Electron Probe
Microanalysis

Advances in
Electronics and
Electron Physics

Edited by
L. MARTON
National Bureau of Standards, Washington, D.C.

Assistant Editor
CLAIRE MARTON

SUPPLEMENT 1
ELECTROLUMINESCENCE AND RELATED EFFECTS, 1963 HENRY F. IVEY

SUPPLEMENT 2
OPTICAL MASERS, 1964 GEORGE BIRNBAUM

SUPPLEMENT 3
NARROW ANGLE ELECTRON GUNS AND CATHODE RAY TUBES, 1968
HILARY MOSS

SUPPLEMENT 4
ELECTRON BEAM AND LASER BEAM TECHNOLOGY, 1968
L. MARTON AND A. EL-KAREH

SUPPLEMENT 5
LINEAR FERRITE DEVICES FOR MICROWAVE APPLICATIONS, 1968
WILHELM H. VON AULOCK AND CLIFFORD E. FAY

SUPPLEMENT 6
ELECTRON PROBE MICROANALYSIS, 1969
A. J. TOUSIMIS AND L. MARTON, EDS.

Electron Probe Microanalysis

Edited by

A. J. TOUSIMIS

Biodynamics Research Corporation Inc.
Rockville, Maryland

and

L. MARTON

National Bureau of Standards
Washington, D.C.

1969

ACADEMIC PRESS New York and London

ACADEMIC PRESS, INC.
111 FIFTH AVENUE
NEW YORK, NEW YORK 10003

United Kingdom Edition
Published by
ACADEMIC PRESS, INC. (LONDON) LTD.
BERKELEY SQUARE HOUSE, LONDON W1X 6BA

Library of Congress Catalog Card Number: 63-12814

PRINTED IN THE UNITED STATES OF AMERICA

List of Contributors

Numbers in parentheses indicate the pages on which the authors' contributions begin.

I. ADLER, Goddard *Space Flight Center, National Aeronautics and Space Administration, Greenbelt, Maryland* (313)

WILLIAM L. BAUN, *Materials Physics Division, Air Force Materials Laboratory, Wright-Patterson Air Force Base, Ohio* (155)

JAMES D. BROWN, *Faculty of Engineering Science, The University of Western Ontario, London, Ontario, Canada* (45)

J. W. COLBY, *Bell Telephone Laboratories, Inc., Allentown, Pennsylvania* (177)

JOHN R. DORSEY,* *Department of Defense, Fort George G. Meade, Maryland* (291)

JOSEPH I. GOLDSTEIN,† *Goddard Space Flight Center, National Aeronautics and Space Administration, Greenbelt, Maryland* (245)

G. HALLERMAN,‡ *Metals and Ceramics Division, Oak Ridge National Laboratory, Oak Ridge, Tennessee* (197)

VICTOR G. MACRES, *Materials Analysis Company, Palo Alto, California* (73)

L. MARTON, *National Bureau of Standards, Washington, D.C.* (1)

CYNTHIA W. MEAD, *U.S. Geological Survey, Washington, D.C.* (227)

P. S. ONG, *Philips Electronic Instruments, Mt. Vernon, New York* (137)

M. L. PICKLESIMER,§ *Metals and Ceramics Division, Oak National Laboratory, Oak Ridge, Tennessee* (197)

E. J. RAPPERPORT, *Ledgemont Laboratory, Kennecot Copper Corporation, Lexington, Massachusetts* (117)

GUNJI SHINODA, *Department of Applied Physics, Osaka University, Yamada, Suita, Osaka-fu, Japan* (15)

J. I. TROMBKA, *Goddard Space Flight Center, National Aeronautics and Space Administration, Greenbelt, Maryland* (313)

RICHARD C. WOLF, *Materials Analysis Company, Palo Alto, California* (73)

HARVEY YAKOWITZ, *Metallurgy Division, Institute for Materials Research, National Bureau of Standards, Washington, D.C.* (361)

* Deceased.

† *Present address:* Department of Metallurgy and Materials Sciences, Lehigh University, Bethlehem, Pennsylvania.

‡ *Present address:* Inland Steel Research Laboratories, East Chicago, Indiana.

§ *Present address:* Southern Research Institute, Birmingham, Alabama.

Foreword

The idea of presenting a collection of reviews on different aspects of electron probe microanalysis originated at the first national conference on this subject, held on the University of Maryland campus in 1966. This book appears well after the inception of this idea, but by rewriting and updating the contributions the volume should prove useful to the growing number of practitioners of this branch of science, as well as to the non-specialist, who needs information on this excellent analytical tool.

A. J. Tousimis
L. Marton

Washington, D.C.
August, 1969

Contents

Backscattered and Secondary Electron Emission as Ancillary Techniques in Electron Probe Analysis

J. W. Colby

The Influence of the Preparation of Metal Specimens on the Precision of Electron Probe Microanalysis

G. Hallerman and M. L. Picklesimer

Electron Probe Microanalysis in Mineralogy

Cynthia W. Mead

Electron Probe Analysis in Metallurgy

Joseph I. Goldstein

Survey of Microanalysis—Interpolation and Extrapolation*

L. MARTON

National Bureau of Standards
Washington, D.C.

When Dr. Tousimis and your committee first did me the honor of inviting me to start off the technical part of this meeting, I surveyed rapidly the field and came to the conclusion that it would be perhaps useful to talk about other things besides electron probe microanalysis. You will be hearing for the next few days nothing but papers on microanalysis by means of electron probes, and therefore, it may be useful to talk about something different. On the other hand, it is not good to go too far afield, and this type of consideration led me to limit myself to a survey of all kinds of methods of microanalysis, chemical or physical. A rapid comparison of this type may serve to put our own field in proper contrast and to emphasize the advantages which can be derived from using the electron probe method. I would like to ask, therefore, for your indulgence in listening to a rapid comparison of some of the types of microanalysis, of their limitations, and of their achievements which are different from the achievements of electron probe microanalysis. By doing this I do not want to imply that chemical identification in the application of our instrument is the only use or useful purpose. There are others and they will be properly emphasized during this meeting. The limitation to the micro-analytical aspect here is merely a convenience for gaining a perspective from which to judge some of our achievements.

Let's start out therefore in enumerating some of the methods which are apt to give us a chemical analysis in extended or limited form and compare their sensitivities and other characteristics. The first which comes to mind is a good old-fashioned chemical analysis. The smallest sample which can be used for analytical purposes is of the order of a microgram. In a microgram sample, as far as I can tell, the smallest amount which can be definitely identified is on the order of one part in a thousand. I think I am not very far off if we set the limit of chemical microanalysis as 10^{-9} grams. The disadvantage of the process is that in doing the analysis the sample is destroyed. The advantage is that it can be extended to practically any substance and that its applicability is extremely wide.

A powerful aid of chemical analysis is the use of spectroscopy. There are two ways of using spectroscopy, either in emission or in absorption. Again, the smallest sample which can be used for analytical purposes in emission spectroscopy is of the order of a microgram or at best $\frac{1}{10}$ of a microgram. The limits of sensitivity, particularly when one is using the so-called method of ultimate lines, can be about one part in ten thousand or maybe one part in one hundred thousand. Thus with some optimistic assumption, we reach the picogram range: 10^{-12} grams. Again the disadvantage of the method is that the sample is destroyed, particularly if it is a very small sample held in place by a large electrode, but the advantage is that it can be applied to very many substances. The sensitivity is a function of the substance used for the analytical determination.

In absorption spectrochemical analysis the situation is more or less comparable. The sensitivity depends very much on the element to be detected as well as on the matrix used for the method. The ultimate sensitivity fluctuates quite a bit for the different elements, and it may go down from $\frac{1}{10}$ microgram to below 10^{-9} grams for selected elements.

Next, perhaps we should look at fluorescence microscopy. Fluorescence microscopy has the advantage of not destroying the sample we are looking at, although in certain cases, modification of the sample is achieved by adding to it a substance which is selectively adsorbed in certain places. This is the so-called method of induced fluorescence, and under very favorable conditions and for selected materials, an estimated limit of 10^{-14} grams can be reached.

Another method for microanalysis is X-ray fluorescence. Its sensitivity is not very notable: the limit appears to be 10^{-7} grams.

Maybe some of you will not agree with me if I list under microanalytical methods the olfactory sense. All of us know how sensitive our noses are for certain smells, particularly the obnoxious ones, and it may be worthwhile in the context of microanalysis to compare the ultimate sensitivity of this method. The objection may be made that it is not applicable to a very wide variety of substances and that is it not giving a chemical analysis. On the other hand, for selected substances, it gives a very specific reaction by means of which a rather positive identification of the substance can be achieved. If in the process, a very small portion of the sample is used up, it is usually so small that the method can be almost described as a nondestructive method. Usually the substances which are rather easily discerned are high molecular weight substances, and there is practically no application of the method to the elements. Assuming, however, that the sample which we inhale is about a hundredth of a liter, we can detect 2×10^{-13} grams of artificial musk, and dogs can do much better than we. Their olfactory limit may be as far as 100 to 1000 molecules. If we assume the molecular weight of 100, this amounts to 10^{-18} grams. Equally sensitive is the olfactory sense of certain flies.

An interesting sidelight is the reaction to the substance which is known to act as a sex-attracting one to the bee [1]. If a bottle containing the substance is uncorked, a bee at a mile distance is going to head for it as soon as it is perceived. Whether it is a stimulator of the olfactory sense or of something else is not known, but assuming isotropic distribution, the total amount can not be more than 10 molecules. I inserted the words about isotropic distribution: of course, the estimate may not be true if the bee is downwind.

Another method we should consider is mass spectrometry [2]. It has the great advantage of being an almost universal method because it determines the individual elements with rather good accuracy. According to the information I have, its limit of sizes of sample is about a microgram, and one part in 10^7 can be detected out of a microgram sample under optimum conditions. This brings us to the sensitivity limit of 100 femtograms.

Table I gives a résumé of these considerations so that you can easily compare what we have with what follows on electron probe microanalysis.

TABLE 1
ULTIMATE SENSITIVITIES OF MICROANALYTICAL METHODS

Chemical	10^{-9} grams
Emission spectroscopy	10^{-12} grams
Absorption spectroscopy	10^{-9} grams
Fluorescent microscopy	10^{-14} grams
X-ray fluorescence	10^{-7} grams
Mass spectroscopy	10^{-13} grams
Olfactory sense	10^{-18} grams
Sex attraction of the bee	10^{-20} grams
Electron probe	10^{-15} to 10^{-16} grams

To make the comparison, I have made certain assumptions and hope that you agree with them. I assume that we have a probe which is one micron square in dimensions and that the penetration volume from which the X rays are issuing is not more than 10 μ^3. Assuming an average density of 10, we have a total mass under investigation corresponding to 10^{-10} grams. Under favorable conditions, one part in a hundred thousand of this amount can be detected, and therefore, we would come to a sensitivity limit of about 10^{-15} grams. In other words, we are very close to the limit of sensitivity of some of the most sensitive methods which are listed above. In addition, the method can be applied to so many elements, that it can compete very advantageously with the most universal methods which we know for microanalysis.

Now that we have convinced ourselves that our method has really a very honorable place in the ranks of quite a number of different methods of microanalysis, let's look briefly at its early history. It is common knowledge that any new development in science has precursors. It is no shame to admit that no method of discovery or invention has sprung ready-made out of the brain of its discoverer, and the father of the electron probe microanalysis, Raymond Castaing, is the first one to pay homage himself to his predecessors. In his review paper published in 1960 [3], he points out how the most important part of his discovery was anticipated by Moseley in 1913 [4]. According to the description given by Castaing, a trolley held different samples and these different samples could be excited to their characteristic X-ray emission by turning the proper handles (Fig. 1).

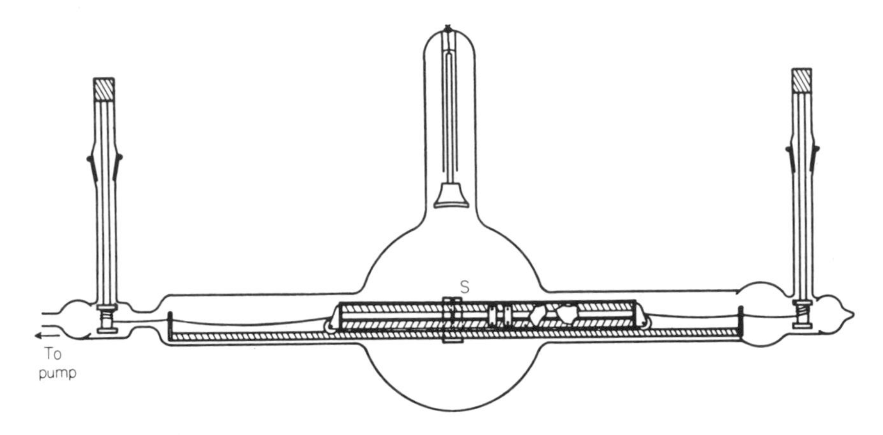

FIG. 1. Moseley's apparatus.

The times were ripe at the time of Castaing's thesis [5] for this type of instrumentation. Hillier applied for a patent in 1943 describing a similar combination of electron probes with X-ray spectography. The patent was granted in 1947 [6], but there is a fundamental difference betwen his and Castaing's method. Whereas Castaing devised means for an absolute calibration, Hillier's patent is essentially a miniaturization of Moseley's work.

Much closer to Castaing's work in date are the efforts to create electron microspectrographs. In 1944 two electron microspectrographs had been described. Both started from Ruthemann's observations of characteristic energy losses [7] as well as of discreet energy losses accompanying emission of X-rays quanta. One was Hillier's microanalyzer, where an electron optical system produced a small probe on the sample as in Castaing's apparatus [8]. The energy losses suffered by the electrons were then analyzed by a magnetic deflection system (Fig. 2). Figure 3 shows my own arrangement [9] which differs from the others in that instead of producing a small probe, a magnified image of the specimen is projected onto a screen provided with a small slit, and only the part of the image falling onto the slit is analyzed. Therefore, at very high magnifications, the total area included in the microanalysis can

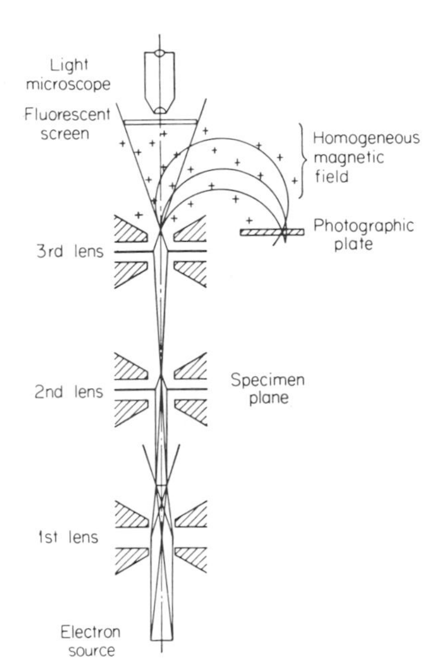

FIG. 2. Microanalyzer of Hillier.

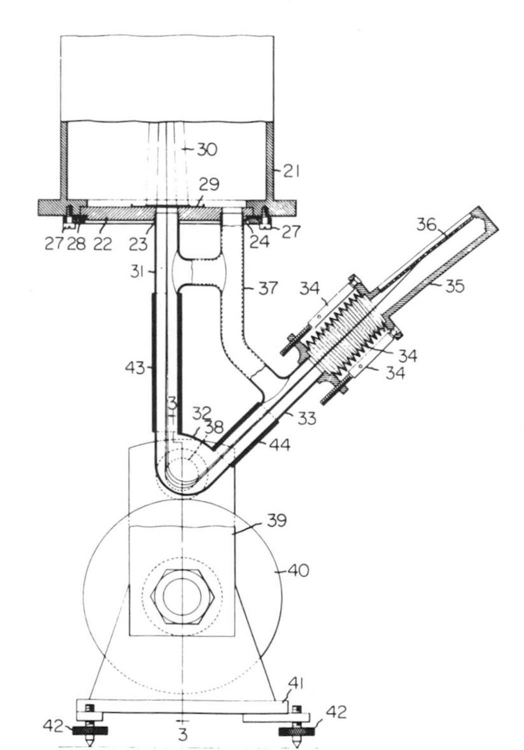

Fig. 3. Microspectroscope of Marton.

be extremely small. Figure 4 shows the appearance of the analyzer part of the instrument at the bottom of a vertical electron microscope.

This may be the place to mention another invention which comes a little closer to some of the modern ideas of Castaing and some recent workers.

FIG. 4. View of Marton's microspectroscope.

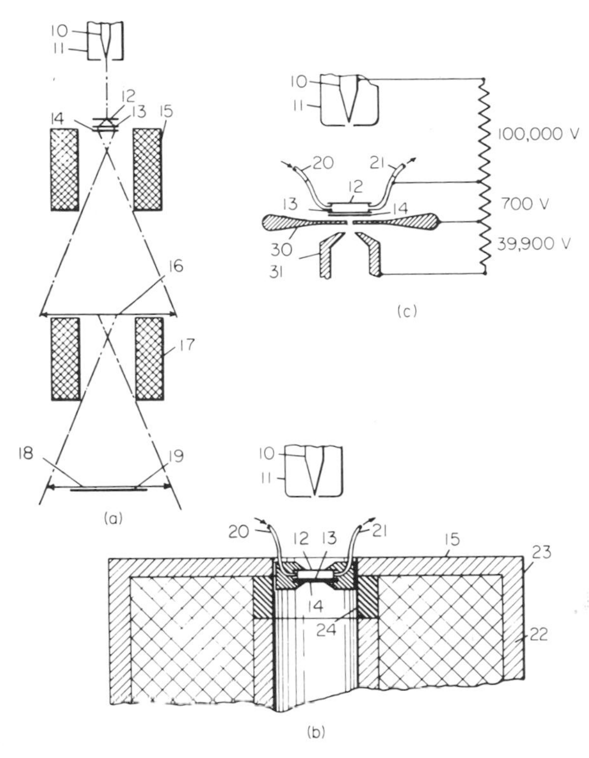

FIG. 5. Proposed microanalyzer based on photoelectric emission excited by X rays.

As you may see from Fig. 5, my patent application of 1946 [10] consisted of the following ideas. A first X-ray tube (Fig. 5a) produces X rays in a thin target. The resulting irradiation of the sample produces photoelectrons which are characteristic for the specimen which is irradiated with the X rays. The photoelectrons are accelerated by a potential difference and used to form an electron microscope image consisting of nothing else but the image produced by the characteristic photoelectrons emitted by the sample. To produce this kind of selective image formation there are several variables. One is the variation of the X-ray exciting potential difference. By operating a little above the K limit of the element which we want to detect, the conditions are reasonably favorable for emitting essentially the K photoelectrons. The other possibility is the use of filters indicated by Number 13 in the figure, which can be exchanged to produce more or less monochromatic X-ray irradiation.

Needless to say, none of these last three instruments has ever been used for practical microanalytical purposes as much as the electron probe micro-analytical method of Castaing is. I am mentioning them merely for historical reasons and not for claiming any kind of results in competition with present methods. In fact, showing these early abortive attempts is a good illustration of how tortuous the path of science can be. Bricks are brought together by many people, some good ones and some poor ones. The good ones serve to erect the edifice, the poor ones are discarded, but not completely. They may serve in a different capacity: to pave a sidewalk, to serve as a temporary support for the scaffolding—they may be ground up for some other purpose. Nothing is completely wasted, however defective a component it may be. But what is needed is the knowledge that such a component or material is available. If it is not known, we are obliged to use a substitute or to create the component which is needed. I shall show later another illustration of this need for information.

In 1943 there appeared a short paper by Professor Andre Guinier titled "Method of Quantitative Analysis Constituents by Means of Absorption Measurements of X-Rays" [11]. Guinier started from considerations of microradiographic images where he demonstrated that in a good microradiograph, quantitative analysis of a thin layer is possible by using not more than one micron square of the surface. As far as I can tell, this was the origin of Guinier's search for a better method to explore quantitatively the chemical constitution of very small samples. Soon thereafter Castaing became Guinier's student, working on his doctorate thesis, and developed the new, very well-known method of electron probe microanalysis. This method today is linked to the name of Castaing, but Castaing with his usual modesty says in the introduction to his thesis [5], "This study was made under the direction of Professor A. Guinier. Professor Guinier had the idea of this new method and proposed that I undertake its realization. I want to thank him and give him the expression of all my gratitude for the interest with which he has oriented and followed my research, indicating to me some of the difficulties which I could not have avoided." There are many examples of this kind in the history of science, and many are the times where the originator of the idea is completely forgotten.

In Castaing, Guinier had an excellent pupil. Not only had he a very good background in physics and mathematics, but he was also full of ideas. His thesis, which appeared in 1952, is still one of the best documents on the origins of the subject. Figure 6 shows a reproduction of the original figure from his thesis, and Fig. 7 shows a somewhat more recent version of the Castaing microanalyzer.

This may be the place to mention the early work of Borovskii. Between 1951 and 1953 he developed an electromagnetic microprobe analyzer, whereas Castaing's original work used an electrostatic instrument [12].

Now we come to a good illustration of how scientific communications are sometimes hampered by lack of distribution. Castaing's drawing shows the incorporation, in the design of the instrument, of a light optical means for

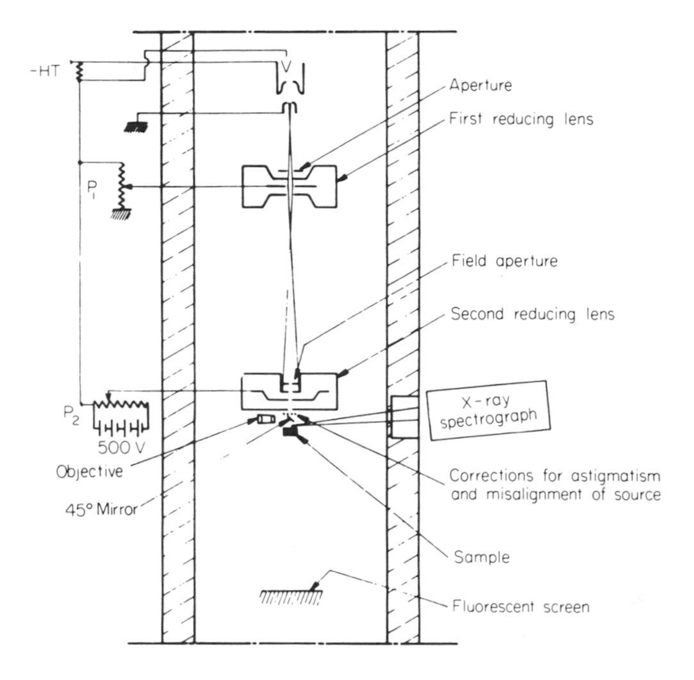

FIG. 6. First version of Castaing's microanalyzer.

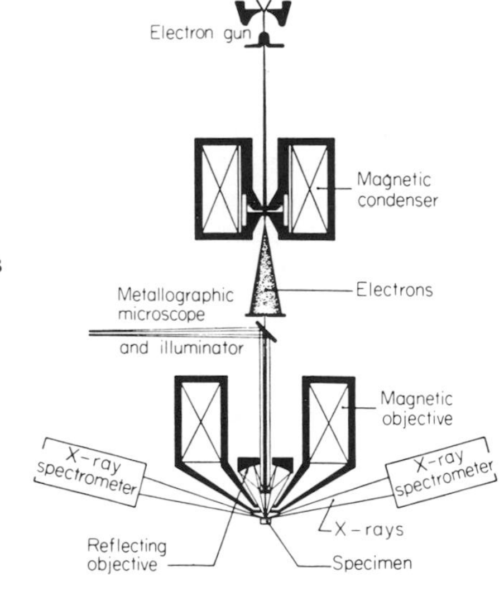

FIG. 7. 1950 Version of Castaing's analyzer.

viewing the specimen while under electron optical observation. It is curious to know that the same problem arose earlier in combination with the electron microscopes, and as an illustration, I would like to show Fig. 8 taken from an early patent application [13]. As can be seen from the figure, the identical problem arose. What was required was a simultaneous observation of a specimen by means of an electron microscope and a light microscope, and

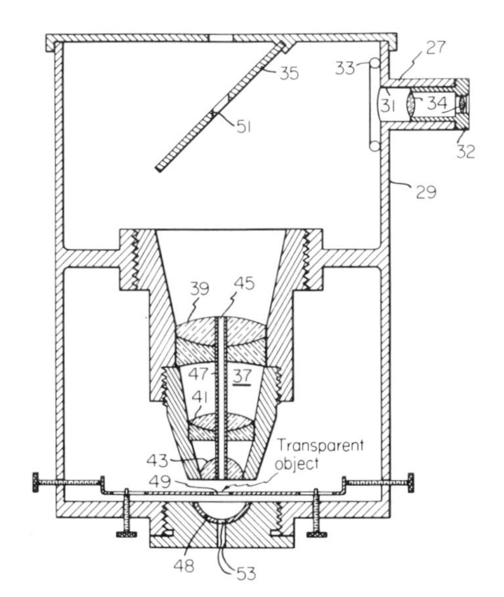

FIG. 8. Combined Light Optical Electron Microscope.

the solution was of necessity very similar—Fig. 9 shows some variations of illuminating and observing means. Castaing could not know about this development because it never appeared in the open literature, and obviously very few people take the trouble of looking at patents issued in different countries. Incidentally, the scheme proved much less practical for electron microscopy than for electron probe microanalysis.

As mentioned before, the greatest advantage of Castaing's new method was that he used pure elements while keeping the conditions of electron bombardment constant. Thus the radiation emitted by the pure element and by the sample were observed under identical conditions.

A very careful survey of necessary corrections was an added help for quantitative analysis. Such corrections are the absorption correction—a correction introduced by distribution in depth of the characteristic emission—and the fluorescence correction. The original thesis of Castaing contains all the basic elements for these important corrections and secured, therefore, a rather permanent base for application of the new method. I do not want to spend much time on sketching more recent history. One item that may be

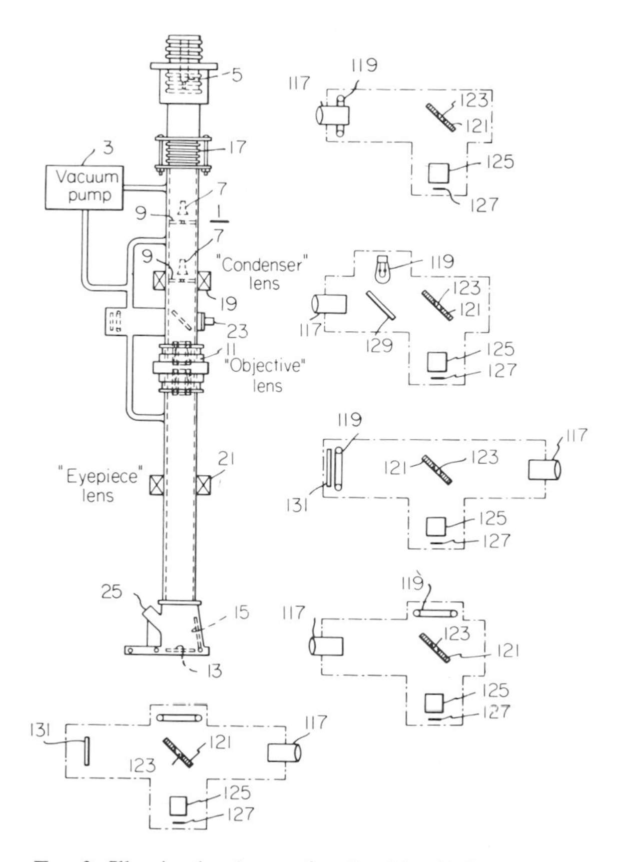

FIG. 9. Illuminating System for Combined Miscroscope.

worth pointing out, though, is that the scanning probe microanalyzer of Duncumb [14] is equally based on much earlier conceptions, particularly on the scanning electron microscope where the form given to it by von Ardenne is the best known [15]. Again, one may easily find precursors to von Ardenne's ideas. All this is to illustrate my point that no idea is without some precursors of its kind.

The title of my talk indicates interpolation and extrapolation. I think we have done quite a bit with the interpolation part. Let us now proceed to the extrapolation. We have phrased our question as follows: In what direction is progress likely for the electron probe microanalysis method, and in what direction are further applications possible? Now you gentlemen have spent much more time than I on debating or at least cogitating on these subjects, and I hope you do not mind my sticking my neck out, perhaps in areas where I may be utterly wrong.

I have spent quite a number of years working on the electron microscope, and resolution was one of the prime considerations which very often guided my thinking and the thinking of those who worked on the electron microscope in those times. I would list, therefore, resolution or improvement of the resolution as one of the main goals for future development, and this is only possible if we can reduce the observed area of the specimen. There are two limitations in this respect. One is the size of the incident beam, and you know very well that this can be reduced considerably. The electron optical system allows reduction without any trouble. However, if we do reduce it, we are hitting two difficulties. One is that the intensity of the outcoming signal is reduced, and it may be reduced to an extent where you may have difficulties recording it. The other limitation is due to the mean free path of the electrons in the material which is irradiated and to the effective volume created by the incident electrons. Let's take the second limitation first because, in principle at least, we may be able to reduce this effective volume from which X rays emerge by reducing the incident electron energy. Assuming that there are no serious limitations on primary energy reduction, we will find that we have reduced even more the total intensity of X rays emerging from the sample.

The intensity limitations are partly due to very poor collection efficiency of the detecting devices. In a well-built conventional electron probe microanalyzer, the spectrograph sees only a very small part of the solid angle comprised by the emerging irradiation. I understand from one publication that it may be as low as 0.07 radian. This collection efficiency may be considerably improved by the pulse height analysis method and in optimal cases, an improvement of a factor of 10 might be reached. What I would like to propose is to increase collection efficiency by a considerably better means, improving the collection efficiency more than tenfold, perhaps even one hundred-fold in some cases. I think this can be done, and the means which I am proposing for it is the creation of X-ray fiber optics [16]. You are probably all familiar with the fiber optics used in light optics; no fiber optics exists at present for X rays. Such fiber optics, according to some of my calculations, could be easily created and reasonably good means achieved for its application in an electron probe microanalyzer. I am very obliged to the organizers of this symposium for asking me to do the work on electron probe microanalyzers because this stimulated me toward thinking of fiber optics for X rays and for their possible applications in fields other than your own. From my limited studies until today I am very optimistic. I think they could be done relatively easily and would find rather wide applications.

It has been pointed out in the past that there exist essentially three methods of microanalysis using electron probes and X rays [17]. The original method of Castaing may be called emission microanalysis, as it involves the emission of characteristic X rays under bombardment with electrons. The absorption method of microanalysis is derived from the early method of microradiography, and we referred to it earlier in this paper. The third method is fluorescence microanalysis, where a primary X-ray beam, limited in area, hits the specimen and excites fluorescent radiation. In the conventional fluorescence microanalysis, as illustrated by the work of Long [18] for instance, the size of the fluorescent spot and the intensity of the fluorescent radiation derived from it are limited by mechanical considerations. One way of improving the efficiency is to have an electron microprobe excite X rays only from a very limited area and to bring that limited target as close as possible to the sample in order to provide only a small area of X-ray illumination. The second method is to aperture the X-ray beam by some mechanical means [19]. I think in this method, too, progress could be achieved if we could increase the intensity of the X-ray illumination reaching the specimen. Perhaps again a judicious application of the fiber optics technique could help in increasing the available intensity without increasing the irradiated volume or perhaps even reducing it.

I do not dare to stick out my neck any further on possible future improvements of the electron probe microanalysis method. It is very difficult to indicate new areas of application, particularly in view of the fact that they range over extremely wide areas. There are, however, two areas which occurred to me and where, to my knowledge there is very little of application. One is very obvious because it is an extension of already existing investigations. I mean the forensic methods. In criminal investigations, all kinds of chemical, metallurgical, and biological investigations are going on, so extension to forensic medicine or forensic analysis is not very new. It may be even newer to extend the application of microanalytic methods to archaeometry. This is a new word, which I just discovered and which in my interpretation, covers the quantitative aspects of archaeology. The subject is extremely fascinating, and, besides the scientific advances, it offers a tremendous esthetic satisfaction to its practitioners.

To finish this long talk I would like to digress a minute to cover another aspect of future development. With the widespread interest in electron probe microanalysis, there should exist a focal point bringing together those interested in the subject. The existing societies can offer only limited hospitality, according to their own orientation. I would like to finish, therefore, by recommending that you devote some thought about the advisability of forming a national society devoted to electron probe microanalysis.

References

1. H. Sleik and A. Turk, "Air Conservation Engineering," 2nd ed. Connor Eng. Corp. Danbury, Connecticut, 1953.
2. See for instance, A. J. Ahearn, ed., " Mass Spectrometric Analysis of Solids," Chapter 5. Elsevier, Amsterdam, 1966.

3. R. Castaing, *Advan. Electron. Electron Phys.* **13**, 317 (1960).
4. H. Moseley, *Phil. Mag.* [4] **26**, 1024 (1913); **27**, 703 (1914).
5. R. Castaing, Doctoral Thesis, Univ. of Paris, 1952. English translation by D. B. Wittry, Rept. No. WAL 142/59-7, Dept. of the Army, December 1, 1955.
6. J. Hillier, Electron probe analysis employing X-ray spectrography, U.S. Patent 2,418,029 (1947).
7. G. Ruthemann, *Naturwiss.* **29**, 648 (1941); **30**, 145 (1942).
8. J. Hillier and R. F. Baker, *J. Appl. Phys.* **15**, 663 (1964).
9. L. Marton, *Phys. Rev.* **66**, 159 (1944).
10. L. Marton, Electron micro-analyzer, U.S. Patent 2,440,640 (1948).
11. A. Guinier, *Compt. Rend.* **216**, 48 (1943).
12. I. B. Borovskii, *Akad. Nauk SSSR, Sb.* 135–139 (1953).
13. L. Marton, Microscope, U.S. Patent 2,301,302 (1942).
14. P. Duncumb and D. A. Melford, *Proc. Intern. Symp. X-Ray Microscopy and X-Ray Microanal., 2nd* p. 358. Elsevier, Amsterdam, 1960.
15. M. v. Ardenne, *Z. Physik* **109**, 553 (1938); *Z. Techn. Phys.* **19**, 407 (1938).
16. L. Marton, *Appl. Phys. Letters* **9**, 194 (1966).
17. V. E. Coslett, P. Duncumb, J. V. P. Long, and W. C. Nixon, Microanalysis by X-ray absorption, fluorescence, emission, and diffraction using ultrafine X-ray sources *in* "X-Ray Analysis" (*Proc. Ann. Conf. Appl. X-Ray Anal., 6th Univ. of Denver*), Vol. 1, p. 329. Plenum Press, New York. 1957.
18. J. V. P. Long and H. Röckert, X-ray fluorescence in microanalysis and the determination of potassium in nerve cells, *in* "X-Ray Optics and X-Ray Microanalysis" (H. H. Pattee, Jr., V. E. Cosslett, and A. Engström, eds.), p. 513. Academic Press, New York, 1963.
19. T. C. Loomis and S. M. Vincent, Trace and microanalysis, *in* "Handbook of X-Rays" (E. F. Kaelble, ed.). McGraw-Hill, New York, 1967.

Behavior of Electrons in a Specimen

GUNJI SHINODA

Department of Applied Physics
Osaka University
Yamada, Suita
Osaka-fu, Japan

I. Introduction

When electrons impinge on the surface of a specimen, some of them are backscattered; and the rest penetrate into the specimen, lose their kinetic energy, and are finally absorbed by the specimen. Still, there is another kind of emitted electrons called secondary electrons which differ from back-scattered electrons only in their energies. Since the energy distributions of backscattered electrons have a wide range, from zero electron volts to that of primary electrons, we cannot distinguish them definitely. Customarily, those having energies less than 50 eV are called secondary electrons (SE) and the others backscattered electrons (BSE).

Electrons that penetrate into the specimen travel in their original directions until a certain depth is attained; then diffusion takes place. During the diffusion process, electrons lose their energy and X-rays are generated. Therefore, the size of the diffused X-ray source is larger than that of the original electron beam. On the other hand, since the absorbed and backscattered number of electrons are a function of the atomic number of the target material, more specifically, the mean atomic number of the surface layer, the beam size corresponding to backscattered electron and specimen current is not much different from that of the original electron beam. These electron-target interreactions are closely related to the problem of resolving power of the electron probe microanalyzer.

In the above statement, we have assumed that backscattered electron and specimen currents are determined by the weight mean atomic number of the target material.* This is true as a first approximation, but since the currents will be influenced by many other factors such as crystal orientations, scattering by phonon, chemical bonding, etc., resolution based on mean atomic number will be confused when the specimen consists of elements having nearly the same atomic numbers.

Also, it is expected that the depth distribution function of the X-ray will be influenced by the orientations of the crystals of the target point if the accelerating voltage is not high compared with the critical excitation potential because the X ray should be generated by electrons whose directions of travel do not deviate greatly from that of the primary electron beam.

These second order effects are troublesome from the standpoint of quantitative analysis in the electron probe microanalyzer. On the other hand, since these effects are closely related to the properties of materials, they are important from the viewpoint of solid state physics.

Whether or not these effects could be partly attributed to secondary electron emission is not known at present. But since the secondary electron emission is much affected by the physical properties and topography of the surface, the secondary electrons will also give us much useful information on the surface state.

Usually, the backscattered ratio η is defined as the ratio of the number of primary backscattered electrons to that of primary electrons, secondary yield δ as the ratio of the number of secondary electrons to that of primary electrons, and absorbed ratio σ as the ratio of the number of absorbed elec-

* Theoretically, it should be atom mean atomic number; however, weight mean atomic number is generally used for practical reasons.

trons to that of primary electrons. Among these ratios the following relation holds

$$\eta + \delta + \sigma = 1. \tag{1}$$

An absorbed electron current is called a specimen current.

II. Backscattering and Diffusion of Electrons

A. Stopping Power and Range of Electrons

When electrons penetrate into the specimen, they interact with outer shell electrons and are scattered by elastic or inelastic collisions. While traveling, electrons lose their energies and are finally stopped or absorbed by the target material. Then quantities such as the range of an electron and the stopping power of the material become important in discussing the behavior of electrons in a material. Of course, these quantities vary with the kinds of material and the energy of the electrons. The semiempirical treatment of this problem by Thomson–Whiddington [1] is well known and is expressed as

$$dE/dx = -\text{const}/E \tag{2}$$

or

$$v_0{}^4 - v^4 = c\rho x \tag{2a}$$

where E is the kinetic energy of the electron, and v_0 and v are initial velocity and that at x, respectively. This formula is correct qualitatively, but it is not so accurate quantitatively. On the other hand, the theory of Bethe *et al.* [2], which had been abandoned because of a difficulty in numerical calculation, becomes usable as the recent developments in electronic computation make it easier.

According to Bethe, the stopping power formula is given by

$$-\frac{dE}{dx} = \frac{4\pi N_A \rho z e^4}{m_0 v^2 A} \ln \frac{m v_0{}^2}{J} \tag{3}$$

where v_0 is the initial velocity of the electron and J is the average excitation energy, usually given by*

$$J = 11.5Z \quad \text{eV}. \tag{4}$$

* According to recent investigations, this value should be revised; c.f. Duncumb and Reed [2a].

From Eq. (3), the Thomson–Whiddington law is derived as its approximate formula, and the stopping power range X_r is given by

$$X_r = 4.24 \times 10^{-10} \frac{ZA}{\rho} \left[E_i\left(2 \ln \frac{0.174\overline{V}}{Z} \right) \right] \tag{5}$$

where $E_i(y)$ is the logarithmic integral.

B. *Theory of Backscattering of Electrons*

When electrons enter the material, their energy and travel direction will be changed by:

(i) direct collision with an atom, with abrupt decrease of velocity;
(ii) single scattering by an atom, with abrupt change in direction;
(iii) interaction with outer-shell electrons, with gradual decrease of velocity;
(iv) multiple scattering by atoms, with gradual change in direction of travel.

The first two are closely related to the generation of X-rays, and the other two play important roles in the dissipation of energy in a specimen. The former two are not only very important in electron probe microanalysis but also, as pointed out by Castaing and Descamps [3], they are important factors in the backscattering of electrons.

1. *Transport Equation. Comparison with Experiment*

Bothe [4] applied the diffusion equation to this problem; but since it is a special case of the transport equation used by Bethe *et al.* [2], we will start from the Boltzmann transport equation. Let an electron be at r and the direction of motion be u, then the electron density function $f(r, u, t)$ is given by

$$\frac{\partial f}{\partial S} = \frac{1}{v^{-1}} \frac{\partial f}{\partial t}$$

$$= u \operatorname{grad} f + \int N_\sigma(\theta, u) \times [f(r, u', t) - f(r, u, t)] \sin\theta \, d\theta \, d\varphi. \tag{6}$$

Where v is the velocity of the electron, $dS = v\,dt$, θ and φ correspond to the direction of the unit vector, and σ is the scattering cross section. The first term shows convection, and the second term is scattering.

If we neglect the convection term and put $F(r, s) = \int du\, f(r, u, s)$, we have

$$\partial F/\partial S = -(\lambda/b)\,\Delta F. \tag{7}$$

In this equation used by Bothe [4], usually called the diffusion equation, λ is the transport mean free path. It is very difficult to obtain a rigorous analytical solution of the Boltzmann transport equation which satisfies boundary conditions appropriate for backscattering phenomena; solutions hitherto obtained are usually approximate, with a term which depends on energy dissipation or small angle scattering. Therefore, in this solution, data related to backscattering will disappear. Then, as far as the effect of backscattering is concerned, Bothe's treatment, which contains certain contribution of multiple scattering, is more valuable. The distribution of energies of backscattered electrons is qualitatively explained by his theory.*

However, theoretical study on single scattering has been done by Everhart [5]. His theory is based on the assumption that (1) backscattering is mainly composed of the Rutherford type of large scattering angle and (2) energy loss in a specimen is expressed by the Thomson–Whiddington law. The backscattered ratio η obtained is given by

$$\eta = (a - 1 + 0.5^a)/(a + 1) \tag{8}$$

where $a = \pi Z^2 e^4 N_A / m^2 c A$, N_A is Avogadro's number, m is mass of the electron, and A is atomic weight. This equation holds for elements with $Z \leq 30$, as shown in Fig. 1. Experimental data are those collected by Burkhalter [6].

In the Everhart theory, since the effect of multiple scattering is neglected, it is quite natural that the equation fits better for a thin film target than for a solid block. Shimizu and Shinoda [7] derived a more generalized equation for the thin film target based on the experimentally obtained result of the effect of incidence angle α on η, namely

$$\eta_t(\alpha) = a[(\tan^2 \alpha) \cdot I(n + 2) + 2I(n + 1) - I(n)] \tag{9}$$

where

$$I(n) = n^{-1}\{1 - (1 - l/\cos \alpha)^n\}$$

and l is a nondimensional quantity L/X_r, L the film thickness and X_r the range of electrons.

Substituting $l = (1 + \sec \alpha)^{-1}$,

* This point has been much improved by numerical computation (not analytical) by Brown and Ogilvie [5a] which gives good results on X-ray production. However, with their boundary conditions, it will be difficult to obtain good results on electron behaviors in the material. This point will become clear if we compare their boundary conditions with the result of our Monte Carlo calculation as will be shown in Fig. 14. For this reason, agreement between their calculated result and the experimental one on backscattering behaviors is not good.

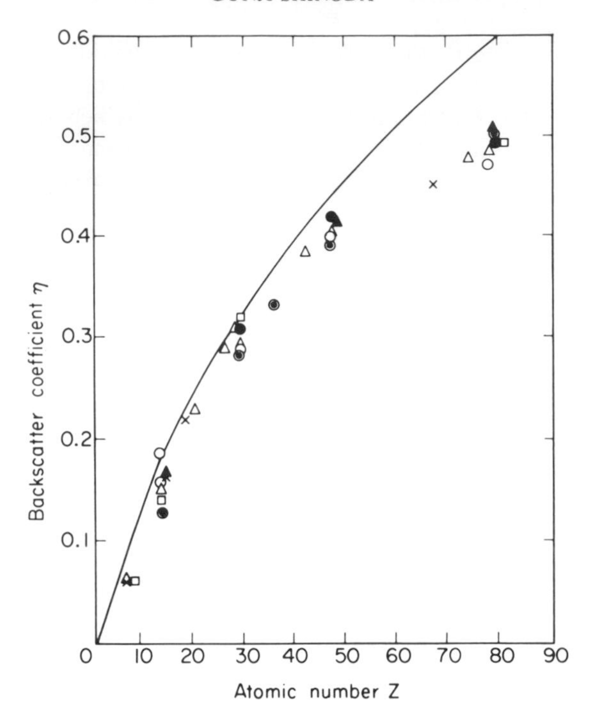

FIG. 1. Comparison of Everhart theory on backscattered ratio η with experiment (— Theory; Shinoda and Shimizu, 15 kV; ▲, Burkhalter, 25 kV; ×, Holliday and Sternglass, 20 kV; △, Palluel, 16 kV; ●, Kanter, 10 kV, ○, Kulenkampff and Spyra, 20–40 kV; ⊙, Schonland, 100 kV).

we have an expression for bulk target. For normal incidence, we have

$$\eta_{(0)} = (a - 1 + 0.5^a)/(a + 1) \tag{10}$$

which is the same result as that of Everhart. Comparing his treatment with experiment, Shimizu [8] found that c, in a of Eq. (8), is no longer a constant, but depends on the depth and is written as $c(x)$; or, since it depends on atomic number, it is better to use the form $c' = \overline{c(xZ)}$. The effect of multiple scattering is included in this treatment. This effect becomes more significant with the increase of the atomic number Z, and the shape of the curve showing angular distribution will be modified.

Theoretically, it is expected that the shape of the curve changes from semicircular to distorted. But, as shown in Fig. 2, the experimental curve does not show much distortion.

In the case of secondary emission as reported by Jonker [9], the cosine law holds for angular distribution. In this case, the effect of multiple scattering will result in loss of the history of electron trajectories, and consequently the

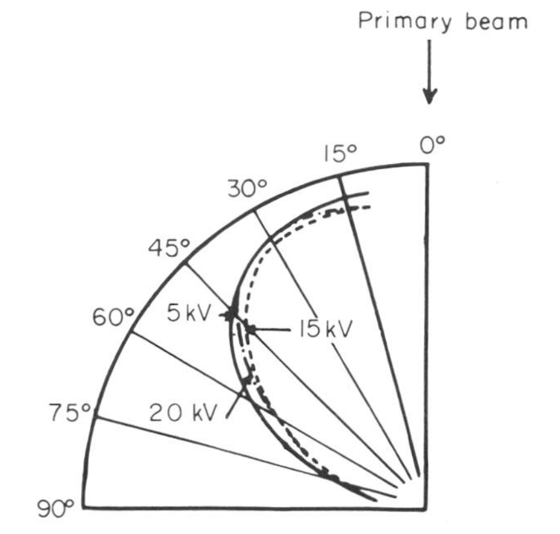

FIG. 2. Angular distribution of backscattered electron for copper.

directions of backscattered electrons will become random. This tendency is significant, especially in high atomic number elements. As shown in Fig. 3, there is a discrepancy between theory and experiment at a high angle of incidence and with high atomic number elements, which provides evidence of multiple scattering of electrons.

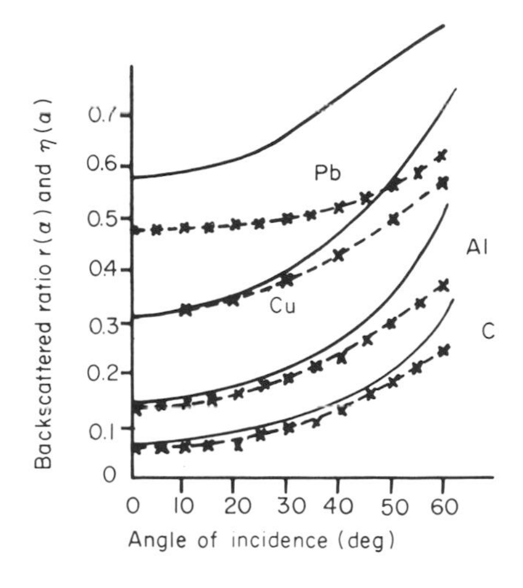

FIG. 3. Backscattered ratio versus angle of incidence; comparison of theory $r(\alpha)$ with experiment (—×—) $\eta(\alpha)$. Experimental curves are corrected for collector emission. Accelerating voltage is 15 kV.

Backscattering from the standpoint of multiple scattering was treated by Archard [10], and combination with single scattering was treated by Dashen [11] using the integral equation and its approximate solution.

2. *Backscattered Electron Image in the Electron Probe*

In the electron probe, the backscattered electron image furnishes very important information. From the standpoint of quantitative analysis, if the take-off angle is great, detection of low energy electrons is preferable in order to increase the contrast because most of these electrons have experienced multiple scattering.

C. Electron Diffusion

From the standpoint of the electron probe, electron diffusion is very important because it is closely related to the generation of X rays, the resolving power, and the specimen current technique. Archard [10] treated this problem using a complete diffusion model, introducing the depth of complete diffusion X_d where the average of the cosine of the angle between the directions of electron travel and the incident direction becomes equal to e^{-1}, according to the definition of Bethe *et al.* [2]. The definition of X_d is given by the following equation

$$X_d = \int_{E_d}^{E_0} \frac{-E\,_d E}{2\pi N e^4 Z \ln(2E/J)} \tag{11}$$

where N is the number of atoms per unit volume J, e the electronic charge, Z the atomic number, J the average excitation potential, E_0 the initial energy of the primary electron, and E_α the solution of

$$\int_{E_d}^{E_0} \frac{dE}{\lambda |dE/dX|} = \frac{1}{2} \tag{12}$$

and

$$\lambda^{-1} = \frac{\pi N e^4 Z^2}{2E^2} \ln \frac{2a_H \cdot m(2E/m)^{1/2}}{\hbar Z^{1/3}} \tag{13}$$

where $\hbar$ is Planck's reduced constant and a_H the Bohr hydrogen radius.

Integration of Eq. (10) gives the expression for X_d, originally obtained by Worthington and Tomlin [12]. According to Archard's expression, it is written as

$$\rho X_d = 4.24 \times 10^{-10} ZA[f(\overline{V}_0) - f(\overline{V}_d) \tag{14}$$

$$f(\overline{V}) = E_i\langle 2 \ln(0.174\, \overline{V}/Z)\rangle \tag{15}$$

where $E_i\langle y\rangle$ is the logarithmic integral. This treatment has been extended to obtain correction formula for quantitative analysis (Archard and Mulvey [13]).

Roughly speaking, as in the Archard model, electrons keep their initial direction until they reach the depth of X_d and then diffuse. But even between the surface and the depth X_d, some scattering will take place. Therefore, we can understand X_d as the depth where all scattering angles are equally probable. From this standpoint, X_d can be determined by measuring angular distribution of transmitted electrons using films of various thickness. It has been observed, however, that the early part of the transmission-film thickness curve does not fit any simple relation, and after some depth exponential law holds. Since the exponential law is related to the diffusion phenomenon, the beginning point of the exponential curve will give the depth of complete diffusion. From these two methods Cosslett and Thomas [14] determined X_d. They compared these values with the theoretically calculated one, but agreement was not good. Even in electron diffusion phenomena, there are still many problems to be solved. In the present state, rigorous theoretical study is very difficult. Therefore, at first the Monte Carlo method is tried; and then, tracing the result of the Monte Carlo calculation, new theoretical considerations should be tried.

The electron diffusion problem is also important in determining the spot size of the electron beam, and then the spot size of the diffused X-ray source can be estimated. Shimizu and Shinoda [15] studied this problem using a wedge-shaped specimen. They assumed that the intensity distribution of the electron probe is expressed by the Gaussian function and the wedge-shaped specimen by a step-function. Then the probability of electron escape $\xi(a)$ from the side surface of the wedge-shaped specimen is given by

$$\xi(a) = \int_{-\infty}^{\infty} \int_{0}^{x_r - x_d} f_0(a - sy) \cdot l(s)\, ds\, dy \tag{16}$$

where

$$l(s) = \frac{1}{2}\left(1 - \frac{S}{X_r - X_d}\right) \quad \text{and} \quad f_0(x, y) = (\delta_0{}^2\pi)^{-1} \cdot \exp\left(-(x^2 + y^2)/\delta_0{}^2\right) \tag{17}$$

and δ_0 is the spot size of the electron probe. The solution of Eq. (16) is as follows

$$\xi(a) = \frac{1}{2}\Bigg[\left(1 - \frac{a}{X_r - X_d}\right)\left\{E\left(\frac{\sqrt{2}a}{\delta_0}\right) - E\left(\frac{\sqrt{2}}{\delta_0}(a - X_r + X_d)\right)\right\}$$
$$+ (X_r - X_d)^{-1} \cdot \frac{\delta_0}{2\sqrt{\pi}}\left\{\exp\frac{-(a + X_d - X_r)^2}{\delta_0{}^2} - \exp\frac{-a^2}{\delta_0{}^2}\right\}\Bigg]$$

$$\text{for} \quad a \geq X_r - X_d$$

$$\xi(a) = \frac{1}{2}\left[\left(1 - \frac{a}{X_r - X_d}\right)\left\{E\left(\frac{\sqrt{2}a}{\delta_0}\right) + E\left(\frac{\sqrt{2}}{\delta_0}(-a + X_r - X_d) - 1\right)\right\}\right.$$

$$\left. + (X_r - X_d)^{-1} \cdot \frac{\delta_0}{2\sqrt{\pi}}\left\{\exp\frac{-(-a + X_r - X_d)^2}{\delta_0} - \exp\frac{-a^2}{\delta_0^{\,2}}\right\}\right]$$

$$\text{for} \quad 0 \leq a \leq X_r - X_d$$

$$\xi(a) = \frac{1}{2}\left[\left(1 - \frac{a}{X_r - X_d}\right)\left\{E\left(\frac{\sqrt{2}}{\delta_0}(-a + X_r - X_d)\right) - E\left(-\frac{\sqrt{2}a}{\delta_0}\right)\right\}\right.$$

$$\left. + (X_r - X_d)^{-1} \cdot \frac{\delta_0}{2\sqrt{\pi}}\left\{\exp\left(\frac{-(-a + X_r - X_d)^2}{\delta_0^{\,2}}\right) - \exp\frac{-a^2}{\delta_0^{\,2}}\right\}\right]$$

$$\text{for} \quad a \leq 0$$

where

$$E(x) \int_{-\infty}^{x} (2\pi)^{-1/2} \operatorname{erf}(-\tfrac{1}{2}x^2)\, dx. \tag{18}$$

Comparing this $\xi(a)$ with experimentally obtained escape electron current, δ_0 can be obtained. Theoretical and experimental results of escape electron current are shown in Fig. 4.

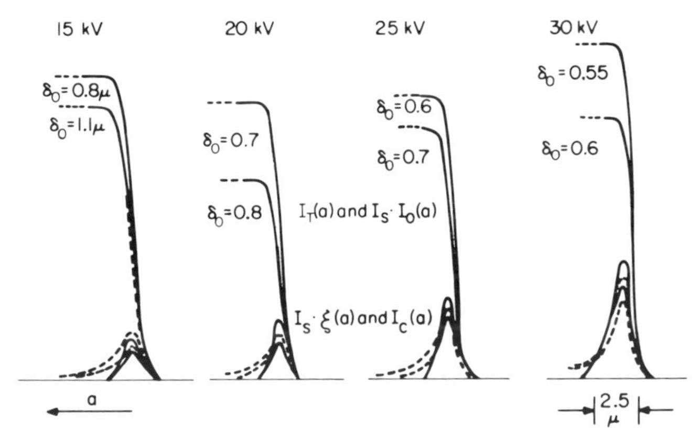

FIG. 4. Comparison of theoretical curves $I_s \cdot I_0(a)$ and $I_s \cdot \xi(a)$ with experimental curves (— — —) $I_T(a)$ and $I_c(a)$ at $\bar{V}_c$ = zero volt for different two values of δ_0 found from $I_T(a)$ curve.

D. Spot Size of Primary Electron Beam

Usually, the spot size of the electron beam is determined by using a knife edge and measuring the change of specimen current with the displacement of the knife edge. But, as electron diffusion occurs in a knife-edge material, the specimen current curve does not take a symmetric form. Such evidence of electron diffusion will be clear from the Fourier transform of the specimen current curve. Correction for the effect of electron diffusion is made by utilizing escape electron current, namely:

$$I(a) = \left\{ I_0\left(\frac{\sqrt{2}a}{\delta_0}\right) - \xi\left(\frac{\sqrt{2}a}{\delta_0}\right) \right\} i_0(1 - \eta) \tag{19}$$

where $I(a)$ is the nondimensional form of the true specimen current and i_0 is incident electron current. Therefore, $I_0(a)$ is an ideal specimen current without electron diffusion and is given by

$$I_0(a) = \begin{cases} E(\sqrt{2}a/\delta_0) & \text{for} \quad a \geq 0 \\ 1 - E(\sqrt{2}a/\delta_0) & \text{for} \quad a < 0. \end{cases} \tag{20}$$

In actual procedure an appropriate value of δ_0 is taken, and $I(a)$ is calculated from Eqs. (19) and (20). Then comparison with experiment is done.

III. Spot Size of Diffused X-Ray Source and Depth Distribution of Characteristic X-Ray

A. *Spot Size of Diffused X-Ray Source*

In the last section, electron diffusion in the specimen was studied. After energy of electrons decreases to that of critical potential of X-ray generation, the electron will soon be stopped. For instance, critical potential is about 6 kV in a copper target, and with this energy the electron can travel only a few tenths of a micron. Thus, the spot size of the diffused electron beam is nearly equal to that of the diffused X-ray source. Ehrenberg and Franks [16] and later Watanabe [16a] measured the spot size of the diffused electron beam from the optical fluorescence due to electron penetration. However, for optical fluorescence, only 10 or 20 eV energy is enough; but for X-ray generation usually several kiloelectron volts are necessary. Consequently, the X-ray source must be somewhat smaller. On the other hand, indirect ionization causes spread of the X-ray source. Therefore, the source size should be determined experimentally. If δd is the spot size of the X-ray source corresponding to an extremely narrow beam and assuming a step function for the wedge-shaped specimen and Gaussian distribution of the X-ray source, the X-ray intensity curve $I_k(a)$ becomes

$$I_k(a) = \int_0^\infty (\pi)^{-1/2} \exp -(x - a)^2 \cdot E(\sqrt{2}kx)\, dx \tag{21}$$

where

$$E(x) = (2\pi)^{-1/2} \int_{-\infty}^{x} \exp\left(-\tfrac{1}{2}t^2\right) dt).$$

In actual experiments, as in Fig. 4, nondimensional variables such as $x = s/\delta_0$, $a = a_0/\delta_0$, and $\delta_0/\delta\alpha = k$ have been used. If we can find the most suitable value of k, the spot size of the characteristic X-ray source is given by $2\delta_T = 2(\delta_0{}^2 + \delta d^2)^{1/2}$.

Figure 5 is the result for copper, a value quite consistent with an expected value from experience and also from the experiment of Ehrenberg and Franks [16]. The shape of the X-ray source has been determined as shown in Fig. 6, but the location of the maximum of X-ray generation is still unknown. A means of clarifying this point is to determine the depth distribution function of X-ray generation.

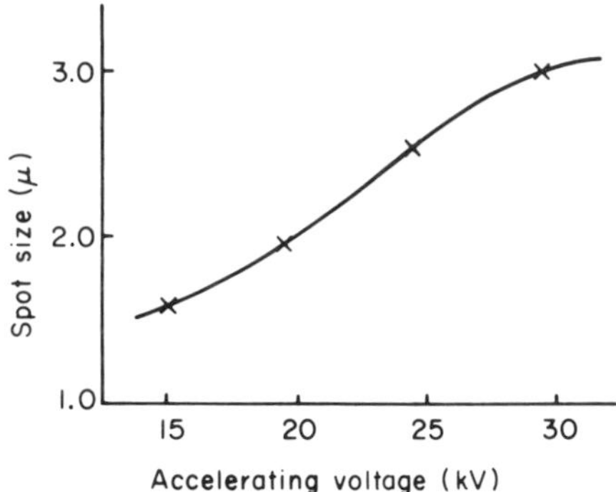

FIG. 5. Change of spot size of X-ray source with accelerating voltage.

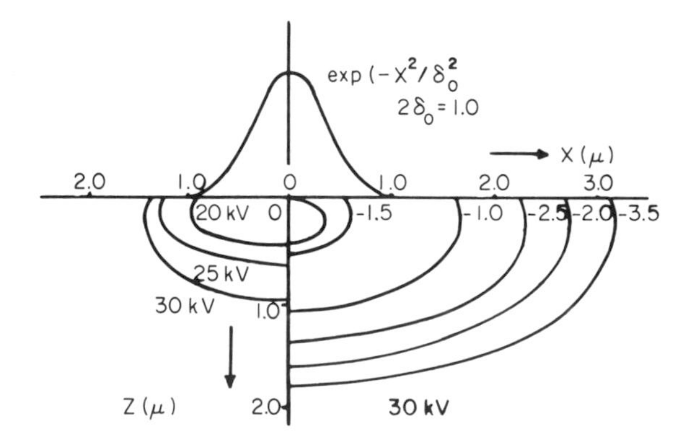

FIG. 6. Shape of X-ray source for copper. Right-hand side is contour of X-ray intensity for accelerating voltage of 30 kV. Left-hand side is locus of (intensity) equals (peak value)/e for various accelerating voltages. For lateral broadening the effect of excitation by general X-ray is included, but for the depth direction only the primary excitation is considered.

B. Depth Distribution of Characteristic X-Ray

The depth distribution function of characteristic X-rays is important for correction of quantitative analysis. More than ten years ago it was determined by Castaing and Descamps [3] with their tracer method. Recently the Monte Carlo calculation and the analysis of angular distribution methods have been developed.

The tracer method is well known. First, a thin film of zinc is evaporated on a copper block, and then several stepwise layers of copper are evaporated. The electron beam is directed to each step, penetrating through the copper steps and generating the characteristic X-ray of zinc. Since the atomic numbers of zinc and copper are neighboring, absorption of Zn K_α in copper is nearly the same as that of Cu K_α in copper. In actual experiment, care must be taken to obtain sound zinc film, and 10–13 steps of 600–800 Å each are suitable [17].

1. *Monte Carlo Calculation*

When an electron of energy strikes the target normally at a point P_0, the electron collides with atoms and dissipates its energy. Some electrons lose all their energies in the target, while others leave the surface of the target with

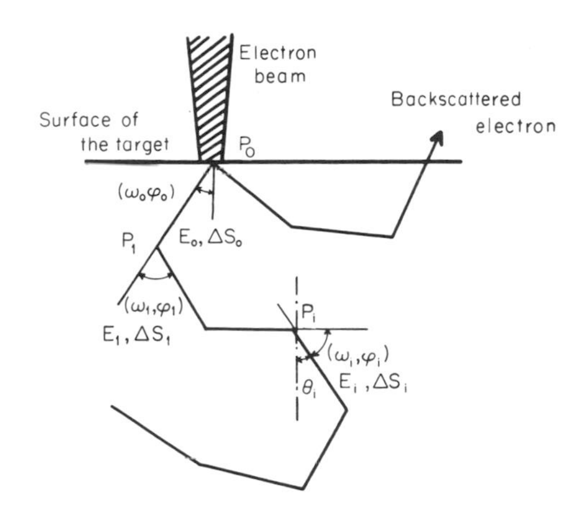

FIG. 7. Simplified model of electron trajectory in Monte Carlo calculation.

considerably high energies. Figure 7 shows a simplified model of an electron trajectory. In the Monte Carlo calculation, it is assumed that an electron travels a small distance, Δs_i, in a straight line with constant energy E_i after collision at P_i, the scattering angle being (ω_i, φ_i) as shown in Fig. 7. Behavior of electrons in the target is determined by the angular distribution of scattering electrons, energy loss, and step length.

Angular distribution of electron scattering is found by following Lewis' solution of the transport equation for an infinite medium

$$f(w) = (4\pi)^{-1} \sum_{l=1}^{\infty} (2l + 1) \cdot P_l(\cos w) \exp\left(- \int_0^s x_l \, ds\right) \tag{22}$$

where

$$X_l = 2\pi N \int_0^\pi \delta(w)[1 - P_l(\cos w)] \sin w \, dw$$
$$= A\left(2\beta(1+\beta)^{-1} - \tfrac{1}{2}(l+1) \int_0^1 \frac{\lambda^l(1-\lambda)^l \, d\lambda}{(\lambda+\beta)^{l+2}}\right) \tag{23}$$

and $P_l(\cos \omega)$, Legendre polynomial; β, screening parameter;

$$\delta(\omega) = \frac{Z^2 e^4}{\rho^2 v^2 (1 - \cos \omega + 2\beta)^2},$$

single scattering cross section; and $A = 2\pi N Z^2 e^4 / \rho^2 v^2$.

For small values of Δs or thin layers $\int_0^{\Delta s} X_l \, ds = X_l \, \Delta s$. The values of β are given by Moliere [18], Wentzel [19], Nigam *et al.* [20], and Cosslett and Thomas [14]. The value of Nigam *et al.* is best suited. The final result is shown in Fig. 8, and agreement is quite good.

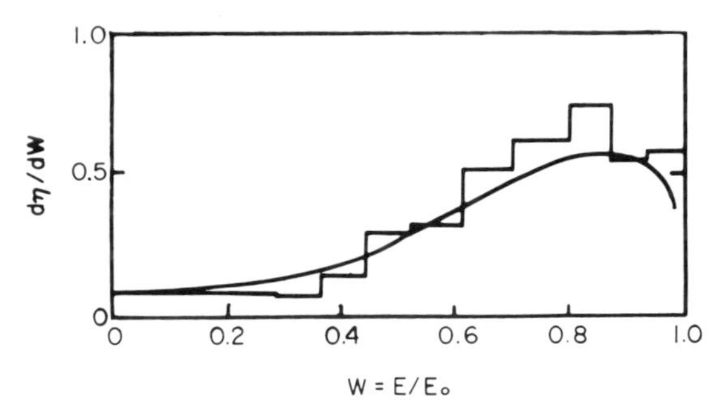

FIG. 8. Energy distribution of backscattered electron. Full curve: Kulenkampff and Spyra [30]. Histogram: result of present calculation (Cu, 30 kV; $\eta = 0.331 \pm 0.01$).

As to the energy loss, the following Bethe relation is used

$$-\frac{dE}{ds} = \frac{2\pi e^4 N Z}{E} \ln \frac{2E}{J} \tag{24}$$

and J is mean excitation potential. Then the energy of the electron at the ith step is given by

$$E_i = E_{i-1} + (dE/ds)\,\Delta S_i - 1. \tag{25}$$

The step length is taken as 1.8×10^{-5} cm for a copper target.

The Monte Carlo calculation for depth distribution of characteristic X rays will be taken from these data. Similar calculation done by Green [21] and Bishop [22] is based on the experimental data on electron scattering, and so it belongs to an empirical method. Even the present calculation method is semiempirical since selection of the most suitable value of β is necessary.

Although difficult, pure theoretical treatment will be required. As we have already found that there is not much difference between empirical and semi-empirical treatment, we cannot expect much new information with such difficult calculation.

2. *Angular Distribution of Characteristic X-Ray*

Although the characteristic X-ray is emitted isotropically, the intensity observed with a certain take-off angle will have a certain angular distribution since the X-ray is generated not only at the surface but also at a certain depth in the target. As this angular distribution is determined by the depth distribution, take-off angle, and absorption of X-ray in the target, we can deduce the depth distribution function by analysis of the angular distribution of the characteristic X-ray.

Let $\varphi(\rho z)$ be a depth distribution of the characteristic X-ray and θ_i a take-off angle; the intensity of X-ray observed is

$$F(X_i) = \int_0^\infty \varphi(\rho z) \exp(-X_i \rho z)\, d\rho z \tag{26}$$

where

$$X_i = \frac{\mu}{\rho} \csc \theta_i$$

According to Kirkpatrick and Hare [23], $F(X_i)$, which will be observed at the take-off angle of θ_i, is written as

$$F(X_i) = \frac{b_0}{X_i + H} + \frac{b_1}{(X_i + H)^2} + \frac{2b_2}{(X_i + H)^3}, \qquad i = 1, 2, \ldots, m+1. \tag{27}$$

provided that functional representation of the depth distribution is given by

$$\varphi(u) \doteqdot \left(\sum_{j=0}^{m} b_j u_i \right) e^{-Hu}. \tag{28}$$

However, if self absorption is small, the solution of (28) becomes inaccurate. Therefore, the minimum seeking method, widely applied in Optimal Control, using an electronic computer, is tried.

It actual procedure, it is convenient to select an appropriate objective function $f(y)$ and to adopt a slope correction method.

As a depth distribution, Gaussian distribution has been assumed, and $f(y)$ to be minimized is given by

$$f(y) = \sum_{j=1}^{m} \left| F(X_j) - \int_0^\infty \exp(-y_2(\rho z - y_1)^2) \exp(-X_i \rho z)\, d\rho z \right|^2 \tag{29}$$

3. *Comparison of Three Methods*

One example of comparison of these three methods is shown in Fig. 9, and good agreement will be seen. This tells us that to obtain depth distribution, any one of the three methods is applicable. The Monte Carlo method is suitable for investigating the behavior of electrons in the specimen. On the

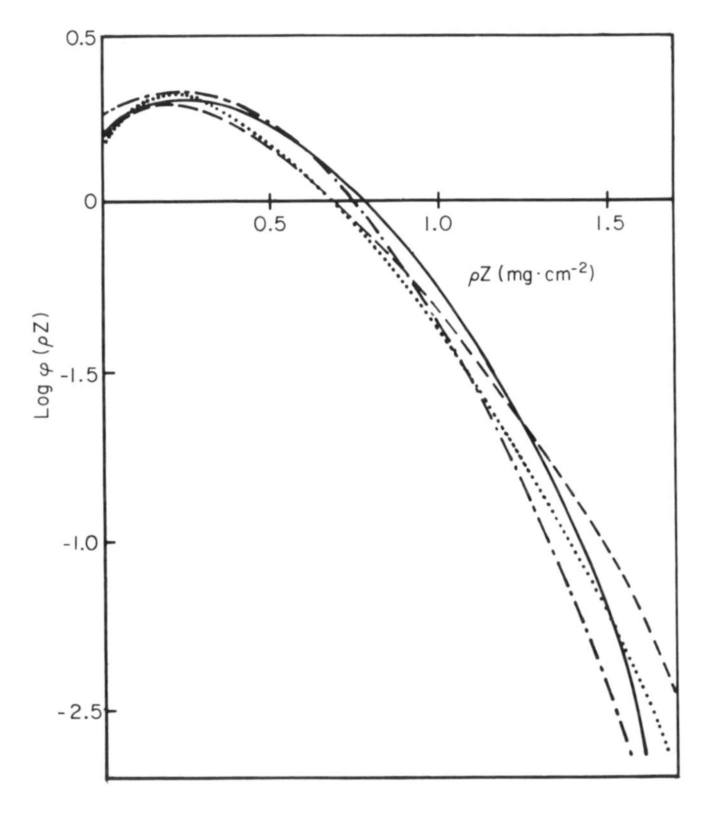

FIG. 9. Comparison of depth distribution curves of Cu K_α obtained by tracer method, Monte Carlo calculation, and angular distribution measurement. Castaing and Descamps' data are added for comparison. (— — —, tracer method, 28.5 kV, $\eta = 90°$; —, Monte Carlo calculation, 30 kV, $\gamma = 90°$; —–— analysis of angular distribution, 30 kV, $\gamma = 90°$; ---, Castaing and Descamps, 29 kV, $\gamma = 80°$; γ, incident angle).

other hand, from angular distribution of the characteristic X-ray, correction for absorption will be made without computing the depth distribution. Until now the depth distribution functions are determined only for several elements such as gold, silver, copper, aluminum, and carbon; so sometimes angular distribution curves provide convenient means of corrections for quantitative analysis. Figure 10 shows depth distribution for aluminum, and the result is quite satisfactory.*

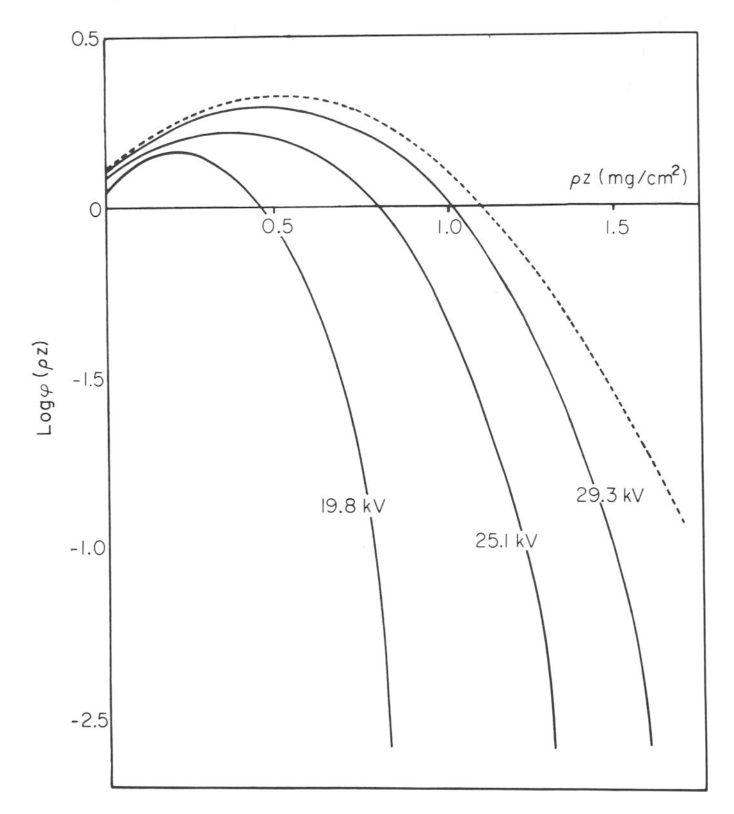

FIG. 10. Depth distribution of Al K_α. (—, Monte Carlo calculation; – – –, analysis of angular distribution; 30 kV, $\mu/\rho = 0.459$.)

IV. Resolving Power of Electron Microprobe

A. Introduction

Recently in the optical field, the response function, which is generally used for evaluating the resolving power of an optical system, has been used to evaluate the resolving power of an optical instrument. Usually, in the optical system we have various kinds of aberrations, and they play a very important

* For aluminum, at different voltages, some apparent discrepancies have been found. However, they are not significant, as shown by Shimizu *et al.* [23a].

role in determining the resolving power. But in the case of the electron microprobe, that which determines the resolving power is not the electron optical aberration but the effect of electron diffusion in a material because, as already stated in Section III, 0.5 $\mu\phi$ of electron beam increases to 3 $\mu\phi$ in the material by electron diffusion. On the other hand, electron optical deficiencies, such as the radius of the circle of confusion, are much smaller than those of the diffused X-ray source.

B. Resolving Power from Spot Size of X-Ray Source

The spot size of the X-ray source $2\delta_T$ is given by

$$\delta_T = (\delta_d{}^2 + \delta_0{}^2)^{1/2}$$

in which $2\delta_0$ is the spot size of the electron probe.

If we denote the radius of the circle of confusion due to spherical aberration of the electron optical system by δ_s, we have

$$\delta_0{}^2 = \delta_a{}^2 + \delta_s{}^2 \tag{30}$$

where δ_a is the spot size when aberration becomes zero, and δ_s is given by

$$\delta s = \tfrac{1}{4}C_s\left(\frac{4}{\sqrt{3}}\frac{J_0}{\pi^2 Cs^2 C_j \overline{V}_0}\right)^{3/4} \tag{31}$$

where C_s is the coefficient of spherical aberration, and C_j is a constant depending on the instrument, and

$$\delta_a{}^2 = \{2 \cdot x(3\pi^2)^{-3/4} I_0^{3/4} C_j^{-3/4} \overline{V}_0^{3/4} C_s^{1/2} \tag{32}$$

For instance, if $C_j = 2$ A/V $\cdot$ cm^2, $V_0 = 25$ kV, $C_s = 2.5$ cm, $I_0 = 0.4 \times 10^{-6}$ A, and $2\,\delta_a = 0.5 \times 10^{-4}$ cm, we have a reasonable value of the spot size, as $\delta_0 = 3.4 \times 10^{-5}$ cm.

Usually δ_d, which corresponds to spot size of the diffused X-ray source when the diameter of the initial electron beam becomes zero, is several times larger than δ_0. Thus, the factor which plays the most important role in resolution is not the aberration originated from electron optics but that due to electron diffusion.

Wittry [24] evaluated the volume from which the X-ray is generated and determined volume resolution, using Castaing and Descamps' depth distribution function and assuming the volume to be spherical. The result is as shown in Fig. 11. The minimum lies in the neighborhood of $u = 1.6$ to 2.3, u being a reduced accelerating voltage (i.e., $u = E_0/E_K$). The volume analyzed by Castaing's probe is thereby decreased to about one-quarter of that which would be obtained by using $\tilde{u} = 3$. But as Wittry's estimation of the volume of the X-ray source is based on the assumption above mentioned, we have

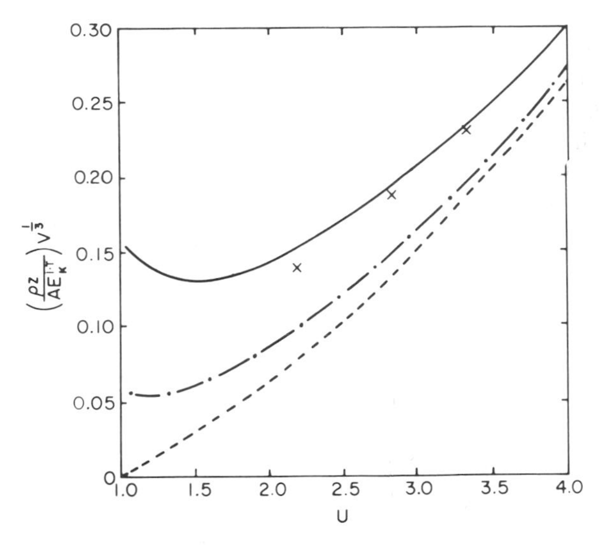

FIG. 11. Comparison of resolution estimated by Wittry and experimentally obtained by Shimizu and Shinoda; ×, Shinoda *et al.*; —, $\alpha p = 246$, Castaing's probe; —·—·— $\alpha p = 8100$, ideal probe; – – –, limiting value).

calculated volume resolution using our experimentally obtained data. The result is inserted in the same figure. The tendency is quite the same.

As pointed out by Bell and Sizmann [25], the emission of the characteristic X-ray depends on the orientation of the surface crystals of the target material, especially for exciting voltages just above the critical potential. From this standpoint, too low an accelerating potential is not preferable. In actual experiment, potentials of 20–25 kV will be reasonable for copper or iron.

Next, we must consider the problem of resolving power in the specimen current and backscattered electron method. Backscattered electrons are mainly composed of those scattered near the surface layer and have no electrons scattered at a depth deeper than X_d. Therefore, the resolution will be determined roughly from the diameter of the electron beam before it reaches the depth of complete diffusion. In other words, it is nearly equal to δ_0, which is much smaller than δ_T. Also, since (specimen current) = (beam current)—(backscattered electron current), the resolution for the specimen current method is nearly the same as that for the backscattered electron method. Figure 12 shows the result of linear scans of the titanium and silver bimetal boundary. In the X-ray curve the shoulder has considerable roundness; however, in the specimen current curve the change of slope occurs abruptly and shows high resolution. The radii of curvatures are 3.2 μ and 1.0 μ, respectively, and their ratio is nearly equal to that of δ_T and δ_0 (26).

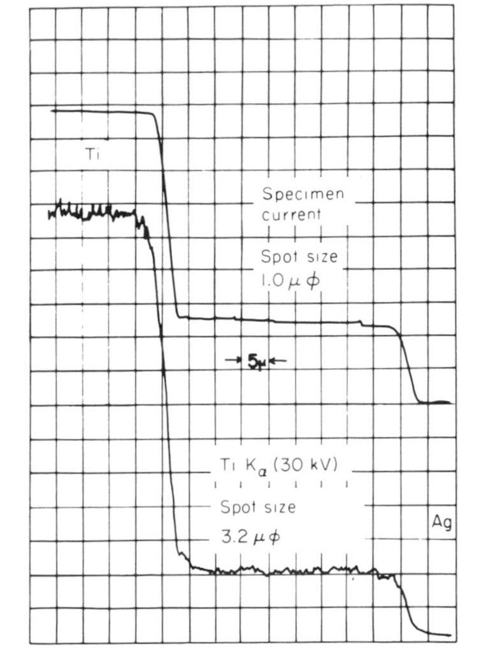

FIG. 12. Comparison of resolutions for X-ray intensity and specimen current in Ti–Ag (diffusion couple) system.

C. Resolving Power from Information Theory

According to the information theory, the optimum condition of the communication system is given by Shannon's capacity of communication C, where

$$C = W \log(1 + P/N) \tag{33}$$

W: frequency band of communication system
P: average power of signal
N: average power of noise.

Applying the sampling theorem, we have a formula for capacity per unit length

$$\bar{C} = \frac{1}{2} \int_{-N_0}^{N_0} \log_2 \left[1 + \frac{|Y(N)^2 \sigma_u{}^2(N)|}{\sigma_n{}^2(N)} \right] dN \tag{34}$$

where $\sigma_u{}^2$ and $\sigma_n{}^2$ are average powers of signal and noise, respectively, and N is spatial frequency. In the case of the electron probe, N means spatial frequency and $C = (n_s/ZN_0)\bar{C}$, n_s being the number of sampling points.

$Y(N)$ is a response function given by

$$Y(N) = I(N) \cdot \Phi(N). \tag{35}$$

Here, $I(N)$ is a response function for the original electron probe and is given by

$$I(N) = \exp(-\pi^2 \cdot N^2 \cdot \delta_0^2). \tag{36}$$

$\Phi(N)$ is related to conversion of the electron beam to X-ray and will be evaluated by the following equation if intensity distribution of the X-ray $\varphi(x)$ is known

$$\Phi(N) = \int_{-\infty}^{\infty} \varphi(x) \exp(-2\pi_i NX)\, dx. \tag{37}$$

But as this calculation is difficult, $\Phi(N)$ must be determined experimentally. Figure 13 shows $I(N)$, $\Phi(N)$, and $Y(N)$ as obtained by Shimizu [8]. When the

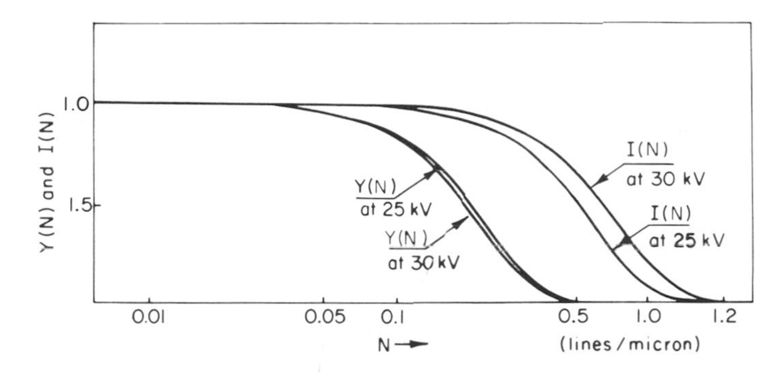

FIG. 13. Response function for electron probe $I(N)$ and diffused X-ray source $Y(N)$ in copper (25×10^{-7} A at 25 kV; 16×10^{-7} A at 30 kV).

accelerating voltage has increased, $Y(N)$ decreases; however, as it corresponds to the widening of the slit in the spectrometer system, the quantity of communication will increase thereby, and σ_u will become large. Then, from Eq. (34), the capacity will change. Since the optimum condition of operation corresponds to the maximum of the capacity, the accelerating voltage must be chosen according to the purpose of experiment and condition of the material.

As shown in Section IV,D, $I(N)$ will correspond to the response for the backscattered electron or specimen current method, and we can see that high frequency response is much improved in these methods.

Thus, conversion of the primary electron beam to X-ray calls for the addition of one converter, which is equivalent to a filter circuit; and it results in much increase of communication capacity and decrease of resolving power and signal to noise ratio.

D. Comparison of Backscattered Electron and Specimen Current Images. Secondary Electron Image. Future Possibilities

Figure 14 is a result of Monte Carlo calculation [26] and shows the relations between the number of electrons backscattered or absorbed and their depth in the specimen. Since the bottom of the histogram of backscattered electrons is shallower than x_d, and since, until this depth, the electron beam does not widen, the response function of the backscattered electron current will be nearly equal to $I(N)$, the function of the primary electron beam. As the value of the specimen current is determined by the backscattered electron current, their response functions must be equal.

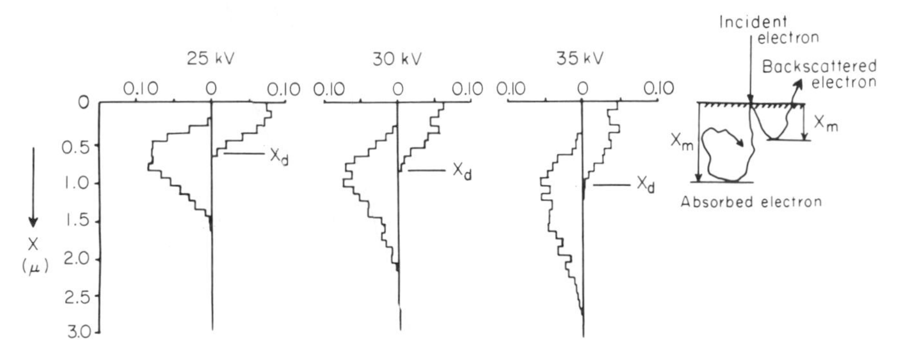

FIG. 14. Population for backscattered and absorbed electrons. Monte Carlo calculation.

In actual experiment, as shown in Fig. 15, the definition of the backscattered electron image is a little better than that of the specimen current. In the former case, only electrons which have been scattered in a certain direction are detected. Also, the effect of the secondary electron will be more significant, and therefore the condition will be slightly different.

However, as surface information is more marked in the backscattered electron method, the image will be much influenced by surface topography. For investigation of inner structure, some procedure to eliminate such effects of surface topography will be necessary. Kimoto and Hashimoto [27] used two detectors, an adder and a subtractor, to discriminate surface topograph and composition image. The specimen current method is suitable for usual metallurgical and mineralogical research purposes.

The backscattered electron method is preferable for the study of surface physics. However, information from backscattered electrons is not due to true surface structure but corresponds to a range at a certain depth. The use of secondary electron emission or photoelectron microscopy is one obvious approach to surface physics. Photoelectron microscopy will be more closely related to the state of the surface. But as it is related to from several to a few

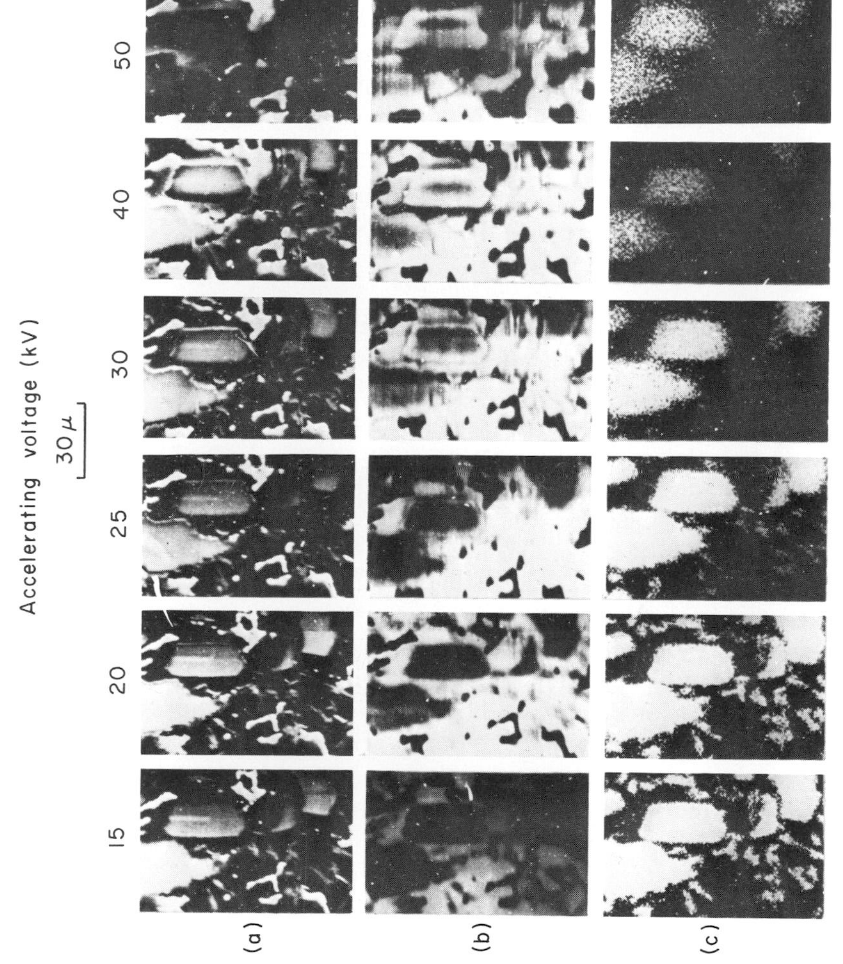

FIG. 15. Backscattered electron and specimen current images for Al–Si alloy, containing 19 wt % Si and a few per cent of Cu, Ni, Cr, Mg, and Ti. (a) Backscattered electron image; (b) absorbed electron image; (c) Si K_α image.

tens of atomic layers, it is very difficult to obtain true surface information, such as given by low energy electron diffraction. To study surface physics is one of the most important current problems; the electron microprobe will fullfill part of the necessary requirements.

Another important problem is how to increase the resolving power. This will be partly fulfilled if we can realize an electron beam having an intensity distribution of $\sin x/x$, instead of Gaussian (Shinoda and Suzuki [28]). Another important improvement will be made possible by the use of an electron optical system used in the scanning microscope. However, information concerning chemical compositions will be lost unless backscattering image is utilized.

V. Application of Backscattered Electron and Specimen Current Methods

A. *Introduction*

Backscattered electron and specimen current methods are used for quantitative analysis, especially for a material composed of elements whose atomic numbers differ greatly. To obtain accurate results, it is necessary that backscattered and absorbed ratios depend on mean atomic number only and do not depend on other physical properties such as crystal orientations (26). Usually these ratios are measured by materials composed of microscopic grains, and the beam diameter is large enough to cover the large numbers of crystal grains. But in the electron microprobe the diameter of the electron beam is usually far less than that of the crystal grain. Therefore, in these ratios the effect of crystal orientation will possibly appear.

This is very important from the standpoint of material research although from that of quantitative analysis, it is undesirable. As stated in the last section, backscattered electron and specimen current methods have an advantage in resolving power. For example, in a binary alloy composed of high and low atomic number elements, quantitative analysis will be done by the specimen current method without the help of X-ray data.

However, for an alloy composed of neighboring elements, combination of the X-ray and specimen current methods is preferable. Accurate compositions may be obtained from X-ray data while specimen current data will provide details of local variation of composition. In short, as each method has its own advantages and disadvantages, the combination of them can provide a better result.

B. *Aluminum–Silicon Alloy*

Although aluminum and silicon are neighboring elements, their contrast backscattered electron and specimen current images is high enough to distinguish both components easily. Usually, in microstructures of hyper-eutectic alloys of this system, polygonal primary silicon crystals are surrounded by eutectics. Since the thickness of these primary crystals is not great, an electron beam having high energy can penetrate through them; and, consequently, the definition of image is not good. However, if the accelerating voltage of electrons is not so high, they cannot penetrate the crystals; and a proper image, bright for silicon and dark for eutectics in the backscattered electron image, will be obtained. With the increase of accelerating voltage, fading of contrast will occur, and at a certain voltage even an inversion of blackening will be expected.

Figure 15 shows an example of this system. A remarkable change in contrast occurs between 40 and 50 kV in the backscattered electron image. In the specimen current image, change of contrast appears partly between 25 and 30 kV. The reason for such discrepancy is not clear; however, it might be explained as follows. In specimen current, electrons are absorbed mainly at a depth which is deeper than that at which backscattering has taken place. Also, while penetrating the specimen, the velocity of electrons decreases, resulting in an increase of the refractive index. Compared to the eutectics, silicon has a somewhat larger atomic number, and the rate of increase of the refractive index must be large. Then, at the interface of silicon and aluminum, total reflection of electrons will take place. It will result in an increase of electron absorption probability near the interface; and, consequently, a brighter rim will appear near the interface. This will be clearly seen in the specimen current images for 30 to 50 kV.

The resolution of backscattered electron images is better than specimen current images. This is because the backscattered image is composed of electrons scattered at a layer just under the surface. Also, in backscattered image, the effect of secondary electron emission and the relation between take-off angle and energy of the backscattered electron must be considered.

C. Copper–Zinc Alloy[5]

Anomalies in specimen current of a copper–zinc alloy [26] were first pointed out by Shirai *et al.* [29]. They found abnormally low specimen current in the γ-phase.

Figure 16 is a microphotograph of a copper–zinc alloy having nearly 50% zinc. The main constituent is β, and α has precipitated at certain grain boundaries. Figure 17 shows the linear scannings for Zn K_α and the specimen current. Apparently, no atomic number relations in specimen current exist between each phase. However, in another part of the specimen, normal atomic number relation holds between each phase. These discrepancies will be due to differences in crystal orientation. Then, even in precipitated α, the effect of original orientation of the matrix crystal may be conceivable, and the trace

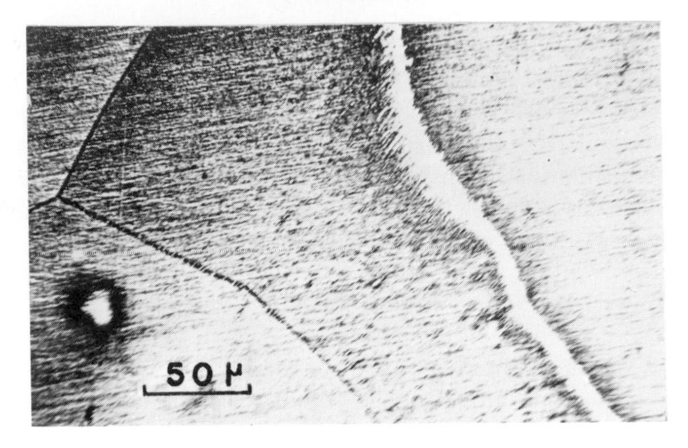

FIG. 16. Optical microphotograph of cast Cu–Zn alloy containing 45 wt % Zn. The α phase precipitation is seen in a certain grain boundary and orientation of each β grain is different.

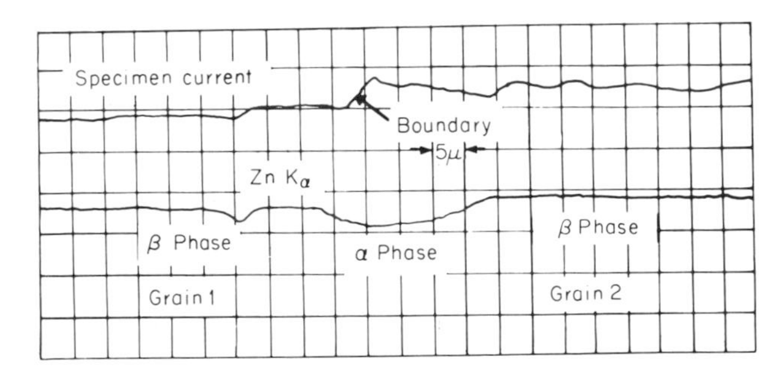

FIG. 17. The Cu K_α X-ray intensity and specimen current for Cu-Zn alloy containing 45 wt % Zn. Cast state.

of the original grain boundary will appear as an inflection point of the curve, as shown in the figure. Many important metallurgical conclusions will be pointed out, such as the fact that the composition of α at the initial state of precipitations is quite close to that of β, that there exist favorable and unfavorable crystal orientation relations for precipitation in the grain boundaries, and that composition of the matrix is not uniform.

D. Copper–Tin Alloy[5]

In tin–bronze casting, so-called cored or dendritic structure appears. The core consists primarily of crystallized and tin-poor crystals. In the grain

[5] See Shinoda [26].

boundaries, the β or decomposed β phase appears. Between these two structures, the filling with gradually changing compositions appears as shown in Fig. 18. Figure 19 is the result of linear scan for specimen current and Cu K_α X-ray. In this case, since the difference of atomic numbers of both elements is great, the specimen current has a nearly linear relation to the composition. Details of change of composition in each phase are clearly seen. For instance, the change of the composition in the filling is not continuous, but stepwise, suggesting spiral-like or multilayered cyrstallization; and precipitates in the

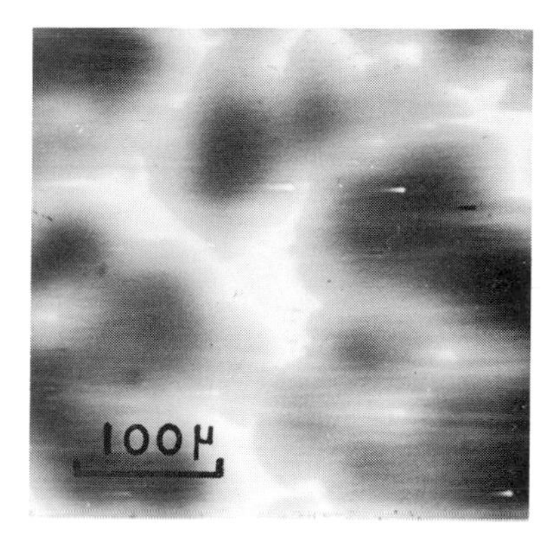

FIG. 18. Specimen current image of bronze casting containing 10 wt % Sn.

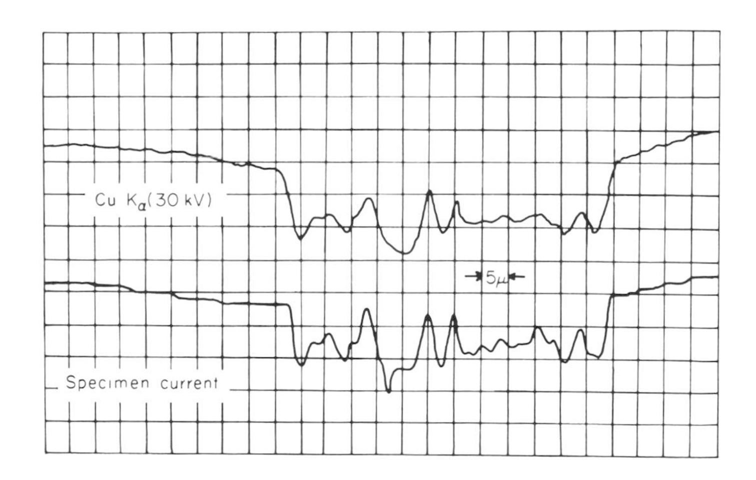

FIG. 19. The Cu K_α X-ray intensity and specimen current curves for bronze casting containing 10 wt % Sn. Around the core region stepwise concentration gradient is seen.

TABLE I
CONCENTRATION, SN WT %

Core, α	Filling α range	β (or γ) and its reactions products		
		β (or γ)	α	δ
$0.3 \sim 5.2$	$2.2 \sim 8.2$	$21.4 \sim 25.6$	$9.2 \sim 11.0$	$33.1 \sim 35.8$

grain boundary have partly decomposed to $\alpha + \delta$, with others remaining in the β or slightly transformed γ state. The composition of each component is as shown in Table I.

REFERENCES

1. R. Whiddington, *Proc. Roy. Soc.* (*London*) *Ser. A* **86**, 360 (1912).
2. H. A. Bethe, M. E. Rose and L. P. Smitt, *Proc. Am. Phil. Soc.* **78**, 573 (1938).
2a. P. Duncumb and S. J. B. Reed, Progress in the calculation of stopping power and backscatter effects, *Wash. Seminar Electron Microprobe Quant. Anal.*; *June 1967, Natl. Bur. Std.*
3. R. Castaing and J. Descamps, *J. Phys. Radium* **16**, 304 (1955).
4. W. Bothe, *Z. Physik* **54**, 161 (1929).
5. J. E. Everhart, *J. Appl. Phys.* **31**, 1483 (1960).
5a. D. B. Brown and R. E. Ogilvie, *J. Appl. Phys.* **37**, 4429 (1966).
6. P. G. Burkhalter, *US Bur. Mines Rept. Invest. 6681* (1965).
7. R. Shimizu and G. Shinoda, *in* "X-Ray Optics and X-Ray Microanalysis" (H. H. Pattee, V. E. Cosslett, and Arne Engström, eds.), p. 419. Academic Press, New York, 1963.
8. R. Shimizu, Thesis. Osaka Univ., Osaka, Japan, 1964.
9. J. L. H. Jonker, *Philips Res.* **9**, 391 (1954).
10. G. D. Archard, *J. Appl. Phys.* **32**, 1505 (1961).
11. R. F. Dashen, *Phys. Rev.* **A134**, 1025 (1964).
12. C. R. Worthington and S. G. Tomlin, *Proc. Phys. Soc.* (*London*) **A69**, 401 (1956).
13. G. D. Archard and T. Mulvey, *J. Appl. Phys.* **31**, 393 (1960).
14. V. E. Cosslett and R. N. Thomas, *Brit. J. Appl. Phys.* **15**, 235, 883 (1964).
15. R. Shimizu and G. Shinoda, *Technol. Rept. Osaka Univ.* **14**, 897 (1964).
16. W. Ehrenberg and J. Franks, *Proc. Phys. Soc.* (*London*) **B66**, 1057 (1953).
16a. H. Watanabe, private communication. Hitachi Central Laboratories, Tokyo.
17. R. Shimizu, H. Kishimoto, T. Shirai, K. Murata, G. Shinoda and M. Miura, *Technol. Rept. Osaka Univ.* **16**, 415 (1966).
18. G. Moliére, *Z. Naturforsch.* **3a**, 78 (1948).
19. G. Wentzel, *Z. Physik* **40**, 590 (1927).
20. B. P. Nigam, M. K. Sundaresan and Ta-You Wu, *Phys. Rev.* **115**, 491 (1959).
21. M. Green, *Proc. Phys. Soc.* (*London*) **82**, 204 (1963).
22. H. E. Bishop, *Proc. Phys. Soc.* (*London*) **85**, 855 (1965).
23. P. Kirkpatrick and D. G. Hare, *Phys. Rev.* **46**, 831 (1935).

23a. R. Shimizu, K. Murata and G. Shinoda, *Technol. Rept., Osaka Univ.* **17**, 13 (1967).
24. D. B. Wittry, *J. Appl. Phys.* **29**, 1543 (1958).
25. F. Bell and R. Sizmann, *Phys. Letters* **19**, 171 (1965).
26. G. Shinoda, H. Kawabe, K. Murata, and T. Shirai, *Technol. Rept. Osaka Univ.* **16**, 423 (1966).
27. S. Kimoto and H. Hashimoto, *in* "The Electron Microprobe" (T. D. McKinley, K. F. J. Heinrich, and D. B. Wittry, eds.), p. 480. Wiley, New York, 1966.
28. G. Shinoda and T. Suzuki, *Technol. Rept. Osaka Univ.* **10**, 681 (1960).
29. S. Shirai, Y. Kawakami, A. Shimizu and Y. Yamada, *Fall Meeting Japan Soc. Appl. Phys. Osaka, 1964*; G. Shinoda, *in* "X-Ray Optics and Microanalysis" (R. Castaing, P. Deschamps and J. Philibert, eds.), p. 97. Hermann, Paris, 1966.
30. W. Kulenkampff and W. Spyra, *Z. Phys.* **137**, 416 (1954).

The Sandwich Sample Technique Applied to Quantitative Microprobe Analysis*

JAMES D. BROWN

Faculty of Engineering Science
The University of Western Ontario
London, Ontario, Canada

I. Introduction

In his thesis, Castaing [1] outlined the basic concepts for quantitative electron probe microanalysis using pure elements as standards. At present, analyses based on these concepts are limited to approximately $\pm 5\%$ of the amount present by inaccuracies in the correction equations, the mass absorption coefficients, and the other parameters used in the equations. Experience of the Washington Area Probe Users Group indicates a limitation of 1 or 2% in the measurement of X-ray intensities with an electron probe microanalyser. Thus, improvement in the correction techniques can result in an improvement of quantitative analysis.

In the quest for better quantitative analysis, several empirical techniques have been published, notably those by Ziebold and Ogilvie [2]. Provided suitable standards are available, these empirical methods are entirely satisfactory.

* Research described in this report was carried out while a member of the staff of the U.S. Department of the Interior Bureau of Mines, College Park Research Center, College Park, Maryland.

However, because of the difficulty of preparing standards for these empirical methods, the full potential of quantitative electron probe microanalysis cannot be realized until quantitative techniques are available for all samples using pure elements as standards.

Castaing proposed the use of relative X-ray intensities for quantitative analysis. The measured intensity of a characteristic line of an element from a sample is compared with the intensity of the same line, measured under identical conditions from the pure element. As a first approximation this ratio is set equal to the concentration of the element in the sample. However, four corrections may be necessary to convert these relative intensities to composition of the sample. These are corrections to the actual intensities emitted from the sample. Thus, it will be assumed that dead time, drift, and background corrections have been made. Definitions for the four major corrections are as follows.

The incident electrons generate X-rays at some depth in the sample. These X-rays must pass through a part of the sample to be detected by the X-ray spectrometer. Some absorption takes place during this process. The magnitude of this absorption depends on the composition of the sample and, therefore, differs between pure element standard and sample. The effect of this difference in absorption on the measured relative intensity is termed the *absorption effect*.

If the energy of a characteristic line of one element is slightly greater than the absorption edge energy of a second element in the same sample, then strong absorption of the characteristic line occurs. The absorbing element is excited and in returning to the ground state it may emit characteristic quanta. These add to the intensity observed for that element. This increase in intensity of the characteristic lines of one element due to absorption of the characteristic lines of another is termed the *fluorescence effect*.

The same effect can occur due to absorption of continuum radiation whose energies are greater than the absorption edge energy. The radiation is absorbed by atoms which can reemit characteristic photons, adding to the observed intensity. This effect is termed *fluorescence due to the continuum*.

Finally, the number of characteristic X-rays produced in the sample per incident electron is a function of the average atomic number of the sample. Thus, the number of Fe K_α X-rays produced in a sample of 1 wt % Fe in Al is quite different from the number produced for the same concentration of Fe in U for the same excitation conditions. Differences in backscattering and stopping power for electrons have been shown to be the cause for this effect, known as the *atomic number effect*.

Equations have been developed for correcting for all of these effects so that measured X-ray intensities can be converted to chemical composition. All of these equations are developed from a model proposed by Castaing

[1] for electron probe microanalysis based on X-ray production and absorption from planes parallel to the surface of the sample. Application of this model to the understanding of the four corrections to microprobe data and the experimental techniques for verifying the results of this model are the subject of this paper.

II. The Absorption Correction

A. Theoretical Derivation

The geometry used in deriving the absorption correction is shown in Fig. 1. Consider a thin slice $d\rho z$ at a depth ρz in a sample of pure element A, where ρ, the density, is introduced to define path and thickness in terms of mass units. Let the characteristic intensity of A in the thin slice relative to

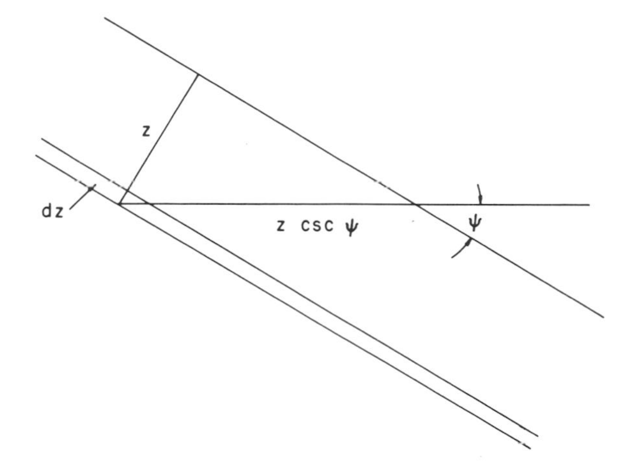

Fig. 1. Geometry of the absorption correction.

the intensity in a slice of identical thickness isolated in space be $\phi_A(\rho z)$. In leaving the sample, X-rays from the layer $d\rho z$ are absorbed along the path $\rho z \csc \psi$ where ψ is the X-ray take-off angle. If the intensity $\phi_A(\rho z)$ is in appropriate units with respect to solid angle, then the intensity which reaches the detector from the thin slice $d\rho z$ is

$$dI_A^{A'}(\rho z) = \phi_A(\rho z) \exp(-\mu_A{}^A z \csc \psi)\, d\rho z \tag{1}$$

where $\mu_A{}^A$ is the linear absorption coefficient of A for its characteristic radiation. The total intensity $I_A^{A'}$ reaching the detector from a sample of pure element A is obtained by integrating Eq. (1) over all depths in the sample, i.e.,

$$I_A^{A'} = \int_0^\infty \phi_A(\rho z) \exp(-\mu_A{}^A z \csc \psi) \, d\rho z. \tag{2}$$

The difficulty with this expression is that it is the intensity $I_A{}^A$ directly excited by the incident electrons within the sample that is related to the concentration of the element, not the measured X-ray intensity. If there were no absorption by the sample, then the intensity measured by the detector would be

$$I_A{}^A = \int_0^\infty \phi_A(\rho z) \, d\rho z. \tag{3}$$

In terms of the measured intensity

$$I_A{}^A = \frac{I_A^{A'} \int_0^\infty \phi_A(\rho z) \, d\rho z}{\int_0^\infty \phi_A(\rho z) \exp(-\mu_A{}^A z \csc \psi) \, d\rho z} = \frac{I_A^{A'}}{f_A(\chi_A)} \tag{4}$$

where $\chi_A = \mu_A{}^A z \csc \psi$, and $f(\chi) = F(0)/F(\chi)$, $F(\chi)$ and $F(0)$ being the integrals in the denominator and numerator of Eq. (4), respectively. For a sample which is not the pure element A, but which has a uniform concentration of A in a matrix, Eq. (4) for the intensity of A generated in the sample is

$$I_A{}^S = \frac{I_A^{S'} \int_0^\infty \phi_S(\rho z) \, d\rho z}{\int_0^\infty \phi_S(\rho z) \exp(-\mu_A{}^S z \csc \psi) \, d\rho z} = \frac{I_A{}^{S'}}{f_S(\chi_S)} \tag{5}$$

where $I_A^{S'}$ is the measured intensity of A from the sample.

Assuming that the relative intensity generated in the sample is equal to the concentration C_A, then

$$C_A = \frac{I_A{}^S}{I_A{}^A} = \frac{I_A^{S'}}{I_A^{A'}} \cdot \frac{f_A(\chi_A)}{f_S(\chi_S)}, \tag{6}$$

and $f_A(\chi_A)/f_S(\chi_S)$ is termed the absorption correction. The value of the function $f(\chi)$ can be determined for any value of χ if the distribution of the generation of X-rays is known as a function of depth. The function $f(\chi)$ can also be determined by measuring the variation in observed intensity from a single sample as a function of take-off angle (and hence χ). Four geometries which have been used for these measurements are shown in Fig. 2. Typical $f(\chi)$ curves are given in Fig. 3 for copper. Although the absorption correction is calculated from the values of $f(\chi)$, nothing is gained toward the determination of the other major corrections to microprobe data. The significance of the $\phi(\rho z)$ distribution function is that once it is known, the other corrections can be calculated from it.

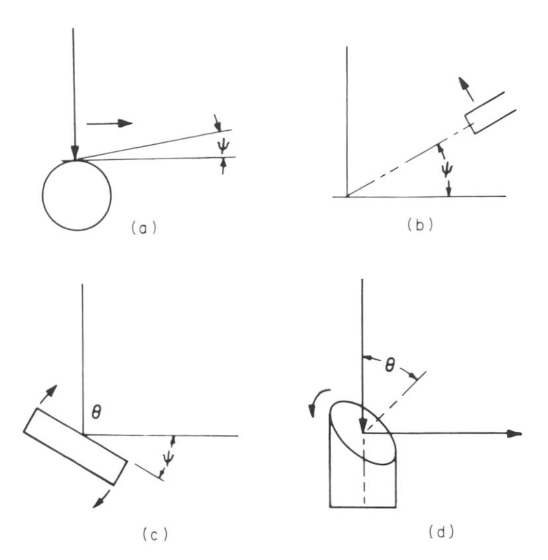

FIG. 2. Methods of measuring $f(\chi)$ curves: (a) Translate a wire sample past a stationary beam, then ψ changes but so does θ (from Castaing [1]). (b) Move detector in circle about focal spot perpendicular to sample surface to change ψ (from Green [3]). (c) Rotate sample about axis in its surface. This changes both ψ and θ, (from Philibert [4]). (d) Rotate sample about axis of electron beam. ψ changes, θ remains constant $\theta \neq 0°$ (from Macres [5]).

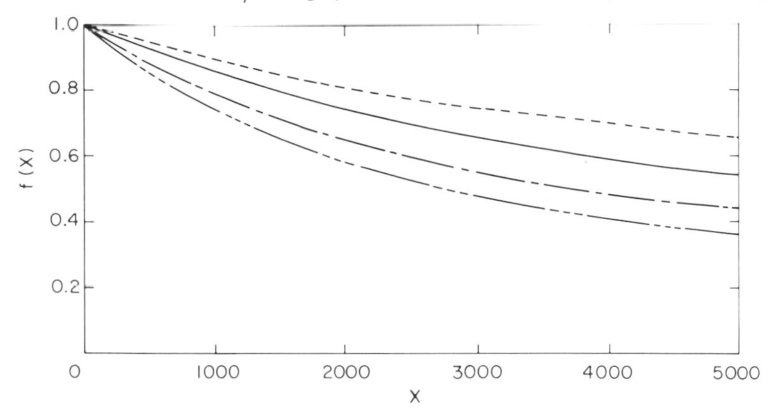

FIG. 3. The $f(\chi)$ curves from measured $\phi(\rho z)$ curves for copper [electron voltage (keV) - - - , 13.4; —, 18.2; - - -, 23.1; - - - - -, 27.6].

Attempts have been made to calculate the $\phi(\rho z)$ function from theoretical models as well as to measure it experimentally by use of tracers. It is also theoretically possible to calculate $\phi(\rho z)$ by an inverse transformation of the $f(\chi)$ curves [6]. The latter method has not been successful, probably because the $f(\chi)$ curves are not too sensitive to the shape of $\phi(\rho z)$ as is evidenced by the good agreement with the $f(\chi)$ curves obtained using the Archard diffusion model [7] although the $\phi(\rho z)$ curves were quite erroneous.

B. Experimental Measurement

Castaing and Descamps [8] first proposed the sandwich sample technique for measuring the distribution of X-ray production as a function of depth in the sample. A cross-sectional view of the sandwich sample is shown in Fig. 4.

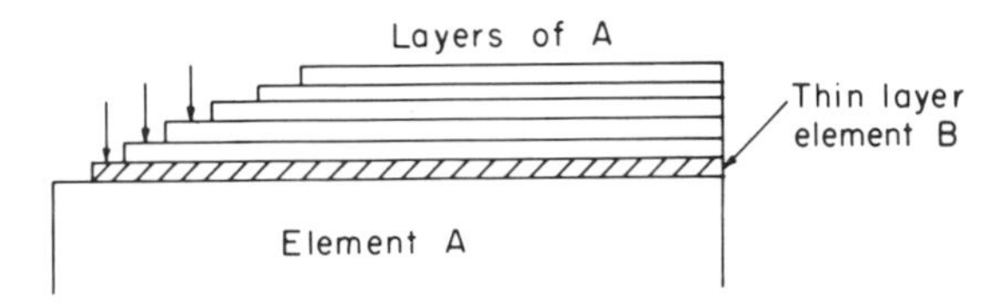

FIG. 4. Sandwich sample used to measure $\phi(\rho z)$.

On a polished block of a pure element A, is deposited a thin layer of a second pure element B. This second element is chosen to be one greater in atomic number than A so that its properties with regard to excitation by electrons will be essentially identical to those of A but the characteristic lines of A are not able to cause fluorescence of the characteristic lines of B. The thin layer B is then covered in part by successively thicker layers of A so that the result is a sample consisting of a thin layer of B buried at various depths in a sample of pure element A.

If the intensity from the B layer is measured at each depth in the sample of A, then a curve is obtained which gives the distribution of production of X-rays in a sample of pure A as a function of depth. To put such measured curves on an absolute basis, Castaing and Descamps measured the intensity from an identical thickness of the tracer element B suspended in space. The ratio of the intensity from the sandwich sample to the intensity from the tracer isolated in space gives the values of $\phi(\rho z)$ at each depth. The measured curves of Castaing and Descamps for Al, Cu, and Au are shown in Fig. 5. These have been replotted on the basis of a linear scale for $\phi(\rho z)$ rather than the logarithmic scale of the original paper. Unfortunately, the details of the preparation of the sandwich samples by Castaing and Descamps are rather sketchy.

Except for some later measurements on the effect of the tracer element on the $\phi(\rho z)$ curves for an aluminum matrix [9], no further measurements of $\phi(\rho z)$ were made for more than ten years. In 1966, I measured $\phi(\rho z)$ curves for copper at four electron voltages [10] (Fig. 6). Details of sample preparation techniques and the methods of measuring sample thickness are given in

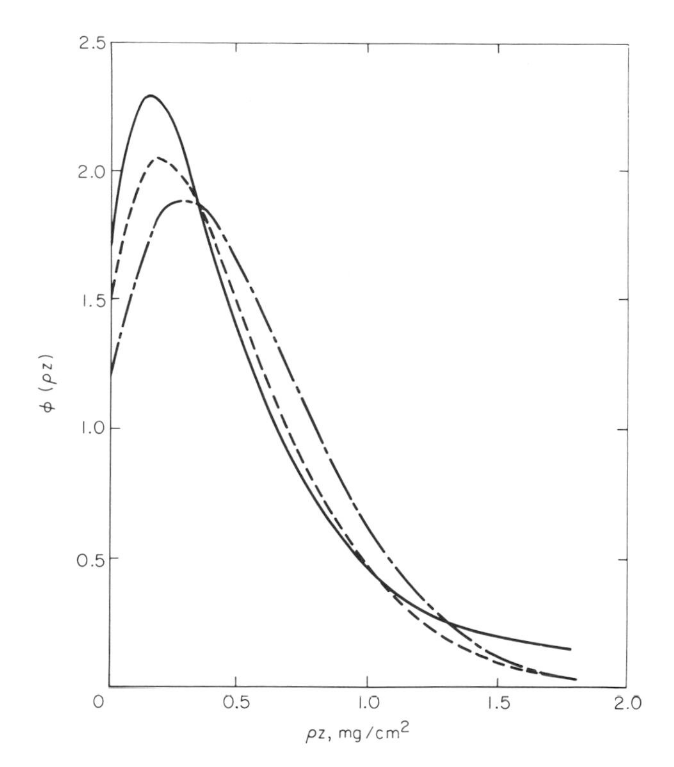

FIG. 5. The $\phi(\rho z)$ curves of Castaing and Descamps measured at 29 keV [—, Au (Bi tracer); – – –, Al (Cu tracer); - - -, Cu (Zn tracer)].

my thesis. The geometry of the instrument used in the measurement was somewhat different from the previous measurements by Castaing and Descamps. The electron beam was incident on the sample at an angle of 30° from the normal rather than the 10° of the earlier measurements. Comparison of the two sets of results is shown in Fig. 7. The difference in depth of production because of the greater inclination of the sample can be removed from the comparison by multiplying the depth of production by the geometric factor sec 30°/sec 10°. The result of this recalculation is also shown in Fig. 7. The remaining difference between the curves can be laid to a difference of 1.4 keV in the electron accelerating voltage.

The curves reproduced in Figs. 5 and 6 represent seven of the ten $\phi(\rho z)$ curves measured by 1966. Recently Vignes and Dez [26] have measured four curves for titanium and two for lead. These 16 curves are the only experimental data presently available for comparison with theoretical calculations of the $\phi(\rho z)$ curves.

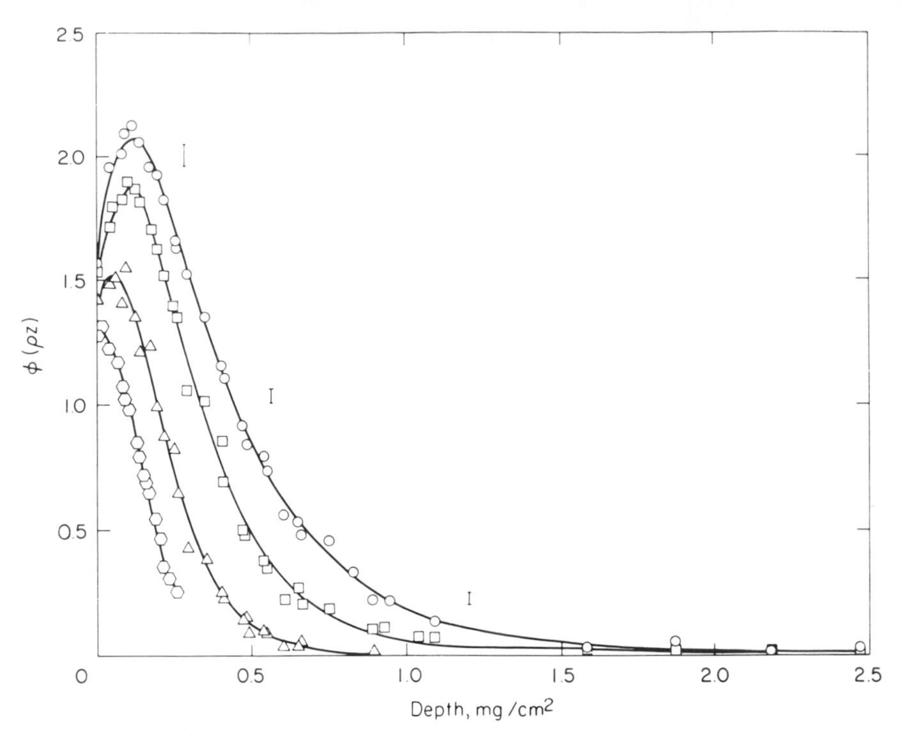

FIG. 6. The $\phi(\rho z)$ curves for copper measured at four electron voltages [electron voltage (keV): ○, 27.6; □, 23.1; △, 18.2; ⬡, 13.4].

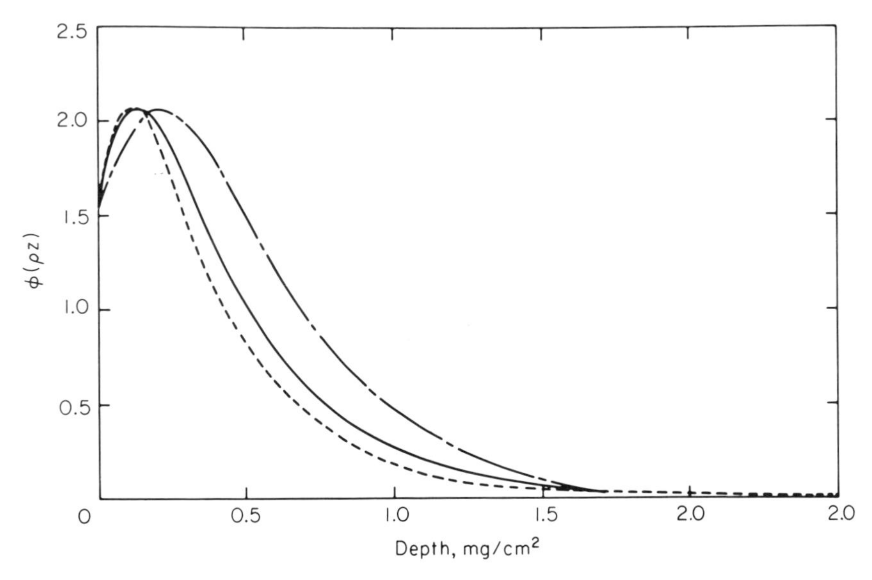

FIG. 7. Comparison of measured $\phi(\rho z)$ [– – –, Castaing and Descamps, 29 keV; - - -, Brown, 27.6 keV; —, Brown, 27.6 keV, corrected for electron incidence].

C. Calculation of $\phi(\rho z)$ Curves

Several models have been proposed in which the basic properties of the interactions of electrons and atoms could be used to calculate the $\phi(\rho z)$ curves and the absorption $f(\chi)$ curves. Archard [7] proposed a model which has come to be known as the Archard diffusion model. In this model the electrons penetrate in a straight line to a depth in the sample corresponding to an electron range. From this point they diffuse equally in all directions and laws of absorption are used to calculate the generation of X-rays. He was able to obtain good agreement with $f(\chi)$ curves for a number of elements, but the $\phi(\rho z)$ curves were quite in error. Monte Carlo methods have been used by Green [11] and Bishop [12] and a transport equation has been used by Brown and Ogilvie [13] for similar calculations. These methods show promise, for not only is good agreement with $f(\chi)$ curves obtained but also the $\phi(\rho z)$ distributions are in good agreement with the measured $\phi(\rho z)$ curves. The major difficulty with these models lies in a lack of reliable input parameters for the electron interactions and a lack of measured curves for a variety of elements and electron voltages for comparison with the results of the calculations.

D. The $f(\chi)$ Curves Derived from Measured $\phi(\rho z)$ Curves

Once $\phi(\rho z)$ curves have been measured, the $f(\chi)$ curves can be calculated directly. Table I gives a comparison of $f(\chi)$ values calculated from the $\phi(\rho z)$

TABLE I
COMPARISON OF $f(\chi)$ VALUES CALCULATED FROM MEASURED $\phi(\rho z)$ CURVES WITH PHILIBERT'S EXPRESSION

	$f(\chi)$ Values							
	27.6 keV		23.1 keV		18.2 keV		13.4 keV	
χ	Meas.	Philibert	Meas.	Philibert	Meas.	Philibert	Meas.	Philibert
100	0.967	0.956	0.974	0.968	0.984	0.980	0.989	0.991
200	0.936	0.915	0.949	0.938	0.968	0.962	0.979	0.983
500	0.853	0.809	0.882	0.857	0.924	0.909	0.949	0.958
800	0.783	0.724	0.824	0.788	0.883	0.861	0.920	0.935
1000	0.742	0.676	0.789	0.747	0.857	0.832	0.902	0.920
1500	0.655	0.578	0.714	0.661	0.799	0.766	0.860	0.884
2000	0.586	0.503	0.652	0.591	0.748	0.709	0.821	0.850
3000	0.484	0.396	0.556	0.484	0.663	0.616	0.754	0.790
4000	0.412	0.329	0.486	0.409	0.595	0.543	0.697	0.737
5000	0.360	0.270	0.432	0.351	0.540	0.484	0.647	0.691

curves of copper (Fig. 6) with values calculated from Philibert's expression for $f(\chi)$ [4] as modified by Duncumb and Shields [14]. The inclined electron beam was taken into account in this comparison by multiplying the value of σ by the geometric factor sec θ. Except at 13.4 keV, the measured $f(\chi)$ values are consistently and significantly higher than those predicted by the Philibert expression. The factor sec θ is therefore not sufficient to account for the decreased depth of production of X rays in an inclined sample. Bishop [12] suggested that by using the expression $\chi' = \chi(1 - b\cos^2\theta)$, where b is a constant, $f(\chi')$ curves for other than normal incidence could be obtained from $f(\chi)$ curves for normal incidence (Table II). The values at 27.6 keV

TABLE II
COMPARISON OF MEASURED AND CALCULATED $f(\chi)$ VALUES WHEN USING THE MODIFICATION OF BISHOP[a]

	$f(\chi)$ Values							
	27.6 keV		23.1 keV		18.2 keV		13.4 keV	
χ	Meas.	Calc.	Meas.	Calc.	Meas.	Calc.	Meas.	Calc.
100	0.967	0.967	0.974	0.978	0.984	0.987	0.989	0.993
200	0.936	0.935	0.949	0.953	0.968	0.973	0.979	0.987
500	0.853	0.855	0.882	0.894	0.924	0.933	0.949	0.970
800	0.783	0.786	0.824	0.838	0.883	0.896	0.920	0.952
1000	0.742	0.747	0.789	0.805	0.857	0.873	0.902	0.941
1500	0.655	0.658	0.714	0.733	0.799	0.820	0.860	0.914
2000	0.586	0.593	0.652	0.670	0.748	0.773	0.821	0.887
3000	0.484	0.485	0.556	0.570	0.663	0.692	0.754	0.840
4000	0.412	0.406	0.486	0.496	0.595	0.626	0.697	0.797

[a] See Bishop [12].

agree to within 1%. At lower voltages the data are overcorrected. It therefore appears that a voltage dependent correction for sample tilt is required.

III. FLUORESCENCE DUE TO CHARACTERISTIC LINES

A. Theoretical Derivation

Several correction procedures have been proposed by Castaing [1], Wittry [15, 16], Birks [17], Criss and Birks [18], and Reed [19]. The usual geometry for calculating the fluorescence correction is shown in Fig. 8. Primary radiation of element B is generated at point P, in the layer $d\rho z$ at a

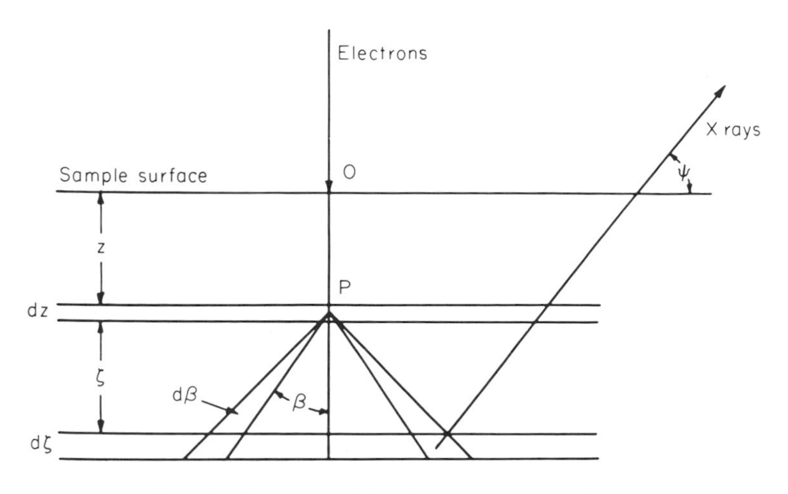

FIG. 8. Geometry of the fluorescence correction.

depth z in the sample. This primary radiation is emitted isotropically from P. The intensity of B radiation, absorbed by A atoms in the layer $d\zeta$ at a depth $z + \zeta$ in the sample is calculated. By integrating over all depths ζ, the contribution due to absorption of B radiation generated at P is obtained. Finally, by integration of all depths z at which primary B radiation is generated, the total contribution of primary B radiation, absorbed by A atoms and re-emitted as A characteristic quanta, is obtained. This same procedure in the calculation of the fluorescence intensity is used by all of the authors listed above. The resultant expressions differ only because different expressions are used to approximate the $\phi(\rho z)$ curve for the primary radiation and to relate the intensities of characteristic lines of different elements. The simplest approximation that has been used is that the primary radiation is generated at a point at the surface of the sample. Castaing assumed that the primary radiation was an exponential function of depth in the sample. Wittry fitted the equation

$$\phi(\rho z) = C\{1 - [\exp(-b(\rho z - \rho z_0)) - 1]^2\} \tag{7}$$

where C and b are arbitrary constants, to the curves of Castaing and Descamps at two arbitrary points. Finally, Criss and Birks used an exponential series

$$\phi(\rho z) = \sum_{i=1}^{N} A_{iA} \exp(-b_{iA}\rho z) \tag{8}$$

where the A_i's and b_i's are constant, to obtain complete agreement with the $\phi(\rho z)$ curves. The important point is that if the $\phi(\rho z)$ curve for the primary radiation is known, then the correction for fluorescence by characteristic lines can be calculated.

B. The Distribution of Secondary Radiation as a Function of Depth

1. *Experimental Measurement*

The distribution of secondary radiation as a function of depth in the sample can be measured using the same sandwich sample technique as Castaing and Descamps [8], but with different tracer elements. If the tracer element is chosen so that the characteristic X-rays of the matrix element can excite it, then the $\phi(\rho z)$ curve measured for such a sample is the sum of contributions due to directly excited radiation, fluorescence by the continuum, and fluorescence by characteristic lines. By subtracting from the $\phi(\rho z)$ curve of this sample, the $\phi(\rho z)$ curve for the primary distribution, the distribution of the secondary radiation is obtained.

This secondary distribution can be determined from a single sample if the fluorescence of L lines of a tracer element by K lines of the matrix element is measured. If the tracer element is chosen so that the K_α line of the matrix element is intermediate in energy between the L_{II} and the L_{III} edges, then selective fluorescence of the L_{α_1} line occurs without equivalent fluorescence of the L_{β_1} line. Such a pair of elements is copper and dysprosium. The copper

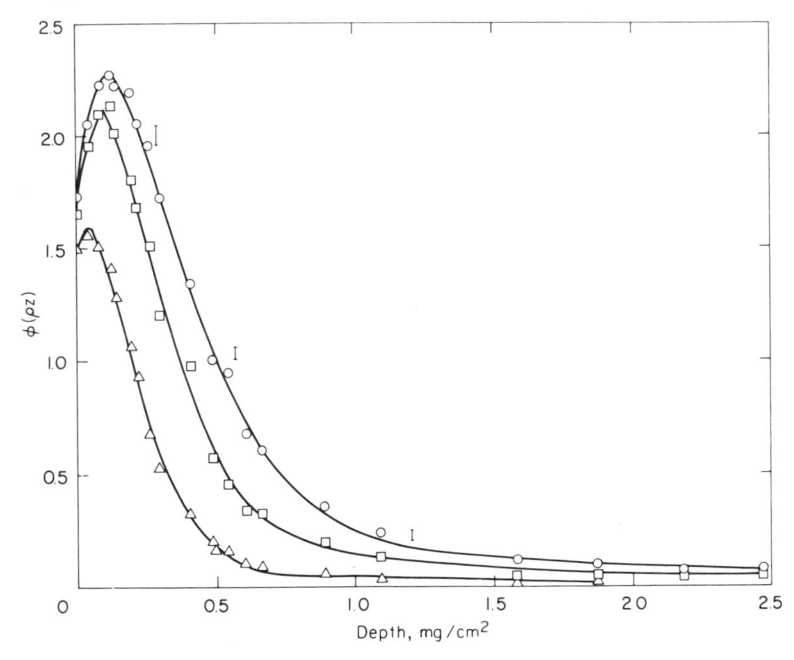

FIG. 9. Measured $\phi(\rho z)$ curves for an iron tracer in copper [electron voltage (keV): ○, 27.6; □, 23.1; △, 18.1].

K_α line has an energy of 8.04778 keV, which is between the L_{II} and L_{III} edge energies of dysprosium, which are 8.5830 and 7.7897 keV, respectively. The difference in the measured distributions for the Dy L_{α_1} and Dy L_{β_1} lines is the distribution of the fluorescence of the Dy L_{α_1} by the Cu K_α line. Differences between the curves related to the difference in excitation potential of the L_{II} and L_{III} edge and the fluorescence due to the K_β lines are not significant.

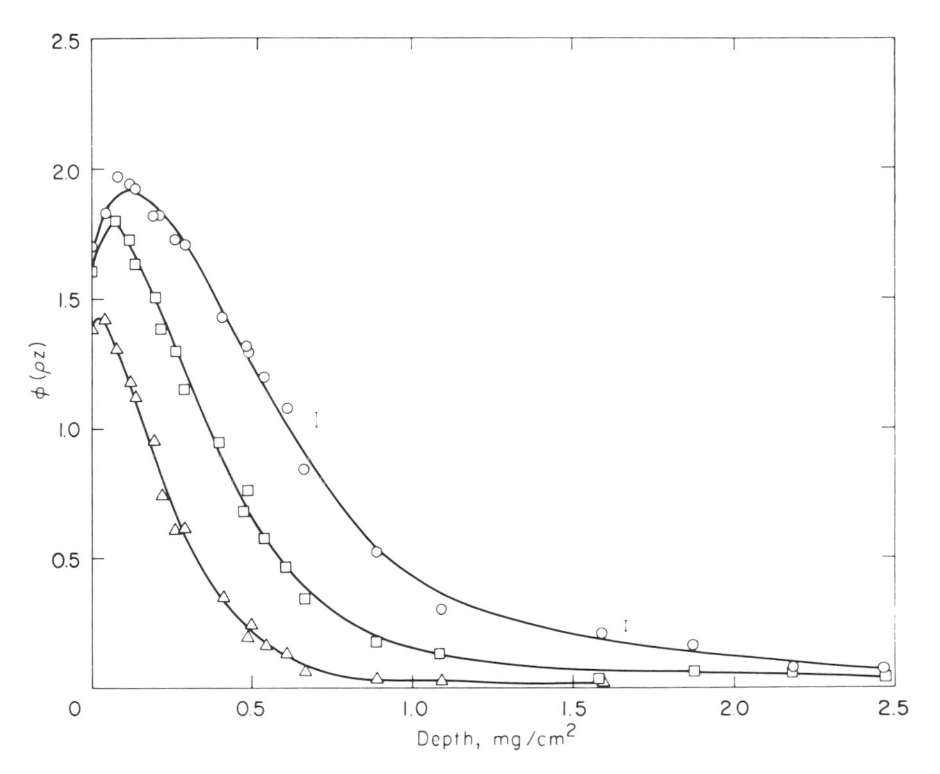

FIG. 10. Measured $\phi(\rho z)$ curves for Dy L_{α_1} in copper [electron voltage (keV): ○, 27.6; □, 23.1; △, 18.2].

Such measurements have been carried out [10]. Sandwich samples of Zn, Ni, Co, Fe, and Dy tracers in a copper matrix were prepared and the $\phi(\rho z)$ secondary distributions were measured at three electron voltages. The resultant curves are shown in Figs. 6, 9, and 10. All of the curves plus the data used in plotting them are given in my thesis [10]. In the instrument used in the measurements, the beam was inclined at an angle of 30° relative to the normal to the sample surface. The curves drawn in Figs. 6, 9, and 10 were fitted to the data points using the following techniques. The expression used was an exponential series suggested by Criss and Birks [18], of the form

$$\phi(\rho z) = \sum_i A_i \exp(-b_i \rho z) \tag{9}$$

where A_i and b_i are constants.

An analog computer was used to generate a five-term series in which the A_i's and b_i's could be adjusted independently. The experimental points were plotted on graph paper and compared with the computer output on an X–Y plotter.

From the computer settings the b's for the exponential series were calculated. A digital computer was then used to adjust the A's to give the best least-squares fit to the experimental data. The resultant curves, which are plotted in Figs. 6, 9, and 10, do not necessarily represent the best fits which could be obtained using a five-term exponential series, since some judgment was involved in determining the b parameters with the analog computer.

Some differences are apparent between the curves of Figs. 9 and 10 that include the secondary contribution and the primary curves of Fig. 6. The most significant difference is the increase in the tail of the curves that include the secondary contribution. This effect is expected since the secondary radiation is generated at relatively greater depths in the sample. For a single sample, the most striking features are the decrease in the value of $\phi(\rho z)$ at the surface of the sample and the rapid decrease in the volume in which the X-rays are produced as the voltage decreases.

The secondary distributions of Fig. 11 are generated by subtracting a $\phi(\rho z)$ curve for the primary radiation from the appropriate $\phi(\rho z)$ curve that includes the secondary radiation. The curves of Fig. 11 represent the difference between the fitted equations—not the difference in experimental data. The reason for this is that although the error for determining the total curves is small, the secondary distribution is the relatively small difference between two curves. Note that the vertical axis of Fig. 11 is 0–0.25 rather than the 0–2.5 for the total $\phi(\rho z)$ curves. For individual measurements of the secondary distribution the errors of measurement are 20–30% of the value of the secondary radiation. By using the calculated curves, the errors in the measurement of individual points are damped out and reasonable curves are obtained. The secondary distributions are all characterized by a hump in the region where the primary radiation is generated, and a slowly decreasing tail. The secondary curves at a given voltage should all have similar shapes, differing only in magnitude since the shape of each curve depends on the absorption properties of the sample which is essentially pure copper. Practical difficulties in the evaporation of the copper limited the maximum depth at which the distributions were measured to 2.5 mg/cm^2.

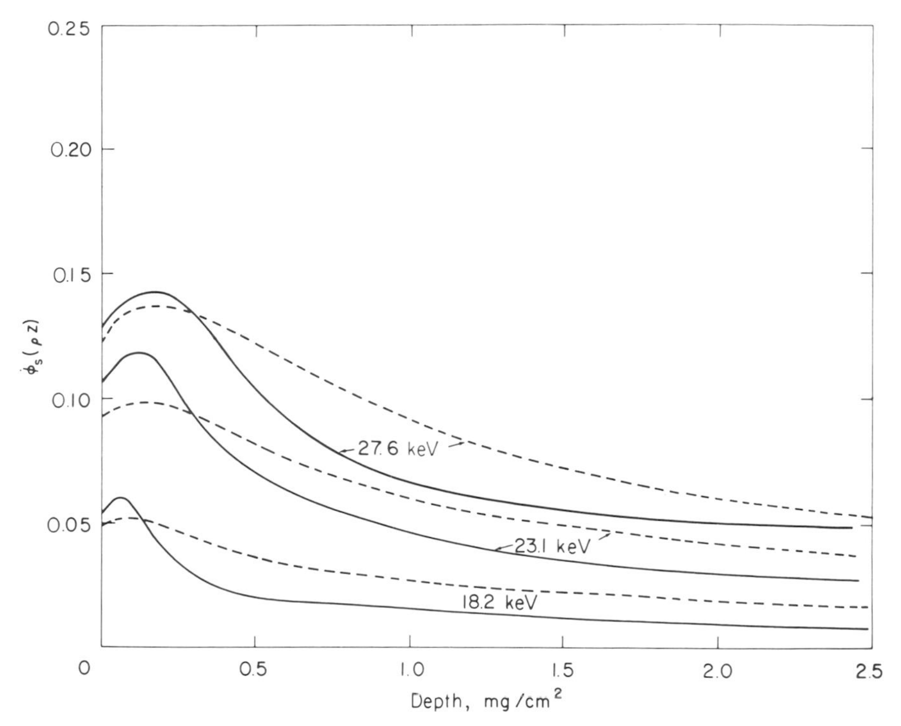

FIG. 11. Comparison of calculated and measured secondary distributions for Fe K_α excited by Cu K_α [—, measured; - - -, calculated].

2. *Theoretical Calculation*

The importance in knowing the distribution as a function of depth for the generation of X-rays in the sample is that corrections for all geometries can be calculated. Birks *et al.* [20] were the first to calculate the distribution of secondary radiation as a function of depth. They used a Monte Carlo technique in which the primary $\phi(\rho z)$ distribution was approximated by 500,000 photons distributed at 23 depths in the sample. The photons proceeded from their starting points in any of 36 directions with equal probability. The distances traveled by the photons before being absorbed were determined by random numbers, taking into account the matrix mass absorption coefficient. The distribution of secondary radiation was determined from the direction and path length of the original photons to their point of absorption.

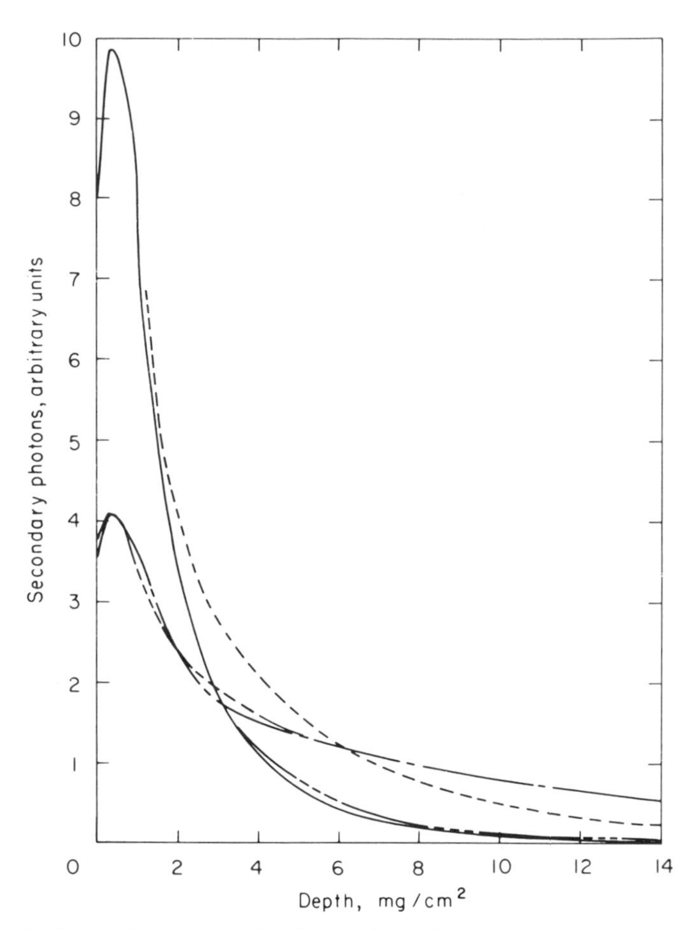

FIG. 12. Comparison of calculated secondary distribution [—, Birks *et al.*, GaAs; - - -, Brown, GaAs generated; - - - - -, Brown, GaAs observed, $\psi = 20°$; – – –, Birks *et al.*, $NbFe_2$; – – – – – –, Brown, $NbFe_2$].

The distributions obtained by Birks *et al.* are shown in Fig. 12. The calculation of each curve required approximately 6 hr on a medium speed computer.

The same distributions can be calculated in a much more straightforward manner using the primary distribution data. The following derivation is condensed from that in my thesis. Consider a sample of elements A and B in which the characteristic X-rays of element B are of greater energy than the absorption-edge energy of a shell in element A. Let the intensity of a specific characteristic wavelength of B from ionizations by electrons at point P at a depth z in the sample be $I_B(z)$. Since the generation of characteristic X-rays is isotropic, the intensity leaving P per unit solid angle is $I_B(z)/4\pi$.

It is important to note that the derivation is for one specific wavelength of the exciting element B. The distribution of secondary radiation of A resulting from the absorption of any other wavelength of B will be somewhat different because of the different absorption coefficient for the B radiation. This is a manifestation of the point made earlier that the distribution of secondary radiation generated in the sample depends only on the absorption characteristics of the sample for the primary radiation.

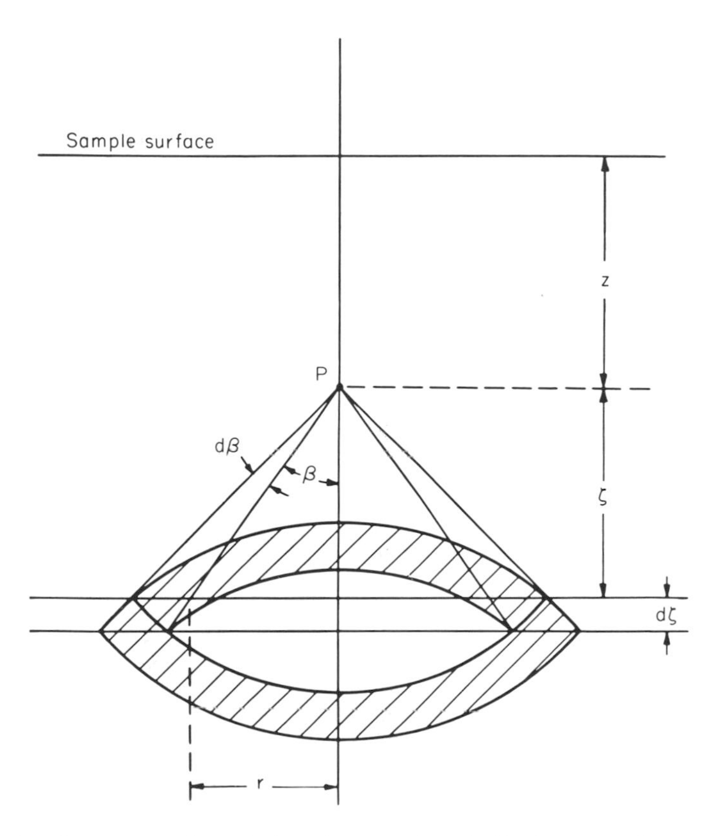

FIG. 13. Geometry used to calculate the secondary distribution from a known primary distribution.

Consider the absorption of the radiation of B in the volume element formed by the intersection of cones of revolution of half-angle β and $\beta + d\beta$ with a layer $d\zeta$ at a perpendicular distance ζ from P (Fig. 13). The solid angle Ω subtended at P by the volume element is

$$\Omega = 2\pi \sin \beta \, d\beta. \tag{10}$$

The intensity absorbed within the volume element by the A atoms is

$$dI_{\text{abs}} = C_{\text{A}} \frac{I_{\text{B}}(z)}{2} \mu_{\text{B}}{}^{\text{A}} \tan\beta \exp(-\mu_{\text{B}}{}^{\text{S}}\zeta/\cos\beta)\, d\beta \cdot d\zeta \tag{11}$$

where C_{A} is the weight fraction of A.

The total intensity $dI_{f\text{A}}(d\zeta)$ of A generated in the layer $d\zeta$ due to absorption of primary radiation of B generated at P is obtained by integrating with respect to β and multiplying by the fraction of ionizations of A which result in the reemission of the specific characteristic line of A. The fraction of ionizations depends on r_{A} the absorption jump ratio, ω_{A} the fluorescent yield and $P_{i\text{A}}$ the probability of emission. Thus

$$dI_{f\text{A}}(d\zeta) = C_{\text{A}} \tfrac{1}{2}(I_{\text{B}}(z)) \frac{r_{\text{A}} - 1}{r_{\text{A}}} \omega_{\text{A}} P_{i\text{A}} \mu_{\text{B}}{}^{\text{A}}\, d\zeta \int_0^{\pi/2} \tan\beta \exp(-\mu_{\text{B}}{}^{\text{S}}\zeta/\cos\beta)\, d\beta. \tag{12}$$

To reduce the integral of this equation to a simpler form, substitute

$$x = \mu_{\text{B}}{}^{\text{S}}\zeta/\cos\beta \tag{13}$$

and

$$dx = (\mu_{\text{B}}{}^{\text{S}}\zeta \sin\beta\, d\beta)/\cos^2\beta \tag{14}$$

so that the limits of integration become $\mu_{\text{B}}{}^{\text{S}}\zeta$ and ∞. Then Eq. (12) can be written

$$dI_{f\text{A}}(d\zeta) = C_{\text{A}} \tfrac{1}{2}(I_{\text{B}}(z)) \frac{r_{\text{A}} - 1}{r_{\text{A}}} \omega_{\text{A}} P_{i\text{A}} \mu_{\text{B}}{}^{\text{A}}\, d\zeta \int_{\mu_{\text{B}}{}^{\text{S}}\zeta}^{\infty} \frac{e^{-x}}{x}\, dx \tag{15}$$

The integral $\int_a^\infty (e^{-x}/x)\, dx$ is known as the exponential integral and cannot be integrated in closed form. Tables of values of this integral for various values of a have been published [21]. The value of the integral depends only on the lower limit of integration, i.e., $\mu_{\text{B}}{}^{\text{S}}\zeta$. Let its value be designated by $\text{S}(\mu_{\text{B}}{}^{\text{S}}, \zeta)$. The equation becomes

$$dI_{f\text{A}}(d\zeta) = C_{\text{A}} \tfrac{1}{2}(I_{\text{B}}(z)) \frac{r_{\text{A}} - 1}{r_{\text{A}}} \omega_{\text{A}} P_{i\text{A}} \mu_{\text{B}}^{\text{A}} \text{S}(\zeta, \mu_{\text{B}}^{\text{S}})\, d\zeta \tag{16}$$

By integrating for the primary radiation generated at each depth in the sample, the total secondary radiation generated in the layer $d\zeta$ can be calculated. Let s be the perpendicular depth in the sample of the slice $d\zeta$, then $s = \zeta + z$ (Fig. 13) and the total secondary intensity of A generated at a depth s in the sample is

$$I_{f\text{A}}(s) = \tfrac{1}{2}(C_{\text{A}}) \frac{r_{\text{A}} - 1}{r_{\text{A}}} \omega_{\text{A}} P_{i\text{A}} \mu_{\text{B}}{}^{\text{A}} \int_0^\infty \phi_{\text{B}}(\rho z) \text{S}(s - z, \mu_{\text{B}}{}^{\text{S}})\, dz \tag{17}$$

where the intensity $\phi(\rho z)$ relative to the intensity from a thin slice isolated in space has been substituted for the total intensity $I_B(\rho z)$. Equation (17) can be evaluated by numerical methods. If the sample is divided into a number of equal depths, and at each depth an appropriate value for $\phi_B(\rho z)$ is assigned, then the equation can be written in terms of the summation

$$I_{fA}(s) = \tfrac{1}{2}(C_A)\frac{r_A - 1}{r_A}\,\omega_A P_{iA} \sum_i \phi_B(\rho z)S(s - z_i, \mu_B{}^S). \tag{18}$$

This equation cannot be evaluated directly because the inverse square law was used to calculate the intensity incident on the thin slice $d\zeta$. If the distance from the point source to the layer $d\zeta$ is large, no difficulty is experienced. However, the product of mass absorption coefficient and thickness for X-ray production is typically 0.01 so that the calculated secondary intensity is greater than the primary intensity since unit dimensions are assumed for the primary source. This problem can be circumvented by considering the meaning of the integral of Eq. (17). Since all other factors in this equation are constants, the integral contains all the dependence of the intensity of secondary radiation on depth. Thus the magnitude of secondary radiation can be calculated if the proper value for the constant which multiplies this integral is derived. The first step toward evaluating the constant is to change the value of the integral so that it represents the fraction of primary radiation absorbed within the sample. Having calculated this fraction, the intensity of secondary radiation resulting from absorption by A atoms is obtained in a straightforward manner.

The fraction of primary radiation absorbed within the sample can be visualized with the aid of Fig. 14. The horizontal axis of this figure represents the magnitude of intensity absorbed at each depth in the sample. If primary radiation is generated at point P, the radiation is absorbed in the sample with the intensity absorbed decreasing according to the value of the exponential integral for both greater and lesser depths as the distance from P increases. The total intensity absorbed in the sample is represented by the shaded area of Fig. 14. The fraction absorbed is this area divided by an area corresponding to the total primary intensity generated at P. This total intensity can be represented by the area under the absorption curve for a sample that extends infinitely in both directions from P. Mathematically this can be expressed as

$$\text{Fraction absorbed} = \frac{\sum_{s=0}^{\infty} S(s - z_i, \mu_B{}^S)}{\sum_{s=-\infty}^{\infty} S(s - z_i, \mu_B{}^S)}. \tag{19}$$

Of the total intensity absorbed in the sample, only that which is absorbed by A atoms and reemitted as characteristic radiation of A contributes to the

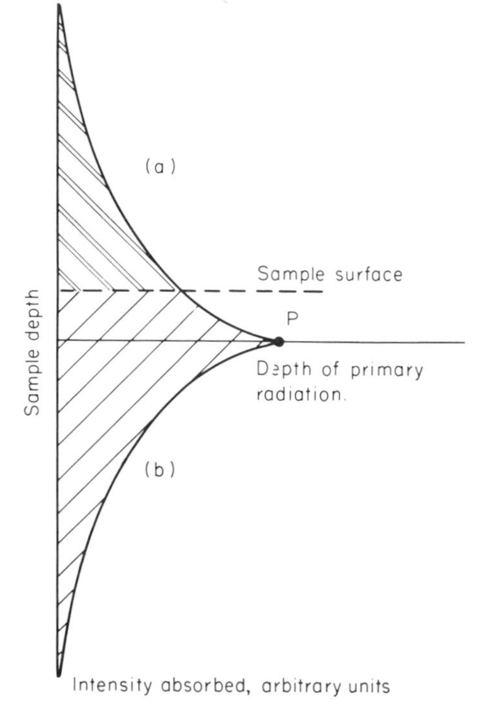

FIG. 14. Fraction of primary radiation absorbed within the sample. (a) Radiation escaping from sample; (b) radiation absorbed in sample.

secondary radiation. Thus, the intensity of secondary radiation at any depth s in the sample is given by

$$I_{f\mathrm{A}}(s) = C_{\mathrm{A}} \frac{r_{\mathrm{A}} - 1}{r_{\mathrm{A}}} \omega_{\mathrm{A}} P_{i\mathrm{A}} \frac{\mu_{\mathrm{B}}{}^{\mathrm{A}}}{\mu_{\mathrm{B}}{}^{\mathrm{S}}} \frac{\sum_{s=0}^{\infty} \phi_{\mathrm{B}}(\rho z_i) \mathrm{S}(s - z_i, \mu_{\mathrm{B}}{}^{\mathrm{S}})}{\sum_{s=-\infty}^{\infty} \mathrm{S}(s - z_i, \mu_{\mathrm{B}}{}^{\mathrm{S}})}. \tag{20}$$

In the summation of Eq. (20), z_i cannot be zero, i.e., the intensity of secondary radiation cannot be calculated at the same depths that have been chosen to represent the primary radiation. This problem can be avoided by calculating the intensity of secondary radiation at points midway between those for the primary distribution. Since Eq. (20) relates the intensity of secondary radiation of A to primary intensity of B, to obtain the increase in A intensity due to absorption of B radiation it is necessary to relate the primary intensities of A and B.

3. *Primary X-Ray Intensities*

Several expressions have been used in the fluorescence corrections for comparing primary intensities of different elements. It is important to consider what intensities must be compared in these expressions, whether directly excited by electrons within the sample or observed at some take-off angle outside the sample. In Castaing's expression and the more complete expressions in Wittry's thesis the comparison is between $I_B(z)$ and $I_A(z)$, the primary intensity of each element generated in a thin slice of the sample for each depth in the sample. Since Rosseland's ionization function [22] is used for comparing these intensities, $I_B(z)$ and $I_A(z)$ differ only by a constant factor regardless of the depth z. Thus, the primary distribution as a function of depth is assumed identical for A and B radiation.

In Wittry's G method [16] and Birk's expression [17], the primary intensities are assumed to originate at the surface of the sample. These authors have used intensities measured outside the sample as the basis for relating intensities from different elements.

An alternate method of comparing intensities in terms of measured $\phi(\rho z)$ curves is suggested by Eq. (17).

The fluorescent intensity of A has been calculated relative to the intensity of B directly excited by electrons within the sample (actually relative to the area under the curve of $\phi_B(\rho z)$). To emphasize the intensities which must be compared, Eq. (17) can be written in terms of the actual intensity of B as

$$I_{fA}(s) = k \int_0^\infty I_B(z) \int_{s-z}^\infty (e^{-x}/x)\, dx\, dz \tag{21}$$

where k is equal to the product of the constant factors. By dividing by the total intensity of B generated in the sample, the fraction of fluorescent radiation of A relative to B is obtained, thus

$$\frac{I_{fA}(s)}{I_B{}^S} = \frac{k \int_0^\infty I_B(z) \int_{s-z}^\infty (e^{-x}/x)\, dx\, dz}{\int_0^\infty I_B(z)\, dz}. \tag{22}$$

To relate the fluorescent intensity of A to the primary intensity of A, multiplication by the ratio $I_B{}^S/I_A{}^S$ is required, and

$$\frac{I_{fA}(s)}{I_A{}^S} = \frac{k \int_0^\infty I_B(z) \int_{s-z}^\infty (e^{-x}/x)\, dx\, dz}{\int_0^\infty I_B(z)\, dz} \frac{\int_0^\infty I_B(z)\, dz}{\int_0^\infty I_A(z)\, dz}. \tag{23}$$

Writing Eq. (23) in terms of the distribution function $\phi(\rho z)$

$$\frac{I_{fA}(s)}{I_A{}^S} = \frac{k \int_0^\infty \phi_B(\rho z) \int_{s-z}^\infty (e^{-x}/x)\, dx\, dz}{\int_0^\infty \phi_B(\rho z)\, dz} \frac{C_B \int_0^\infty \phi_B(\rho z)\, dz}{C_A \int_0^\infty \phi_A(\rho z)\, dz} \frac{I_A{}^0}{I_B{}^0}. \tag{24}$$

Equation (24) consists of three factors. The first is the distribution of the secondary radiation as a fraction of the primary radiation. A method for calculating this distribution has already been described. The second factor is the ratio of the areas under the curves of $\phi(\rho z)$ versus depth for the primary radiation of A and B. The only data from which such areas can be determined are those of Castaing and Descamps [8]. The calculated areas are given in Table III in which the area for the copper curve is set equal to 1.0. In view of the scarcity of data and the fact that a copper tracer was used for the Al curve, no general statements can be made concerning the total area under the $\phi(\rho z)$ curves.

TABLE III
AREA UNDER $\phi(\rho z)$ CURVES OF CASTAING AND DESCAMPS[a]

Element	Tracer	Relative area	Excitation potential for tracer (keV)
Cu	Zn	1.000	9.66
Au	Bi	1.068	13.43
Al	Cu	1.040	8.98

[a] See Castaing and Descamps [8].

The third factor is the ratio of the intensities generated in thin isolated samples of the pure element. The determination of intensities from thin samples should be easier to treat both theoretically and experimentally than such determinations for thick samples.

4. *Calculations of Secondary Distributions*

Birks *et al.* [20] used a Monte Carlo technique for calculating the distribution of secondary radiation as a function of depth in the sample. These calculations required several hours on a medium speed computer for each distribution determination. The same distributions can be calculated using Eq. (20) in approximately 1 min using an equivalent computer. The results of the calculation of secondary distributions based on my measurements of the primary distributions for copper are shown in Fig. 15. The calculations are based on a composition of 0.1% iron in copper and are calculated on the same relative scale as the primary distributions. Similar distributions as would be observed outside the sample at various take-off angles are shown in Fig. 16 for 27.6 keV. The decrease in depth of sample which significantly contributes to the observed intensity as the take-off angle decreases is readily apparent. Although a significant intensity of secondary radiation is generated at depths at up to 20 mg/cm^2, even for a take-off angle of 90° little contribution to the observed intensity originates from a depth greater than 10 mg/cm^2. At a take-off angle of 10°, the maximum depth is 4 mg/cm^2, not very much greater than the depths at which primary radiation is generated.

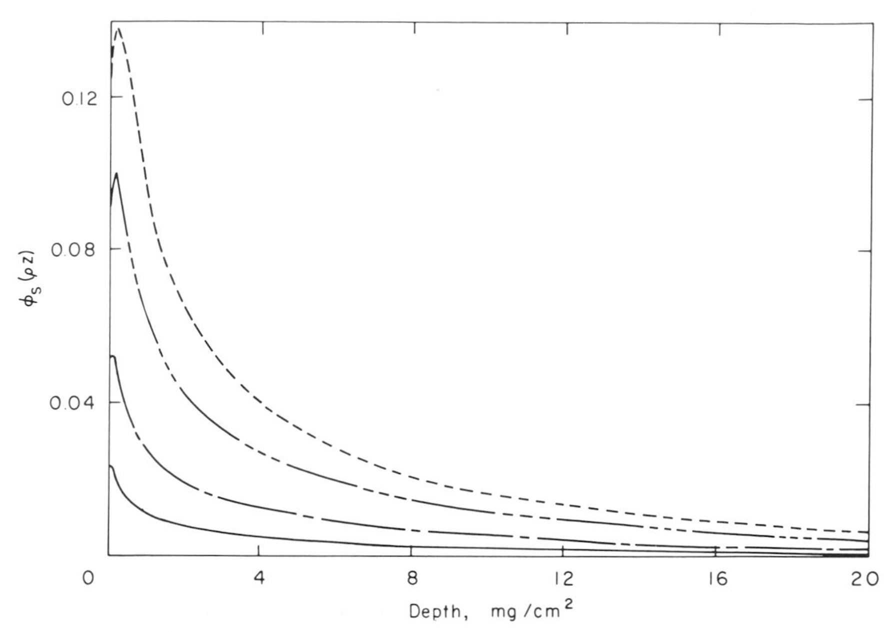

FIG. 15. Calculated secondary distributions for Fe K_α excited by Cu K_α [electron voltage (keV): - - -, 27.6; - - - - -, 23.1; – – –: 18.2; —, 13.4].

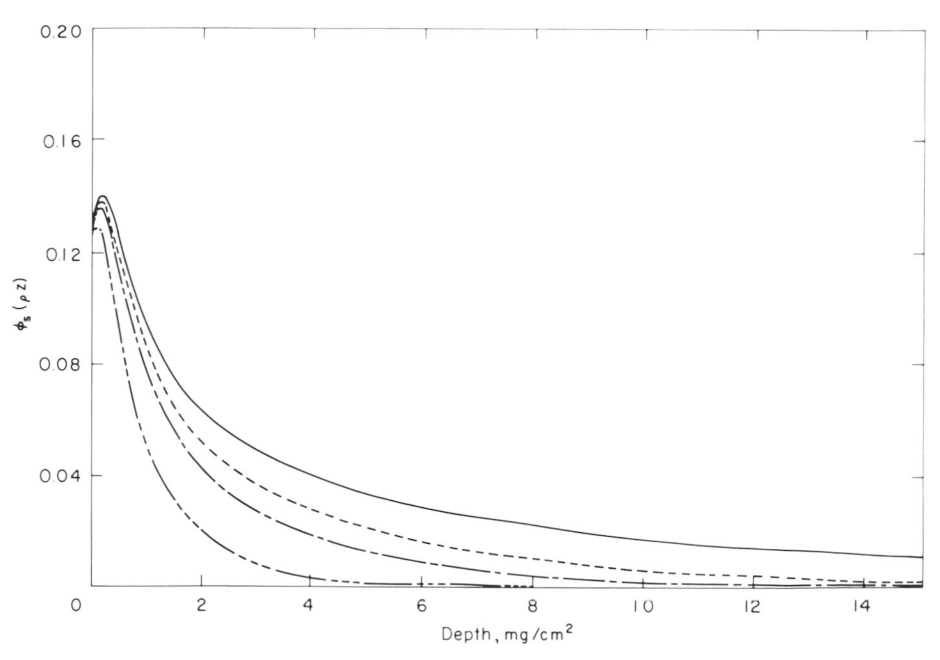

FIG. 16. Effect of absorption within the sample on the secondary distribution [0.1% Fe in Cu: —, generated; - - -, observed at $\psi = 90°$; – – –, observed at $\psi = 30°$; - - - - -, observed at $\psi = 10°$].

Secondary distributions were also calculated for GaAs and $NbFe_2$. Good agreement was obtained with the results of Birks *et al.* [20] except that the curve for GaAs of Fig. 2 ([20] p. 205), is not the distribution of secondary X-rays generated in the sample but is the distribution observed at a take-off angle of 20°.

IV. The Atomic Number Effect

Several equations have been proposed as a correction for the atomic number effect, notably those of Thomas [23], Archard and Mulvey [24], and Long and Reed as reported by Smith [25]. All of these reflect the original formulation of Castaing. In his approach the atomic number effect is considered as a separate correction that is applied to remove any remaining discrepancy between calculated and measured compositions after all other corrections are made. Unfortunately, because of this approach, any errors resulting from inaccurate mass absorption coefficients or other parameters complicate any evaluation of the atomic number correction. Differences between the calculated and known composition that is ascribed to the atomic number effect, may partly or completely result from incorrect mass absorption coefficients. The atomic number correction usually includes any correction due to fluorescence by the continuum since the continuum fluorescence has not been removed from the experimental data before applying the atomic number correction. However, separation of the atomic number correction and fluorescence due to the continuum may neither be desirable nor necessary.

Another approach to a study of the atomic number correction is possible using the sandwich sample technique. Consider the differences in the $\phi(\rho z)$ curves measured with sandwich samples in which a single tracer material is embedded in low, medium, and high atomic number matrices. In the absence of fluorescence effects the differences in the areas under the curves are directly the result of the change in energy lost from the sample because of backscattered electrons and in the stopping power of the matrix for electrons. These two factors are exactly what is defined as the atomic number effect. Thus measurement of $\phi(\rho z)$ curves for carefully selected sandwich samples yields a direct measurement of the atomic number effect.

Several remarks should be made about sandwich sample measurements of the atomic number effect. Even in measurements of this kind, errors in mass absorption coefficients still play a role since a correction for absorption in the sandwich sample must be made. At the same time, however, such measurements do not depend on errors in the model used for the absorption correction as is the case for corrections to total intensity from a microprobe sample. The correction for absorption in a sandwich sample is straightforward. Vacuum evaporation techniques have reached the point where alloys of accurately controlled composition can be evaporated. A sandwich sample with an alloy matrix could settle the question of which $f(\chi)$ curve should be used for a sample intermediate in composition between pure element end members. Finally, since the atomic number correction can be written in terms of the depth distribution of the production of X-rays, combining the atomic number and the absorption corrections into a single correction would seem an appropriate approach to quantitative electron microprobe analysis.

V. A Model for Quantitative Electron Probe Microanalysis

Criss and Birks [18] have proposed a method of quantitative analysis in which the primary intensity distribution is approximated by an exponential series and the area under the curve is set equal to unity. Because of the form of the series used, integration is possible so that the correction for fluorescence due to absorption of characteristic lines is exactly defined. The atomic number correction is treated as a separate multiplicative factor.

A modification of this model would lead to a general correction procedure which could be evaluated using sandwich sample techniques. The coefficients of the exponential power series would be obtained by fitting to measured $\phi(\rho z)$ curves, but these would not be normalized to unit area since this normalization removes the atomic number correction from the coefficients. The dependence of the coefficients on atomic number, composition, excitation potential and electron voltage could be determined by measurements of a single tracer in a number of matrix elements. This dependence should be a smooth function of these variables. Quantitative analysis would require a computer program in which the primary distribution was adjusted on each iteration by adjustment of the coefficients in the exponential series according to an estimated composition. Absorption, fluorescence, and atomic number corrections could then be applied with no further assumptions. The correction for fluorescence due to the continuum could be treated as a separate correction to be applied to the measured intensities before comparing with a calculated intensity or it could simply be viewed as part of the primary $\phi(\rho z)$ distribution. Mathematically, this approach can be described by the equation relating relative intensity to concentration as

$$\frac{I_{\mathrm{A}}^{\mathrm{S}'}}{I_{\mathrm{A}}^{\mathrm{A}'}} = C_{\mathrm{A}} \frac{\int_0^\infty \phi_{\mathrm{S}}(\rho z) \exp(-\mu_{\mathrm{A}}{}^{\mathrm{S}} \csc \psi)\, d\rho z}{\int_0^\infty \phi_{\mathrm{A}}(\rho z) \exp(-\mu_{\mathrm{A}}{}^{\mathrm{A}} \csc \psi)\, d\rho z} (1 + k_f). \tag{25}$$

The integrals of Eq. (25) contain the atomic number, absorption and, if desired, fluorescence due to the continuum corrections. The value of k_f for the fluorescence correction can be obtained by integration as has been described by Criss and Birks [18]. However, since the area under the curves for the primary radiation represents the total intensity generated in the sample, intensities from different elements are related by the ionization function and not the total intensity from massive targets.

At the present time, quantitative analysis is being studied by the analysis of standard samples of uniform, accurately known composition. In many cases the correction procedures are justified on the basis of the analysis of samples that are not well characterized, and in which the various errors which could exist have not been carefully examined. Much confusion exists because of such evaluations. The sandwich sample technique provides an alternate and independent route to a new model of quantitative analysis as well as toward providing data for significant tests of the accuracy of the present correction procedures.

Acknowledgments

Much of the work described in this paper is based on a thesis submitted to the faculty of the Graduate School, University of Maryland, in partial fulfillment of the requirements of the degree of Doctor of Philosophy. I would like to thank Dr. Ellis R. Lippincott of the Chemistry Department, University of Maryland, for his constant encouragement during the course of my studies and Dr. William J. Campbell of the U.S. Bureau of Mines, College Park, Maryland, for the many helpful discussions. I am grateful for the permission granted by the U.S. Bureau of Mines and the University of Maryland to conduct this cooperative research program.

References

1. R. Castaing, Ph.D. thesis. Univ. of Paris (1951).
2. T. O. Ziebold and R. E. Ogilvie, *Anal. Chem.* **36**, 322 (1964).
3. M. Green, Ph.D. thesis. Univ. of Cambridge, pp. 30–59 (1962).
4. J. Philibert, *in* "X-Ray Optics and X-Ray Microanalysis" (H. H. Pattee, V. E. Cosslett, and A. Engstrom, eds.), p. 379. Academic Press, New York, 1963.
5. V. G. Macres, Procedure for preliminary evaluation of a chemical analysis by the electron microprobe. *Ann. Conf. Application X-Ray Analysis, 12th, Denver, 1963.*
6. P. Kilpatrick and D. G. Hare, *Phys. Rev.* **46**, 831 (1934).
7. G. D. Archard, *J. Appl. Phys.* **32**, 1505 (1961).
8. R. Castaing and J. Descamps, *J. Phys. Radium* **16**, 304 (1955).
9. R. Castaing, *Advan. Electron. Electron Phys.* **13**, 317 (1960).
10. J. D. Brown, Ph.D. thesis. Univ. of Maryland (1966).
11. M. Green, *Proc. Phys. Soc.* (*London*) **82**, 204 (1963).
12. H. E. Bishop, *Proc. Phys. Soc.* (*London*) **85**, 855 (1965).
13. D. B. Brown and R. E. Ogilvie, *J. Appl. Phys.* **37**, 4429 (1966).
14. P. Duncumb and P. K. Shields, *in* "The Electron Microprobe" (T. D. McKinley, K. F. J. Heinrich, and D. B. Wittry, eds.), p. 217. Wiley, New York, 1966.

15. D. B. Wittry, Ph.D. thesis. California Inst. Technol. (1957).
16. D. B. Wittry, Univ. of Southern California, Rept. 84–204 (1962).
17. L. S. Birks, *in* "Electron Probe Microanalysis," p. 118. Wiley (Interscience), New York, 1963.
18. J. Criss and L. S. Birks, *in* "The Electron Microprobe" (T. D. McKinley, K. F. J. Heinrich, and D. B. Wittry, eds.), p. 217. Wiley, New York, 1966.
19. S. J. B. Reed, *Brit. J. Appl. Phys.* **16**, 913 (1965).
20. L. S. Birks, D. J. Ellis, B. K. Grant, A. S. Frish, and R. B. Hickman, *in* "The Electron Microprobe" (T. D. McKinley, K. F. J. Heinrich, and D. B. Wittry, eds.), p. 199. Wiley, New York, 1966.
21. G. Placzek, "Tables of Functions and Zeros of Functions," pp. 51–111. *Natl. Bur. Std.* (*U.S.*) *Applied Math. Ser.* No. 37, 1954.
22. S. Rosseland, *Phil. Mag.* **45**, 65 (1923).
23. P. M. Thomas, *UK At. Energy. Authority Res. Group Rept.* R.4593 (1964).
24. G. D. Archard and T. Mulvey, *Brit. J. Appl. Phys.* **14**, 626 (1963).
25. J. V. Smith, *J. Geol.* **73**, 830 (1965).
26. A. Vignes and G. Dez, *Brit. J. Appl. Phys. Ser.* 2, **1**, 1309 (1968).

Quantitative Microprobe Analysis: A Basis for Universal Atomic Number Correction Tables

RICHARD C. WOLF and VICTOR G. MACRES

Materials Analysis Company
Palo Alto, California

I. Introduction

The absolute method of quantitative chemical analysis by emission X-ray spectroscopy, first proposed by Castaing [1], is based on the comparison of primary emitted intensities of the same characteristic radiation from an unknown and from a pure element standard. This simple ratio of two intensities, often referred to as the emission-concentration proportionality law, gives a first approximation to the concentration of an element in the unknown.

In addition to being nondestructive in nature, this method has the advantage of yielding an *in situ* chemical analysis of microscopic regions by the use of a focused electron beam.

Experimentally, it is not possible to measure primary emitted intensities since X-rays emerging from an unknown will, in general, be absorbed differently from those emerging from the pure element standard. Procedures for the correction of these differential absorption effects have been pursued by several investigators [1–11]. In addition, the measured X-ray intensities may also include contributions from secondary emission (fluorescence) caused by higher energy characteristic radiation generated in the sample by the electron beam or by the higher energy portion of the continuous spectrum. Procedures for correction of fluorescence effects have been given in a number of publications [1, 9, 12–16].

It is of interest to note that due to the complex calculations and uncertainties involved in applying correction formulas, some investigators prefer to use a comparison method based on calibration standards to determine compositions of unknowns [17]. Although this method may be useful in some cases, its application is limited by the availability of homogeneous standards whose compositions are accurately known. Furthermore, it does not recognize the full potential of quantitative chemical analysis by primary emission spectroscopy.

The emission-concentration proportionality law, as originally derived, is based on a somewhat simplified electron deceleration law by Williams [18]. A more rigorous treatment requires the use of a refined electron deceleration law to account for differences in mass penetration as well as a factor taking into account backscatter effects from targets of different atomic numbers. The use of a deceleration law by Webster *et al.* [19] in deriving the emission-concentration proportionality relation recognizes the fact that the loss of energy per unit length traveled by an electron decreases with increasing atomic number Z, and that this variation may be roughly approximated by use of the ratio Z/A, where A is atomic weight. The relation based on the Webster law is given by

$$I_1/I(1) = C_1(Z/A)_1/\sum C_i(Z/A)_i \tag{1a}$$

where I_1 and $I(1)$ are primary excited intensities from the unknown and pure element standard, respectively, and C is weight concentration. This relation is still somewhat simplified and does not account for differences in backscatter.

Recognizing the shortcomings of these electron deceleration laws, Castaing [20] proposed a relation involving a coefficient α_i, which represents the "specific decelerating power" of element i,

$$I_1/I(1) = C_1\alpha_1/\sum C_i\alpha_i \, . \tag{1b}$$

The α coefficient takes into account both the mass penetration and backscatter effects due to atomic number differences. Castaing has shown that estimates for the α coefficients may be determined from Z/A ratios and knowledge of the depth distribution of primary emission. However, this method is limited by the availability of emission-distribution curves and by experimental uncertainties inherent in their determination. The α coefficient may also be determined experimentally by measurements from alloys of known composition, but this method is also limited by experimental uncertainties; in particular, corrections for primary emission absorption and fluorescence effects.

In order to examine the physical basis of primary emission analysis from first principles, Archard and Mulvey [21] derived a relation between intensity and concentration based on Bethe's [22] more refined electron deceleration relation. Their objective was to determine under what circumstances, if any, the relation between concentration and X-ray intensity remains linear. The method for determining atomic number corrections in the present investigation is based upon this more rigorous approach of Archard and Mulvey. This method consists of a model for electron penetration from which primary emitted intensity ratios are determined. Effects of atomic number differences on the linearity of the emission-concentration proportionality law are then determined from the calculated intensity ratios. All variables on which atomic number effects depend are accounted for in the method and include: (1) atomic number, (2) Z/A ratios, (3) electron accelerating voltage, (4) critical excitation voltage, (5) electron beam-target incident angle, and (6) composition. In addition, the model allows complete separation of electron backscatter and mass penetration components inherent in the overall atomic number effect and is applicable to multicomponent systems. Since use of the model requires a large number of numerical calculations, the equations involved have been written into a computer program that is used to calculate mass penetration, electron backscatter, and the total atomic number effect for specific conditions. Most important, the program is capable of generating simplified tabulations in terms of a factor λ', which is independent of Z/A, but which incorporates all other variables. The λ' factor, used in conjunction with a readily determined $(Z/A)'$ factor, allows calculation of the atomic number correction. Tabulations of this type are universal and provide a correction procedure that is based on a rigorously defined model and is applicable from a practical viewpoint for routine calculations.

In the following presentation, the development of the diffusion model will be carried out in a manner similar to that of Archard and Mulvey [21] but incorporating a more exact determination of the depth of complete electron diffusion, a refined integration procedure, and the exact definition of effective atomic number. The correction formula developed will be discussed and evaluated in terms of mass penetration and backscatter effects and dependence on experimental variables, and will be compared with other correction procedures. The basis for obtaining universal correction tables is set forth and examples illustrating application of the correction procedure to selected alloy systems are presented.

II. Formulation of Atomic Number Correction

A. *Diffusion Model*

When a sample target is exposed to an electron beam, the incident electrons undergo a series of elastic and inelastic collisions with the target atoms. Inelastic collisions result in a deceleration and loss of energy of the incident electrons and a subsequent emission of characteristic X-rays by excited target atoms. During the deceleration process, an electron may change direction many times and may escape from the target before its energy is expended, thus contributing to backscatter losses. Figure 1 is a simplified schematic

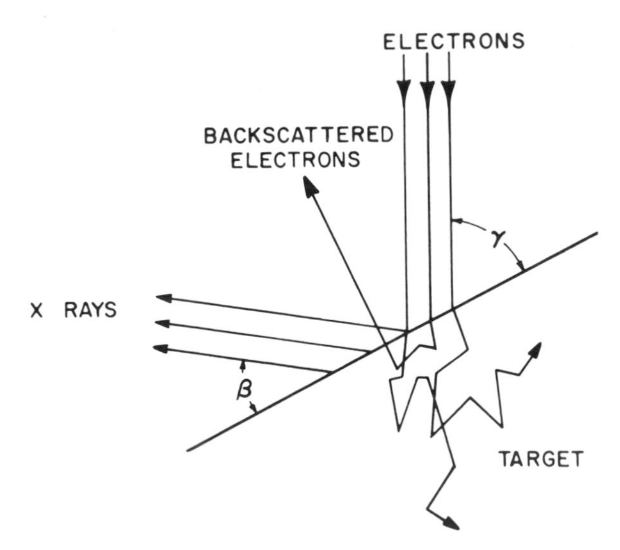

Fig. 1. Schematic representation of penetration and backscattering of electrons.

representation of electron paths illustrating diffusion and backscatter of electrons in a target material. In order to evaluate the electron deceleration process, it is necessary to obtain an expression for the electron path during which primary emission occurs. Bethe *et al.* [23] approached the problem by assuming that electrons travel straight into the target to a given depth after which they diffuse randomly in all directions. This approach, often referred to as the electron diffusion theory, assumes elastic reflections, which an electron may undergo before reaching this depth of complete diffusion, are negligible. Archard [24] has pointed out that for atomic numbers greater than 6, the contribution from elastic scattering is small. In such cases, therefore, the fraction of electrons backscattered may be calculated using the Bethe approach.

The loss of energy by electrons in passing through a solid target of a pure element has been expressed by Archard and Mulvey [21] using Bethe's deceleration law,

$$d\,\mathrm{eV}/dx = -(2\pi NZe^3/V)\ln(2\,\mathrm{eV}/I) \tag{2}$$

where e is electron charge in electrostatic units, V is accelerating voltage, Z is atomic number of target, N is number of atoms per cubic centimeter, I is mean ionization potential of target atoms ($I = 11.5Z$ eV), and x is the path length traveled by electrons in centimeters. For a multicomponent system, NZ may be replaced by the summation $\sum N_i Z_i$ where i denotes the elements present.

Integration of Eq. (2) gives a form of Bethe's law which describes the depth of electron penetration in centimeters for a multicomponent target

$$x = \frac{7.68 \times 10^{12}}{\sum N_i Z_i}[V_0{}^2F(V_0) - V^2F(V)] \tag{3}$$

where,

$$V_0 = \text{initial voltage, in volts,}$$
$$F(V) = 1/y(1 + 1/y + 2!/y^2 + \cdots),$$
$$y = 2\ln(0.174\,V/Z).$$

To determine $F(V)$, the semiconvergent series is taken only as far as its smallest term, and the multicomponent target is assumed to have an effective atomic number $\bar{Z}$. The correct definition of $\bar{Z}$ is

$$\bar{Z} = \sum n_i Z_i^2 / \sum n_i Z_i \tag{4}$$

where n_i is the atomic fraction of element i. By substituting the relation $N_i = \rho C_i N_A/A_i$ in Eq. (3), the relation for the electron penetration can be expressed in terms of ρx,

$$\rho x = \frac{7.68 \times 10^{12}}{N_A \sum (C_i Z_i/A_i)}[V_0{}^2F(V_0) - V^2F(V)] \tag{5}$$

where ρ is density in grams per cubic centimeter, A_i is atomic weight of element i, C_i is weight fraction of element i, and N_A is Avagadro's number.

Worthington and Tomlin [25] have assumed that electrons effectively come to rest when the function $F(V) = 0$ rather than when $V = 0$ since $F(V)$ becomes indeterminate at $V = 0$, and such a condition does not correspond to a physical reality. From Eq. (5) the full range of electron penetration, ρx_R, then becomes

$$\rho x_R = \frac{7.68 \times 10^{12} V_0^{\,2}}{N_A \sum (C_i Z_i / A_i)} F(V_0). \tag{6}$$

However, in calculating the intensity of characteristic radiation excited in a target, the electron penetration of primary interest is that which occurs between electron energies eV_0 and eV_c where V_c is the critical excitation voltage required to excite characteristic radiation. After the electron energy falls below eV_c, it is unable to excite the characteristic radiation, and therefore it is not of interest in the intensity calculations. The range of electron energies, eV_0–eV_c, over which primary emission occurs is often referred to as the effective range of electrons, and the effective depth of penetration, ρx_r, as determined by Eq. (5) becomes

$$\rho x_r = \frac{7.68 \times 10^{12}}{N_A \sum (C_i Z_i / A_i)} [V_0^{\,2} F(V_0) - V_c^{\,2} F(V_c)]. \tag{7}$$

The effective penetration ρx_r may be considerably less than the full penetration of electrons ρx_R, especially for low accelerating voltages where the ratio V_0/V_c becomes small.

In formulating a model to describe electron penetration from the Bethe diffusion theory, it is necessary to determine the depth of penetration at which complete diffusion occurs. Bethe *et al.* [23] define this depth as that at which the average cosine between the actual direction of motion of an electron and the direction of the primary electron beam becomes e^{-1}. Under this condition, the Bethe deceleration equation may be expressed as

$$\int_{Vd}^{V_0} \frac{Z \ln(0.54 V^{1/2} Z^{-1/3})}{4 \ln(0.174 V/Z)} \, dV = \frac{1}{2} \tag{8}$$

where eV_d is the electron energy at ρx_d, the depth of complete diffusion. Archard [24] used an approximation of this relation to determine ρx_d, given by

$$\rho \chi_d / \rho x_R = 40/7Z. \tag{9}$$

In the present investigation, the more rigorous form, Eq. (8), is used to determine V_d, and this result is then used in Eq. (5) to determine ρx_d.

Figure 2 compares electron paths determined from the diffusion model for elements differing greatly in atomic number when the accelerating voltage is

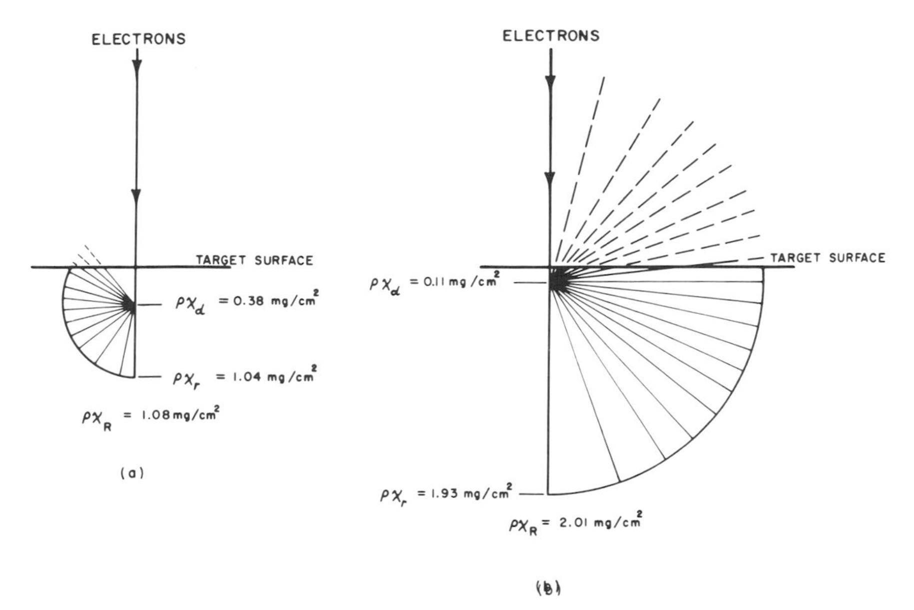

FIG. 2. Comparison of diffusion model for high and low atomic number targets; $V_0 = 20$ kV. (a) Aluminum (at. no. 13) K_α, $V_c = 1.56$ kV; (b) gold (at. no. 79) M_α, $V_c = 2.24$ kV.

held constant and the critical excitation voltages are similar. The depth ρx_d at which complete diffusion occurs is much greater for low atomic numbers, whereas the effective range of electron penetration ρx_r and the full range of electron penetration ρx_R are greater for high atomic numbers.

Application of the diffusion model can be carried out for any electron beam-target geometry. As shown in Fig. 3, the effect of inclined

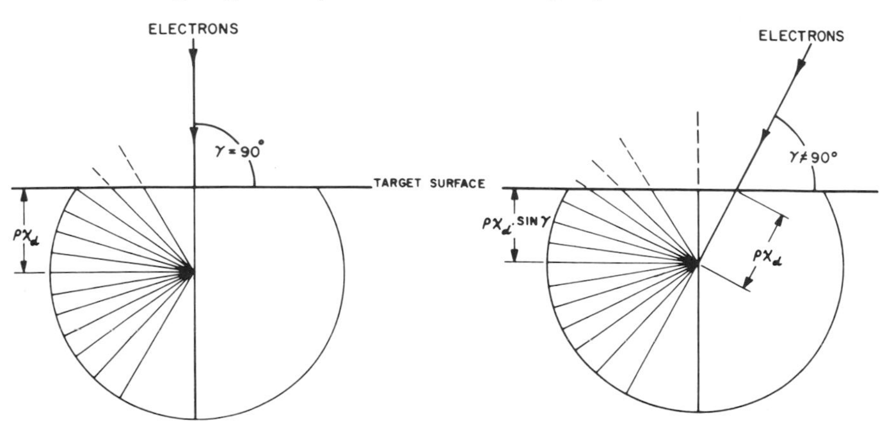

electron incidence on the diffusion model is that the center of the sphere of electron diffusion is positioned closer to the surface of the target as the electron incident angle γ decreases. The values of the electron penetration parameters ρx_R, ρx_r, and ρx_d determined by the model are not affected by inclination of the electron beam. As will be shown later, the portion of atomic number effect resulting from differences in electron penetration is not dependent on electron incident angle, whereas the portion resulting from backscatter losses increases as the electron incident angle decreases.

It is important to consider the extent to which the model is applicable to the determination of electron backscatter effects for various values of ρx_r and ρx_d. For this purpose Fig. 4 shows three different cases that must be considered:

Case 1: $\rho x_r \leq \rho x_d$ For targets of atomic number less than 6, the electron backscatter is primarily due to large single elastic scattering not included in the Bethe approach. According to the model, no sphere of electron diffusion exists and no electron backscatter is predicted since electrons lose their critical energy before complete diffusion can occur. The diffusion theory does not provide a physically realistic model, and therefore it is not applicable.

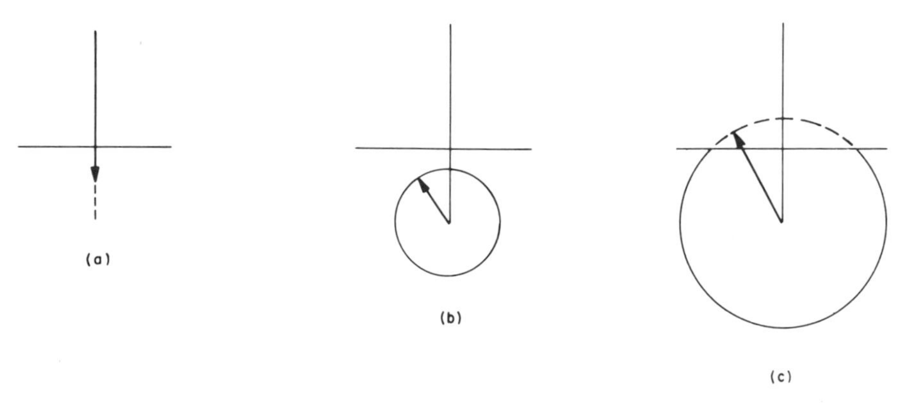

FIG. 4. Applicability of diffusion model for determining electron backscatter effects. (a) case 1: $\rho x_r \leq \rho x_d$; (b) case 2: $\rho x_d < \rho x_r \leq 2\rho x_d$; (c) case 3: $\rho x_r > 2\rho x_d$.

Case 2: $\rho x_d < \rho x_r \leq 2\rho x_d$. For targets of low atomic number, but greater than 6, and for ratios of V_0/V_c approaching unity, the sphere of electron diffusion lies entirely below the surface of the target, and no electron backscatter is predicted since electrons lose their critical energy before they can escape from the target. Although in this case contributions from large single elastic scattering are small, a model based on the diffusion theory is still physically unrealistic. However, use of such values of the ratio V_0/V_c

are not practical in quantitative microprobe analysis since the emission-concentration proportionality law is not applicable.

Case 3: $\rho x_r > 2\rho x_d$. For targets of atomic number greater than 6 and ratios V_0/V_c such that $\rho x_r > 2\rho x_d$, the sphere of electron diffusion intersects the target surface. In this case, the contributions from large single elastic scattering are sufficiently small to be considered negligible and the diffusion model becomes physically realistic thus allowing backscatter effects to be determined. This case describes the physical conditions encountered in quantitative microprobe analysis and provides the basis in this paper for applying the electron diffusion model to the determination of the atomic number correction.

B. Calculation of Intensity Ratios

Having established a model for electron penetration in a target material, it is now possible to express the intensity for a particular characteristic radiation excited from the target. Electrons are assumed to travel in straight lines into the target until reaching the depth of complete diffusion, and thereafter travel with equal probability in all directions. The intensity of characteristic quanta emitted in any electron path increment may be expressed in terms of the ionization cross section. The total intensity I of characteristic radiation excited in a target of a single element is

$$I = \sum \int NQ\, dx \tag{10}$$

where N is number of atoms per cubic centimeter, Q is ionization cross section, and dx is increment of path length traveled by an electron. By combining the differential form of Eq. (3) with Eq. (10), the intensity from a pure element target becomes

$$I(1) = \sum \int \frac{7.68 \times 10^{12} Q\, d(V^2F(V))}{Z_1}, \tag{11}$$

and that for element 1 in a multicomponent target becomes

$$I_1 = \sum \int \frac{7.68 \times 10^{12} (C_1/A_1) Q\, d(V^2F(V))}{\sum (C_i Z_i/A_i)}. \tag{12}$$

The ratio of intensities from a multicomponent target and a pure element standard is, thus

$$\frac{I_1}{I(1)} = \frac{C_1 Z_1/A_1}{\sum C_i Z_i/A_i} \frac{\sum \int Q\, d(V^2F(V))_{\text{sample}}}{\sum \int Q\, d(V^2F(V))_{\text{standard}}}. \tag{13}$$

This intensity ratio is equivalent to that determined using the approximation of Castaing, Eq. (1b), and represents the total correction required for effects of atomic number differences as well as a measure (to the extent that the model is accurate) of the nonlinearity of the emission-concentration proportionality law.

Mott and Massey [26] give an expression for the ionization cross section Q for K radiation as

$$Q = \frac{4.56 \times 10^{-14}}{V/V_c} \ln \left[\frac{4V/V_c}{1.65 + 2.35 \exp(1 - V/V_c)} \right]. \tag{14}$$

Burhop [27] has shown that the Q–V curve for L radiation has the same form as the expression given by Mott and Massey except for the numerical constant. However, in the intensity equations, the absolute value of Q is not required since it cancels when determining intensity ratios. The present investigation also assumes that the Q–V relation for M series radiation takes the same form as the relation for K radiation with the exception of the numerical constant.

In order to evaluate the integrals given in Eq. (13), a numerical integration procedure is employed. A schematic diagram illustrating how this integration is used in conjunction with the diffusion model is given in Fig. 5. A sphere in which electron diffusion occurs and is capable of exciting characteristic radiation is constructed using the effective range of electrons ρx_r defined earlier. The sphere has a radius equal to $\rho x_r - \rho x_d$, and its center is a distance $\rho x_d \cdot \sin \gamma$ below the surface of the target. The sphere is divided into radial increments $d\rho x$ such that the dependence of ρx on V^2 is taken into account. The manner in which the radial increments are chosen is important since the ionization of intensities is more efficient near the center of the sphere where the electron energy eV is large. The sphere is also divided into conical segments such that each contains an equal number of electron paths, thus an equal number of ionizations occur in each segment. Selection of segments in this manner has the advantage that each is weighted equally with respect to total intensity from the sphere. The method of integration employed consists of evaluating numerically the integral of Eq. (11) or (12) in a radial direction along the mean electron path of each conical segment $d\beta$. The integration extends from the center of the sphere in increments of $d\rho x$ until the electron has reached the surface of the target or until $\rho x = \rho x_r$, whichever occurs first. As shown by the unshaded area of Fig. 5, a portion of electrons will not reach their effective range ρx_r within the target. Since these electrons will not contribute to X-ray excitation after leaving the target, they represent losses due to backscatter and must be accounted for by integrating only that portion of each conical segment lying below the surface

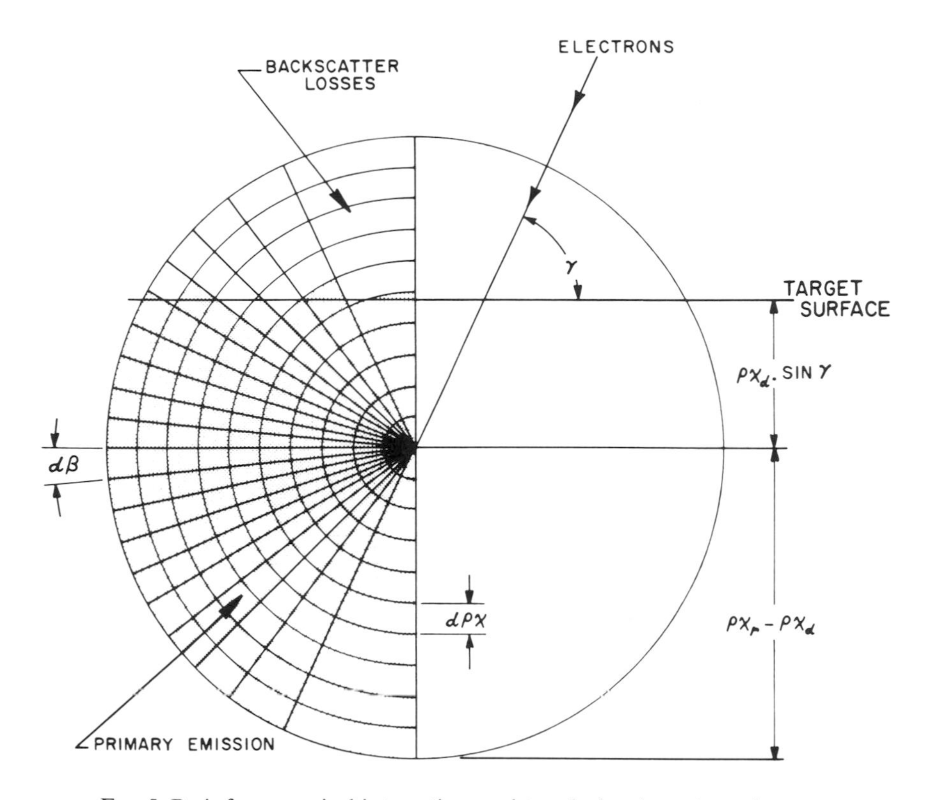

FIG. 5. Basis for numerical integration used to calculate intensity ratios.

of the target. The intensities from each conical segment are then summed to determine the total intensity excited in the sphere. In addition to the integration described, it is necessary to account for the intensity of characteristic radiation which electrons excite before reaching the depth of complete diffusion. This determination may be done by evaluating Eq. (10) from $\rho x = 0$ to $\rho x = \rho x_d$ in increments of $d\rho x$. It is only necessary to integrate the intensity for one electron path since the electron diffusion model assumes that all electrons travel identical paths until reaching the depth of complete diffusion. This intensity is then added to the intensity excited within the sphere.

Using this numerical integration procedure, the intensities $I(1)$ for the standard and I_1 for the sample are evaluated, and the intensity ratio (Eq. (13)) is determined. Thus, the correction factor for atomic number effect can be expressed by the relation

$$\text{at. no. corr. factor} = C_1/I_1/I(1). \tag{15}$$

III. Discussion and Evaluation of the Correction Formula

A. *Mass Penetration and Backscatter Effects*

As mentioned earlier, the diffusion model is unique in the determination of atomic number corrections because it allows separation of mass penetration and backscatter effects. Mass penetration effects may be determined separately by assuming that no electron backscatter losses occur. This determination is accomplished by integrating the intensities given in Eq. (13) from all portions of the conical segments shown in Fig. 5, including the unshaded areas above the target surface. Thus, the only difference between the standard and sample is the size of the electron diffusion sphere, determined by $\rho x_r - \rho x_d$, which varies with the atomic number and Z/A ratio of the target. Although such a calculation is physically unrealistic since backscatter effects are ignored, it can be used to gain insight into that portion of atomic number effects attributable solely to mass penetration effects. The correction factor for these effects may be determined from Eq. (15), using the intensity ratio obtained by this procedure. Although an intensity ratio for backscatter effects cannot be determined directly from Eq. (13), the backscatter correction factor may be obtained from a ratio of total atomic number and mass penetration correction factors,

$$\text{bksc. corr. factor} = \frac{I_1/I(1) \quad \text{at. no.}}{I_1/I(1) \quad \text{mass penet.}}. \tag{16}$$

Using the relations developed above, evaluation of the correction factors was first carried out to determine their sensitivity to the total number of volume elements N_T used in the numerical integration, where N_T is the product of the number of radial increments and the number of conical segments. The nature of the numerical integration procedure used was such that the ratio of conical segments to radial increments varied between 1 and 2 depending upon the relative values of ρx_r and ρx_d. For the purpose of this evaluation, the Cu–Al system was chosen. Results show that for values of $N_T > 270$, the calculated mass penetration correction remained constant. On the other hand, as shown in Fig. 6, values of $N_T > 2700$ were required before the calculated backscatter correction remained constant. This behavior is reasonable since the sensitivity of both the mass penetration and backscatter factors is dependent upon the precision with which ρx_d and ρx_r are determined. However, the sensitivity of the backscatter factor is also greatly dependent upon the number of volume increments intersecting the target surface. To determine the dependence of the atomic number correction on N_T in a case in which an extreme difference in atomic number exists, the thorium ($Z = 90$)–magnesium ($Z = 12$) system was tested for values ranging

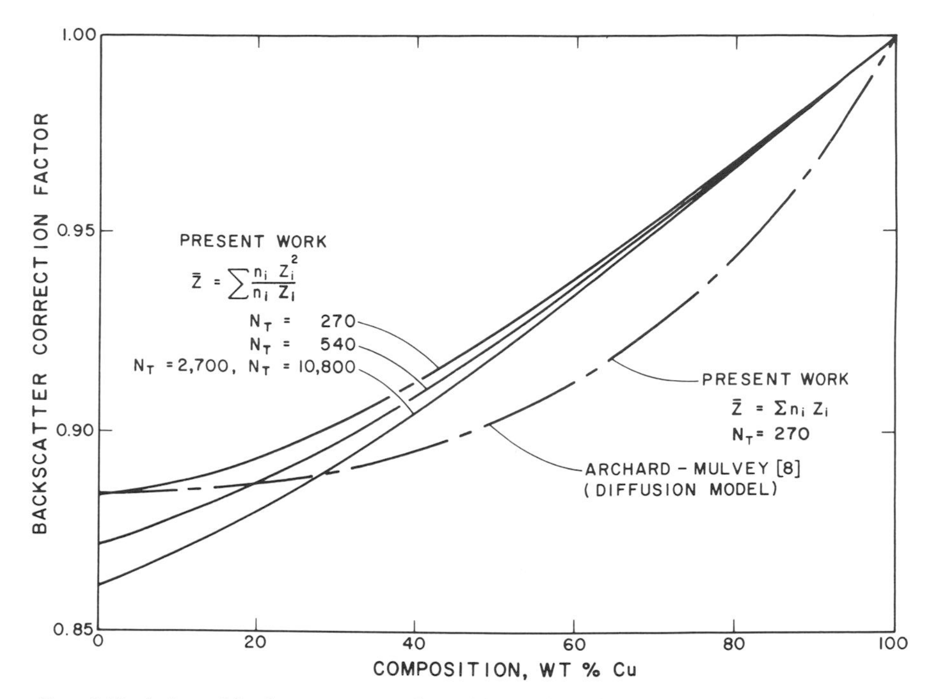

FIG. 6. Variation of backscatter correction with number of numerically integrated volume elements N_T and effective atomic number $\bar{Z}$ for Cu K_α from Cu–Al system: $V_0 = 29$ kV, $V_c = 8.98$ kV, $\gamma = 90°$.

up to $N_T = 28{,}000$. Results of these tests show that for $N_T > 6000$, the atomic number corrections for Th M_α and Mg K_α excited from Th–Mg alloys at 10 kV do not vary significantly. In view of these results, values of $N_T > 10{,}000$ were used for all data reported in the present investigation.

Figure 6 also shows that backscatter correction results from the present work differ greatly from those determined by Archard and Mulvey [8]. As can be seen in the figure, this difference is evidently due to the use of an improper relationship for evaluating the effective atomic number $\bar{Z}$, and an insufficient number of volume elements in the integration. If the relationship

$$\bar{Z} = \sum n_i Z_i \tag{17}$$

where n_i is the atomic fraction of element i, is used for determining effective atomic number instead of Eq. (4), and the value of N_T is set equal to 270 in the integration process, exact coincidence is obtained with the Archard and Mulvey results. The large difference in results obtained by using Eq. (17) instead of Eq. (4) for determining $\bar{Z}$ is illustrated by the two curves in the figure obtained for $N_T = 270$; the percent difference being as large as 80% of the backscatter factor for alloys containing greater than 80% copper.

To illustrate the difference in the two equations, an effective atomic number of 20 would correspond to copper weight concentrations of 45.1 and 64.7% using Eqs. (4) and (17), respectively. Furthermore, in comparing backscatter results, attention must also be given to the method for calculating ρx_d. Archard and Mulvey [8] used an approximation (Eq. (9)) which gives larger values for ρx_d than those determined by the more rigorous method used in the present investigation. A larger value for ρx_d causes the diffusion sphere to be smaller and to lie farther below the target surface, thus yielding smaller backscatter losses. As a result, relative backscatter differences between targets of different atomic number become greater, and thus backscatter corrections become greater. This consideration indicates that backscatter results of Archard and Mulvey (Fig. 6) which correspond to $N_T = 270$ (this work) actually represent an $N_T < 270$ for their work. The resultant effect of these differences on the total atomic number correction will be shown later for the Cu–Al system.

For the purpose of illustrating the separation of mass penetration and backscatter effects, results for Al K_α and Au M_α from the Au–Al system are shown in Figs. 7 and 8 along with the total atomic number correction. These effects are expressed in terms of the numerical factor required to correct

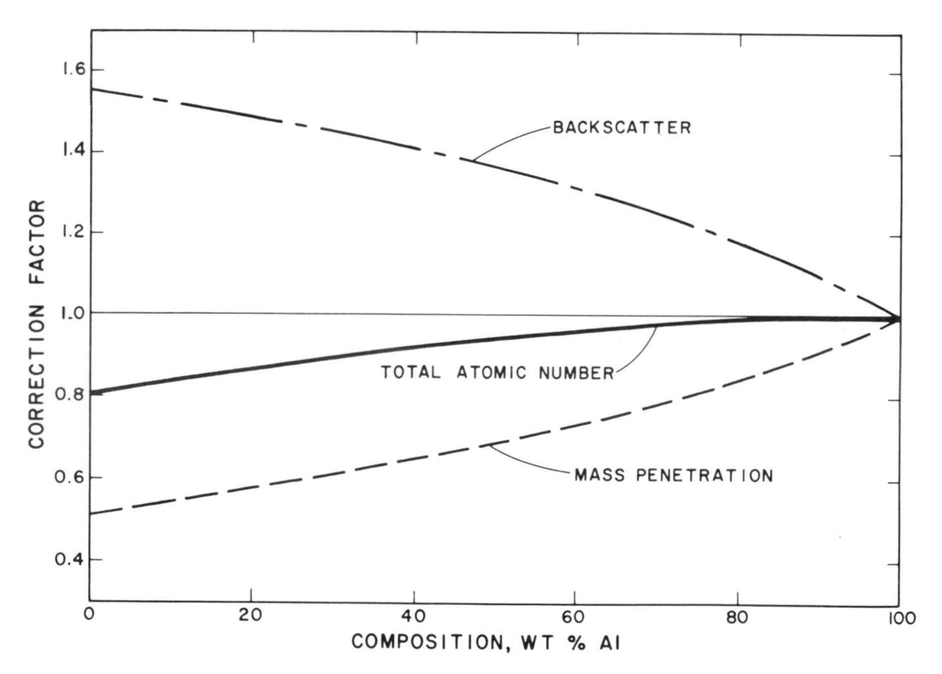

FIG. 7. Separation of atomic number correction into mass penetration and backscatter components for Al K_α from Au–Al system; $V_0 = 20$ kV, $V_c = 1.56$ kV, $\gamma = 62.5°$.

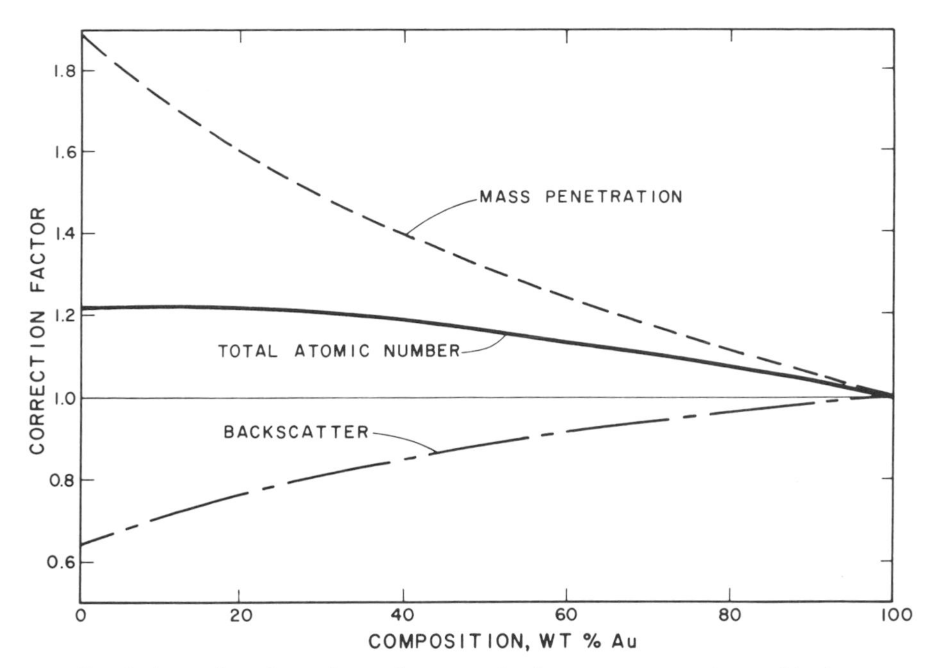

FIG. 8. Separation of atomic number correction into mass penetration and backscatter components for Au M_α from Au–Al system; $V_0 = 20$ kV, $V_c = 2.24$ kV, $\gamma = 62.5°$.

measured intensity ratios for which primary emission absorption has been accounted. In the case of Al K_α, the penetration effect enhances these ratios whereas the backscatter effect reduces them. However, in the case of Au M_α, they are reduced by the penetration effect and enhanced by the backscatter effect. This example typifies the general manner in which these effects influence measured intensity ratios. That is, for analysis of low Z elements in a high Z matrix, the mass penetration effect enhances, and the backscatter effect reduces these ratios. For analysis of high Z elements in a low Z matrix, the reverse is true. Since mass penetration and backscatter effects generally act in opposite directions, they tend to cancel one another. The extent to which cancellation is incomplete results in a residual atomic number correction usually due to the greater effect of mass penetration. The magnitude of this residual correction is dependent upon atomic number differences between the element analyzed and the target material. For large atomic number differences, the correction required may be very significant. For example, in the case of the Au–Al system, this correction approaches 15–20% of the amount present for small amounts of Au in Al.

It could be further generalized that in the analysis of low Z elements in a high Z matrix that the net of the atomic number effects is to enhance the measured intensity ratios; whereas in the analysis of high Z elements in a low Z matrix, the intensity ratios are reduced. This generalization is not a completely accurate statement for the correction procedure described herein. For example, in the case of Al K_α from Au–Al alloys (Fig. 7) for compositions greater than 83 wt % Al, the total atomic number correction becomes slightly greater than unity, e.g., the correction factor for 90 wt % Al equals 1.002. This effect is caused by backscatter differences given by the diffusion model which are greater than those that are experimentally observed for elements of low atomic number. This effect results in a larger backscatter correction, and hence too great a cancellation of mass penetration effects. Although this generalization applies to large differences in the Z values of the element analyzed and the target, it is not necessarily true for small Z differences. This fact relates to the dependence of Z/A on atomic number. Although Z/A typically increases over large differences in atomic number, localized variations occur in which the reverse is true. For example, in the Cu–Ni system, Cu ($Z = 29$) has a considerably lower Z/A value than Ni ($Z = 28$) and gives rise to a significant atomic number correction [20]. In this case, the atomic number effect enhances the measured intensity ratio of copper, the higher Z element.

B. Dependence on Experimental Variables

As mentioned earlier, the atomic number correction determined from the Archard diffusion model may be expressed as a function of several easily determined variables.

$$\text{at. no. corr.} = f(Z, Z/A, V_0, V_c, \gamma, C). \tag{18}$$

Dependence on Z, Z/A, and C have been discussed in the previous section. Dependence on the remaining variables will be discussed below.

In order to evaluate the dependence of atomic number correction on accelerating voltage V_0, it is worthwhile considering separately the dependence of mass penetration and backscatter effects. Columns 1 and 2 of Table I show the dependence of these effects on a change in accelerating voltage from 10 to 30 kV for Au M_α from Au–Ag alloys containing infinitely small amounts of Au. As V_0 increases, the depth of penetration (ρx_r) and the depth of complete diffusion (ρx_d) increase for both Au and Ag, the mass penetration in Au always exceeding that in Ag. However, the relative increase for Ag is greater than that for Au, thus causing the percent relative difference in mass penetration between Au and Ag to decrease. This effect results in a decrease in the mass penetration correction factor as V_0 increases.

Also shown in Table I are values for the relative difference between Au

TABLE I

DEPENDENCE OF CORRECTION FACTORS ON ACCELERATING VOLTAGE AND EXCITATION VOLTAGE ($\gamma = 62.5°$)

Accelerating voltage, V_0 (kV):	10		30		30	
Excitation voltage, V_c (kV):	2.24 (Au M_α)		2.24 (Au M_α)		11.92 (Au L_α)	
Target:	Au	Ag[a]	Au	Ag[a]	Au	Ag[a]
Depth of penetration, ρx_r (mg/cm^2):	0.575	0.440	3.92	3.14	3.12	2.53
Depth of complete diffusion, ρx_d (mg/cm^2):	0.031	0.046	0.23	0.33	0.23	0.33
Percent difference in mass penetration between Au and Ag[b]	−23.5		−19.9		−18.9	
Mass penetration corr. factor for Au:	1.309		1.256		1.236	
Percent difference between Au and Ag in portion of diffusion sphere above surface:	−6.8		−6.3		−8.3	
Backscatter corr. factor for Au:	0.888		0.900		0.879	
Total at. no. corr. factor for Au:	1.162		1.130		1.087	

[a] For infinitely dilute solutions of Au in Ag, the target may be assumed to be pure Ag.
[b] Minus signs denote decreases from Au to Ag.

and Ag in the portion of the diffusion sphere above the target surface. These values are due to relative changes in the size of the sphere of diffusion as well as its depth below the surface, as determined by ρx_r and ρx_d, and are indicative of relative backscatter differences. As the results show, backscatter differences between Au and Ag decrease, and thus the backscatter differences between Au and Ag decrease, and thus the backscatter correction decreases for an increase in V_0. As mentioned earlier, mass penetration and backscatter effects generally act in opposite directions, and the residual atomic number correction is usually due to the larger effect of mass penetration. In the present example, the decrease in mass penetration is greater in magnitude than the decrease in backscatter, thus the total atomic number correction decreases with increases in V_0. This variation of atomic number correction is shown in Fig. 9 for Au M_α from Au–Ag alloys at several accelerating voltages.

Dependence of atomic number correction on accelerating voltage is an important consideration in selecting operating conditions for analysis since it is desirable to minimize the magnitude of total corrections. It is significant to note that the voltage dependence of atomic number effects is opposite to that associated with primary emission absorption effects which generally increase with increases in V_0. Therefore, the magnitude of the absorption correction must be compared with the atomic number correction when it is desirable to select operating conditions such that total corrections are minimized.

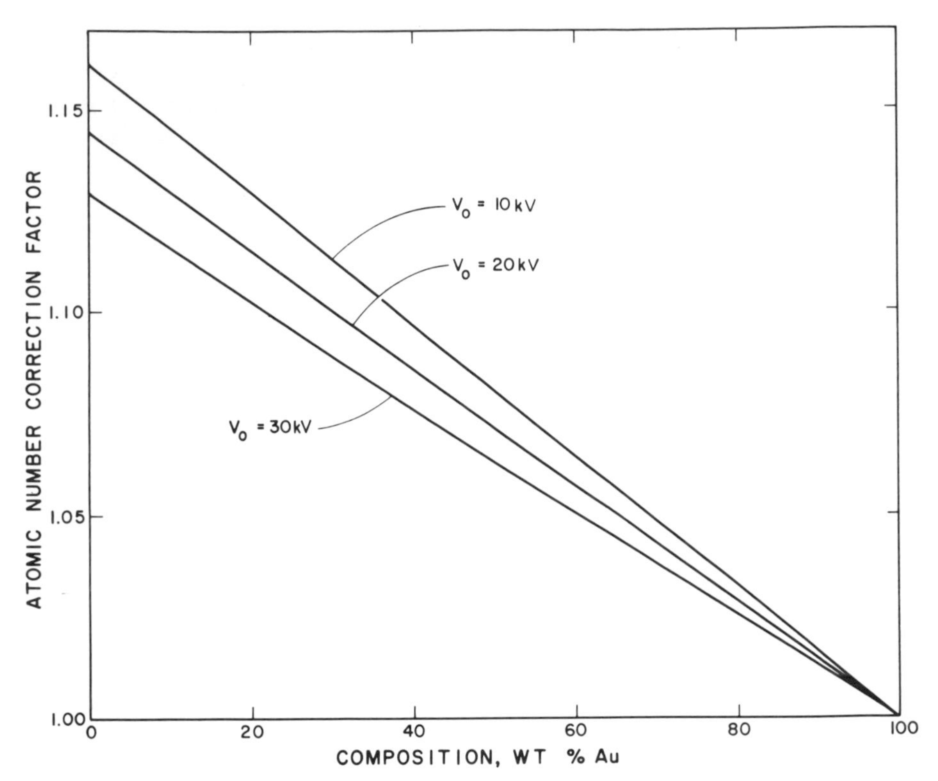

FIG. 9. Dependence of atomic number correction on accelerating voltage for Au M_α from Au–Ag system; $V_c = 2.24$ kV, $\gamma = 62.5°$.

The atomic number correction is also greatly dependent on the critical excitation voltage V_c. This dependence can be evaluated by considering changes in V_c while holding Z and V_0 constant. Columns 2 and 3 of Table I show the dependence of atomic number effects for a change in V_c from 2.24 kV (Au M_α) to 11.92 kV (Au L_α) for Au–Ag alloys. Under these conditions, as V_c increases, ρx_r for Au and Ag decreases, whereas ρx_d remains constant since it is not dependent on V_c. The relative decrease in depth of penetration for Au is greater than that for Ag, resulting in a decrease in the relative mass penetration difference between Au and Ag, and hence a decrease in the mass penetration correction.

The amount of backscatter for Au and Ag also decreases as V_c increases. This effect is due to a decrease in the size of the diffusion sphere $(\rho x_r - \rho x_d)$ and its constant depth below the surface which cause a relative decrease in the portion of the sphere above the surface. However, as shown in Table I, the relative backscatter decrease is greater in Ag than in Au, causing an increase in relative backscatter difference and an increase in the backscatter correction. In this case, therefore, the mass penetration correction decreases and backscatter increases, resulting in a decrease in the residual atomic number correction as V_c increases. Figure 10 shows the dependence of atomic

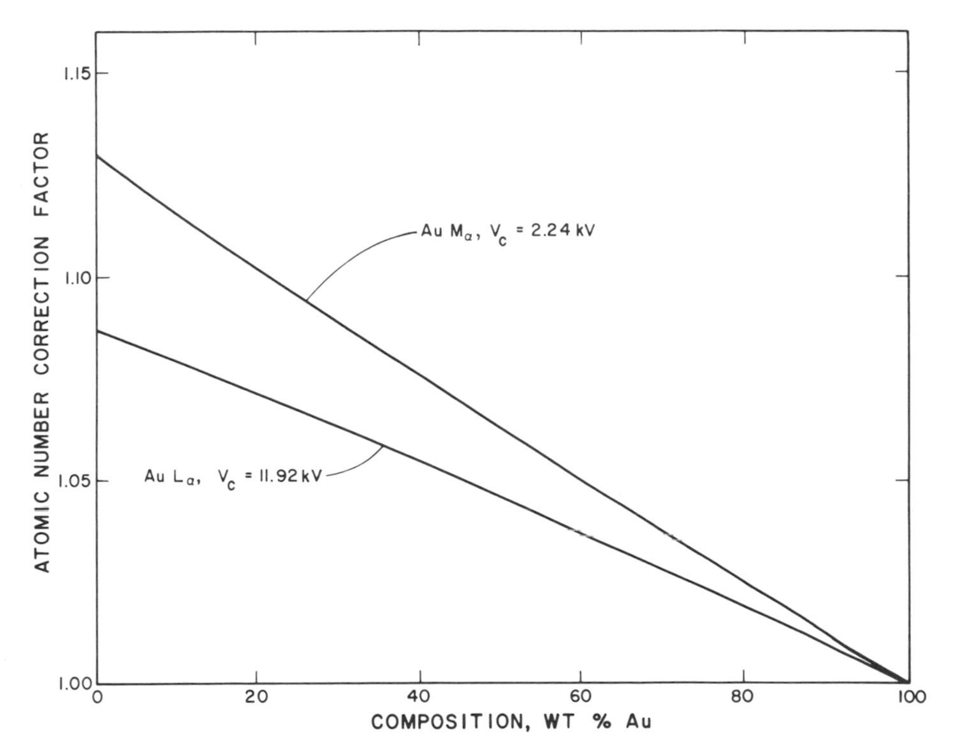

FIG. 10. Dependence of atomic number correction on critical excitation voltage for Au M_α and Au L_α from Au–Ag system; $V_0 = 30$ kV, $\gamma = 62.5°$.

number correction on critical excitation voltage for Au M_α and Au L_α from Au–Ag alloys. This large dependence makes the critical excitation voltage an important consideration in selecting characteristic X-ray lines for analysis. For example, where a choice of characteristic X-ray lines exists, e.g., between K and L or L and M series, it is desirable with regard to minimizing atomic number correction, to select the shorter wavelength, i.e., largest critical excitation voltage. This criterion is consistent with primary emission absorption effects which also decrease significantly for shorter X-ray wavelengths.

The diffusion model is applicable to any electron beam-target geometry. Magnitudes of the electron penetration parameters ρx_r and ρx_d, and thus the size of the diffusion sphere determined by the model, are not affected by inclination of the electron beam. Therefore, mass penetration is not dependent on electron beam-target geometry. However, the position of the diffusion sphere below the target surface is determined by the parameter $\sin \gamma \cdot \rho x_d$, thus causing backscatter to be dependent on electron beam-target geometry. As the electron incident angle decreases, the portion of the diffusion sphere above the target surface increases, increasing backscatter losses and thus decreasing relative backscatter differences for targets of different atomic number. This effect results in a decrease in backscatter correction and less complete cancellation of the larger mass penetration correction. Thus the effect of decreasing the electron incident angle is to increase the total atomic number correction, a conclusion opposite to that made from previous work on the Cu–Al system [8]. Dependence of atomic number correction on electron incident angles is shown in Fig. 11 for Au M_α from Au–Ag alloys. The magnitude of this effect is less than 5% of the total atomic number correction for incident angles varying from 60 to 90°.

In the preceding discussions of the dependence of atomic number correc-

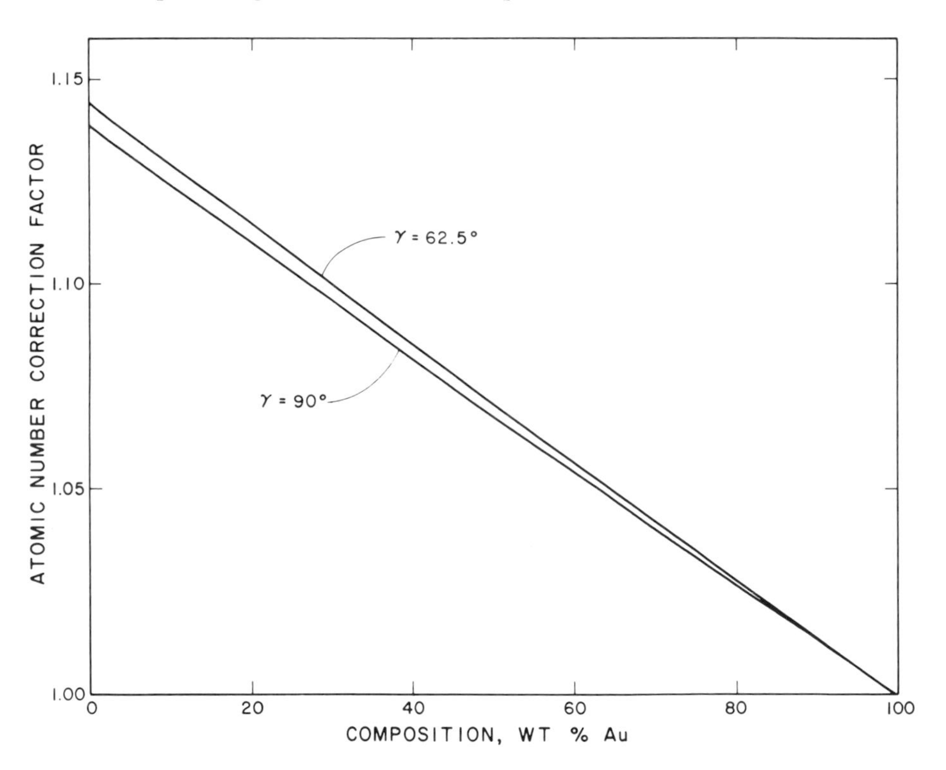

FIG. 11. Dependence of atomic number correction on electron incident angle for Au M_α from Au–Ag system; $V_0 = 20$ kV, $V_c = 2.24$ kV.

tion on V_0, V_c, and γ, only results for a heavy element in a light matrix were considered. For the case of a light element in a heavy matrix, the dependence of atomic number correction on V_0, V_c, and γ would be the same with the exception that the total atomic number correction would be less than unity, and that mass penetration would enhance and backscatter would reduce intensity ratios.

C. Comparison with Other Correction Procedures: Copper–Aluminum System

Evaluation of the atomic number correction procedure set forth above requires a comparison with other correction procedures [8, 20, 21, 28–30]. Results of this comparison are shown in Fig. 12 for analysis of Cu K_α from Cu–Al alloys at 29 kV and 90° electron incidence.

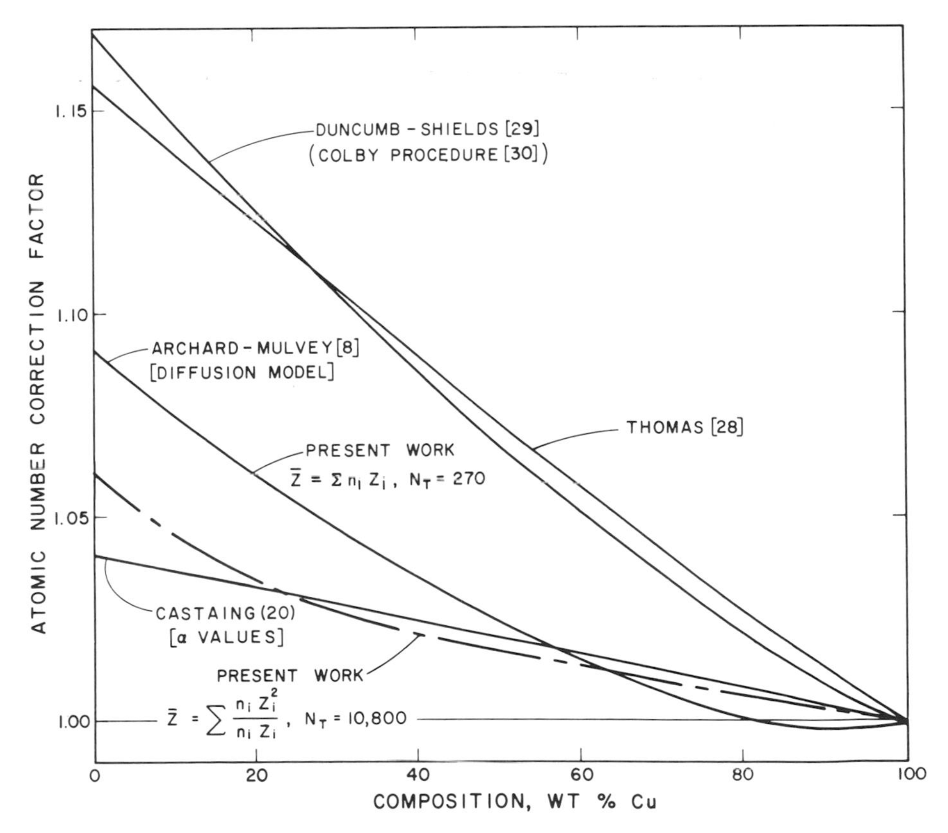

FIG. 12. Comparison of several atomic number corrections for Cu K_α from Cu–Al system; $V_0 = 29$ kV, $V_c = 8.98$ kV, $\gamma = 90°$.

The Archard–Mulvey results show a behavior substantially different from that determined in the present investigation. Of particular significance is the greater correction for compositions of less than 60% Cu. The difference is due to two factors stated previously, namely, (1) the manner in which effective atomic number $\bar{Z}$ is calculated, and (2) the smaller number of volume elements N_T used in the numerical integration of the diffusion model. Use of Eq. (17) to calculate $\bar{Z}$ instead of Eq. (4) has the effect of displacing calculated correction values to higher copper concentrations. As shown in Fig. 6, if the number of volume elements is too small, the calculated backscatter correction is too small, resulting in less cancellation of mass penetration, and thus yielding a larger residual atomic number correction. These effects are partially offset, however, by an increase in backscatter correction resulting from the larger values of ρx_d obtained by use of Eq. (9). In addition, for Cu–Al alloys above 82 wt % Cu, Archard–Mulvey give atomic number corrections which are slightly less than unity. This behavior is a result of too few volume elements used in the numerical integration.

Corrections obtained from α values given by Castaing compare exceptionally well for compositions greater than 20% Cu; the difference between corrections becoming increasingly larger as the Cu composition decreases below 20%. These α values are the product of the Z/A ratio and a λ coefficient determined from empirical emission distribution curves. Since the α values are determined for each of the pure constituents of the compound sample and assumed to vary linearly with composition, the correction curve shown in Fig. 12 varies linearly with weight concentration. On the other hand, correction curves given by the diffusion model are nonlinear.

The corrections calculated using the Thomas and Duncumb–Shields procedures agree very well with each other, but agreement with the results of this work is substantially poorer than either the Archard–Mulvey or Castaing procedures. The Thomas correction accounts for backscatter with an effective current factor R, calculated from data on the energy distribution of electrons backscattered from pure element targets, and the target stopping power S, taken from the tabulations by Nelms [31]. The Duncumb–Shields correction is that described by Colby [30] in which the Bethe expression is used to calculate target stopping power as given by Nelms [31] and backscatter losses are determined from fifth degree polynomials which have been fitted to backscatter yields of Bishop [32] as a function of overvoltage ratio V/V_c. The Duncumb–Shields method involves many complex calculations which do not lend themselves to practical tabulations, and therefore without the use of a computer is not easily applicable to routine calculations. Comparisons of the correction method described herein with the Thomas and Duncumb–Shields methods will be made in Section V,C for experimental data from the Au–Al system.

IV. Basis for Universal Correction Tables

The relationship between the intensity ratio $I_1/I(1)$ and composition as given by Eq. (13) is such that the complex numerical integration process described above need not be carried out for individual analyses in order to apply an atomic number correction. On the contrary, evaluation of the integrals need be carried out only once and tabulated, the tabulation accounting for variation in γ, V_0, V_c, and $\bar{Z}$. The tabulation may then be used to correct for atomic number effect in any analysis problem. Furthermore, the procedure for using such tables is simpler than that required for determining absorption corrections from tables such as those prepared by Adler and Goldstein [33] based on Philibert's empirical equation [3].

The basis for this tabulation procedure involves separating the expression in Eq. (13) into two factors which are designated as $(Z/A)'$ and λ'. The $(Z/A)'$ factor is dependent only on the composition and the Z/A ratios of the components of the sample, and it can be calculated for any single component of a multicomponent system by the expression,

$$(Z/A)' = \frac{\sum C_i Z_i/A_i}{Z_1/A_1} \tag{19}$$

where C_i is the best known value of weight concentration for each component. It should be noted that the $(Z/A)'$ factor is equivalent to the atomic number correction obtained by use of the Webster law for electron deceleration (Eq. (1a)). The λ' factor is dependent on γ, V_0, V_c, and $\bar{Z}$ and is evaluated from the expression

$$\lambda' = \frac{\sum \int Q\, d(V^2F(V))_{\text{standard}}}{\sum \int Q\, d(V^2F(V))_{\text{sample}}}. \tag{20}$$

Since the weight concentration C_1 of element 1 appearing in the numerator of Eq. (13) represents the measured intensity ratio $I_1/I(1)$, corrected for primary emission absorption and secondary emission effects, the atomic number correction is obtained by multiplying $I_1/I(1)$ by $(Z/A)'$ and λ'.

It should be noted that the $(Z/A)'$ and λ' factors do not represent separate mass penetration and backscatter effects and bear no relation to the earlier discussion on these effects. The λ' effect given by Eq. (20) must also be distinguished from λ values, defined by Castaing [20], which represent a "specific decelerating power" for each component. The $(Z/A)'$ factor is a partial expression of the mass penetration effect while the λ' factor includes the total backscattering effect plus the remaining portion of the mass penetration effect not included in $(Z/A)'$.

A separation of the atomic number correction into $(Z/A)'$ and λ' factors

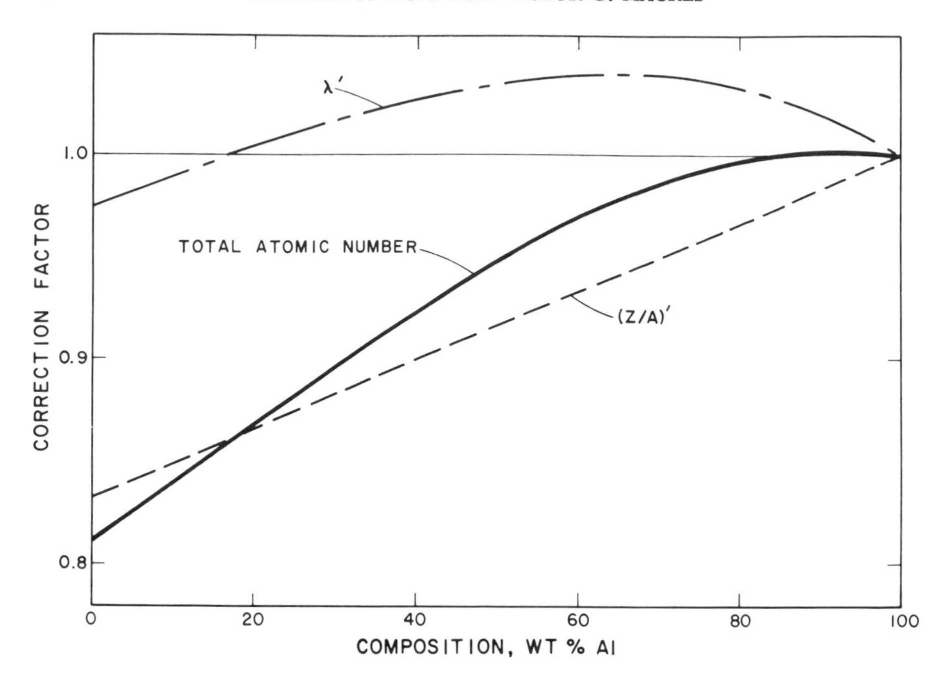

FIG. 13. Separation of atomic number correction into $(Z/A)'$ and λ' factors for Al K_α from Au–Al system; $V_0 = 20$ kV, $V_c = 1.56$ kV, $\gamma = 62.5°$.

is given in Figs. 13 and 14 for Al K_α and Au M_α, respectively from Au–Al alloys. The $(Z/A)'$ factor varies linearly as a function of weight concentration whereas the λ' factor varies in a nonlinear manner due to its dependence on the Bethe relation.

In the cases of both Al K_α (Fig. 13) and Au M_α (Fig. 14), the magnitude of the λ' factor is smaller than the $(Z/A)'$ factor. The portion of mass penetration effect not included in $(Z/A)'$, but included along with the backscatter effect in λ', can be seen by comparing Figs. 13 and 14 with Figs. 7 and 8, respectively. In the case of Al K_α, the portion not included in $(Z/A)'$ causes the λ' factor to act in the same direction as $(Z/A)'$ for alloys less than about 17% Al. In the Au M_α case, this effect is sufficiently large to cause λ' to act in the same direction as $(Z/A)'$ over the entire composition range. The resultant atomic number corrections for Al K_α and Au M_α are the same as those set forth in Figs. 7 and 8, respectively, but are included for reference. It should be noted, in particular, that the difference between the $(Z/A)'$ and atomic number correction curves directly illustrates the difference in the atomic number correction using Webster's law (Eq. (1a)) and the diffusion model.

Separation of Eq. (13) into two parts is significant since it removes the composition dependent component $(Z/A)'$ and permits values of λ' to be

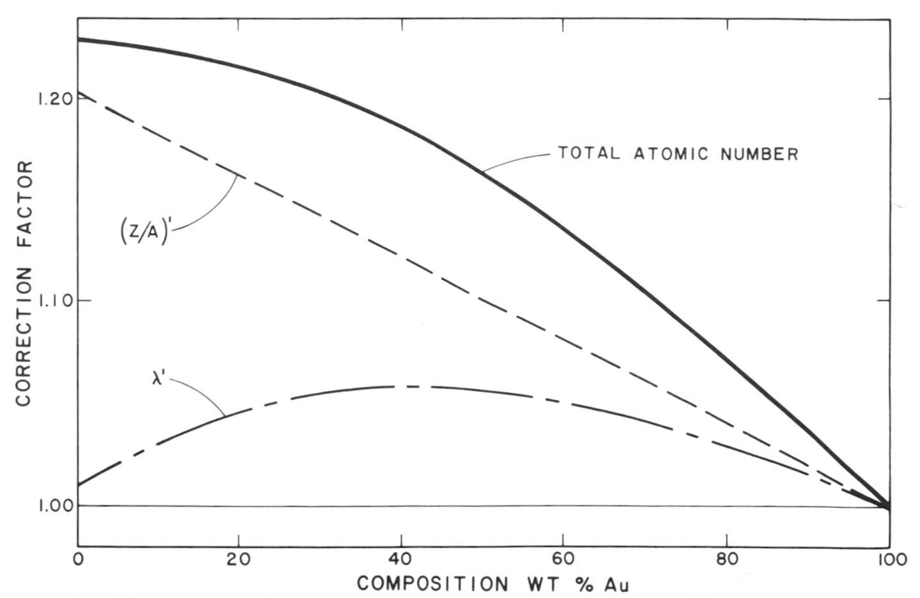

FIG. 14. Separation of atomic number corrections into $(Z/A)'$ and λ' factors for Au M_α from Au–Al system; $V_0 = 20$ kV, $V_c = 2.24$ kV, $\gamma = 62.5°$.

computed versus effective atomic number $\bar{Z}$. A computer program has been written which calculates values of λ' for specific values of electron accelerating voltage V_0, critical excitation voltage V_c, electron incident angle γ, and for effective atomic numbers, $\bar{Z}$, from 10 through 100 in intervals of 5. The program has been written in the ALGOL language for use on a Burroughs 5500 binary computer. By generating λ' values for a number of characteristic X-ray lines (different values of V_c) and for various accelerating voltages V_0, λ' tabulations have been compiled. Appendix I shows a typical set of λ' tabulations for Mg, Al, and Si K_α lines and Pt, Au, and Hg M_α lines for an accelerating voltage of 20 kV and an electron incident angle of 62.5°.

Tabulations of λ' can be used to calculate atomic number corrections for any multicomponent system regardless of the number of elements present. An example calculation is given in Appendix II. The effective atomic number $\bar{Z}$ of the target is determined from the best known compositions using Eq. (4). A value for the λ' factor is obtained, using interpolation where necessary, from applicable tabulated values (such as those given in Appendix I) for the $\bar{Z}$ calculated, the characteristic wavelength analyzed (V_c) and conditions used (V_0 and γ). The $(Z/A)'$ factor is calculated from Z/A ratios for the components and the best known composition of the target, using Eq. (19). The product of λ' and $(Z/A)'$ gives the correction factor which may then be applied to measured intensity ratios to account for effects of atomic number. It should be noted that the measured intensity ratio must be corrected for deadtime, drift, background, fluorescence, and absorption effects prior to applying this atomic number correction.

This procedure for atomic number correction may easily be incorporated into a reiterative process of successive approximations for determining the composition of an unknown. Furthermore, its use in a computer program for the reduction of microprobe intensity data to chemical compositions may be facilitated by fitting λ' values with polynomial equations. This technique is presently being investigated.

V. Application to Experimental Data

A. Analytical Procedures

In order to illustrate the applicability of the proposed atomic number correction, a number of alloy systems were analyzed and the correction applied. Before discussing the results of this experimental work, it is worthwhile to review some of the problems which must be considered. A major problem concerns the uncertainties arising in corrections for deadtime, drift, background, fluorescence, and primary emission absorption effects, all of which must be applied prior to the atomic number correction. It is possible to minimize corrections for deadtime, drift, and background by careful selection of operating conditions such as accelerating voltage, analyzing crystal, primary and secondary slits, pulse height selector settings, and count rates. Fluorescence corrections caused by other characteristic radiations usually are significant only when the exciting element and the fluoresced element are relatively close in atomic number. Therefore, selection of alloy systems may be made in which these effects are negligible. In most cases, fluorescence caused by the continuum is negligible compared to the uncertainties of other corrections and measured data. The absorption of primary emission usually accounts for the largest percentage of the total correction. The accuracy with which this effect can be corrected is limited to uncertainties in both the $f(\chi)$ values and mass absorption coefficients on which this correction is based. Although these uncertainties can be less than 1% in certain cases, an uncertainty in extreme cases of 50% is not unusual. Consequently, the validity of an atomic number correction procedure is not easily ascertained.

Ideally, to study atomic number effects, it is desirable to select alloy systems in which all effects other than atomic number are negligible. Although this selection is possible to a limited extent, it is generally difficult to find alloy systems in which absorption effects are small but in which significantly large atomic number effects are present. Alloy sytems which exhibit large atomic number effects are usually those in which the constituent elements vary widely in atomic number. Unfortunately, these systems often have large absorption effects associated with one or more of the constituents. In some situations, however, absorption effects can be minimized by analyzing K and L series characteristic X-ray lines of shorter wavelengths. This alternative will be shown later for the case of Au L_α. In the case of lighter elements, e.g., Mg, Al, Si, in which even the K series lines are often highly absorbed, this alternative is not always possible. In addition, the selection of X-ray lines is also dependent on the types of analyzing crystals available, 2θ range of the X-ray spectrometer and peak-to-background ratios attainable. For example, although the absorption effects of Mo K_α (0.710 Å, LiF crystal) are usually very low, the Mo L_α (5.406 Å, PET crystal) is preferred for most analyses since the peak-to-background ratio attainable is considerably higher. Another difficulty is that of obtaining alloys for which the true composition or stoichiometry are well known. This requirement is particularly important in microprobe analysis since the composition and homogeneity must be known on a micron scale and not merely in terms of bulk chemical analysis.

In the present investigation, quantitative analyses were carried out on alloy systems selected to evaluate the applicability of the correction procedure for a wide range of atomic number differences and experimental conditions. Prior to applying the atomic number correction, microprobe X-ray data were corrected for drift, background, fluorescence by characteristic X-ray lines (where these effects were significant), and primary emission absorption effects. X-ray counting rates did not exceed 5000 counts per second and effects of deadtime were negligible for the counting system used, therefore obviating the need for a deadtime correction. The drift correction was made by assuming that drift occurred linearly between any two successive standard intensity measurements. In all cases drift was less than 1%. Background intensity I^B was determined from the following relation:

$$I^B = \sum C_i I_i^{\,B} \tag{21}$$

where $I_i^{\,B}$ is the background intensity measured from the pure element standard of i without changing the spectrometer setting. In the case of the element being analyzed, an equivalent background intensity was interpolated from pure element standards of neighboring elements. Where significant fluorescence by characteristic X-ray lines occurred, the correction derived by Castaing [1] was applied. With the exception of the tetrahedrite mineral, no alloy systems were studied which required a fluorescence correction greater than 0.1 wt % of the compositions analyzed.

In all cases a correction for absorption effects was made using the empirical equation given by Philibert [3]. The effect of critical excitation potential on the absorption correction was included by using the Duncumb and Shields

[10] modification for sigma given by the relation

$$\sigma = \frac{2.39 \times 10^5}{V_0^{1\cdot 5} - V_c^{1\cdot 5}}. \tag{22}$$

Values for mass absorption coefficients, given in Appendix III, were taken from tables by Heinrich [34], Frazer [35], and from recent measurements by Bucklow [36]. The absorption parameter χ was determined using the relations

$$\chi = (\mu/\rho) \cdot \chi_T \tag{23}$$

$$\chi_T = \csc \beta \cdot \sin \gamma \tag{24}$$

where β is the X-ray take-off angle. The geometric factor χ_T is an exact definition of the X-ray absorption path for any incident beam–X-ray take-off geometry. Interpretation of Monte Carlo results by Bishop [37] indicates agreement to within 1% using this relation for χ_T (i.e., within experimental error of $f(\chi)$ curves obtained for electron incident angles ranging from 60 to 90°). The instrument used in the present investigation (Materials Analysis Company model 400S) had an electron incident angle of 62.5° and an X-ray take-off angle of 33.5° ($\chi_T = 1.607$).

B. Effect of Experimental Variables: Gold–Silver System

The Au–Ag system was chosen to study the effect of accelerating voltage and choice of spectral line on the applicability of the diffusion model for atomic number correction. Eight carefully prepared Au–Ag alloys of well-known compositions ranging from 11.5 to 80.5 wt % Au were available for analysis. The alloys had been prepared from spectrographically pure Au and Ag by fire assaying. Although the alloys were postannealed, slight inhomogeneities were still observed in the specimens containing 20.8 and 30.9 wt % Au. To minimize the effect of inhomogeneity in all specimens, a beam diameter of 30 μ was used and about 10 different areas were analyzed on each alloy. Analyses were made for Ag L_α and Au M_α at 10 and 20 kV and for Au L_α at 30 and 40 kV. The Au L_α and M_α analyses provided a basis for evaluating the applicability of the atomic number correction when used in conjunction with L and M series lines. However, Ag L_α was not analyzed above 20 kV because of the large absorption which occurs and the consequent uncertainty of the absorption corrections to be applied. For example, the magnitude of the absorption correction for small amounts of Ag in Au at 30 kV exceeds 100% of the measured intensity ratio. Even at 20 kV the absorption correction for high Au alloys is 60% of the measured intensity.

Average results for 8 to 12 analyses on each alloy are given in Tables II–IV. The tabulated measured intensity ratios have been corrected for drift

TABLE II
GOLD–SILVER SYSTEM: Au L_α ANALYSES (CONCENTRATION, Wt % Au)

	Meas. intensity ratio ($\pm\sigma$)	After absorption corr.	After at. no. corr.	True concentration
30 kV	10.7 ± 0.3	10.8	11.6	11.5 ± 0.1
	19.6 ± 0.4	19.7	21.1	20.8 ± 0.3
	29.0 ± 0.4	29.1	30.9	30.9 ± 0.2
	39.1 ± 0.3	39.2	41.3	41.2 ± 0.2
	48.5 ± 0.3	48.6	50.8	50.9 ± 0.2
	58.0 ± 0.4	58.1	60.2	60.5 ± 0.1
	68.4 ± 0.3	68.5	70.4	70.6 ± 0.1
	78.9 ± 0.3	79.0	80.5	80.5 ± 0.2
40 kV	10.3 ± 0.4	10.4	11.2	11.5 ± 0.1
	19.8 ± 0.8	19.9	21.3	20.8 ± 0.3
	29.1 ± 0.8	29.3	31.1	30.9 ± 0.2
	39.2 ± 0.6	39.4	41.5	41.2 ± 0.2
	49.0 ± 0.4	49.2	51.4	50.9 ± 0.2
	58.2 ± 0.6	58.4	60.5	60.5 ± 0.1
	68.8 ± 0.3	69.0	70.8	70.6 ± 0.1
	78.6 ± 0.2	78.8	80.2	80.5 ± 0.2

TABLE III
GOLD–SILVER SYSTEM: Au M_α ANALYSES (CONCENTRATION, Wt % Au)

	Meas. intensity ratio ($\pm\sigma$)	After absorption corr.	After at. no. corr.	True concentration
10 kV	18.8 ± 0.5	19.2	21.6	20.8 ± 0.3
	28.2 ± 0.6	28.6	31.8	30.9 ± 0.2
	37.7 ± 0.9	38.2	41.8	41.2 ± 0.2
	47.3 ± 0.4	47.9	51.7	50.9 ± 0.2
	57.2 ± 0.4	57.8	61.5	60.5 ± 0.1
	67.5 ± 0.5	68.0	71.2	70.6 ± 0.1
	79.3 ± 0.3	79.4	81.8	80.5 ± 0.2
20 kV	9.81 ± 0.3	10.3	11.6	11.5 ± 0.1
	18.4 ± 0.8	19.3	21.5	20.8 ± 0.3
	27.4 ± 0.4	28.4	31.2	30.9 ± 0.2
	37.2 ± 0.7	38.4	41.7	41.2 ± 0.2
	46.1 ± 0.5	47.3	50.6	50.9 ± 0.2
	56.1 ± 0.3	57.2	60.4	60.5 ± 0.1
	66.5 ± 0.3	67.4	70.2	70.6 ± 0.1
	76.8 ± 0.3	77.3	79.5	80.5 ± 0.2

TABLE IV
GOLD–SILVER SYSTEM: Ag L_α ANALYSES (CONCENTRATION, Wt % Ag)

Meas. intensity ratio ($\pm\sigma$)	After absorption corr. Heinrich [34]	After absorption corr. Frazer [35]	After absorption corr. Bucklow [36]	After at. no. corr. Heinrich [34]	After at. no. corr. Frazer [35]	After at. no. corr. Bucklow [46]	True concentration
			10 kV				
76.8 ± 1.2	80.6	81.1	81.3	78.7	79.2	79.3	79.2 ± 0.3
65.2 ± 1.1	70.2	70.7	70.8	67.6	68.1	68.2	69.1 ± 0.2
56.0 ± 0.6	61.7	62.2	62.3	58.7	59.2	59.3	58.8 ± 0.2
45.9 ± 0.7	51.7	52.2	52.4	48.5	49.0	49.2	49.1 ± 0.2
36.7 ± 0.5	42.3	42.7	42.9	39.2	39.6	39.8	39.5 ± 0.1
27.1 ± 0.5	31.8	32.3	32.4	29.1	29.6	29.8	29.4 ± 0.1
17.7 ± 0.2	21.1	21.6	21.7	19.1	19.5	19.6	19.5 ± 0.2
			20 kV				
82.5 ± 1.2	88.5	89.7	90.0	87.4	88.5	88.8	88.5 ± 0.1
70.4 ± 1.2	80.5	81.8	82.0	78.7	79.9	80.1	79.2 ± 0.3
58.0 ± 1.6	70.1	71.7	71.9	67.6	69.2	69.4	69.1 ± 0.2
45.3 ± 0.9	58.1	59.7	59.9	55.4	57.0	57.2	58.8 ± 0.2
36.9 ± 0.8	49.8	51.3	51.5	46.9	48.3	48.5	49.1 ± 0.2
28.5 ± 0.7	40.3	41.7	41.9	37.5	38.8	39.0	39.5 ± 0.1
19.8 ± 0.2	29.3	30.4	30.6	27.0	28.0	28.2	29.4 ± 0.1
12.7 ± 0.1	19.7	20.4	20.7	17.9	18.5	18.8	19.5 ± 0.2

and background effects. The standard deviation in the intensity ratios is best for the Au L_α data and poorest for the Ag L_α data. This result is apparently related to the higher absorption of the Ag L_α and its consequent greater sensitivity to specimen inhomogeneity and perfection of the polished surface. Since large uncertainties exist in absorption data for Ag L_α in Au, e.g., Heinrich and Bucklow (Metals Research) data differ by about 10% (see Appendix III), intensity data were corrected using absorption coefficients from the three different sources. Weight concentrations after absorption and after atomic number correction are tabulated to show the magnitude of each correction as well as the effect of the different absorption data. Correction for fluorescence of Au M_α by Ag L_α is not included. The effect of this fluorescence would be to increase the calculated Au compositions, but its magnitude is estimated to be less than 0.1 wt % for the compositions analyzed.

Results for the Au L_α analyzes at 30 and 40 kV are given in Table II. However, Ag L_α was not analyzed at these voltages because of the high absorption it undergoes. Hence, in applying corrections to the Au L_α data,

Ag was assumed to make up the composition balance. Since absorption corrections for Au L_α are extremely small (less than 1% of the intensity ratio even at 40 kV), high accuracy in the absorption correction procedure is not required. The minimal significance of the absorption correction provides an excellent opportunity to check the validity of the atomic number correction method. The magnitude of the required correction varies from 8% of the intensity ratio at low Au compositions to about 2% at high compositions, the corrections being almost exactly equal at both 30 and 40 kV. The agreement between calculated and actual compositions is excellent for the entire range for both sets of data, the average agreement being about 0.5 and 1.1% for the 30 and 40 kV data, respectively.

In Table III the results of the Au M_α analyses are given. Although the absorption of Au M_α is larger than that for Au L_α, it is still relatively small at both 10 and 20 kV. Thus, high accuracy in the absorption correction is not required to test the validity of the atomic number correction procedure. Compared to the Au L_α analyses, the correction is slightly greater, and varies from about 13% of the intensity ratio for low Au compositions to about 3% for high compositions, the corrections being slightly greater for the 10 than for the 20 kV analysis. The agreement between calculated and actual compositions is very good for the entire range for both sets of data, the average agreement being about 2 and 1.2% for the 10 and 20 kV data, respectively. The better average agreement at the higher rather than the lower kilovolt value is the reverse of the Au L_α case.

In the reduction of Ag L_α intensity data, the correction for absorption is large, amounting to about 20 and 60% of the intensity ratios for low Ag alloys at 10 and 20 kV, respectively. As a consequence, the validity of the atomic number correction is dependent upon the accuracy with which the absorption correction can be made. Unfortunately, the reliability of the Philibert equation decreases substantially when the absorption to be corrected is large. In Table IV the results for the Ag L_α analyses are presented using absorption coefficient data from Heinrich, Frazer, and Bucklow. For the 10 kV data, the atomic number correction ranged from about 2.5 to 10% from high to low Ag compositions while the absorption correction varied from 5 to 23%. In spite of these corrections, the agreement between calculated and actual compositions is excellent for all alloys analyzed. The best average agreement is obtained using Frazer and Bucklow absorption coefficient data and is about 0.8%. Using data from Heinrich, the agreement is about 1.2%, the calculated compositions all being lower than the actual values. At 20 kV, the atomic number corrections are about the same, but the absorptions increase considerably, varying from about 9 to 60% from high to low Ag compositions. In this case, the agreement between actual and calculated composition is poorer and is about 1.8% using Frazer and Bucklow absorption coefficient data and 4.5% using Heinrich data. It should be noted that the agreement generally becomes poorer as the Au concentration increases, lending support to the argument that the absorption correction is in error. Further support is obtained by comparison of the Au and Ag analyses which show that the combined total of each alloy analysis deviates further from 100% for the low than for the high Ag alloys.

In summary, the Au L_α and M_α data show excellent agreement with actual compositions, the L_α data being somewhat better, and hence lending strong support to the validity of the diffusion model for atomic number correction. Although in the case of Ag L_α data, the accuracy of the correction is somewhat masked by uncertainty in the absorption correction, its agreement with actual compositions is also impressive. The uncertainty in the absorption correction, which is likely due to limitations of the Philibert equation, perhaps causes the 10 kV data to appear fortuitously in better agreement than the 20 kV data in contrast to the Au M_α analyses in which the opposite is true.

C. Examples of Binary Systems

1. *Gold–Aluminum System*

The gold–aluminum system represents an extreme difference in atomic number of the constituents and would be expected to show large atomic number effects. Three compounds in this system were analyzed: $AuAl_2$, AuAl, and Au_2Al. These compounds had been formed in a diffusion couple between pure Au and pure Al and were assumed to be stoichiometric in composition in accordance with the equilibrium phase diagram.

Analyses of the phases were made at an accelerating voltage of 20 kV for the Al K_α and Au M_α spectral lines. Results of these analyses corrected for effects of absorption and atomic number are given in Tables V and VI. Correction for fluorescence of Al K_α by Au M_α is not included. Although this effect increases the Al K_α intensity and gives higher calculated Al concentrations, estimates showed that its magnitude does not exceed 0.1 wt % for the alloys analyzed. Considerable disagreement exists in absorption coefficient data for both Al K_α and Au M_α given by Heinrich, Frazer, and Bucklow. As shown in Appendix III, these differences are as great as 13% for Au M_α in Au, 8% for Al K_α in Au, 5.5% for Au M_α in Al and 5.5% for Al K_α in Al. Due to these relatively large variations in absorption data and to the magnitude of the absorption effect, especially in the case of Al K_α, absorption corrections for the Au–Al compounds were determined separately using absorption data from the three different sources. In addition, results from each

TABLE V

GOLD–ALUMINUM SYSTEM: Al K_α ANALYSES (20 kV, CONCENTRATIONS, Wt % Al)

Compound	Meas. intensity ratio (±σ)	Absorption data used	After absorption corr.	At. no. corr. procedure [a–c]	After at. no. corr.	Stoichiometric (assumed) composition
Au Al_2	15.6 ± 0.4	Bucklow [36]	26.8	W–M	23.4	21.48
			26.8	D–S	19.3	
			—	T	27.7	
		Heinrich [34]	28.2	W–M	24.6	
			28.2	D–S	20.3	
			—	T	29.6	
		Frazer [35]	27.9	W–M	24.4	
			27.9	D–S	20.1	
			—	T	29.3	
Au Al	8.81 ± 0.2	Bucklow [36]	15.8	W–M	13.4	12.03
			15.8	D–S	11.0	
			—	T	16.8	
		Heinrich [34]	16.9	W–M	14.3	
			16.9	D–S	11.8	
			—	T	18.0	
		Frazer [35]	16.6	W–M	14.1	
			16.6	D–S	11.6	
			—	T	17.7	
Au_2Al	4.49 ± 0.2	Bucklow [36]	8.15	W–M	6.73	6.40
			8.15	D–S	5.64	
			—	T	8.93	
		Heinrich [34]	8.85	W–M	7.36	
			8.85	D–S	6.13	
			—	T	9.64	
		Frazer [35]	8.71	W–M	7.18	
			8.71	D–S	6.03	
			—	T	9.45	

[a] W–M: Wolf and Macres, present work.

[b] D–S: Duncumb and Shields [10], Colby procedure [30].

[c] T: Thomas [28].

set of absorption data were corrected using three different atomic number corrections [present work, 28, 30], so that results obtained under identical conditions could be compared. Thus, nine values of the corrected concentration are given for each of the compounds analyzed.

The results of the Al K_α analyses are shown in Table V. The magnitude

TABLE VI
GOLD–ALUMINUM SYSTEM: Au M_α ANALYSES (20 kV, CONCENTRATIONS, Wt % Au)

Compound	Meas. intensity ratio ($\pm\sigma$)	Absorption data used	After absorption corr.	At. no. corr. procedure [a–c]	After at. no. corr.	Stoichiometric (assumed) composition
Au Al_2	67.6 ± 0.4	Heinrich [34]	72.7	W–M	78.3	78.52
			72.7	D–S	75.2	
			—	T	78.8	
		Frazer [35] or Bucklow [36]	73.5	W–M	79.1	
			73.5	D–S	77.1	
			—	T	79.9	
AuAl	80.7 ± 0.3	Heinrich [34]	83.7	W–M	87.4	87.97
			83.7	D–S	85.9	
			—	T	88.3	
		Frazer [35] or Bucklow [36]	84.1	W–M	87.8	
			84.1	D–S	86.3	
			—	T	88.8	
Au_2Al	89.8 ± 0.5	Heinrich [34]	91.8	W–M	93.9	93.60
			91.8	D–S	93.0	
			—	T	94.0	
		Frazer [35] or Bucklow [36]	92.0	W–M	94.1	
			92.0	D–S	93.3	
			—	T	94.2	

[a] W–M: Wolf and Macres, present work.
[b] D–S: Duncumb and Shields [10], Colby procedure [30].
[c] Thomas [28].

of absorption correction is large, varying between 70 and 100% of the measured intensity ratio for the compositions analyzed. Compositions after absorption correction vary as much as 5–6% depending on which absorption data is used. Data by Heinrich gives the greatest correction whereas corrections using Frazer's or Bucklow's data are smaller. Comparison of various atomic number corrections also shows a wide range of values. For example, results for the $AuAl_2$ compound range from 19.3 wt % (Bucklow absorption data, Duncumb–Shields atomic number correction) to 29.6 wt % (Heinrich absorption data, Thomas atomic number correction). Results for Al K_α given by the Thomas atomic number correction were always greater (undercorrected) than stoichiometry by as much as 29–51%. On the other hand, results with the Duncumb–Shields correction were somewhat less than stoichiometry (overcorrected) by 2–12%. Results for Al K_α with the present work were somewhat greater than stoichiometry (undercorrected) by 5–13%.

Table VI summarizes results for Au M_α analyses. The magnitude of absorption correction for Au M_α is considerably less than that for Al K_α, varying from 2–9% of the measured intensity ratio for the compounds analyzed. Absorption data of Frazer and Bucklow (Metals Research) give results which are identical within ± 0.1 wt % whereas results using Heinrich's data are slightly lower. The Thomas correction tends to overcorrect the Au M_α data slightly, varying from 0.5 to 1.4% greater than stoichiometry. On the other hand, the Duncumb–Shields method tends to undercorrect the data from 0.3 to 4.2% less than stoichiometry. Results of the present work show only a slight overcorrection from 0.3 to 0.8% greater than stoichiometry.

In general, corrected results for Au M_α are in much better agreement with each other and with the assumed stoichiometry than results for Al K_α. This behavior is most likely due to the smaller absorption correction required, and hence a lesser dependence on the accuracy of the absorption coefficient data and absorption correction procedure. This same behavior was encountered in the Ag–Au system. It is important to note, however, that even though large atomic number corrections are required in the Au–Al system, the corrected concentrations, using the procedures of this work or Duncumb–Shields, agree very favorably with the stoichiometric values. The uncertainty in the absorption correction does not allow a decisive evaluation to be made on the accuracy of the two procedures. However, the simplicity with which the correction can be made using tabulations based on the diffusion model, with no attendant loss in accuracy, is an important implication.

2. *Selected Binary Compounds*

To demonstrate further the applicability of the diffusion model to the correction of the atomic number effect, three additional high-purity binary compounds were analyzed including $ThMg_3$, V_3Si, and Bi_2Te_3. The results of these analyses are given in Tables VII–IX.

The Th–Mg system, like the Au–Al system, represents an extreme difference in atomic number. Average results of 6 analyses on this compound for an accelerating voltage of 10 kV are given in Table VII. Since no absorption data was available for Mg K_α in Th, values for μ/ρ were extrapolated from the tables of Heinrich and Frazer. Since these values differed by about 10% (see Appendix III), measurements were corrected using both values and are shown for purposes of comparison. The absorption correction for Mg K_α at 10 kV is very large, about 70% of the measured intensity ratio, and the atomic number correction is also large, about 22% of the absorption corrected ratio. In spite of these large corrections, the results agree very favorably with the stoichiometric composition. The agreement of the Th M_α analysis with

TABLE VII
The $ThMg_3$ Compound: Mg K_α and Th M_α Analyses (10 kV, Concentration, Wt %)

	Meas. intensity ratio ($\pm\sigma$)	After absorption corr.		After at. no. corr.		Stoichiometric composition (assumed)
		Heinrich [34]	Frazer [35]	Heinrich [34]	Frazer [35]	
Mg K_α	18.0 ± 0.5	30.7	29.6	24.1	23.2	23.91
Th M_α	68.5 ± 1.2	68.7	69.0	76.0	76.3	76.09

TABLE VIII
The V_3Si Compound: Si K_α and V K_α Analyses (Concentrations, Wt. %)

	Element and spectral line	Meas. Intensity ratio ($\pm\sigma$)	After absorption corr.	After at. no. corr.	Stoichiometric composition (assumed)
20 kV	Si K_α	10.1 ± 0.2	16.8	15.8	15.52
	V K_α	82.7 ± 0.2	83.6	84.8	84.48
30 kV	Si K_α	7.62 ± 0.2	16.4	15.5	15.52
	V K_α	82.1 ± 0.2	83.9	84.7	84.48

TABLE IX
The Bi_2Te_3 Compound: Te L_α and Bi M_α Analyses (15 kV, Concentrations, Wt %)

	Meas. intensity ratio ($\pm\sigma$)	After absorption corr.	After at. no. corr.	Stoichiometric composition (assumed)
Te L_α	41.25 ± 0.3	48.6	47.2	47.8
Bi M_α	48.54 ± 0.3	50.4	52.7	52.2

the stoichiometric value is excellent and is particularly important since the absorption correction is approximately 1% of the intensity ratio while the atomic number correction is 11% of the ratio corrected for absorption. As a result, it is an effective demonstration of the accuracy with which the present work can correct for atomic number differences.

The V–Si system, unlike the Th–Mg or Au–Al systems, represents a relatively small difference in atomic number of the constituents. Nevertheless, it is sufficiently large that in order to obtain accurate quantitative analyses an atomic number correction must be applied. Average results of 10 analyses at accelerating voltages of 20 and 30 kV are given in Table VIII. Corrections for absorption were made using absorption coefficients from Heinrich (Frazer's show negligible differences). The absorption corrections for Si K_α are very large, amounting to 68% of the measured intensity ratio at 20 kV and 115% at 30 kV, while the atomic number correction is only 6.6% at 20 kV and 5.6% at 30 kV. On the other hand, the absorption correction for V K_α is very small, amounting to 1% of the measured intensity ratio at 20 kV and 2.3% at 30 kV. The atomic number correction is also small, being less than 1.4% of the absorption corrected ratio at 20 kV and 1.0% at 30 kV. The agreement between calculated and assumed stoichiometric compositions is excellent for both V K_α and Si K_α. Although it would not be surprising to find poorer agreement in the 30 kV Si K_α data because of the magnitude of the absorption effect occurring, the reverse is true and hence may be fortuitous.

The results of analyses on the binary compound, Bi_2Te_3, are given in Table IX. The analyses were obtained using an accelerating voltage of 15 kV, and the data represents averages of 28 analyses. Corrections for absorption were made using absorption coefficients from Heinrich (Frazer's data show negligible differences). The absorption correction for Te L_α is 18% of the measured intensity ratio, whereas that for Bi M_α is only 4%. The atomic number correction amounted to 3.5% of the absorption corrected ratio for Te L_α and 4.5% for Bi M_α. The agreement between calculated and stoichiometric compositions cannot be easily evaluated since measurements of thermal, electrical, and mechanical properties on Bi_2Te_3 compounds indicate that a range of homogeneity exists [38]. However, this system represents an atomic number difference of the constituents which is similar to that in the Au–Ag system. Since excellent agreement was obtained in the study of the Au–Ag alloys and since the data is an average of 28 analyses, the results should be quite accurate. Assuming, however, that the stoichiometric composition is valid, the agreement is good to about 1%.

D. Example of Multicomponent System: Tetrahedrite Mineral

The significance of the atomic number correction procedure set forth in this paper is the relatively routine manner in which it can be carried out using simple tables. The use of the procedure is not restricted to binary systems. To demonstrate its applicability to multicomponent systems a tetrahedrite mineral was analyzed. This mineral is a sulfosalt containing major constituents of Cu, Ag, Zn, Sb, and S with minor amounts of Fe, As, and Se occurring in substitution for major constituents. The stoichiometry of this mineral is well established and is given by the formula $(Cu, Ag, Fe, Zn)_{12}$ $(Sb, As)_4(S, Se)_{13}$.

Average results for analyses of 29 areas are given in Table X as well as

TABLE X
ANALYSIS OF TETRAHEDRITE MINERAL

Element analyzed:	Cu	Ag	Fe	Zn	Sb	As	S	Se
Spectral line:	K_α	L_α	K_α	K_α	L_α	K_α	K_α	K_α
Accelerating voltage (kV):	30	20	30	20	20	35	20	35

	Average cation and anion subtotals		
Stoichiometry:	$(Cu, Ag, Fe, Zn)_{12}$	$(Sb, As)_4$	$(S, Se)_{13}$
Sum of meas. intensity, ratios (wt %):	44.85	20.98	16.95
After fluorescence and absorption corr. (wt %):	49.90	24.83	25.16
After at. no. corr. (wt %):	50.53	26.85	24.08
After at. no. corr. (at. % $(\pm\sigma)$):	40.8 ± 0.42	13.7 ± 0.17	45.5 ± 0.49
Stoichiometric composition (at. %):	41.38	13.79	44.83

the spectral lines analyzed and accelerating voltages used. Subtotals for elements of each cation or anion group were totaled for each area analyzed prior to determining average values for each group. In this manner, slight compositional variations due to elemental substitutions within each group were also taken into account. Subtotals given in Table X show effects on composition for absorption and fluorescence as well as atomic number corrections. Absorption coefficients of Heinrich were used and are included in Appendix IV. Subtotal concentrations for each of the groups did not vary greatly thus indicating a high degree of stoichiometry for the areas analyzed. The standard deviation (one sigma) of the cation group (Cu, Ag, Fe, Zn) was about ± 0.5 wt % whereas standard deviations of the anion groups were about ± 0.30 wt %. Agreement with stoichiometry can be evaluated by expressing the atomic percent represented by each group of the chemical formula and comparing them with atomic percents for each group calculated from analyses. Results of this procedure are included in Table X and show very good agreement. Analyses of the (Cu, Ag, Fe, Zn) group are slightly lower, whereas the (S, Se) group are slightly higher than that predicted from stoichiometry.

Although the required atomic number corrections are fairly small, their inclusion provide improved agreement of the results with stoichiometry. In particular, deviations of the calculated subtotals are reduced from 2.6, 8.1, and 6.0% to 1.4, 0.07, and 1.6% for the $(Cu, Ag, Fe, Zn)_{12}$, $(Sb, As)_4$, and $(S, Se)_{13}$ groups, respectively. Since atomic number corrections using other procedures would be much more complex and time consuming for this system, the usefulness of the procedure of this paper is clearly illustrated.

VI. Summary and Conclusions

A careful study of the general procedure for atomic number correction first proposed by Archard and Mulvey and based on the electron diffusion theory of Bethe has shown that the correction can be expressed in terms of $(Z/A)'$ and λ' factors, the λ' factor essentially expressing a refinement in the atomic number correction obtained using the Webster law for describing electron deceleration. The λ' factor is significant in that it provides a basis for universal tables which can be used to obtain this refinement. Such tabulations provide a method by which the correction can be carried out routinely for multicomponent as well as simple binary systems, the procedure for correction being simpler than that typically used to correct for primary-emission absorption. In view of this finding, the evaluation of the diffusion model was pursued in a more rigorous manner with respect to the determination of the depth of diffusion ρx_d, the number of volume elements required to obtain an accurate evaluation of the required numerical integration, and the procedure for determining effective atomic number.

Application of this correction to experimental analysis data from a wide range of selected alloys and compounds and for a wide range in experimental conditions show exceptionally good agreement with known compositions. In some cases the accuracy of the agreement is somewhat masked by uncertainties in the procedures used for making absorption corrections. Comparison in the Cu–Al and Au–Al systems of this procedure with those of Thomas and Duncumb–Shields, however, show differences which are in need of explanation. Because of the demonstrated accuracy of the atomic number correction using this procedure, as well as its simplicity, the utility of tables which cover the range of variables encountered in microprobe analysis is evident. Preparation of these tables is presently being pursued. It is suggested that such universal tables be adopted for general use for correction of atomic number effects in quantitative microprobe analysis.

Another important application of the model is that it provides a basis for generating $f(\chi)$ curves for use in determining corrections for primary emission absorption effects. Although Archard and Mulvey [8, 21] have generated such curves, their accuracy may be improved by use of the more rigorous calculation of ρx_d as well as a more exact evaluation of the numerical integration. The use of a larger value of ρx_d, as was the case with Archard and Mulvey, would yield larger absorption corrections for a given value of χ. This result is verified when comparison is made with $f(\chi)$ curves of Green [4] or Castaing and Descamps [2].

Further improvement in the procedure might be obtained by using experimental data on electron backscatter to determine the depth of complete diffusion. In addition, more refined values for the mean ionization potential I, such as those compiled by Duncumb and Reed [39], might also be included. These modifications are presently being investigated.

ACKNOWLEDGMENTS

The authors express their appreciation to Dr. C. M. Taylor of Materials Analysis Company for use of his Au–Ag alloys and for microprobe analysis data which he supplied on the Au–Ag system, Bi_2Te_3 compound, and tetrahedrite mineral. They also express appreciation to Peter Romans of the US Bureau of Mines for supplying the V_3Si, $ThMg_3$, and Bi_2Te_3 compounds analyzed.

APPENDIX I

TABULATION OF λ' VALUES: $V_0 = 20$ kV, $\gamma = 62.5°$

$\bar{Z}$	Spectral line: / V_c (kV):	Mg K_α 1.30	Al K_α 1.56	Si K_α 1.84	Pt M_α 2.15	Au M_α 2.24	Hg M_α 2.32
10		0.993	0.989	0.984	0.997	1.000	1.002
15		1.011	1.008	1.003	1.016	1.019	1.022
20		1.027	1.023	1.019	1.033	1.035	1.038
25		1.039	1.033	1.029	1.045	1.048	1.051
30		1.044	1.039	1.035	1.052	1.055	1.057
35		1.045	1.040	1.037	1.054	1.057	1.060
40		1.041	1.038	1.036	1.053	1.056	1.059
45		1.035	1.033	1.031	1.050	1.053	1.057
50		1.027	1.026	1.025	1.044	1.048	1.051
55		1.019	1.019	1.018	1.038	1.041	1.043
60		1.010	1.011	1.010	1.031	1.033	1.035
65		1.001	1.002	1.002	1.023	1.024	1.027
70		0.991	**0.992**	0.993	1.014	**1.016**	1.018
75		0.981	**0.983**	0.984	1.005	**1.007**	1.009
80		0.970	0.973	0.974	0.996	0.998	1.000
85		0.959	0.963	0.964	0.986	0.988	0.990
90		0.948	0.952	0.954	0.976	0.978	0.980
95		0.937	0.941	0.944	0.966	0.968	0.970
100		0.925	0.930	0.933	0.955	0.958	0.960

APPENDIX II
EXAMPLE CALCULATION USING UNIVERSAL ATOMIC NUMBER CORRECTION TABLES[a]

	Al K_α		Au M_α
Atomic Number	13		79
Effective atomic number, Eq. (4)	—	7.47	—
λ', Appendix I (interpolated)	0.984		1.008
Z/A	0.482		0.401
$(Z/A)'$, Eq. (19)	0.843		1.014
Atomic number correction: $\lambda' \cdot (Z/A)'$	0.838		1.023
Measured intensity ratio: $100 \cdot I_1/I(1)$[b]	8.15		92.0
Corrected concentration: $100 \cdot I_1/I(1) \cdot \lambda' \cdot (Z/A)'$	6.73		94.1

[a] Al K_α and Au M_α from Au_2Al compound: assumed stoichiometric composition 93.60 Wt% Au, 6.40 Wt % Al.

[b] Measured intensity ratio corrected for deadtime, drift, background, and primary emission absorption effects (Bucklow [36], Metals Research absorption coefficients used).

APPENDIX III
MASS ABSORPTION COEFFICIENTS[a]

Spectral line	Wavelength (Å)	Absorber	Source of data: Heinrich [34]	Frazer [35]	Bucklow [36]
Mg K_α	9.889	Mg	464	509	N.A.[c]
		Th	4800[b]	4400[b]	N.A.
Al K_α	8.337	Al	386	404	408
		Au	2494	2463	2300
Si K_α	7.126	Si	328	332	N.A.
		V	1708	1734	N.A.
Au M_α	5.840	Au	1131	1101	989
		Ag	1265	1239	N.A.
		Al	2009	2118	2010[d]
Bi M_α	5.118	Bi	942	892	896
		Te	1236	1217	N.A.
Ag L_α	4.154	Ag	522	516	525
		Au	1959	2109	2150

(continued)

APPENDIX III *(continued)*

Spectral line	Wavelength (Å)	Absorber	Source of data: Heinrich [34]	Frazer [35]	Bucklow [36]
Th M_α	4.138	Th	733	633	N.A.
		Mg	620	634	N.A.
Te L_α	3.289	Te	392	391	N.A.
		Bi	1523	1542	N.A.
V K_α	2.504	V	98	97	N.A.
		Si	235	234	N.A.
Au L_α	1.276	Au	128	130	129
		Ag	130	129	N.A.

[a] Values used for correction of data from various binary systems.
[b] Extrapolated.
[c] N.A. indicates none available.
[d] Interpolated.

APPENDIX IV

MASS ABSORPTION COEFFICIENTS[a]

Spectral line	Wavelength (Å)	Absorber: S	Ag	Sb	Fe	Cu	Zn	As	Se
S K_α	5.373	239	1019	1319	1158	1621	1798	2389	2601
Ag L_α	4.154	1375	522	676	573	803	891	1183	1288
Sb L_α	3.439	816	1635	413	342	479	532	707	769
Fe K_α	1.937	167	404	502	71	100	111	147	161
Cu K_α	1.542	89	218	271	311	54	60	79	86
Zn K_α	1.436	73	179	224	256	44	49	65	71
As K_α	1.177	42	105	131	149	197	214	38	41
Se K_α	1.106	36	88	111	126	167	181	32	35

[a] Values from Heinrich [34] used for correction of data from tetrahedrite mineral.

REFERENCES

1. R. Castaing, Thesis, Univ. of Paris, ONERA Pub . No. 55 (1951).
2. R. Castaing and J. Descamps, *J. Phys. Radium* **16**, 304 (1955).
3. J. Philibert, *in* "X-Ray Optics and X-Ray Microanalysis" (H. Pattee, ed.), p. 379. Academic Press, New York, 1963.

4. M. Green, *in* "X-Ray Optics and X-Ray Microanalysis" (H. Pattee, ed.), p. 185. Academic Press, New York, 1963.
5. R. Theisen, Euratom Rep. EUR-I-1 (1961).
6. D. M. Poole and P. M. Thomas, *J. Inst. Metals* **90**, 228 (1962).
7. D. M. Poole and P. M. Thomas, *in* "X-Ray Optics and X-Ray Microanalysis" (H. Pattee, ed.), p. 411. Academic Press, New York, 1963.
8. G. D. Archard and T. Mulvey, *Brit. J. Appl. Phys.* **14**, 626 (1963).
9. L. S. Birks, "Electron Microprobe Analysis," Chap. 9. Wiley, New York, 1964.
10. P. Duncumb and P. K. Shields, *in* "The Electron Microprobe," p. 284. Wiley, New York, 1964.
11. T. Taylor, Thesis, DMS Rep. No. 65–4, Stanford Univ., Stanford, California (1964).
12. D. B. Wittry, Thesis, California Inst. of Technol. (1957).
13. D. B. Wittry, USCEC Rep. 84–204, Univ. of Southern California Eng. Center (1962).
14. P. Duncumb and P. K. Shields, *in* "X-Ray Optics and X-Ray Microanalysis" (H. Pattee, ed.), p. 329. Academic Press, New York, 1963.
15. J. Henoc, F. Maurice, and A. Kirianenko, CEA Rep. 2321, Comm. l'Energie Atomique, France (1964).
16. S. J. B. Reed, *Brit. J. Appl. Phys.* **16**, 913 (1965).
17. T. O. Ziebold and R. E. Ogilvie, *Anal. Chem.* **36**, 323 (1964).
18. E. J. Williams, *Proc. Roy. Soc.* (*London*) **A130**, 310 (1932).
19. D. L. Webster, W. W. Hansen, and F. B. Duveneck, *Phys. Rev.* **43**, 839 (1933).
20. R. Castaing, *Advan. Electron, Electron Phys.* **13**, 317 (1960).
21. G. D. Archard and T. Mulvey *in* "X-Ray Optics and X-Ray Microanalysis" (H. Pattee, ed.), p. 393. Academic Press, New York, 1963.
22. H. A. Bethe, *Ann. Physik* **5**, 525 (1930).
23. H. A. Bethe, M. E. Rose, and L. P. Smith, *Proc. Am. Phil. Soc.* **78**, 573 (1938).
24. G. D. Archard, *J. Appl. Phys.* **32**, 1505 (1961).
25. C. R. Worthington and S. G. Tomlin, *Proc. Phys. Soc.* (*London*) **A69**, 401 (1956).
26. N. F. Mott and H. S. W. Massey, "Theory of Atomic Collisions." Oxford Univ. Press, London and New York, 1949.
27. E. H. S. Burhop, *Proc. Cambridge Phil. Soc.* **36**, 43 (1940).
28. P. M. Thomas, AERE-R 4593, United Kingdom Atomic Energy Authority (1964).
29. P. Duncumb and P. K. Shields, *Brit. J. Appl. Phys.* **14**, 617 (1963).
30. J. Colby MAGIC—a computer program for quantitative electron microprobe analysis, Bell Telephone Lab. (1967).
31. A. T. Nelms, *Natl. Bur. Std.* (*U.S.*), *Circ.* No. 577 (1956); *Suppl. Circ.* No. 577 (1958).
32. H. E. Bishop, *in* "X-Ray Optics and Microanalysis" (R. Castaing, ed.), p. 112. Hermann, Paris, 1966.
33. I. Adler and J. Goldstein, NASA-TN-D-2984, Natl. Aeron. Space Admin. (1965).
34. K. F. J. Heinrich, *in* "The Electron Microprobe" (T. D. McKinley, ed.), p. 296. Wiley, New York, 1966.
35. J. Z. Frazer, SIO Ref. No. 67–29, Inst. for the Study of Matter, Univ. of California (1967).
36. I. A. Bucklow, Metals Res. Ltd., private communication (1966).
37. H. E. Bishop, *Proc. Roy. Soc.* (*London*) **A85**, 855 (1965).
38. M. Hansen, "Constitution of Binary Alloys," 2nd ed., p. 339. McGraw-Hill, New York, 1958.
39. P. Duncumb and S. J. B. Reed, *in* "Quantitative Electron Probe Microanalysis," p. 133. Spec. Publ. 298, Natl. Bur. Std., Washington, D.C., 1968.

Deconvolution: A Technique to Increase Electron Probe Resolution

E. J. RAPPERPORT

Ledgemont Laboratory
Kennecot Copper Corporation
Lexington, Massachusetts

Summary

Ordinarily, the electron probe analyzer is used to gather quantitative information from volumes whose minimum projected surface dimensions are no smaller than the minimum diameter of the impinging electron beam: 0.5–1.0 μ, approximately. In some instances, however, it may be possible to increase the effective resolution of an electron probe to provide composition information on a distance scale much finer than the actual electron beam diameter.

This may be done by examining in detail the interaction of a known (measured) incident electron spatial profile with an unknown (desired) impurity concentration profile to yield a known (measured) transmitted or backscattered electron spatial profile. Analyzing the interaction of these three distributions as a superposition problem, one finds the profile of the observed backscattered (or transmitted) current distribution $h(x)$ as the convolution of the impurity concentration profile $g(x)$ and the incident electron profile $f(x)$. Using x as a distance variable and t as an auxiliary distance variable, one may write

$$h(t) = K \int_{-\infty}^{\infty} f(x) g(t - x) \, dx.$$

The deconvolution of the impurity concentration profile from the other two is a mathematical problem involving the solution of the given integral equation relating the three profiles. Solutions of this equation are detailed with examples of simulated and real data.

I. Introduction

Commercial electron probe microanalyzers can commonly produce beams as small as 0.5–1.0 μ in diameter. Lens aberrations and electron interactions have made this size a reasonable lower limit for the beam currents and voltages required for most work using the instrument. Studies of features smaller than the beam size have been limited to semiquantitative or qualitative inferences of the composition—distance relations that were present in the specimens. Thus, certainly in some instances, the usefulness of the instrument has been resolution limited to one degree or another.

We can define resolution as the minimum dimension of a region over which quantitative concentration information can be obtained. If electrons are used as the concentration index, by measurement of either the back-scattered or transmitted current, the physical electron beam diameter at the specimen surface is considered as the spatial resolution.

In considering the spatial resolution with respect to X-ray measurements using an electron probe, one must be cognizant of the extremes in matrix absorption and fluorescence that may be encountered. These, of course, drastically influence the total X-ray emitting volume that is being analyzed from a given electron beam geometry. Thus, it is possible in some cases in which the matrix has a low X-ray absorption to get X-ray intensity contributions from regions 200 μ, or farther, from a 1.0 μ diameter impinging electron beam. In these cases, the wavelength of the primary X rays generated and the atomic number of the matrix may well play a controlling role, with the actual primary electron beam affecting an almost negligible volume in comparison to the total X ray contributing volume. It is certainly possible to obtain X-ray contributions from a region one million times the volume actually subjected to the incident primary electron beam [1].

In some recent work at our laboratories concerning grain boundary segregation, the ability to obtain quantitative concentration information on a much finer spatial scale was desired. To this end, the deconvolution technique described below was conceived and is currently in the process of development. This deconvolution approach has the potential, in favorable cases, to generate concentration profiles on a distance scale well below the actual electron beam diameter.

II. Theory of the Deconvolution Method

A physical picture of the problem may be obtained from a consideration of the gross interactions that occur between an electron beam and a specimen whose surface has a concentration discontinuity. This situation is shown schematically in Fig. 1. In this figure a concentration step, denoted as the

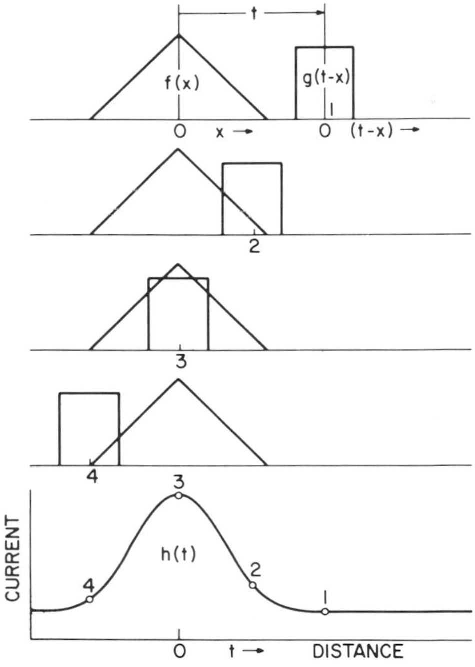

FIG. 1. Development of specimen current (or backscattered current) versus distance curve $h(t)$ as a square-wave concentration discontinuity $g(t-x)$ is translated through an impinging electron density profile $f(x)$.

concentration function $g(t-x)$, is translated through a triangular shaped incident electron profile $f(x)$ with the backscattered or transmitted electron current monitored as a function of position to yield the current-distance response function $h(t)$. The change in current with distance is shown with intermediate numbered points corresponding to several positions of the concentration step during its translation through the beam.

One may assign a backscatter coefficient σ_M to the matrix and another, σ_I, to the small impurity. Let us now consider the mathematics relating the measured backscattered current $h(t)$ to the impurity distribution function $g(t-x)$ and to the electron density generating function $f(x)$.

If the profile $f(x)$ lies entirely on the matrix (position 1 of Fig. 1), the backscattered current $h(t)$ is a constant whose value is

$$h(t) = \sigma_M \int_{-\infty}^{\infty} f(x)\,dx. \tag{1}$$

If all or part of the discontinuity $g(t-x)$, lies within the electron profile $f(x)$, as it does in the situations numbered 2, 3, and 4 of Fig. 1, the backscatter coefficient is a distance variable that may not be removed from the integral. Instead, Eq. (1) must be written

$$h(t) = \int_{-\infty}^{\infty} g(t-x) f(x)\,dx \tag{2}$$

where the concentration function $g(t-x)$ may be related to the backscatter coefficient function; i.e., in this case

$$g(t-x) = \sigma_M \qquad \text{off the discontinuity;} \tag{3}$$

$$g(t-x) = \sigma_I \qquad \text{on the discontinuity.} \tag{4}$$

In the general case, of course, $g(t-x)$ is a more complex function than the simple rectangle chosen for this example.

Thus, the fundamental mathematical relationship between the observed electron current versus distance curve $h(t)$ and the incident electron beam density profile $f(x)$, as the beam is swept over a discontinuity $g(t-x)$, is the integral equation (2). A few comments may be made on the factors in Eq. (2) before considering its solution. The physical meanings of the probe distribution function $f(x)$ and the current versus distance response function $h(t)$ are mentioned above. The function $g(t-x)$ in the cases considered here may be viewed as the spatial distribution of electron transmittance (if specimen current is the index) across the concentration discontinuity. Thus, as the concentration changes at the discontinuity, so also does the electron backscattering coefficient and, hence, the electron transmittance. If backscattered current or secondary electron current, instead of specimen current, is used in the measurement of $h(t)$, then $g(t-x)$ will be related to a backscatter coefficient profile or to a secondary electron source profile across the concentration discontinuity.

After the deconvolution analysis described below, one is left with detailed knowledge of $g(t-x)$, the electron transmittance, reflectance, or emission profile across the composition discontinuity. The problem of converting such information into a concentration profile requires a knowledge of the relationship between the current measured and concentration. Although in many cases this relationship is currently imperfectly understood, there is a large body of binary systems which can be satisfactorily handled [2, 3].

In many instances, we are able to obtain the $h(t)$ and $f(x)$ distributions experimentally and are left with the problem of the deconvolution of Eq. (2) to determine $g(t-x)$.

The deconvolution of Eq. 2 is practical if electron currents were monitored as the probe-distribution profile $f(x)$ is of fixed geometry and is independent of the concentration of the target. For the X-ray case this is not true; in that case the $f(x)$ curve is related to a generated X-ray photon intensity profile which can be a strong function of the specimen geometry and composition, as well as the incident electron density. Differences in compositional configuration may give rise to drastic differences in $f(x)$ for the X-ray case. The absence of a unique distribution function makes the deconvolution of X-ray data exceedingly difficult compared to that of electron current measurements.

III. Mathematics of the Deconvolution Method

A. Fourier Transforms

Given the integral equation

$$h(t) = K \int_{-\infty}^{\infty} f(x) g(t-x)\, dx \tag{2a}$$

where $h(t)$ is the response function of a convolution of the probe distribution function $f(x)$ with the composition dependent function $g(t-x)$, we wish to find $g(t-x)$ when $h(t)$ and $f(x)$ are given, usually as a series of measurements. The determination of the curves of $h(t)$ and $f(x)$ are examined in detail below.

To solve the integral equation (2a) take the Fourier transform of both sides. Denoting the transforms by capital letters, we have

$$\int_{-\infty}^{\infty} e^{-it\xi} h(t)\, dt = \int_{-\infty}^{\infty} e^{-it\xi}\, dt \int_{-\infty}^{\infty} f(x) g(t-x)\, dx \tag{5}$$

or

$$H(\xi) = \int_{-\infty}^{\infty} e^{-it\xi}\, dt g(t-x) \int_{-\infty}^{\infty} f(x)\, dx. \tag{6}$$

Letting $t - x = w$,

$$H(\xi) = \int_{-\infty}^{\infty} f(x)e^{-i\xi x}\, dx \int_{-\infty}^{\infty} e^{-i\xi w} g(w)\, dw, \tag{7}$$

$$H(\xi) = F(\xi)G(\xi). \tag{8}$$

This is the well-known property of the Fourier transform of a convolution integral. Solving Eq. (8) for $G(\xi)$, we have

$$G(\xi) = H(\xi)/F(\xi). \tag{9}$$

The desired concentration function $g(t - x)$ may be related to the transform $G(\xi)$ by

$$g(t - x) = (2\pi)^{-1} \int_{-\infty}^{\infty} e^{i\xi w} G(\xi)\, d\xi, \tag{10}$$

thus allowing the determination of the function $g(t - x)$, if the Fourier transform $G(\xi)$ is known or can be obtained from $H(\xi)$ and $F(\xi)$.

In a fashion similar to that developed above, Stokes [4] has expanded the functions $h(t)$, $f(x)$, and $g(t - x)$ into Fourier series with summations replacing the integrals and has limited the distance range to one that encompasses the entire convolution, conveniently. In so doing he arrives at a relation for the complex Fourier series coefficients that is analogous to that given in Eq. (9). Thus, to paraphrase Stokes [4], the procedure when using Fourier series expansions is as follows.

Find the Fourier coefficients of the functions $h(t)$ and $f(x)$: Divide each coefficient of $h(t)$ by the corresponding coefficient of $f(x)$, and use the resulting quotients in a Fourier synthesis to find $g(t - x)$.

The actual mechanical steps in such an unfolding are discussed in detail by Stokes [4] and more recently in an excellent summary by Cohen [5]. Both of these references make use of Lipson–Beevers strips [6] to facilitate the tedious summations involved. In the calculations of this work, however, the determinations were made on digital computers.

In general, rectangular step functions for $g(t - x)$ have not been recoverable with a reasonable number of intervals and harmonics using Fourier analysis. This experience parallels that of Ergun [7], who reports similar difficulties using a somewhat different deconvolution technique.

B. Modified Transforms

The method that we are currently exploring was based on the published work of Rice [8]. His modification of technique lies principally in the obtaining of a series of n simultaneous equations involving a lesser number m different values of the unknown function $g(t - x)$, and the computation of the best of these values using some reasonable criterion, such as either minimizing the maximum deviation or minimizing the sum of the squares of the deviations. The values thus obtained represent a highly selective solution of a series of points on the curve of the unknown function.

In his development, Rice uses the relationship given in Eq. (8) for Laplace transforms of both sides of Eq. (2a) to obtain

$$H(s) = F(s)G(s). \tag{8a}$$

Following Rice, we use the Dirichlet series representations to replace the Laplace transforms

$$\begin{aligned} F(s) &= \sum_{\nu=0}^{m} A_\nu e^{-\nu(\Delta t)s} \\ G(s) &= \sum_{\nu=0}^{N} X_\nu e^{-\nu(\Delta t)s} \\ H(s) &= \sum_{\nu=0}^{n} B_\nu e^{-\nu(\Delta t)s} \end{aligned} \tag{11}$$

where A_ν, X_ν, and B_ν represent the areas under the curves $f(t)$, $g(t)$, and $h(t)$, respectively, for the νth equal interval Δt, and where it is assumed that $A_\nu = 0$ for $\nu > m$, $X_\nu = 0$ for $\nu > N$, and $B_\nu = 0$ for $\nu > n$. Letting a_ν, x_ν, and b_ν represent midpoint amplitude values of $f(t)$, $g(t)$, and $h(t)$, and taking $\Delta t = 1$, and substituting $u = e^{-s}$, the equations of (11) reduce to

$$\begin{aligned} F(u) &= \sum_{\nu=0}^{m} a_\nu u^\nu \\ G(u) &= \sum_{\nu=0}^{N} x_\nu u^\nu \\ H(u) &= \sum_{\nu=0} b_\nu u^\nu. \end{aligned} \tag{12}$$

If the convolution is represented as polynomial multiplication of the equations in (12), and if coefficients of like terms on either side of the equality are equated, the following series of linear simultaneous equations in the unknown amplitudes x_i is obtained:

$$\begin{aligned} &a_0 x_0 &&= b_0 \\ &a_1 x_0 + a_0 x_1 &&= b_1 \\ &a_2 x_0 + a_1 x_1 + a_0 x_2 &&= b_2 \\ &\vdots &&\vdots \\ &a_m x_0 + a_{m-1} x_1 + \cdots + a_0 x_m &&= b_m \\ &a_m x_1 + a_{m-1} x_2 + \cdots + a_1 x_m + a_0 x_{m+1} &&= b_{m+1} \\ &\qquad\vdots &&\vdots \\ &\qquad a_m x_{n-m} + \cdots + a_0 x_n &&= b_n. \end{aligned} \tag{13}$$

Rice points out that a solution of these equations, in general, requires the use of an additional constraint, e.g., minimizing the maximum deviation or minimizing the sum of the squares of the deviation. He then uses matrix algebra to compute a solution using a least-squares criterion.

As noted, in Section IV, Rice's solution yielded unacceptable results when even modest amounts of noise were present in the input. Thus, a modification of this technique is required in order to process real data.

Morris, of the Kennecott Scientific and Engineering Computer Center, has applied ridge analysis [9, 10] to the solution proposed by Rice. The difficulty with the deconvolution procedure, as proposed by Rice, appears to lie in its very exactness. Ridge analysis, however, introduces additional statistical constraints in order to determine how stable the analytic solution is and simultaneously shows the best compromise if it proves unstable [10]. The ridge analysis in this instance was applied by examining the effects of slight perturbations in the solutions of Rice's simultaneous equations. These modified solutions for ordinates on the unknown $g(t-x)$ curve were folded with the known $f(x)$ curve to yield a derived $h'(t)$ curve. Points on the derived $h'(t)$ curve were then compared with the measured $h(t)$ curve. Actually, the ridge analysis in this application yields a series of solutions which trace a path of the minimum least-squares deviation between the $h'(t)$ derived from calculated $g(t-x)$ values, and the given data of $h(t)$. In following the path of this minimum, one gets a number of valid solutions to the matrix Eq. (13), and one may then choose the solutions which minimize the squares of the deviations from the actual data or which satisfy some other statistical criterion.

IV. Experimental Measurements

A. Specimen Current versus Distance across a Concentration Discontinuity: Response Function $h(t)$

There are several methods that may be used to obtain this data. Perhaps the easiest for most instruments is the use of electron scanning with the specimen current (or backscattered current) used to modulate one ordinate on an oscilloscope while the other has a constant scan rate. Thus, a plot of specimen current versus distance may be made on the oscilloscope and photographed to allow subsequent measurement.

Figure 2 is a photograph of such a trace. In this case a beam approximately 1 μ broad and 40 μ long was aligned parallel to a linear concentration discontinuity (a brass foil approximately 1500 Å thick) in aluminum. The beam was then swept across the discontinuity and the specimen current-distance plot obtained.

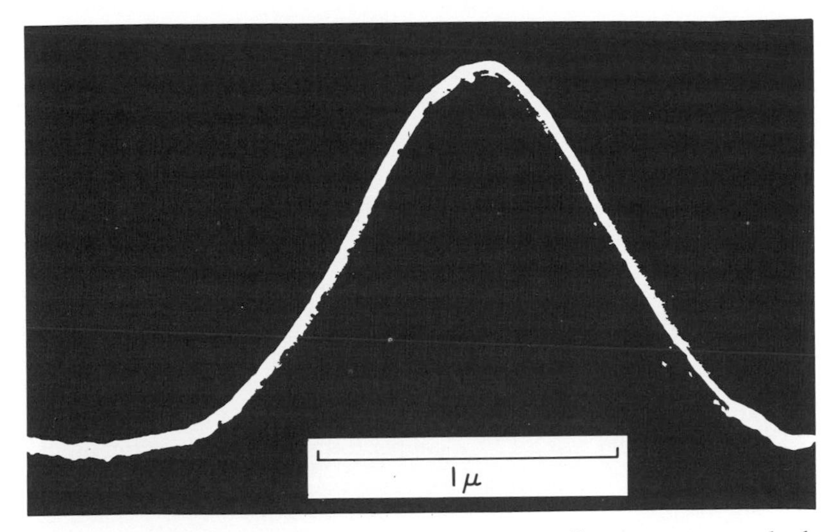

FIG. 2. Oscilloscope trace of specimen current versus distance, as rectangular beam $\sim 1 \times 40\ \mu$ swept across a parallel 1500 Å thick layer of brass in aluminum (25 kV and 0.07 μ a specimen current on aluminum).

A second technique to generate the $h(t)$ curve involves the movement of the beam in small known steps, using the manual position adjustment on the control oscilloscope, with measurement of the desired electron current at each position. This can be done with very fine distance increments as may be seen in Fig. 3 which is an example of this technique. In this case the specimen

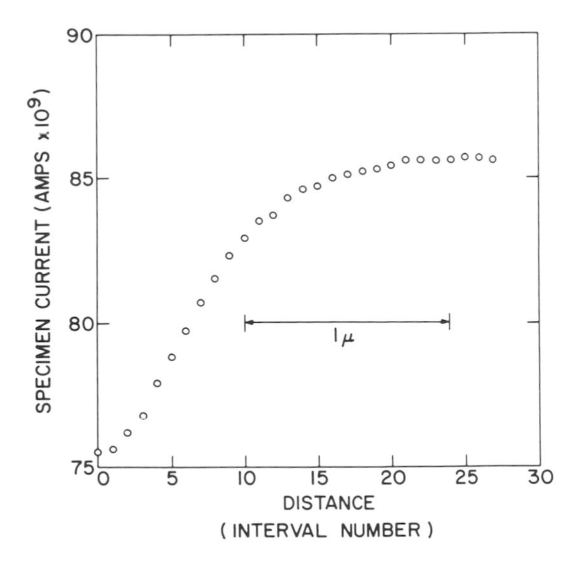

FIG. 3. Plot of specimen current versus distance as rectangular beam $\sim 1 \times 40\ \mu$ was moved in 720 Å increments from copper into aluminum with the beam parallel to the concentration step (25 kV).

current was read using a digital voltmeter across a known resistance on the line to ground. The intervals were approximately 720 Å apart.

The geometry of the beam and specimen used to generate Fig. 3 is similar to that for Fig. 2; i.e., a linear concentration discontinuity, with a parallel, narrow rectangular beam moved perpendicularly across it. In the example of Fig. 3, the concentration discontinuity is simply a step function, and the $h(t)$ curve generated is sigmoidal rather than bell-shaped as in Fig. 2.

A third technique used in this work employed the mechanical drive of the microprobe specimen stage to move the concentration discontinuity through a stationary beam. An example of data collected in this fashion is plotted as Fig. 4. In this instance, the beam was circular in cross section and larger than either of the two previous examples. Again, as in the example of Fig. 2, the concentration discontinuity is a narrow band, compared to the beam. In the case of Fig. 4, the band was a 12.5 μ copper foil sandwiched in aluminum.

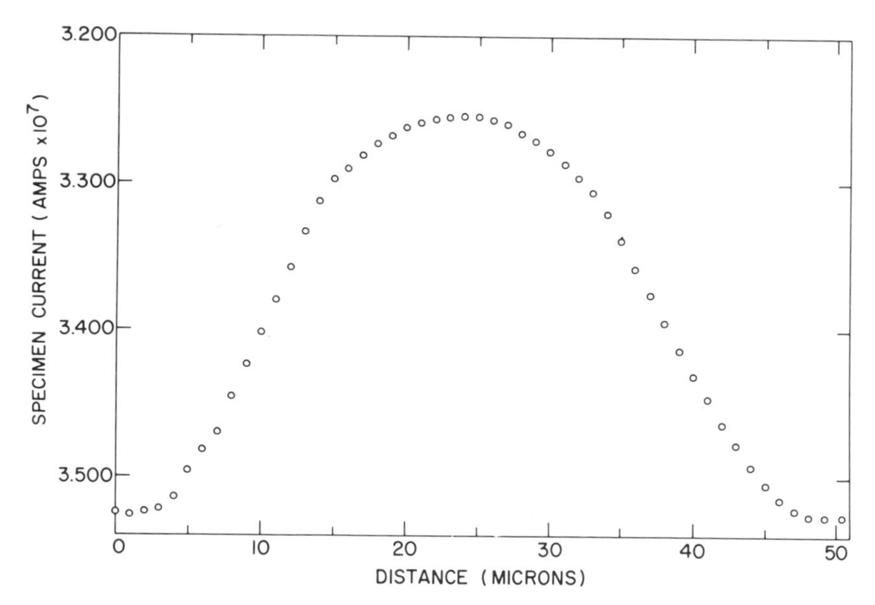

FIG. 4. Response curve $h(t)$ as 12.5 μ copper sheet sandwiched in aluminum was moved incrementally through a 36 μ diameter electron beam.

B. *Generation of the Beam Profile: Beam Generating Function $f(x)$*

In the solution of Eq. (2) for the desired concentration related profile $g(t-x)$, it is necessary to know both $f(x)$ and $h(t)$. The experimental determination of $f(x)$ is similar in mechanics to that of $h(t)$ described above.

The techniques used in this case involved moving the beam across a known composition step, either with an oscilloscope sweep (Fig. 5) or in small measured increments (Fig. 3). In both of these examples, as in Fig. 2, a rectangular beam with the long axis parallel to a concentration discontinuity was taken across that discontinuity using the driving oscilloscope in order to obtain a plot of specimen current versus distance. In like fashion, the stage may be used to translate the discontinuity through the beam to obtain the curve of Fig. 6.

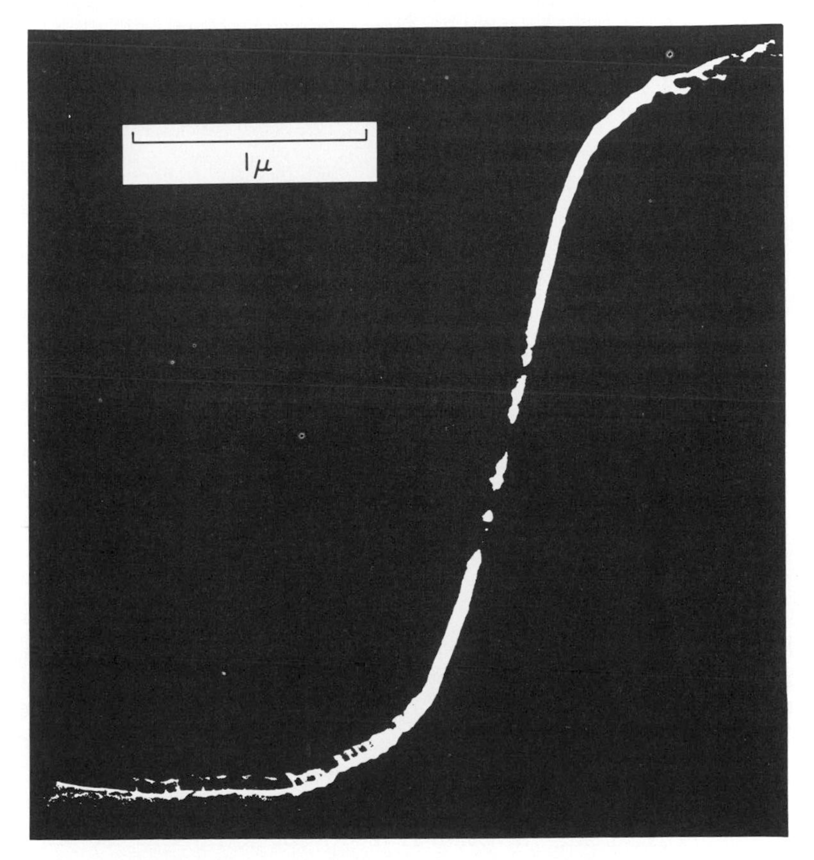

FIG. 5. Oscilloscope trace of specimen current versus distance as rectangular beam swept across a parallel interface from copper into aluminum (25 kV).

The curves obtained in Figs. 3, 5, and 6 are easily related to the desired $f(x)$ curve. In these instances, the portion of the curves rising above their background level are proportional to the integral of $f(x)$ with respect to x to the particular x value of interest. This is made evident on examination of

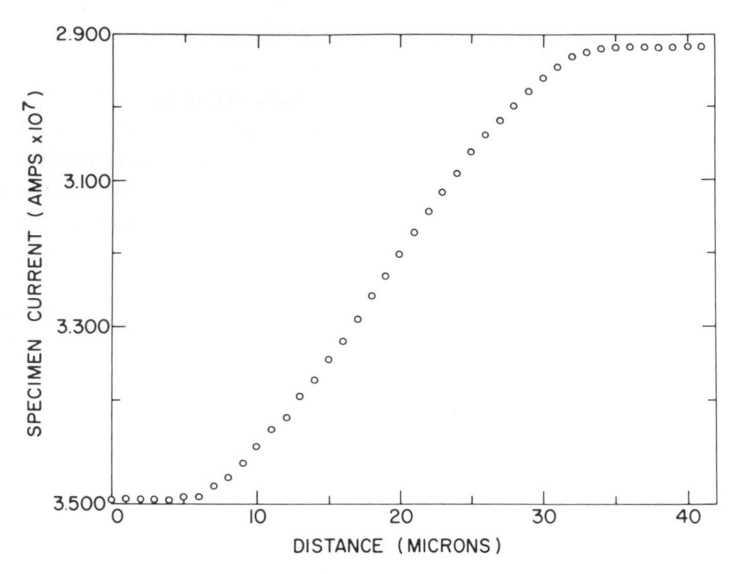

FIG. 6. Plot of specimen current versus distance as a sandwiched copper–aluminum specimen was moved in 1.0 μ increments through a 36 μ diameter beam (25 kV).

Eq. (2) for the case in which the concentration profile undergoes a step to a new constant level. As the beam penetrates the discontinuity, only the difference of backscatter coefficient between the two compositions is a factor. This constant difference may be taken outside the integral to yield

$$h(t) = K' \int_{-\infty}^{t} f(x)\, dx. \tag{14}$$

In this case, a point by point first derivative curve of $h(t)$ with respect to distance will yield the desired $f(x)$ profile multiplied by some constant factor.

The probe distribution function $f(x)$ may be determined independently and directly by an alternative technique. In this instance, it is necessary to prepare a specimen that is a sandwich of an extremely thin layer of a high (or low) atomic number element between thick sheets of a low (or high) atomic number element. The beam is then made to traverse the sandwich with the appropriate electron current monitored to produce the desired probe distribution, directly. This process may be visualized by picturing the production of the $h(t)$ curve of Fig. 1 in the case in which the $g(t - x)$ curve is reduced in breadth by a factor of 50 or 100. Although a plot of $h(t)$ is actually produced, if the sandwiched layer is thin enough relative to the beam dimension, a close approximation to the $f(x)$ profile will be obtained.

V. Reduction of Data

A. Data Reduction by Approximate Techniques

In many instances, it is possible to get useful spatial and concentration information without resorting to deconvolution. For a broad range of discontinuity sizes, it might be sufficient simply to compare the total breadth of the $h(t)$ curve with that of the corresponding $f(x)$ curve or with the breadth of the integral of $f(x)$ versus x curve, e.g., Fig. 5 or 6. The $h(t)$ curve will be broader by the added breadth of the desired $g(t - x)$ curve. If the difference in breadth is measurable, then the breadth of the discontinuity is immediately available as this difference, even if the actual distribution is still unknown within that breadth.

If measurements of $h(t)$ and $f(x)$ have been made, it may be possible to make some approximate concentration estimates for the discontinuity with very little additional work. As has been noted, the breadth of the concentration profile is equal to the difference in breadth between $h(t)$ and $f(x)$. In addition, an estimate of the total surplus (or deficiency) of second element content within the discontinuity may be obtained from a comparison between the peak $h(t)$ value and the integral of $f(x)$ with distance; i.e., comparing the entire monitored beam current on the matrix with that measured when the beam is centered on the discontinuity. The difference between these two is, of course, due to a backscattering difference between the discontinuity and the matrix. Since the breadth of the discontinuity and the probe distribution function are both known, one may estimate the fraction of the incident beam that impinges on the discontinuity under these conditions. With these estimates in hand, one may relate the observed current difference to an integrated second element content within the discontinuity. Thus, the discontinuity breadth and the total second element content of a sub-beam sized concentration fluctuation are available without actual deconvolution, in some cases at least.

An example of this sort of reasoning may be made using the simulated problem of Fig. 7. In this example the base of the triangular $f(x)$ curve is 6 intervals wide and the base of the bell-shaped $h(t)$ curve can be estimated at about 8 intervals. The difference between these two values, 2 intervals, is the breadth estimated for the impurity discontinuity. The difference in the measured specimen current (or backscattered current) when the beam is on the matrix and when the beam is centered on the narrow discontinuity is due to changes in electron backscatter attributable to the discontinuity. In this instance, a centrally located discontinuity of 2 intervals intercepted five-ninths of the incident electron beam. This information is sufficient to compute

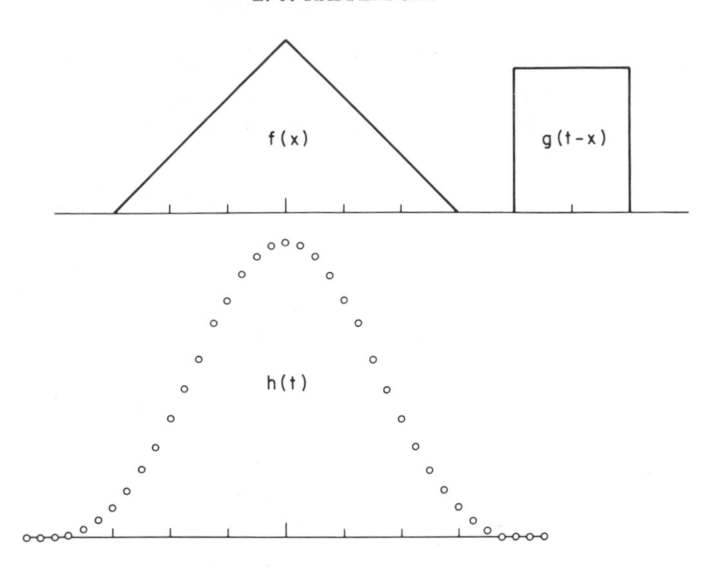

FIG. 7. Assumed triangular probe distribution function $f(x)$ traversing an assumed square-wave concentration discontinuity $g(t-x)$ to produce a calculated current-distance response curve $h(t)$.

an integrated second element content within the discontinuity. If desired, one may assume a gross shape for the concentration profile (rectangle, sine wave, gaussian distribution, etc.) and compute an approximate concentration distribution consistent with the measured breadth, the measured electron currents, and the assumed profile shape. One may even simplify a little further by assuming a probe distribution function $f(x)$ of some analytic form, rather than actually reducing the measurements to a profile.

In this simple model it might be advantageous to utilize X-ray intensity measurements of the second element in the discontinuity compared to a suitable standard. These measurements, coupled with a knowledge of the electron fraction intercepted, might allow a composition inference to be made although fluorescence and matrix absorption effects could be troublesome.

A clue to the shape of the concentration profile is contained in the change in the tails of the $h(t)$ curve with distance, for a sharp change in concentration at the discontinuity boundary will change the value of $h(t)$ more quickly than will a gradual concentration change. It is obvious that this profile determination can be made quantitative by detailed consideration of the slope changes in the tails of the $h(t)$ curve, but this is a very difficult region in which to work.

B. Data Reduction by Exact Techniques

Although the approximate methods described above are useful and may, in many cases, offer a considerable resolution improvement, the exact solution of Eq. (2) for the concentration function $g(t-x)$ is an extremely attractive prospect. At this juncture, the deconvolution of Eq. (2) offers particular promise in the improvement of microprobe resolution if $h(t)$ and $f(x)$ are known sufficiently well.

The major effort of this work, therefore, has been devoted to attaining reliable $h(t)$ and $f(t)$ curves for cases of known $g(t-x)$ and to developing reasonably good deconvolution techniques. An insight for the capabilities of the method and for the limitations imposed by inaccuracies in the data was obtained by detailed examination of simulated problems.

The most useful simulation is that depicted in Fig. 8. In this figure, the

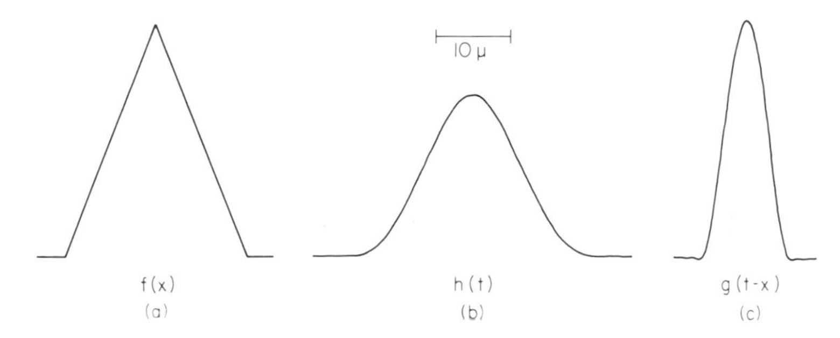

FIG. 8. Simulated deconvolution problem. Triangular generating function $f(x)$ was folded with concentration function $g(t-x) = \frac{1}{4}x^4 - 6x^3 + 36x^2$, to give points on response function $h(t)$. The computed points of $h(t)$ were then processed with the given $f(x)$ to produce the $g(t-x)$ curve shown in (c), which is indistinguishable from the initial $g(t-x)$ input.

beam distribution function $f(x)$ was analytically fixed as a triangle of base 25 μ and height 50 current units. The concentration discontinuity function $g(t-x)$ was assigned the function

$$g(x) = \tfrac{1}{4}x^4 - 6x^3 + 36x^2. \tag{15}$$

These two functions when folded together according to Eq. (2) yielded the bell-shaped response curve $h(t)$ of Fig. 8b. The $h(t)$ data set, thus obtained, was then used as input data along with the exactly known triangular $f(x)$ curve in order to compute the simulated composition discontinuity, $g(t-x)$.

The results of this computation, using Rice's deconvolution technique on exactly known input data points on the curves of $h(t)$ and $f(x)$, is given in Fig. 8c as the composition discontinuity function $g(t-x)$. This computed curve is equivalent to the initially assumed $g(t-x)$ function within the width of the drawn curve. Thus, it may be seen that Rice's technique is able to perform a deconvolution extremely well with exact input.

The effect of slight errors in the data on Rice's solution, as well as the usefulness of ridge analysis is shown in Fig. 9. The $f(x)$ used in this simulated problem is again the triangle used in Fig. 8; however, the response curve $h(t)$ of Fig. 9 has 0.5% of the peak amplitude introduced as random noise in the actual $h(t)$ values used as input. When Rice's solution is used for this case, the $g(t-x)$ that is obtained is so noisy as to be virtually useless. However, ridge analysis is able to generate a reasonable set of data for $g(t-x)$ as may be seen in the progression of the $g(t-x)$ curves of Fig. 9.

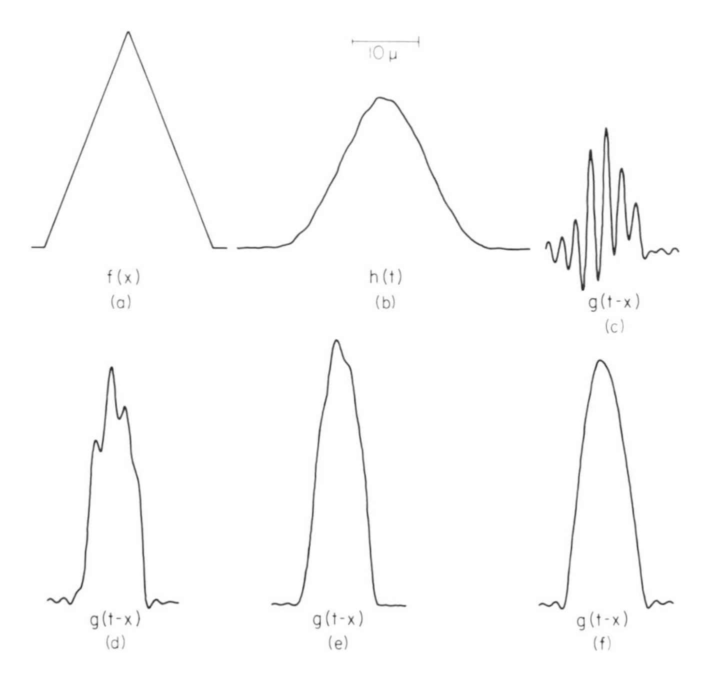

FIG. 9. Deconvolution given response curve $h(t)$ of Fig. 8 with 0.5% peak amplitude random noise and triangular generating function $f(x)$. Rice's solution for concentration function $g(t-x)$ is given in (c) with the ridge analysis solutions explored in (d)–(f). Actual $g(t-x)$ is that of Fig. 8.

Ridge analysis can, in fact, process fairly noisy $h(t)$ curves if $f(x)$ is well known. An example is given in Fig. 10, in which the conditions are the same as those in Figs. 8 and 9 except that random noise equal to 10% of the peak amplitude has been introduced into the $h(t)$ value. The $g(t-x)$ curve that is produced by the deconvolution techniques that have been described may be seen to be a reasonable approximation to the true $g(t-x)$ curve of Fig. 8, in view of the poor response curve $h(t)$ used as input.

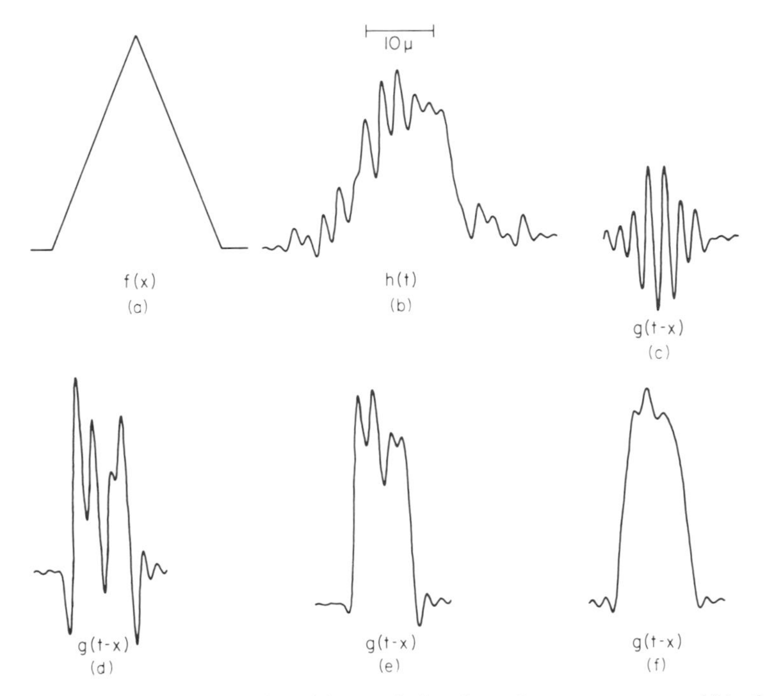

FIG. 10. Deconvolution, using ridge analysis, given the response curve $h(t)$ of Fig. 8 with 10% of peak amplitude introduced as random noise and given triangular generating function $f(x)$. Actual $g(t-x)$ is that of Fig. 8.

VI. DISCUSSION

The preceding text has described the folding of two functions, the generating function $f(x)$, and the concentration function $g(t-x)$ to produce a measurable response function $h(t)$. The unfolding of the given $h(t)$ with a deducible $f(x)$ to produce the desired $g(t-x)$ also has been elaborated in detail. Illustrations of the deconvolution procedures on simulated problems with various degrees of noise have been explored. Additionally, techniques to obtain concentration information on subbeam geometries without deconvolution have been described briefly.

A specially constructed test sample was fabricated in order to allow the collection of real data from a specimen of known geometry. This sample consisted of a 12.5 μ sheet of copper sandwiched between two thick sheets of aluminum. An additional thick sheet of copper was adjacent to one of the thick aluminum sheets. Figure 4 shows the experimental set of data points on the response function $h(t)$ as the 12.5 μ concentration rectangular wave,

$g(t - x)$, was incrementally moved through a 36 μ diameter beam when the specimen was examined on a plane normal to that of the sandwiched sheets.

The data needed to obtain the generating function $f(x)$ was taken from a traverse across the composition step from pure aluminum to pure copper. Figure 6 shows this set of experimental points. The problem of reducing the data of Fig. 6 to an electron density profile has been programmed for computer application using a "hanning" filter [11] along with differentiation. At this stage of development the technique reduced the data of Fig. 6 to an electron profile shown in Fig. 11.

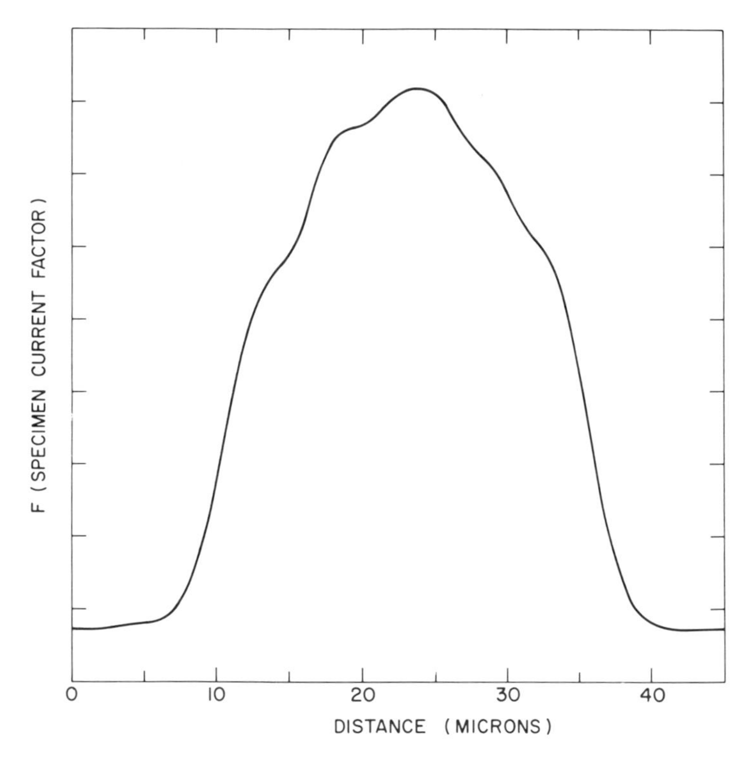

FIG. 11. Generating function $f(x)$ as determined from data of Fig. 6.

Although the profile of Fig. 11 would appear to need refinement, it was used as input along with the response function profile $h(t)$ of Fig. 4. The deconvolution procedures described above, i.e., the superposition of ridge analysis on Rice's solution, were applied to these profiles to produce the desired concentration profile $g(t - x)$, shown in Fig. 12. The profile breadth is nearly correct for this profile, but there is unacceptable noise on the function. Efforts are being made to improve the mathematical operations with

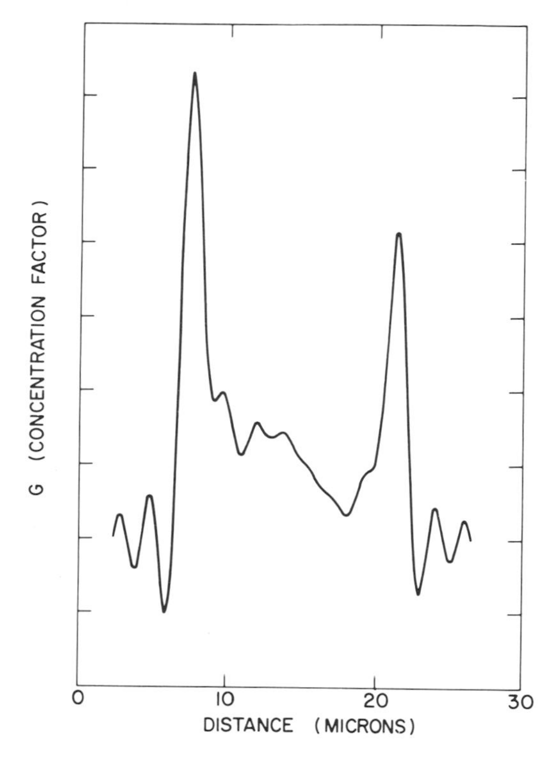

FIG. 12. Concentration profile $g(t-x)$ derived by deconvolution using profiles of Figs. 4, 6, and 11. The actual $g(t-x)$ was a square wave of about 12.5 μ breadth.

smoothing and filtering functions in order to obtain better $f(x)$ profiles, and thereby reduce this problem.

The real problem chosen for a "proof test" analysis is a very special case. The square wave concentration discontinuity introduced by sandwiching sheets of copper and aluminum is particularly difficult mathematically, by our techniques as well as others [7]. It is quite possible, additionally, that the interface geometry is not the step function we have postulated but has a small degree of smear or gapping, which would tend to confuse the analysis. The main advantage of the geometry chosen is that it was relatively easy to prepare, and on the scale used, it was possible to evaluate the specimen quality to some degree with light microscopy. We feel, however, that the results of deconvolution on concentration functions that are more likely to occur in nature would tend to be more satisfactory.

Judging from the results to date, it would appear that the use of deconvolution techniques does offer an increase in probe resolution. The extent of this improvement, as well as the accuracy required in data measurement, is somewhat uncertain at this point. However, the work done on the simulated problem of Figs. 8–10 indicates that some reasonable data inaccuracy can be tolerated. The initial real data results (Figs. 4, 6, 11, and 12) show a resolution potential about equal to the data point interval. Thus, a resolution of one-fifth to one-tenth the beam size would appear attainable.

Acknowledgment

The sustained patient efforts of D. Morris of the Kennecott Scientific and Engineering Computer Center in the mathematics and computer reduction of the bulk of this analysis is gratefully acknowledged.

Additionally, I wish to express appreciation to P. C. Clapp and D. H. Howling for critical discussions of this work, and to R. D. Leite for his painstaking specimen preparation and aid in data collection.

References

1. R. Thiesen, "Quantitative Electron Microprobe Analysis," p. 8. Springer, New York, 1965.
2. J. W. Colby, Microprobe analysis of binary systems containing uranium. *Advan. X-Ray Anal.* **8**, 352 (1965).
3. K. F. J. Heinrich, Interrelationships of sample composition, backscatter coefficient, and target current measurement. *Advan. X-Ray Anal.* **7**, 325 (1964).
4. A. R. Stokes, *Proc. Phys. Soc.* (*London*) **A61**, 382 (1948).
5. J. B. Cohen, "Diffraction Methods in Materials Science," pp. 80–105. Macmillan, New York, 1966.
6. C. A. Beevers, *Acta Cryst.* **5**, 670 (1952).
7. S. Ergun, Direct method for unfolding the convolution products—its application to X-Ray scattering intensities. *Ann. Pittsburgh Diffraction Conf., 24th, 1966* and *J. Appl. Cryst.* **1**, 19 (1968).
8. R. B. Rice, *Geophysics* **27** (1), 4 (1962).
9. A. E. Hoerl, *Chem. Engr. Prog.* **55** (11), 69 (1959).
10. A. E. Hoerl, *Chem. Engr. Prog.* **58** (3), 54 (1962).
11. C. F. George, H. W. Smith, and F. X. Bostick, Jr., *Proc. I.R.E.* (*Inst. Radio Engrs.*) **50**, 2315 (1962).

Analysis for Low Atomic Number Elements with the Electron Microprobe

P. S. ONG

Philips Electronic Instruments
Mt. Vernon, New York

I. INTRODUCTION

In 1963, Henke [1] showed the feasibility of extending the analytical range of the X-ray fluorescent technique to include the elements sodium, fluorine, oxygen, nitrogen, carbon, and boron. A high-power thin-window soft X-ray tube, originally designed for microradiography, was used to excite the K radiation of the above-mentioned elements. The crystals used for dispersing the radiation include a multilayered lead stearate film which has a $2D$ spacing of approximately 100 Å. Excellent analytical sensitivity was obtained by optimizing the excitation parameters as well as the operational condition of the detector.

Previously, electron excitation of the light elements was studied by Dolby [2] using a nondispersive technique, by Holliday [3] using mostly diffraction gratings, and by Nicholson and Wittry [4] who compare the relative efficiencies of stearate films and gratings, among others. The electron beam power employed in the dispersive type analyses, however, were usually high (in the order of 100 mW) as compared to that normally used in the electron microprobe (in the order of 1 mW).

The first report on the use of stearate film "crystals" to detect carbon in a microprobe came from Merritt *et al.* [5]. Although the counting rate was low (60 counts/sec for 0.1 μA at 6 kV), it was nevertheless 25 times larger than obtained by Holliday using line gratings.

Meanwhile, the author was also working on the problems of light element analysis in a microprobe with the cooperation of Henke (Pomona College, Claremont, California), who dipped our first three crystals. By employing a variably bent goniometer [6] which allowed him to optimize the focusing geometry, the author obtained bettter than 20 times more signal as compared to Merritt's results on carbon under essentially the same condition. The signal-to-noise ratio was also improved by optimizing the detector operating parameters. These results, including the detection sensitivities on the elements fluorine, oxygen, nitrogen, and boron, were presented at the 1964 Pittsburgh Conference on Analytical Chemistry and Applied Spectroscopy [7].

"Light element kits" which consist of the necessary components to extend the range of detectability of the microprobe with the elements fluorine ($Z = 9$) through boron ($Z = 5$) became commercially available. Since then, light element analysis has been carried out routinely.

Carbon contamination of the sample under electron beam bombardment is a phenomenon of which microprobe users are aware. But not until the use of the long wavelengths associated with the light element does this contamination become a serious problem. Ways and means to eliminate or at least to reduce its effects are being studied.

Quantitative interpretation of the observed intensities measured at complex sample is a technique which one wishes to extend into the low atomic number elements. Unfortunately, high absorption and uncertainties in the values of absorption coefficient make correction difficult and unreliable. The extension of the wavelength range by the use of stearate crystals is also a welcome feature to study the L and M lines of heavier elements. These lines are more readily affected by the chemical state of the atoms. Thus, line shifts due to chemical binding becomes more pronounced.

II. The Stearate Crystal

A. The Combination Mica-Stearate

The heart of the light element spectrometer is the dispersing element. Ruled line grating can give better resolution than the stearate crystal because of the greater number of diffracting elements (1500 lines in an inch-long grating versus 50–100 double layers in a stearate crystal). Nicholson and Hasler [8] have shown that the overall efficiency of the grating exceeds that of the stearate. The ease with which the stearate can be used in an existing spectrometer and its ready availability make the use of these crystals very attractive in commercial instrumentations. In this article, we will limit ourselves to the use of stearate crystals only.

The technique of dipping stearate crystals, based on the Langmuir–Blodgett [9, 10] method has been described in full detail by Henke [1]. Also described in the same paper was a technique to make stearate-decanoate crystals which have proved to be appreciably better than the pure stearate. In all our experimental and commercial instrumentation, we use lead stearate-decanoate deposited on mica or glass. Mica as a substrate gives some complication in the surface preparation prior to the deposition of the stearate, mainly because of the softness of the material. It has, however, some unique advantages. The most obvious one is that mica itself can be used as a diffracting crystal for soft and hard radiation while, simultaneously, the stearate can be used for the ultrasoft radiation. If third and fifth order diffractions are used, such a mica-stearate crystal can be used to cover a wavelength region between 0.4 and 70 Å in an angular range of $10° < 2\theta < 90°$ (see Fig. 1).

Consequently, the spectrometer covers the element range from boron ($Z = 5$) through uranium ($Z = 92$) without changing the crystal. Undesired orders can easily be separated by using pulse high discriminations. The same is true for distinguishing between diffraction from the mica and the stearate crystal. Gas discrimination can also be employed.

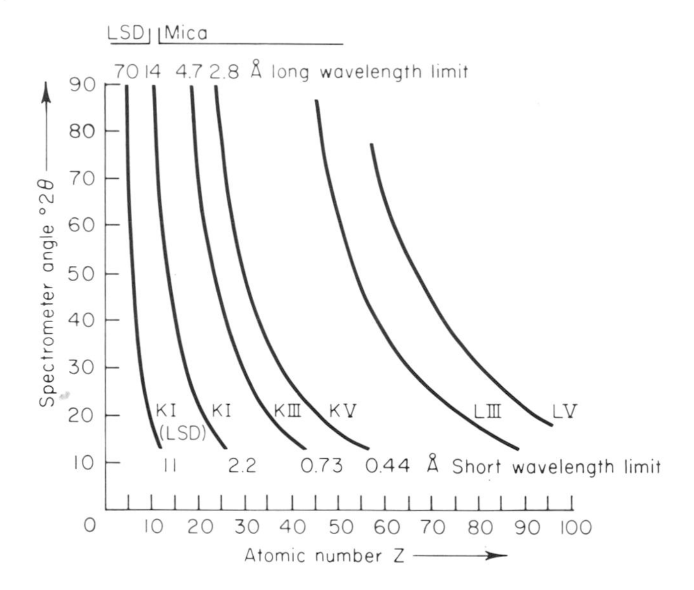

FIG. 1. Spectral range of a mica-stearate crystal.

Another advantage of using mica as a substrate is that aligning the scanning spectrometer for the stearate is simplified and can be carried out in air using short wavelength radiations. Because the diffracting planes of the stearate and the mica crystals are parallel, optimizing the spectrometer can be done on either of the two crystals. With the stearate crystal alone, it is very difficult to check whether a scanning spectrometer "tracks," i.e., whether the spectrometer remains optimally aligned when it is scanned from low angle to high angle. This tracking can easily be tested on the mica. It must be emphasized here that proper tracking of the spectrometer is essential for judging the quality of the stearate crystal by its peak-to-background ratio. A nontracking spectrometer gives a maximum efficiency at the wavelength at which it is aligned.

B. The Quality of the Crystal

The quality of the crystal can be characterized by three numbers, i.e.:

1. *The resolution.* This can conveniently be expressed as the reciprocal value of the relative half-width of the diffracted peak of a monochromatic radiation. Thus, we can write $R = \lambda/\Delta\lambda$. In optics, the resolution equals the number of effective lines in a line grating. In our case, the resolution equals the number of effective double stearate films. This seems to be true as long as the crystal is perfect. However, by increasing number of layers, the imperfection builds up, and it becomes more difficult to obtain the theoretical resolution. Of course, to test the resolution one has to be certain that the absorption of the radiation by the crystal does not limit the use of the deeper layers. Figure 2 illustrates the resolution of an experimental stearate crystal deposited on mica. We see in the right-hand section of Fig. 2 the first order reflection of aluminum K by the mica, and adjacent to it the fifth order reflection on the stearate. The $2D$ spacing of the stearate is just slightly larger than five times that of mica, with the result that the peak appears on the low angle side. The left-hand section shows the resolution of the stearate crystal in the fourth order reflection with aluminum K radiation. The crystal used was an experimental one in an attempt to deposit 200 layers or 100 double layers. In a perfect case, one would expect a resolution in the fourth order of 400. Because of the built-up imperfection, only a resolution of 310 was obtained.

2. *The diffraction efficiency.* The peak intensity is determined by the window transparency, the solid angle at which the X-ray source sees the crystal and also the diffraction efficiency of the stearate. Efficiencies can easily be compared if the same instrument is used. The efficiency increases with the number of layers and also with the perfection. In practice, however, perfection decreases with the number of layers so that an optimum number of layers is

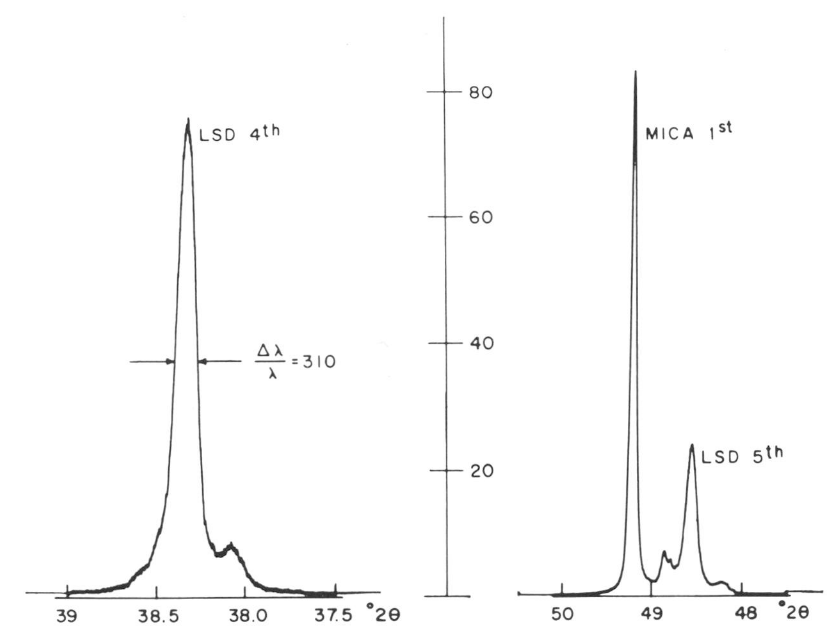

FIG. 2. Resolution of a stearate crystal for aluminum radiation (8.32 Å).

obtained. Also increased absorption of the radiation tends to set a limit to the efficiency.

Figure 3 shows the performance of an experimental crystal for carbon K radiation. The dipping process was interrupted after the deposition of 30, 55, 86, and 115 layers, and the crystal used to scan the carbon line. The graph inserted in Fig. 3 shows a linear response of the total intensity as the number of $2N$ increases. It is, however, difficult to deposit many more than $2N = 100$ layers because the stearate starts to crack. This results in a loss of intensity, and of course, loss of resolution also. Figure 3 also shows the relative half-width which changes as the number of layers increases. The crystal with the 30 layers deserves special attention. In addition to the normal carbon peak, one observes a series of maxima and minima similar to the optical thin film interference pattern. In fact, the phenomenon is identical to that of the thin film diffraction. The rays reflected by the top surface of the stearate and by the mica interfere with each other, resulting in destructive and constructive interference. Because of the long wavelength of the carbon radiation, many optical phenomena are observed here.

3. *Peak-to-background ratio.* The background can be measured either at some angle or wavelength other than that of the peak (in this case we will call it the off-peak background), or it can also be measured at the peak angle

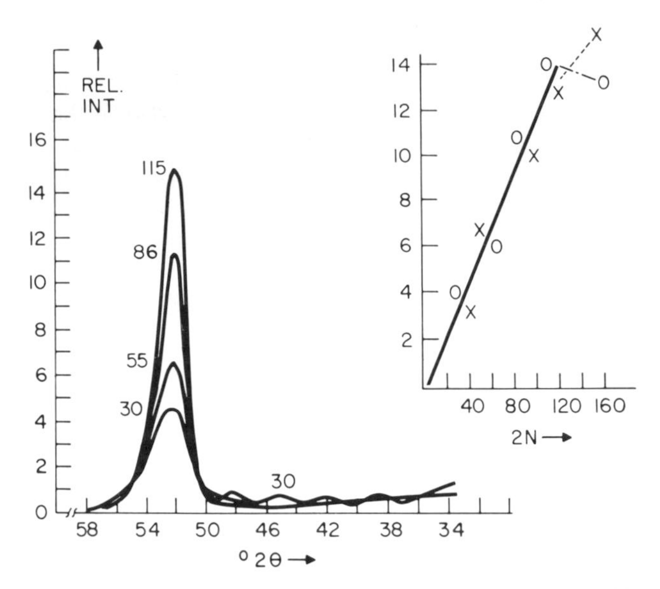

FIG. 3. The efficiency of a typical stearate crystal versus the number of layers. Carbon radiation (44.6 Å).

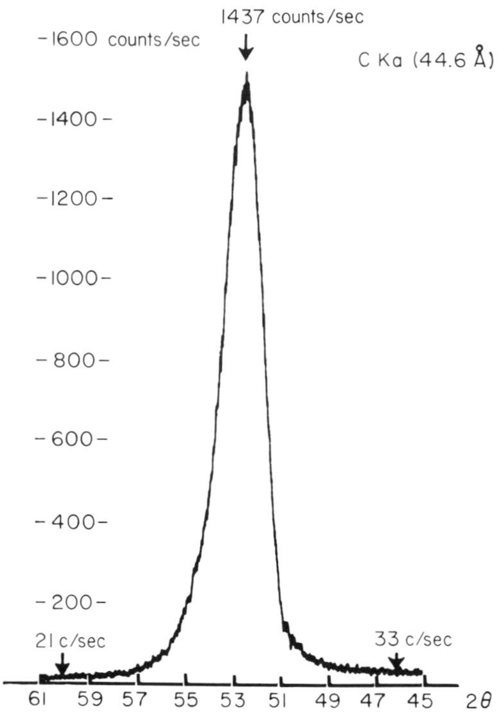

FIG. 4. Typical carbon spectrum as obtained with a stearate crystal, in 1963. High background on the low angle side is caused by specular reflection.

in which the sample is replaced by another element that does not contain the measured radiation. This on-peak background is very important in the determination of limits of detectability. To discuss the off-peak background, we can refer to Fig. 4 as an example.

Figure 4 shows a carbon scan which was typically obtained with a stearate crystal approximately three years ago [7]. We see the familiar asymmetrical carbon K band, and typically we have a high background on the low angle side, increasing toward the low angle. It is obvious that with this type of curve we cannot define a specific angle at which the background should be measured. The same is true for the higher angle close to the peak. One can, of course, arbitrarily agree upon a fixed angle at which the low angle background can be measured, for example, the minimum value. What about the high angle side? Here the background will keep decreasing, and when we assign a particular angle at which the background should be measured, one may run into the difficulty that what was considered background is, in fact, part of the carbon spectrum.

Figure 5 shows a carbon curve obtained with the better stearate. For convenience in studying the shape, the sensitivity of the recorder is increased for the background. First, we see here that on the low angle side the background is improved as compared to the earlier crystal. Here, we estimate a P/B ratio

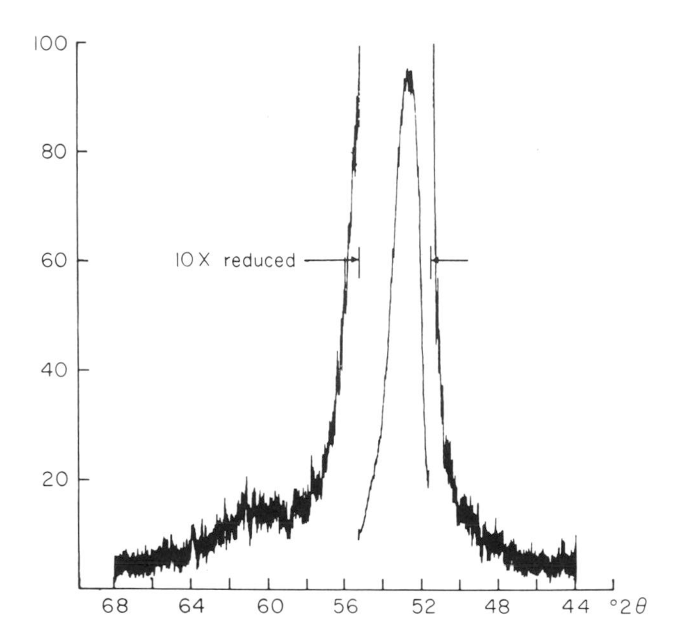

FIG. 5. Newer type crystal showing a better peak-to-background ratio on the low angle side. The sensitivity of the recording is increased by a factor 10 for the background.

close to 200 to 1. If we continue measuring the background on the high angle side at 60°, we will get a constant P/B ratio no matter how good our crystal is; in this case, we again measure a P/B ratio of about 70 to 1. We feel that it is more realistic to measure the background at a larger angle because we still do not know how far the carbon band extends.

C. Background Measurement

Background measurement should be done with great care if an honest representation of the crystal performance is to be obtained. This is especially true when one makes the crystal, judges the performance, and compares it with other crystals made under different circumstances. Perfect alignment and "tracking" of the spectrometer over the range in which the background is going to be measured is, of course, of the greatest importance. Optimizing the instrument for a desired wavelength is not difficult when enough intensity is obtained. This is usually true for the peak. But for the background, this becomes more difficult unless one uses another element which has a peak at the angle in which the background is going to be measured. When a mica substrate is used, aligning is much simplified. Geometrically at least, we are then sure that the spectrometer is optimum in the range of interest.

Another source of error which can easily be overlooked is the peak shift of the pulse high distribution due to intensity differences, as was pointed out by Bender and Rapperport [11]. This is certainly not an imaginary case if one considers that the peak is two orders of magnitude larger than the background. If a narrow window is used to improve the P/B ratio and the PHA is adjusted for the peak, it may not be true for the background.

An error which tends to decrease the peak intensity and not affecting the background is the dead time of the detector. It is, therefore, generally desirable to reduce the total intensity in this type of measurement.

III. Reduction of Carbon Contamination

A. The Problem of Carbon Contamination

Carbon contamination has been known to users of electron microprobes, but it has never put severe limitations on the use of the instrument in the conventional range of elements for which the instrument was originally designed. With the availability of means to extend the useful range to include the element boron, however, this is no longer the case.

Carbon contamination of the sample results in increased absorption (with time) of the ultrasoft radiation and an increase of the measured carbon content. If a small spot size is not required, one can defocus the electron beam. By doing this, the carbon is deposited over a larger area, and thus the effect becomes less. Moving the sample, and thus continuously exposing a clean area, can also be resorted to. However, if microanalysis has to be carried out, these methods cannot be used. In that case, corrections can be made for the carbon contamination. This can, for example, be done by plotting the X-ray intensity as a function of time and then extrapolating the results to the starting time of the experiment. This is, however, time consuming and inaccurate.

There are several ways of reducing carbon contamination, i.e.:

1. by heating the sample;
2. by cooling the surrounding;
3. by ion bombardment;
4. by the use of a gas (air or argon) jet.

Here, we will discuss only the second method.

B. The Use of a Cold Plate

As Heide [12] showed with an electron microscope, carbon contamination can be almost completely eliminated by surrounding the sample with a cooled surface. To prevent undesired carbon removal, the sample should be kept at room temperature while the surroundings should have a temperature of at least 130°C below zero.

Campbell and Gibbons [13] have shown that carbon contamination dropped by an order of magnitude for every drop of 50°C of surrounding tempertúre. The sample here was also completely surrounded by a cold surface. Such an arrangement cannot easily be realized in an existing microprobe without cutting down the sample size drastically. But if we are satisfied with a partial but still significant reduction of carbon contamination, this method can easily be applied to most existing probes [14].

In the first approximation, the efficiency of a cooled surface is proportional to the solid angle at which the sample sees the cooled surface. In a large and thick sample, this means that cooling the top hemisphere will result in 100% efficiency. In a thin, electron transparent sample, as in an electron microscope, both sides of the sample should be surrounded by a cooled surface. If we use a cold plate inserted between sample and electron lens and provided with the necessary holes for the light and X rays, such a plate can easily cover an area of approximately 30% of the hemisphere. Let us assume that the cooled plate temperature is 100°C below room temperature: according to Campbell and Gibbons [13], this can result in a 100-fold reduction of contamination for the totally enclosed specimen. In the case of the cold plate, reductions of approximately 30-fold can therefore be expected. Further, smaller improvement can still be obtained by using extreme precaution in vacuum system design.

Figure 6 shows the cross-sectional drawing of the cold plate. It is mounted on a copper rod and can be moved in its own plane for alignment. A height adjustment is provided to locate the plate with respect to the sample and the lens. A collapsible tube makes it possible to install and remove the device easily. The presence of the cold plate does not in any way interfere with the normal performance of the microprobe.

Figure 7 shows the device: the cold plate, the connecting rod, the glass-to-metal seal, and the soft copper beard extending into the liquid nitrogen flask. The part of the conductive rod which is exposed to the air is insulated with styrofoam and is enclosed in a metal shield. The function of this electrical wire will be explained later. This odd shape of the plate happens to have the largest surface area that can easily be taken in and out of our microprobe without interfering with major components. The two cuts on the side were made to prevent the plate from interfering with the X-ray beam.

Figure 8 shows some results. It is a reproduction of the strip chart recorder, recorded for four elements. The spectrometer was tuned for carbon

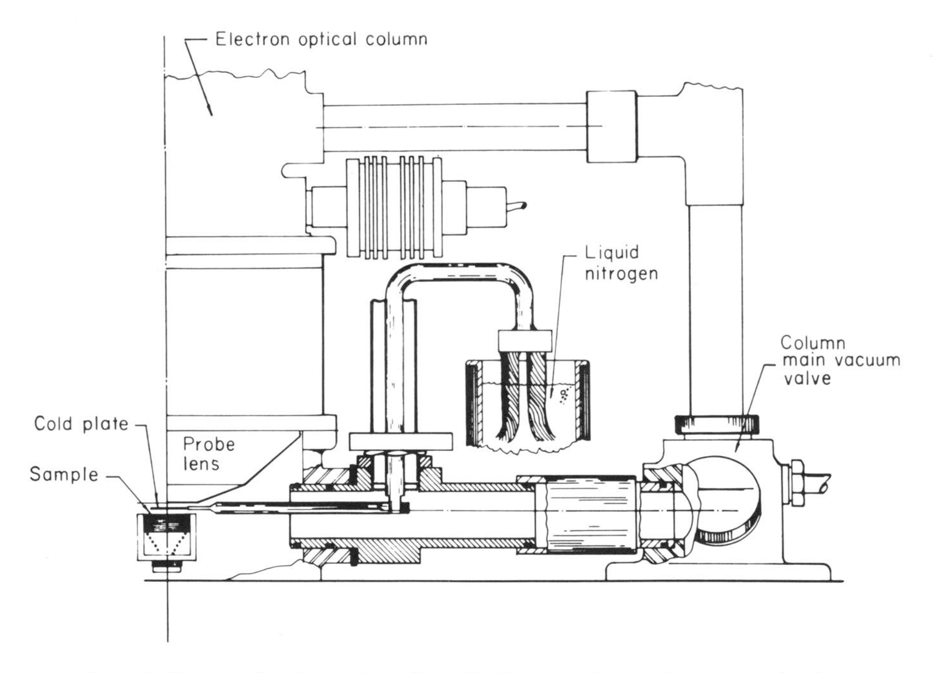

FIG. 6. Cross-sectional drawing of a cold plate to reduce carbon contamination.

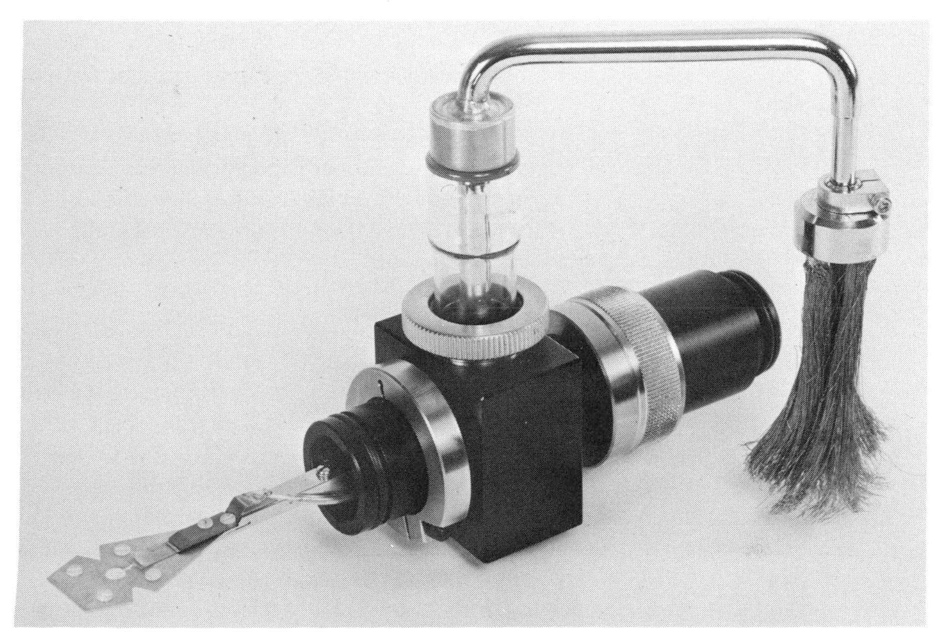

FIG. 7. A cold plate assembly.

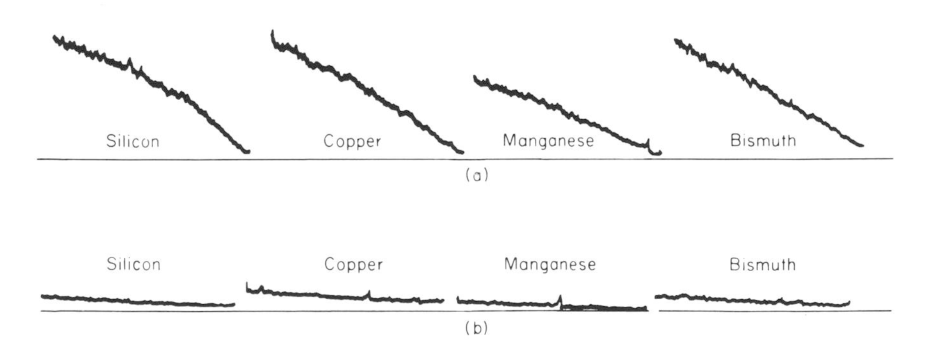

FIG. 8. Carbon contamination as function of time (time scale reads from right to left) (a) with and (b) without cold plate. (a) Contamination rate traces; 1000 counts/sec full scale; excitation potential, 10 kV; time interval, 5 min. (b) Contamination rate traces; cold finger; liquid nitrogen temperature; 1000 counts/sec full scale; excitation potential, 10 kV; time interval 5 min.

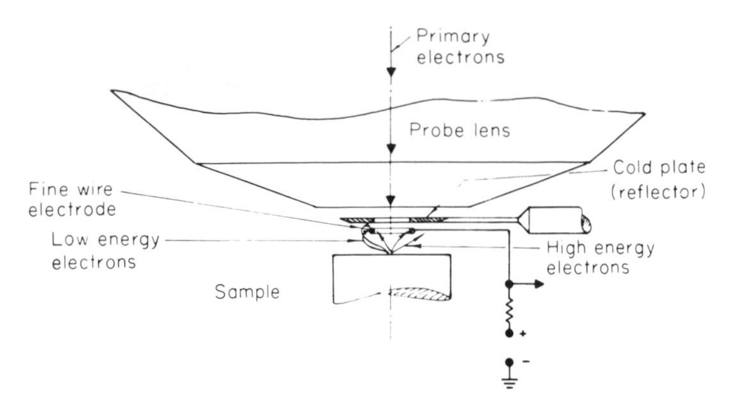
Primary
electrons
Probe lens
Cold plate
(reflector)
Fine wire
electrode
Low energy
electrons
High energy
electrons
Sample

FIG. 9.

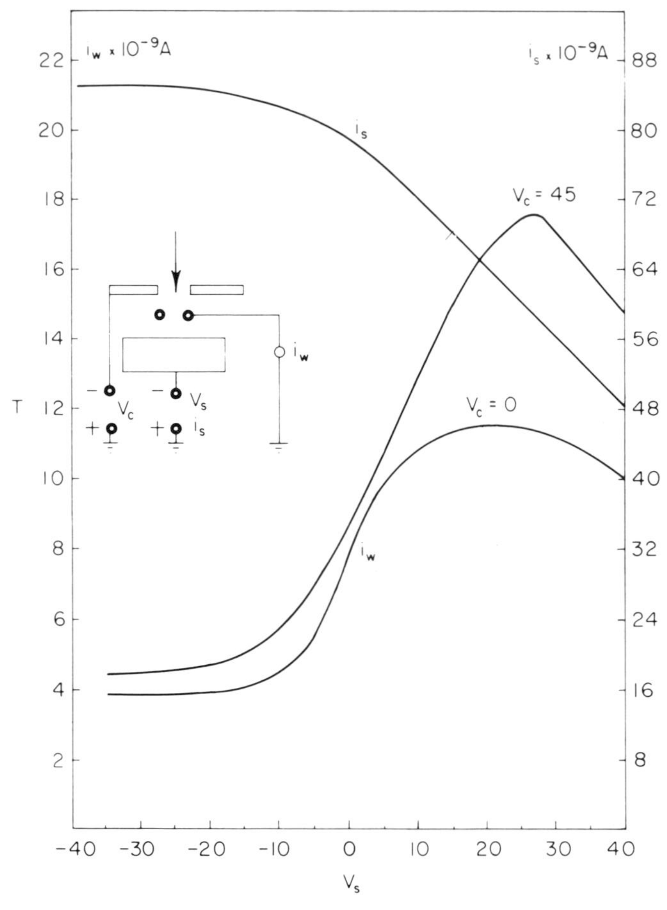
$i_w \times 10^{-9}A$
$i_s \times 10^{-9}A$
i_s
$V_c = 45$
$V_c = 0$
i_w
V_c
V_s
i_s
T
V_s
22
20
18
16
14
12
10
8
6
4
2
88
80
72
64
56
48
40
32
24
16
8
-40
-30
-20
-10
0
10
20
30
40

FIG. 10.

radiation and the beam left stationary for some period of time. Figure 8a with the cold plate at room temperature; Fig. 8b with the liquid nitrogen charge. Equilibrium is usually reached approximately 15 min after filling the Dewar with liquid nitrogen. By measuring the slopes on the curves, we find that the reduction of contamination is somewhere between a factor of 10 and 30.

The cold plate is electrically insulated from ground potential, and thus can also be used for measuring the backscattered electrons irrespective of their energy. Because of its location and size, it will show a minimum of topographical dependence. In addition, we can add another electrode that will collect essentially the low energy electrons. This will be illustrated in Fig. 9 in which we see again the lens, the sample, and the cold plate. Just above the sample is placed a very thin wire in the shape of a small loop. It is attached to the plate but electrically insulated (see also Fig. 7). Because of its small dimension, it will collect hardly any electrons emitted by the sample. By applying a positive voltage with respect to the sample, we can increase the collecting efficiency for the low energy electrons. A further increase of the efficiency can be obtained by making the plate negative with respect to the wire. The results are shown in Fig. 10. For reference, the sample current on a different scale is included in this curve.

IV. Analysis of the Light Elements

A. *Soft X-ray Detector*

Of the various types of X-ray detectors, the flow proportional counter seems to be the most suitable one for the detection of X rays in the 10–100 Å wavelength region. Apart from the high counting efficiency, its energy resolution is excellent and the pulse height distribution, even for boron K radiation (67 Å), is well above the electronic noise. The relative half-width of the pulse height distribution decreases with the square root of the photon energy. Mosley's law, however, states that the energy of the K radiation of the elements decreases also according to a square root function with atomic number. As a result, the degree of separation of the pulse height distribution for the K radiation of the elements is independent of the atomic number. This means that the proportional counter can resolve any two elements that have a difference of 2 in the atomic number scale.

A good resolution cannot be obtained, however, by an inappropriate

---

FIG. 9. Cross-sectional drawing of a wire loop electrode arrangement to collect low energy electrons.

FIG. 10. Current–voltage characteristics of the wire electrode for two different values of cold plate voltage. Wire electrode current versus sample bias voltage. Sample: platinum.

design of the detector or by the use of an inadequate detector gas. Detector geometry and gas mixture density should therefore be adapted to each other. Too low a gas density may result in losses because of the transparency. High density tends also to result in losses because the photoelectrons generated close to the entrance of the detector cannot reach the center wire without undue losses. The gas density is controlled by either the gas pressure or the gas material. A mixture of methane–argon with various compositions can be made to obtain a certain gas density. Figure 11 shows an example of the pulse height distribution of the K radiation of fluorine, oxygen, nitrogen, carbon, and boron, as obtained with a gas mixture of 75% methane and 25% argon.

B. Qualitative Analysis

The light element microprobe is able to detect the light elements even if present in low concentration. Two-dimensional distribution of that particular element can be obtained by scanning the beam over the sample and displaying the X-ray intensity on a display tube. Because of the relatively good sensitivity, such X-ray images made on a routine basis have proved to be useful. Figure 12 shows an example.

C. Quantitative Analysis

Quantitative analysis can be approached by using standards to establish a calibration curve. The composition of the standards should come close to the composition of the sample. Figure 13 shows an example of the detection of carbon in steel [15]. In many cases, however, it is very difficult to obtain standards of sufficient homogeneity and known compositions, and attempts have been made to relate the measured intensities to the weight percentage. Corrections are severe and unreliable in most cases because of:

1. the high absorption of the ultrasoft radiation,
2. the uncertainty of the values of the absorption coefficients.

A phenomenon which complicates quantitative analysis is the wavelength shift due to chemical bonding. Another complication arises from carbon contamination which was dealt with in the previous section.

Atomic number and absorption corrections have been approached by Duncumb and Melford [16], using a "thin film" model. The assumption that the radiation is generated only in a very thin film on the surface holds true for low take-off angles and high absorption matrix.

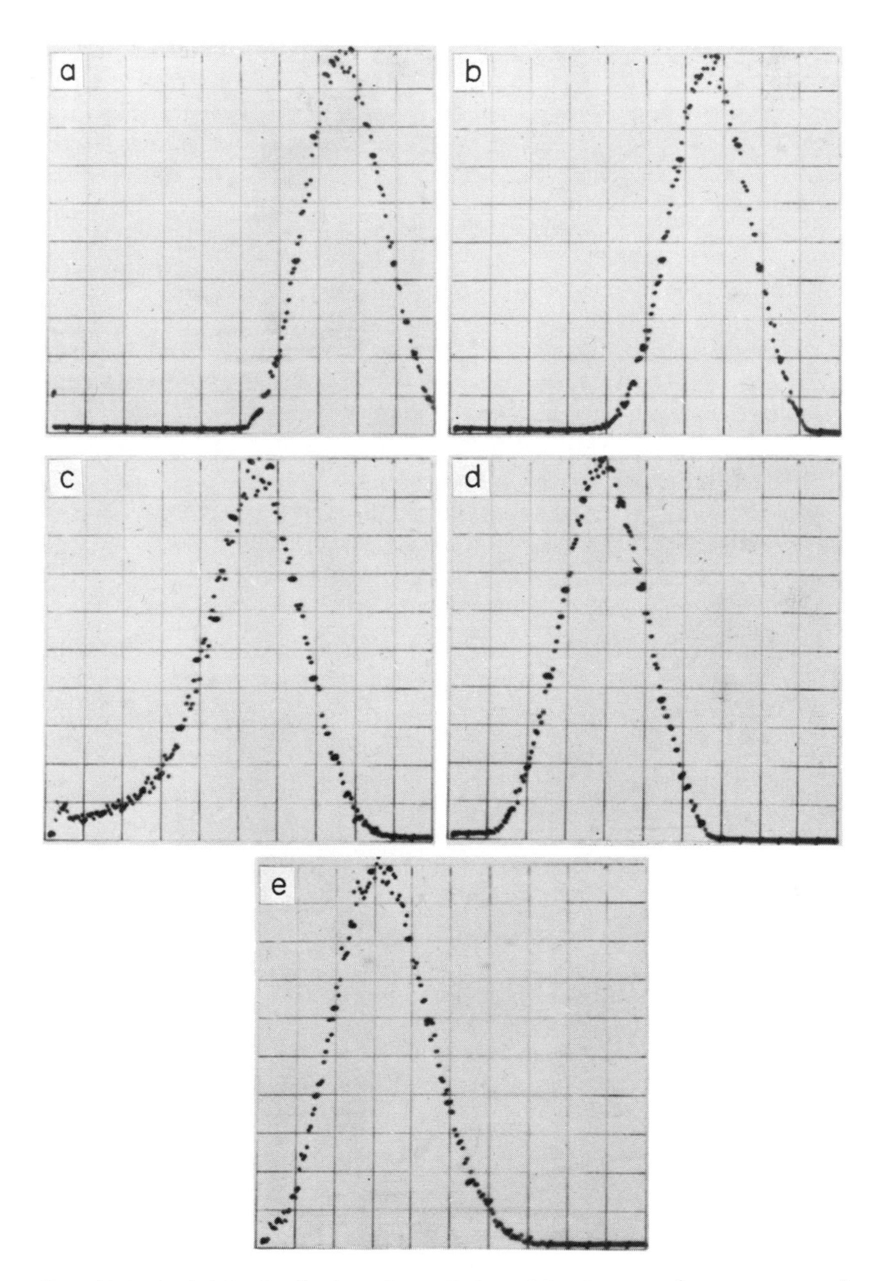

FIG. 11. Pulse height distribution of K radiation of fluorine (18.3 Å), oxygen (23.6 Å), nitrogen (31.6 Å), carbon (44.6 Å), and boron (67.8 Å), using a mixture of 75% methane–25% argon as detector gas.

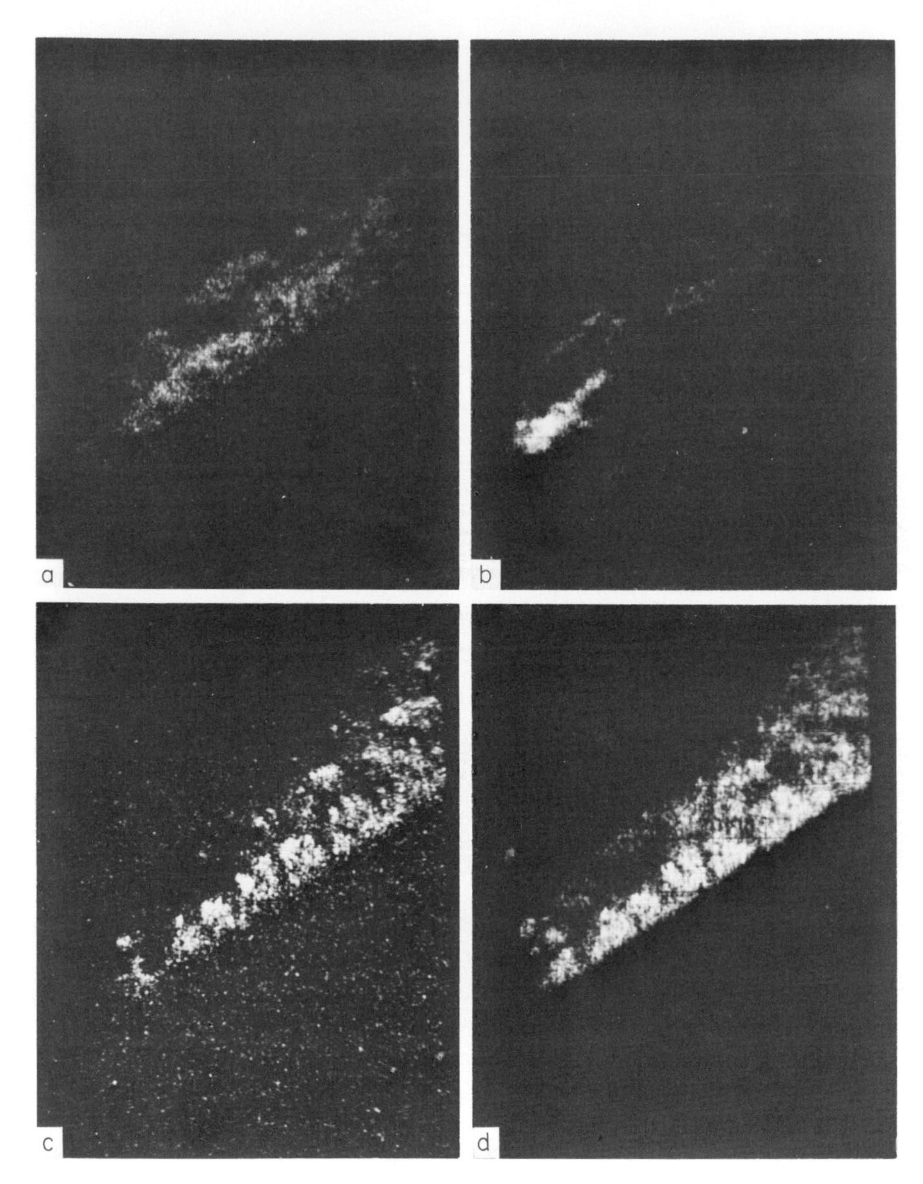

FIG. 12. X-ray scanning images of an inclusion (welding slag), showing the distribution of (a) silicon, (b) carbon, (c) fluorine, and (d) calcium.

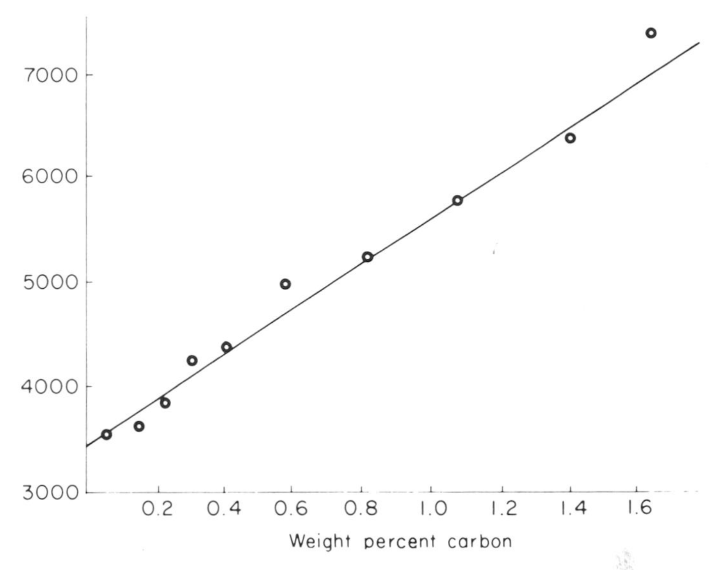

FIG. 13. Typical carbon intensities as obtained from low carbon steel. The fluctuations are mainly due to local variations. A smoother curve can be obtained when larger areas are analyzed. Counts in 90 sec; 0.1 μA, 10 kV. [15]

REFERENCES

1. B. L. Henke, *Advan. X-Ray Anal.* **7**, 460 (1964).
2. R. M. Dolby, *J. Sci. Instr.* **40**, 345 (1963).
3. J. E. Holliday, *Rev. Sci. Instr.* **31**, 891 (1960).
4. J. B. Nicholson and D. B. Wittry, *Advan. X-Ray Anal.* **7**, 497 (1964).
5. J. Merritt, C. E. Muller, W. M. Sawyer, Jr., and Tefler, *Anal. Chem.* **35**, 2209 (1963).
6. H. A. Elion and R. E. Ogilvie, *Rev. Sci. Instr.* **33**, 753 (1961).
7. P. S. Ong, *Conf. Anal. Chem. Appl. Spectroscopy, Pittsburgh, Pensylvania, 1964.*
8. J. B. Nicholson and M. F. Hasler, *Advan. X-Ray Anal.* **9**, 421 (1966).
9. I. Langmuir, *J. Franklin Inst.* **218**, 153 (1934).
10. K. B. Blodgett, *J. Am. Chem. Soc.* **56**, 495 (1934).
11. S. L. Bender and E. J. Rapperport, *in* "The Electron Microprobe," p. 405. Wiley, New York, 1966.
12. H. G. Heide, *Proc. Intern. Conf. Electron Microscopy, 5th, Philadelphia, Pennsylvania.* Academic Press, New York, 1962.
13. A. J. Campbell and R. Gibbons, *in* "The Electron Microprobe," p. 75. Wiley, New York, 1966.
14. P. S. Ong, *Intern. Conf. X-Ray Optics X-Ray Microanal., 4th, Orsay, France, 1965.*
15. J. F. Moskal, P. S. Ong, and P. Crean, *Nat. Meeting Soc. Applied Spectroscopy, 4th, Denver, Colorado, 1965.*
16. P. Duncumb and D. A. Melford, *Intern. Conf. X-Ray Optics X-Ray Microanal., 4th, Orsay, France, 1965.*

Changes in X-Ray Emission Spectra Observed between the Pure Elements and Elements in Combination with Others to Form Compounds or Alloys

WILLIAM L. BAUN

Materials Physics Division
Air Force Materials Laboratory
Wright-Patterson Air Force Base, Ohio

I. INTRODUCTION

A. History and Reviews

For a time following the discovery of X rays, it was believed that characteristic X rays were simple functions of the emitting atoms and that combining the atoms with dissimilar atoms did not change the characteristic spectrum. However, it was predicted by Swinne [1] in 1916 that under certain conditions characteristic X rays would be affected by chemical combination. The first changes observed were not in emission lines but in K absorption edges by Bergengren [2], in 1920. Following this discovery, Lindh and Lundquist [3] and several others found changes due to chemical combination in emission line energies. Shortly afterward, there were many reports of changes in emission spectra, especially in third period elements. The early literature was reviewed in Siegbahn's book [4], in 1931. A complete review and reference list through 1951 was prepared by Herglotz [5]. Faessler's chapter in the Landolt–Bornstein tables [6] emphasizes experimental absorption and emission data for many pure elements and compounds. The "Encyclopedia of Physics," Vol. XXX, "X Rays" includes a section by Sandström [7] on experimental methods of X-ray spectroscopy and one by Tomboulian [8] on soft X-ray spectroscopy and the valence band spectra of light elements Blokhin's book [9] translated from Russian in 1961 and available from the Department of Commerce, contains much useful data and also comments on effects of chemical combination, especially in Chapters 8 and 9 that deal with fine structures in emission and absorption spectra. Shaw's excellent review [10] on the X-ray spectroscopy of solids contains a bibliography of 355 references as well as some unpublished data by Shaw and his co-workers. An annotated bibliography by Yakowitz and Cuthill [11] contains over 500 references and emphasizes the ten year period 1950–1960. Only a few early references are included since reviews such as those by Herglotz [5] and Faessler [6] cover the early literature quite well. Several sessions at meetings have been devoted to the effect of chemical combination on X-ray spectra. A conference was held on the applications of X-ray spectroscopy to solid state problems, in Madison, Wisconsin, in 1950 [12]; on the physics of X-ray spectra at Cornell University in 1965 [13]; and also in 1965, a complete program on the chemical effect was presented at Leipzig's Karl Marx University [14]. A session on the chemical combination effect on Applications of X-Ray Analysis [15] was organized by this author at the 1965 meeting in Denver.

B. General Effects Observed in X-Ray Emission Spectra

1. *Shifts*

Wavelength shifts to both longer and shorter wavelengths have been observed for various elements upon chemical combination. These shifts result from energy level changes due to electrical shielding or screening of the electrons when the valence electrons are drawn into a bond. The greatest shifts are seen in spectra of low atomic member elements. Generally, the so-called last or highest energy member of a given series is most affected by chemical combination. For instance, K bands (K_β) of oxides of Mg, Al, and Si shift more than 4 eV compared to the pure metal while the K_α doublet shifts less than 1 eV (in the opposite direction). Satellite lines shift more than the parent line. X-ray shifts are not usually as great as the actual energy level shifts as measured by electron spectroscopy [16] because in X-ray spectroscopy, we measure only energy level differences. Where each level, say the K and L, shift about the same amount and in the same direction, as is often the case, the resulting K_α line will show little shift with chemical combination.

Although the shifts are sometimes small, Sanner [16a] concluded in his dissertation that the wavelength shift of the K_α doublet can be followed to 25 Mn, that of K_{β_1} to 27 Co, and the quadrupole line K_{β_5} to 28 Ni. Similar measurements by other authors have also established practical shift detection limits for the L and M series (see Sandström [7]).

2. *Shape Changes*

The shape of a band gives some indication of the energy distribution of the electrons occupying positions in or near the valence shell. It was once thought that the electron density of states could be deduced directly from band spectra, but it is now obvious that many factors govern band shapes, and that corrections are necessary to determine the actual electron population. However, there are gross band shape changes with chemical combination, particularly in low Z elements, that can be attributed to combination of electrons in the valance band.

3. *Intensity Changes*

Large intensity changes are seen in certain lines and bands due to changes in excitation probabilities of the electrons undergoing transitions. Several phenomena combine to affect transition probabilities. One possibility is that with chemical combination, there is some change in electron character which allows either an increased or decreased number of electrons to undergo a given transition. Another possibility is the electron depopulation of a given level or state, and the subsequent formation of "exciton" or excitation states. A third contributing factor is the ocurrence of nonradiative transitions. These nonradiative transitions increase with decreasing atomic number, and undoubtedly contribute to the intensity changes observed in low Z elements with chemical combination. The last member of a series shows large intensity changes along with satellite lines as was mentioned earlier concerning wavelength shifts. The large intensity changes in satellite lines have not yet been adequately explained.

4. *Appearance and Disappearance of Lines and Bands*

Certain lines or bands appear or disappear with chemical combination and the origin of the transitions producing these lines is not usually clear. Some of these lines, the ones not explained by satellite production caused by multiple ionized states, have been ascribed to forbidden transitions that are allowed in the chemically combined atom because of a change in electron character caused by bonding. Some are said to be caused by the criss-cross transition of an electron from the outer level of one atom to an inner level of another atom of different character from the first, and still others are attributed to transitions from excitation states.

C. Scope of the Review

In the following section, some of these changes due to chemical combination are summarized for a number of elements with emphasis on low atomic number elements. The wavelength range to be discussed is rather arbitrary. The long wavelength cutoff is placed at beryllium K (116 Å) since wavelengths to 120–130 Å appear to be a practical limit for microprobes employing curved crystal optics and flow proportional counter detection. Naturally, when (and if) microbeam probe equipment is developed using grating spectrometers and windowless detectors, this long wavelength limit will be extended to perhaps 500 Å, which will allow measurements of Li K and Mg L spectra, for instance.

This review is not meant to be a complete coverage of the literature. Rather, specific examples are shown based primarily on the author's interests and the practical use to which the spectra may be put for materials characterization. X-ray absorption spectra are not discussed because the microbeam probe is not normally used for such measurements.

II. X-Ray Emission Spectra

A. K Series

It was mentioned in the Introduction that the most noticeable effects are seen in spectra from low atomic number elements. Figure 1 shows some of the effects of chemical combination on K spectra of very light elements beginning with beryllium. The Be K spectrum in the metal consists of just one asymmetrical band and illustrates an effect of chemical bonding even in the pure element. The $K_{\alpha_{12}}$ line originates from the transition $L_{II,\ III} \rightarrow K$. In beryllium, the $L_{II,\ III}$ level is empty, having only two electrons in the L_I level. Therefore, K_α appears sooner than the first electron appears in the $L_{II,\ III}$ level of the free atom. This indicates the role of chemical bonding in the solids where the external electrons of the atom are excited to the next optical level where these electrons then can complete the transition to the vacancy in the K level. With chemical combination to the oxide the K_α line shifts significantly to longer wavelengths and becomes quite symmetrical, and a new line is observed on the short wavelength side of the K_α line. Probably a new line or band also appears on the long wavelength side, but with the lead lignocerate analyzer, it is not possible to scan to a long enough wavelength to observe it. The Be K

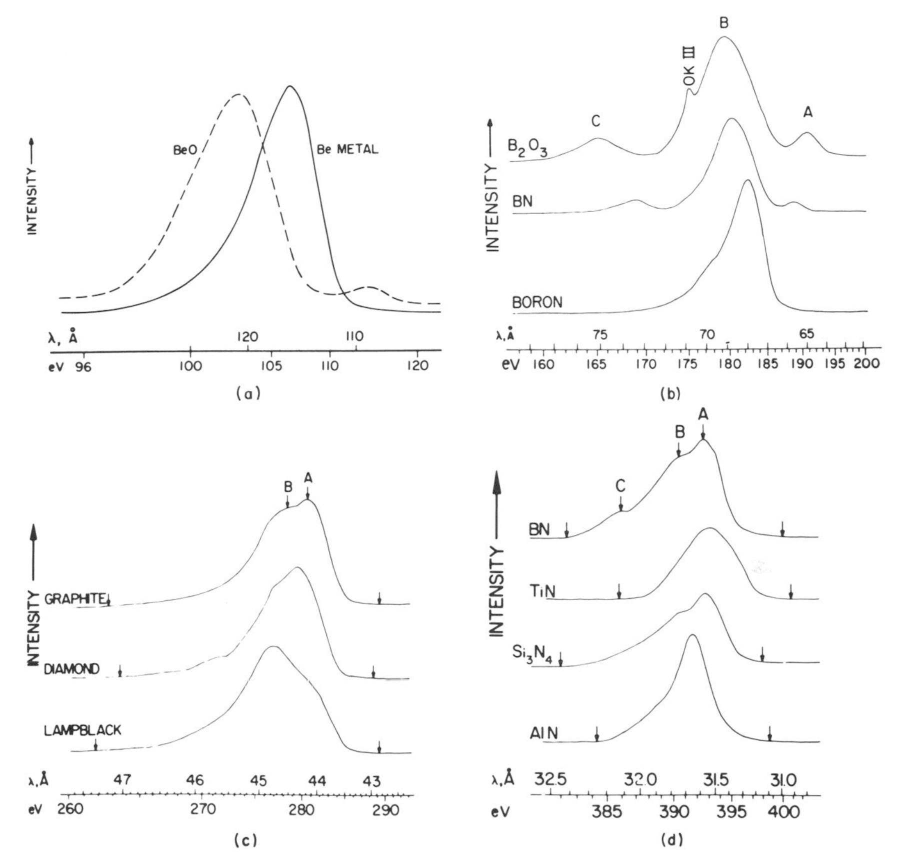

FIG. 1. The K spectra from low atomic number elements: (a) Be K, (b) B K, (c) C K, (d) N K.

spectra are unpublished results obtained in the author's laboratory by D. W. Fischer.

The boron K spectrum is very similar to beryllium K, both in the pure element and in the oxide as is seen in Fig. 1. The B K spectrum is also shown for BN for comparison with the pure element and oxide. The spectra for B, BN, and B_2O_3 illustrate each of the effects due to chemical combination mentioned in the introduction. The wavelength shift between B and B_2O_3 is almost 3 Å. There is a dramatic shape change, two new bands appear, and the intensities of these new bands depend on chemical combination. Data for the B K spectrum [17] from a number of compounds are shown in Table I. The table includes the B K wavelength, uncorrected half-bandwidths, relative intensity, and the peak intensity at 5 kV and 3 mA using primary excitation and a flat crystal spectrometer.

TABLE I
WAVELENGTH, HALF-BANDWIDTHS, AND INTENSITIES FOR BORON K EMISSION BANDS[a]

Target	Band	λ (Å)	$W_{1/2}$(eV)	Relative intensity	Peak intensity (counts/sec) (5 kV, 3 mA)
Boron	B	67.90	5.0		12,000
BN	A	65.56	3.0	12	
	B	68.65	5.9	100	2700
	C	73.35	3.7	9	
B_2O_3	A	64.96	4.1	18	
	B	69.01	7.9	100	2200
	C	74.94	4.2	20	
B_4C	B	67.94	5.8		7500
BP	B	67.93	5.2		900
TiB_2	B	68.16	4.9		3500
ZrB_2	B	68.03	4.8		1700
NbB_2	B	68.12	5.0		1300
W_2B_5	B	67.99	5.1		1400
SiB_6	B	67.92	4.9		1800
AlB_{12}	B	67.94	5.0		3600

[a] See Fischer and Baun [17].

The carbon K emission band has been studied extensively, probably because the carbon K absorption edge in most counter windows and film emulsions allows for easy recording of this radiation. Most of the data in the literature have been recorded using grating instruments and film methods. The C K curves in Fig. 1, however, are second order reflections from lead stearate and were recorded with a thin-window flow proportional counter [18]. A wide variety of peak shapes have been recorded for the C K band especially for graphite. Most of the differences obtained by various authors have been attributed to the use of photographic film techniques and the subsequent corrections required to translate blackening values into true intensity. However, recent data for graphite, shown by Holliday [19] and obtained using a grating and flow counter, do not agree with the curve for graphite, shown in Fig. 1. Probably this disagreement is due to the dispersion method, but it is possible that excitation conditions such as greater voltage above threshold could lead to such discrepancies. At any rate, regardless of the correct* band

* The "correct" shape is arbitrarily defined as the pure band emission at or near threshold voltage.

shape, there are significant differences in the spectra from different forms of carbon and from carbides. Table II gives wavelengths, half-bandwidths ($W_{1/2}$) and full bandwidths (W) for three forms of carbon and four carbides.

TABLE II
WAVELENGTHS AND BANDWIDTHS FOR CARBON K BANDS[a]

Target	A (λ, Å)	B (λ, Å)	$W_{1/2}$ (eV)	W (eV)
Graphite	44.18	44.62	8.4	27
Diamond	44.36	44.80	7.4	25
Lampblack	43.97	44.77	9.0	28
SiC	—	44.45	5.2	21
Al_4C_3	—	44.49	5.9	18
TaC	44.08	44.64	4.5	18
NbC	44.10	44.62	4.6	19

[a] See Fischer and Baun [18].

The nitrogen K band has not been studied in recent years, except by Fischer and Baun [18] and Holliday [19]. The N K band lies on the short wavelength or absorbing side of the C K absorption edge, and intensities generally are low. The N K bands from four nitrides are shown in Fig. 1. These curves are third order reflections from lead stearate and cover an angular range of 135–150° 2θ. There are some changes in the spectra with chemical combination especially in the number of maxima on the main band. Nitrogen-containing compounds which would be expected to show larger changes, such as nitrates and nitrites, are unstable in the electron beam.

Fischer [20] investigated a large group of oxides to determine whether there are any periodic trends in oxygen K band shapes and shifts. He found that the O K band could be classified into three categories; those showing high energy structure on the band, those showing low energy structure, and finally, curves showing no structure on either the high or low energy side of the band. No specific correlation with a property such as crystal structure could be found. The O K band shows a very curious splitting that is dependent on the crystal that is used to disperse the radiation. This band-splitting phenomenon was investigated by Mattson and Ehlert [21]. No satisfactory explanation has been proved, but apparently the splitting is an absorption process of some sort. These same authors also investigated O K spectra from gases and found large differences in various oxygen-containing gases and significant changes from that observed in solids.

In addition to investigating oxygen, Mattson and Ehlert investigated fluorine from a large number of fluorides and miscellaneous compounds such as SF_6 and Teflon. They grouped their spectra from fluorides into crystal structure types and got a fair correspondence between structure and spectral features. Large changes are observed in the intensity, separation, and position of the strong (up to 30% of K_α) K_{α_3} and K_{α_4} short wavelength satellites. Mattson and Ehlert also showed that there was little difference between fluorine K spectra excited by electrons or X rays.

Sodium K has not been studied extensively to determine changes in the spectrum due to chemical combination. Sodium K_α at 11.909 Å is too short for grating studies and somewhat too long for most crystal spectrometers, mainly because a dispersing crystal having sufficient resolution for Na K lines is not available. Also, because sodium is so active, it is difficult to obtain the spectrum of the pure metal. Preliminary studies in the author's laboratory, however, have shown that the satellites K_{α_3} and K_{α_4} are very dependent on chemical combination. Sodium is the first element in which K_β exists. Sodium is similar to beryllium, as discussed earlier, in that a member of the series

TABLE III
ALUMINUM K LINES AND BANDS FROM ALUMINUM AND Al_2O_3[a]

Line	λ (Å)	E (eV)	Intensity[b]	ΔE (eV)[c]
		Aluminum metal		
$\alpha_1\alpha_2$	8.3393	1486.3	1000	
α'	8.3080	1492.0	13	
α_3	8.2854	1496.0	78	
α_4	8.2744	1498.0	39	
α_5	8.2284	1506.3	5.0	
α_6	8.2098	1509.8	3.9	
β	7.9590	1557.3	6.5	
β''	7.8330	1582.4	<0.5	
β'''	7.8048	1588.1	<0.3	
		Aluminum oxide, Al_2O_3		
$\alpha_1\alpha_2$	8.3380	1486.6	1000	+0.3
α'	8.3037	1492.7	18	+0.7
α_3	8.2820	1496.6	64	+0.6
α_4	8.2718	1498.6	60	+0.6
α_5	8.2240	1507.2	4.6	+0.9
α_6	8.2050	1510.7	3.5	+0.9
β'	8.0618	1537.5	1.3	—
β	7.9819	1552.9	7.0	−4.4
β''	7.8522	1578.6	<0.3	−3.8
β'''	7.8204	1585.0	<0.2	−3.1

[a] See Baun and Fischer [22].
[b] Peak intensity.
[c] Shift between metal and oxide.

becomes visible sooner than an electron appears in the level involved in the transition.

The K spectra of Mg, Al, and Si are all quite similar. Baun and Fischer [22] have studied these elements in the form of pure elements, oxides, and various compounds and tabulated wavelengths, energies, intensities and energy differences for diagram and satellite lines. Table III gives some of the data tabulated and is included to show lines and bands existing in the Al K series and the changes which occur in going from metal to oxide. From this table, it is seen that the largest changes due to chemical combination are in K_α satellite lines, the K emission band, and in satellites of the K band (it has not been positively established that all extra lines near the K band are true satellites). The three satellites lines observed closest to the K_α line, $K_{\alpha'}$, and K_{α_3} and K_{α_4} are shown in Fig. 2 for the pure element, and the oxide for Mg, Al, and Si. With oxidation and, in general, with any form of chemical combination, the satellites shift to shorter wavelengths and the intensity relationships change significantly. The K band (K_β) shows similar gross changes as seen in Fig. 3. In all three elements, an asymmetrical band is obtained from the pure metal much the same as shown earlier for pure Be and

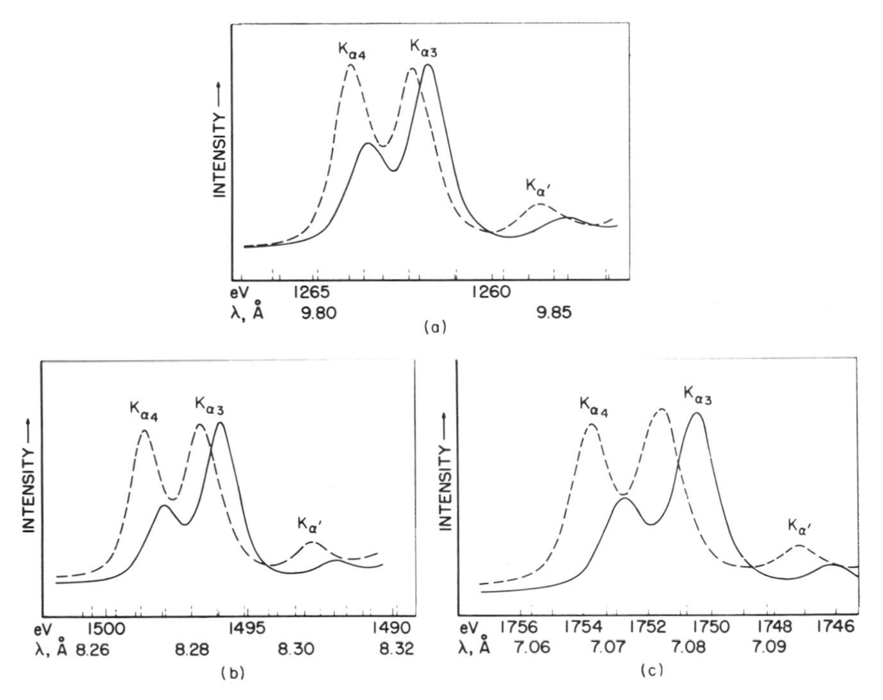

FIG. 2. The $K_{\alpha'}$, K_{α_3}, and K_{α_4} satellite lines (—, metal; - - -, oxide) for (a) magnesium, (b) aluminum, (c) silicon, and their oxides.

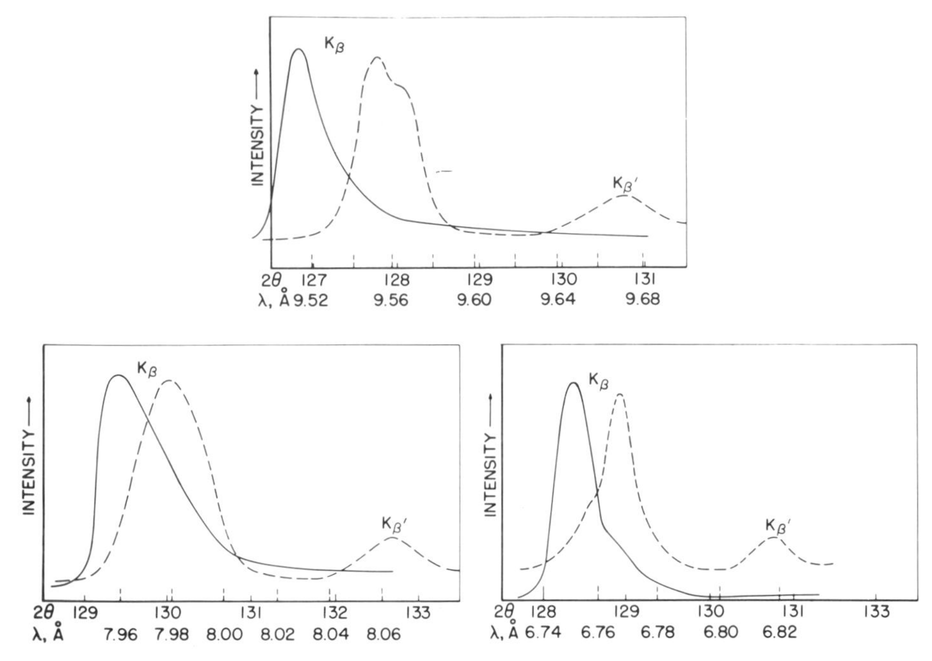

FIG. 3. The K emission band for (a) magnesium, ADP (101) crystal, (b) aluminum, EDDT crystal, (c) silicon, ADP (200) crystal, and their oxides (—, metal, – – –, oxide).

B. With oxidation, the band shifts more than 4 eV and becomes very symmetrical. In addition, $K_{\beta'}$ appears in the oxide. The exact origin of $K_{\beta'}$ is still somewhat in doubt. This is one of the lines or bands that has been assigned at various times to (1) a forbidden transition that is now allowed because of a change in electron character, (2) the criss-cross transition of an outer electron of one member to an inner hole in the other member, and (3) a transition from an excited state.

Alloying also has a significant effect on the X-ray emission spectra of low atomic number elements. In the author's laboratory, a program on the effect of alloying on the Al K spectrum has been carried out on several binary systems with a total of over 200 alloys. Figure 4 shows an example of the change in the Al K band for pure aluminum and some Al–Ni alloys [23]. There is a linear shift in wavelength and a change in band shape. There is also a linear change with composition in the intensity ratio of the satellite lines K_{α_3} and K_{α_4}. The $K_{\alpha_4}/K_{\alpha_3}$ ratio changes from 0.48 in pure aluminum to 0.84 in 4 Al–96 Ni. Several two-phase alloys were investigated using the microprobe, and it was found that the $K_{\alpha_4}/K_{\alpha_3}$ ratio measured by our macroarea excitation techniques was an average of the specific ratio for each pure phase. A comparison of results on K_{α_3} and K_{α_4} using micro- and macroarea excitation is shown in Table IV [23].

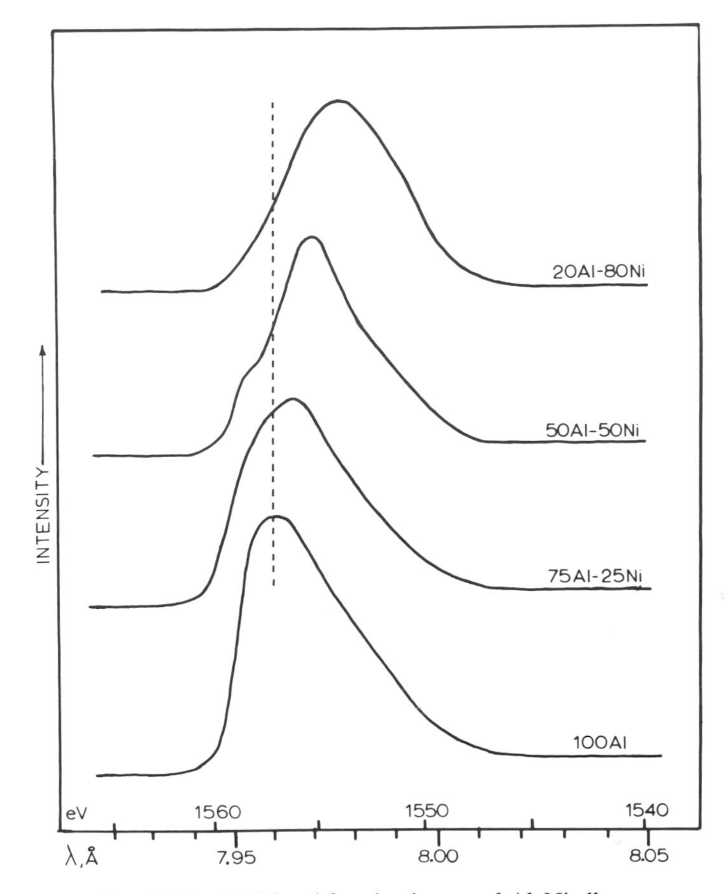

FIG. 4. The Al K band for aluminum and Al–Ni alloys.

TABLE IV
COMPARISON OF AL $K_{\alpha 4}/K_{\alpha 3}$ INTENSITY RATIO IN THE Al–Ni SYSTEM USING MACRO- AND MICROAREA EXCITATION[a]

Sample	$K_{\alpha 4}/K_{\alpha 3}$ macroarea excitation	$K_{\alpha 4}/K_{\alpha 3}$ microprobe
Pure Al	0.48	0.48
85 Al–15 Ni (two-phase)	0.54	0.50 Al-rich
		0.55 Ni-rich
75 Al–25 Ni (Al_3Ni)	0.57	0.57
65 Al–35 Ni (two-phase)	0.61	0.60 Al-rich
		0.63 Ni-rich
50 Al–50 Ni (AlNi)	0.66	0.66

[a] See Fischer and Baun [23].

In other alloy systems, far different results are obtained. In the Al–Mg system, for instance, there is virtually no shift in the K band nor change in the intensity of the K_α satellites. In Al–Cu, however, and in the spectra of nearly all polyvalent metals alloyed with monovalent elements, there are significant changes, especially in the K band. In the Al–Cu system the Al K band splits into two components as shown in Fig. 5 for pure aluminum and

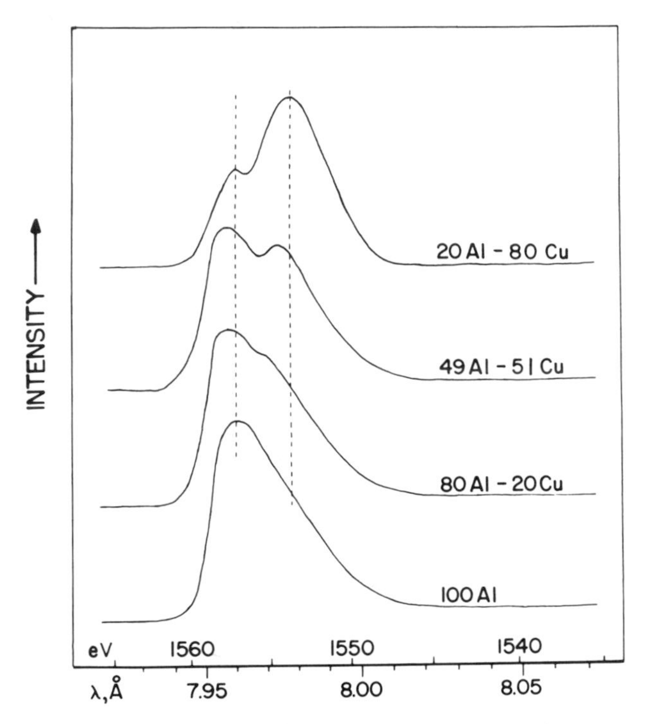

FIG. 5. The Al K band for aluminum and Al–Cu alloys.

some Al–Cu alloys. In this system, the short wavelength component retains the sharp short wavelength limit and shifts only slightly with alloying, while the long wavelength component shifts linearly with composition toward longer wavelengths.

Still other alloys behave differently. A summary of the systematic nature of Al K spectral changes has been prepared and will be published [24].

The K spectra of the elements phosphorus, sulfur, and chlorine exhibit many of the same changes observed for the previous elements, but sometimes K_α shifts are equal to, or greater than, K_β shifts. Schnell [25] showed for Cl K a successive shift to shorter wavelengths in going from KCl to $KClO_2$ to $KClO_3$ and $KClO_4$. He found also that K_β shifts to shorter wavelengths for the same compounds and $K_{\beta'}$ appear for all chlorine atoms coordinated with oxygen atoms. In a later paper, Schnell [26], using secondary excitation, investigated K_α and K_β from pure elements and compounds of elements from magnesium to chlorine in the periodic table. In this work his emphasis was on the variation of K_β intensity with chemical combination. His curves show that K_β occurs in every element between aluminum and chlorine when the element is combined with oxygen, and that the intensity of both K_β and $K_{\beta'}$ varies from compound to compound throughout the series. Schnell's K_α curves show that the unresolved satellite lines have approximately the same intensity as that in work in which primary excitation was used. A very complete report on the relationship between chemical bonding and the X-ray spectrum of sulfur was published recently by Wilbur [27]. In this work K_α and K_β energies were determined for a number of compounds, and the intensity dependence of K_β on chemical binding was investigated. Wilbur found good relationship of K_β intensity and energies for higher sulfur valences, but within

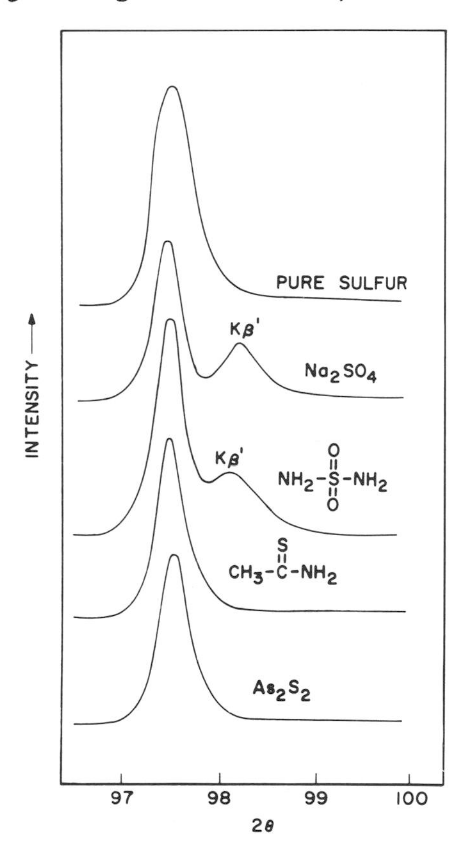

FIG. 6. The S K band for sulfur and some sulfur compounds (Wilbur [27]).

the 0 and -2 states, there was no particular correlation. The K_β energies and widths exhibited large differences, but there was no correlation with bonding. A representative set of K_β profiles from Wilbur's report is shown in Fig. 6. Especially notable are Wilbur's conclusions concerning $K_{\beta'}$. He found that $K_{\beta'}$ occurred only when sulfur was bonded to at least two oxygen atoms. The intensity of $K_{\beta'}$ was found to be negatively correlated with the K_β band reaching a maximum of one-third of the intensity of K_β in Na_2So_4. Wilbur finds most acceptable the idea that $K_{\beta'}$ originates from a 3s → 1s transition, with the intensity correlated with the amount of p character in the 3s band. Increasing intensity with greater bonding involvement would mean that more p character is being given to the 3s band in these situations.

There have been numerous investigations of line and band shifts and changes in the shapes of lines and bands in K spectra of higher atomic number elements. There are especially plentiful data for the first transition series. In general, however, measurements on K spectra of these elements require the use of a two-crystal spectrometer, and the changes due to chemical combination would not be large enough to concern the microbeam probe operator. Therefore, no work in this area will be reviewed.

B. L Series

There are L spectra that show changes as large as, or even larger than, K spectra with chemical combination of the elements. Examples of L spectra from compounds containing chlorine, sulfur, and potassium are shown in Fig. 7 [28]. These curves appear as recorded, with no corrections applied. The chlorine K spectrum consists of one main band having structure on the low energy side. The main band shift may be correlated with the electronegativity of the cation as was shown by Fischer and Baun [28] or with an estimation of the percent of ionic character of a single bond as has been done in Fig. 8, using Pauling's electronegativity values [29]. This plot shows that the greater the ionic character of the bond, the shorter the wavelength of the main L band in simple chlorides.

The sulfur L spectrum undergoes extremely large changes with chemical combination of sulfur. Elemental sulfur gives one main broad band (3s) with a strong component (3p) which is separated from (3s) by about 4 Å. Sulfides, on the other hand, show the main band (3s) and a weaker short wavelength component (3d) that averages nearly a 5 Å separation from the main band. Table V shows wavelengths and relative intensities for sulfur L components for sulfur, sulfides, and sulfates.

Sulfates give a still different spectrum as can be seen from Table V and the spectrum from $BaSo_4$ in Fig. 7. Spectra from the compounds $KHSO_4$

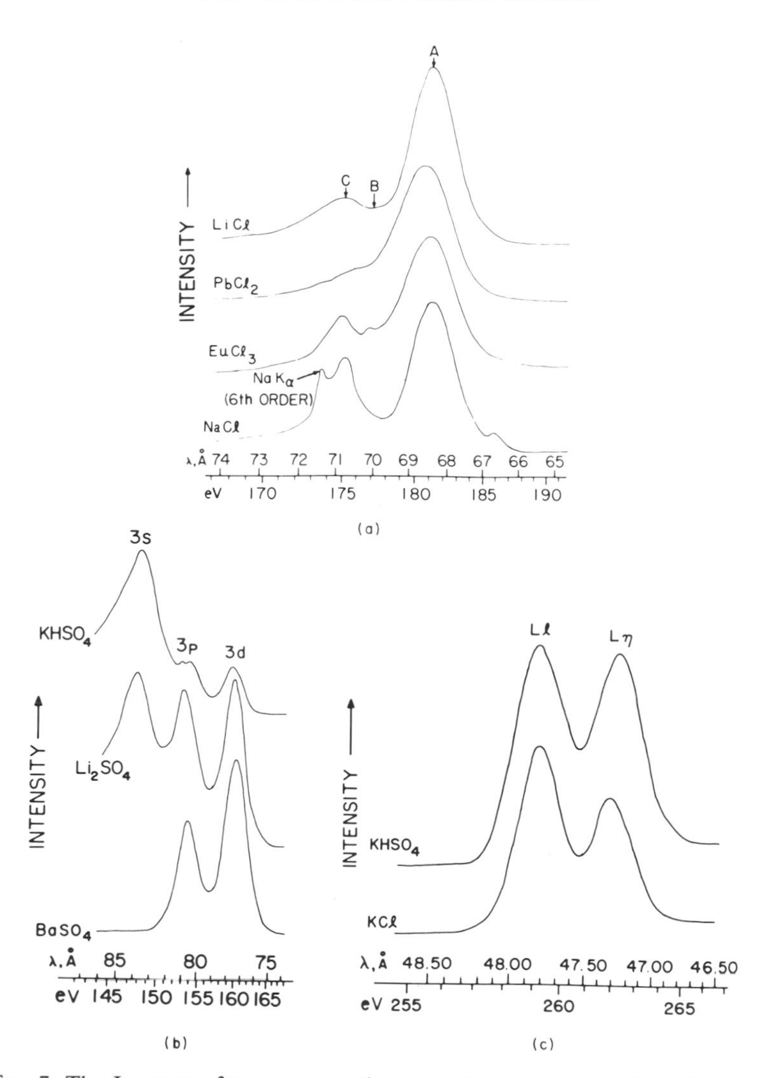

FIG. 7. The L spectra from compounds containing (a) chlorine, (b) sulfur, and (c) potassium.

and Li_2SO_4 are included in Fig. 7 to show the danger of using direct electron excitation on unstable materials. It appears that both of these compounds have partially decomposed to elemental sulfur in the electron beam.

The titanium L spectrum shows significant effects due to chemical combination. Figure 9, from the work of Holliday [19], shows Ti $L_{II, III}$ emission bands ($3d + 4s \rightarrow 2p$ transition) from Ti, TiC, and TiO. There are large shifts and band shape changes observed in these spectra. In recent work in

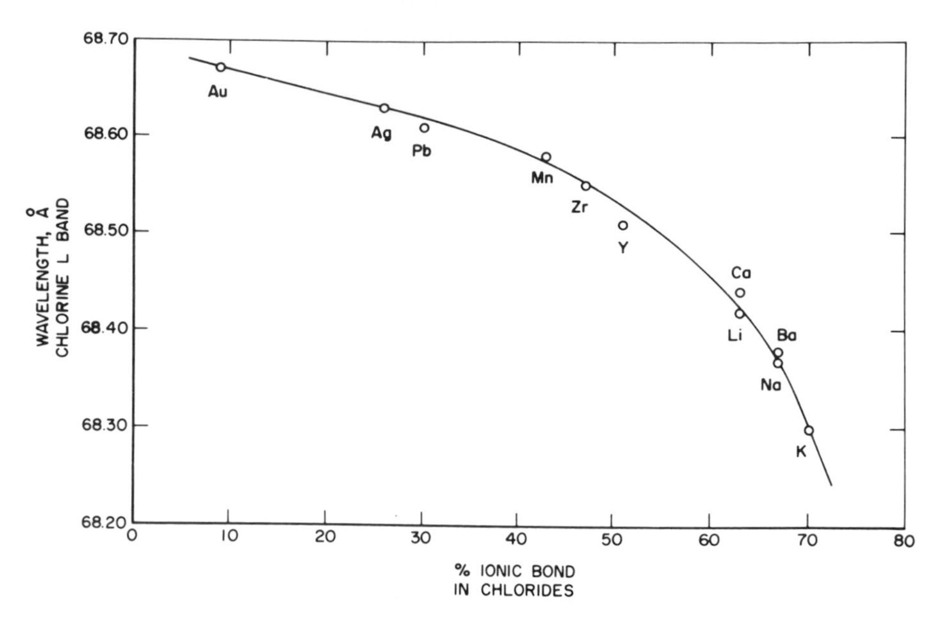

FIG. 8. Plot of wavelength of Cl L band versus % ionic bond in simple chlorides.

TABLE V
WAVELENGTHS AND INTENSITY RELATIONSHIPS FOR SULFUR L BANDS[a]

Target	3s (Å)	3p (Å)	3d (Å)	Relative intensity 3s : 3p : 3d
S	84.10	80.42	—	100 : 40 : 0
ZnS	84.55	—	80.14	100 : 0 : 15
Cu_2S	84.65	—	79.21	100 : 0 : 15
FeS	84.65	—	79.53	100 : 0 : 15
CdS	84.72	—	80.07	100 : 0 : 20
PbS	84.49	—	78.88	100 : 0 : 10
MoS_2	84.41	—	79.55	100 : 0 : 15
Bi_2S_3	84.16	—	79.84	100 : 0 : 20
$FeSo_4$	—	80.95	77.67	0 : 60 : 100
$BaSo_4$	—	80.68	77.63	0 : 45 : 100
Ag_2SO_4	—	80.76	77.69	0 : 50 : 100
$PbSo_4$	—	80.77	77.52	0 : 60 : 100
$HgSO_4$	—	80.90	77.46	0 : 60 : 100

[a] See Fischer and Baun [28].

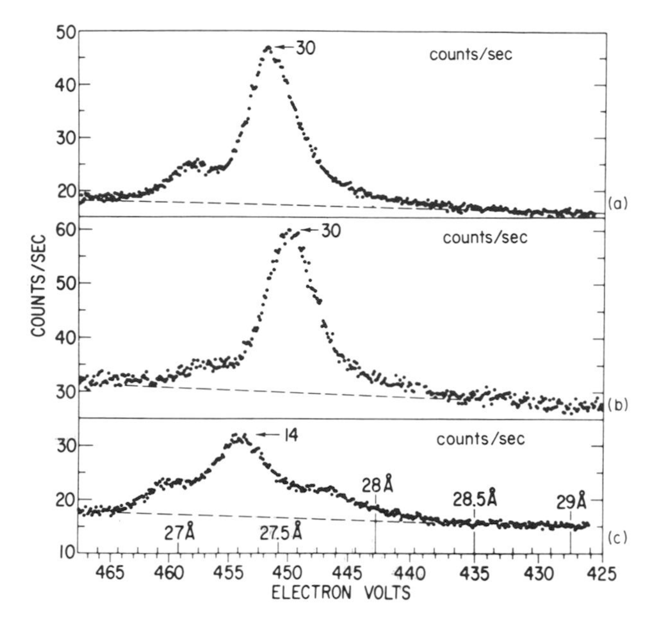

Fig. 9. The Ti $L_{II,III}$ from emission bands (a) Ti, (b) TiC, and (c) TiO (Holliday [19]).

which Fig. 9 was also shown, Holliday analyzes the origin of band features and satellites and discusses the significance of changes in shapes and intensities in Ti L spectra for various compounds [30].

Similar changes take place in the L spectra of other elements of the first transition series and of elements nearby in the periodic table. Fischer [31] has shown that the L_{III} band from oxides of elements from 24 chromium to 30 zinc shifts predictably with respect to the pure element. Shifts are to higher energy in oxides of the transition elements $Z = 24–28$ and to lower energy in Cu and Zn oxides. Intensities were shown to change significantly between the pure elements and oxides as seen in Table VI for 6 kV excitation.

TABLE VI
THE L_{II}/L_{III} BANDS FOR PURE ELEMENTS AND OXIDES AT 6 kV

Element	L_{II}/L_{III}	Element	L_{II}/L_{III}
Mn	0.33	Ni	0.20
MnO_2	0.66	NiO	0.29
Fe	0.17	Cu	0.18
Fe_2O_3	0.49	CuO	0.30
Co	0.16	Zn	0.25
Co_3O_4	0.30	ZnO	0.35

The voltage must be specified since absorption effects cause intensities to vary in both the pure element and the oxides with changes in excitation voltages.

The shape of Cu L_{III} changes significantly with oxidation [32] as shown in Fig. 10 in which uncorrected recorder tracings of Cu L_{III} are presented for Cu, Cu_2O, and CuO. With oxidation, the band shifts to longer wavelengths and the band becomes much narrower. These curves are, of course, normalized to the intensity of L_{III} from the pure metal and do not represent actual intensities at the same power input. The half-width of Cu metal was measured to be 3.5 eV which compares very favorably to the curve shown by Shaw [10], using a two crystal spectrometer.

There are some changes with alloying in L spectra of the first transition elements, but these changes are not as large as in K spectra of low atomic number elements described earlier. In the Al–Ni system [23], the Ni L_{III} band shifts less than 0.20 eV between pure nickel and 10 Ni–90 Al, and L_{II}–L_{III} intensity relationships remain virtually constant. There is a significant change in the shape of the Ni L_{III} band, however, as shown in Table VII.

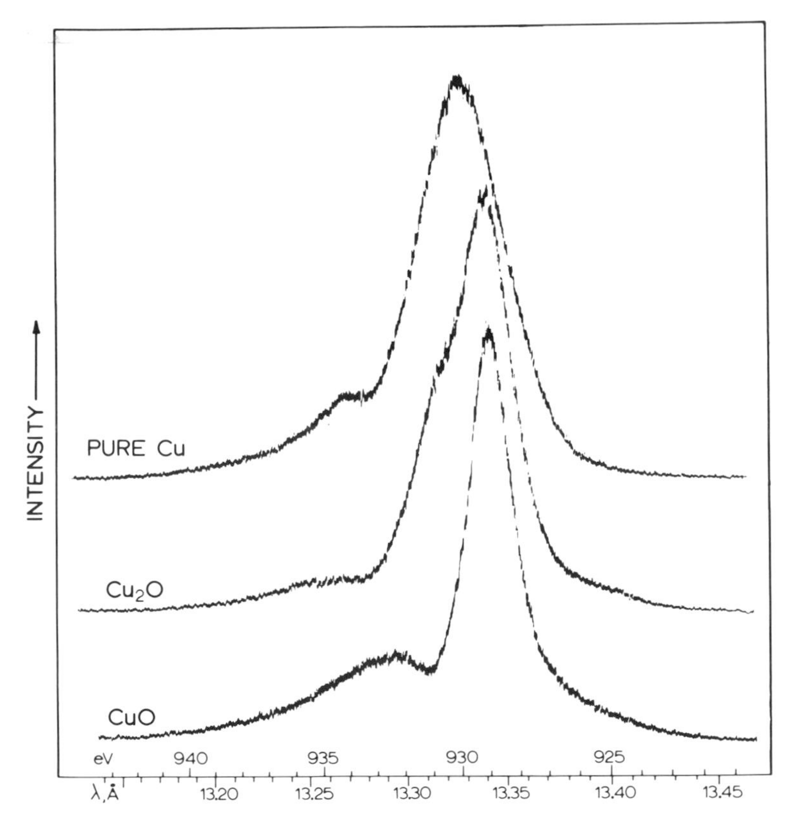

FIG. 10. The Cu L_{III} emission band from Cu, Cu_2O, and CuO.

TABLE VII
UNCORRECTED Ni L_{III} EMISSION BAND CHARACTERISTICS IN Al–Ni SYSTEM[a]

Target	Half-bandwidth (eV)	High energy edge width (eV)	Asymmetry index
100 Ni	2.43 ± 0.08	1.73 ± 0.08	1.74 ± 0.04
80 Ni–20 Al	2.66	1.76	1.72
75 Ni–25 Al	2.60	1.78	1.70
65 Ni–35 Al	2.40	1.86	1.60
60 Ni–40 Al	2.31	1.93	1.50
50 Ni–50 Al	2.34	2.09	1.30
40 Ni–60 Al	2.34	2.25	1.15
35 Ni–65 Al	2.41	2.34	1.08
25 Ni–75 Al	2.54	2.67	1.00
15 Ni–85 Al	2.59	2.94	0.98
10 Ni–90 Al	2.64	3.14	0.95

Throughout the complete alloy series, there are only small changes in the band width at half-maximum, but large changes are seen in the short wavelength edge width and the asymmetry index for the band. The band changes from an asymmetrical band in Ni metal to a completely symmetrical band in 25 Ni–75 Al and then continues to change to a slightly asymmetrical band with the asymmetry on the long wavelength side as compared to short wavelength asymmetry seen in the metal.

Two recent reviews summarize the state of the art of X-ray spectra from alloys [33, 34]. These reviews as well as the papers referenced in them should be consulted for an in-depth explanation of alloy effects. Such an explanation is beyond the space limitations of this review.

C. M Series

A complete review of changes in M spectra due to chemical combination will not be attempted. Very little recent M spectral data exist, especially when the emphasis of the work was on effects of chemical combination. The M bands in the long wavelength region are, in general, quite weak, which probably accounts for the paucity of data on M bands. Furthermore, the majority of M spectra are beyond the wavelength capability of most microbeam probe instruments. It seems unlikely that the microbeam probe could be used for M spectra of elements below about 39 yttrium; and for very high Z elements, there is little effect on M lines due to chemical combination. As an example of the kinds of changes induced in M spectra by chemical combination, Fig. 11 is shown from work of Holliday [19]. Compared to Zr metal, the

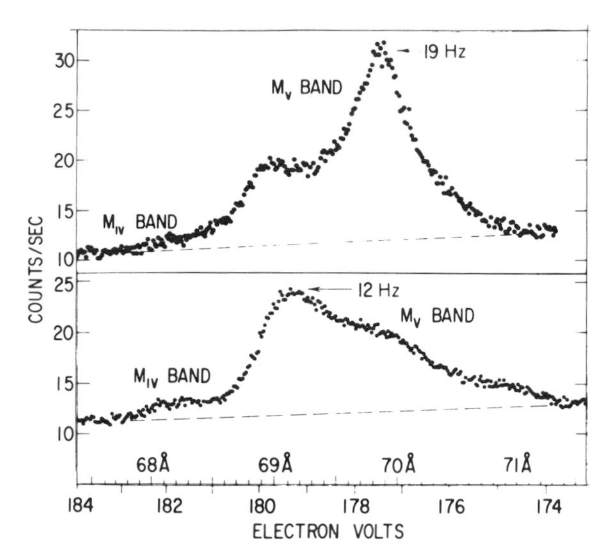

FIG. 11. The Zr M_{IV} and M_V bands from (a) ZrC and (b) Zr (Holliday [19, 30]).

shape of the M_V band is significantly different in ZrC. Holliday [30] has indicated that the bond between Zr and C should be largely covalent. Because of this, the shape of the metal and carbon emission bands should be the same. However, Holliday finds that the C K and Zr M_V bands do not have the same shape and that inner level smearing cannot account for this anomaly. It is evident from just this one example that a better understanding of the relation between bonding and band shape will have to be obtained in order to make theoretical use of such data.

III. APPLICATIONS

Certainly, the full potential of the effect of chemical combination on X-ray spectra has not yet been realized. Probably the greatest potential use of the technique lies with poorly crystalline and glassy materials that are difficult to characterize by conventional techniques. Other possible uses include characterization of thin films, oxidation and reduction phenomena, and valence and coordination determination. In the author's laboratory, the effect of chemical combination on X rays has been used to identify anodized films, for compound identification in diffusion couples, for Al_2O_3 particles in coatings, to determine valency of iron in iron-ore with oxides, and in hypervelocity impact crater studies.

One example of the practical application of the chemical effect on X-ray spectra is illustrated in Fig. 12. In this work by White and Roy [35] the authors set out to find out whether "SiO" is short range ordered silicon monoxide,

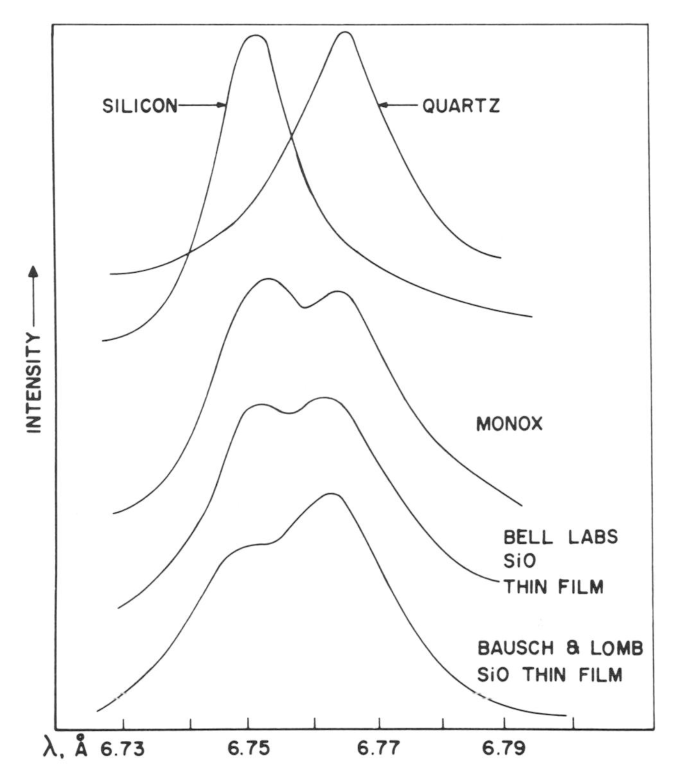

FIG. 12. The Si K emission band from silicon, quartz, and three "SiO" samples (White and Roy [35]).

or whether it is an intimate mixture of Si + SiO_2. An electron microbeam probe was used to obtain the K emission band from pure Si and SiO_2 and three "SiO" samples, two of which were very thin ~ 1 μ films on plastic substrates (bottom two curves). The results of the study are obvious from only a glance at the figure. The "SiO" consisted of a mixture of Si and SiO_2. Furthermore, some indication of relative amounts of both components may be made using the K band. This measurement may be made with such ease that Si metal to Si^{4+} ratios in thin films could be determined routinely as a function of deposition conditions or treatment.

REFERENCES

1. R. Swinne, *Phys. Z.* **17**, 481 (1916).
2. J. Bergengren, *Z. Physik* **3**, 247 (1920).
3. A. E. Lindh and O. Lundquist, *Arkiv. Mat. Astron. Fysik* **18** (14), 3 (1924).
4. M. Siegbahn, "Spektroskopie der Röntgenstrahlen." Springer, Berlin, 1931.

5. H. Herglotz, "Einflüsse der Bindung auf das Rontgenspektrum" ("Influence of Bonding on the X-Ray Spectrum"). *Transl. Documentation Center Technol. Econ.*, Vol. 13, Vienna ,1955, in German).
6. A. Faessler, 1508—Röntgenspektrum und Bindungszustand, *in* "Landolt–Bornstein Tables," Zahlenwerte und Funktionen, 6th ed., Vol. I, Pt. 4, pp. 769–808, 1955.
7. A. Sandström, Experimental methods of X-ray spectroscopy: ordinary wavelengths *in* "Encyclopedia of X-Rays," Vol. XXX, pp. 78–245. Springer, Berlin, 1957.
8. D. H. Tomboulian, The experimental methods of soft X-ray, *in* "Encyclopedia of X-rays," Vol. XXX, pp. 246–304. Springer, Berlin, 1957.
9. M. A. Blokhin, "The Physics of X-Rays," 2nd ed. State Publ. House of Tech.-Theoret. Lit., Moscow, 1957.
10. C. H. Shaw, Theory of alloy phases, *in* "The X-Ray Spectroscopy of Solids," pp. 13–62. *Am. Soc. Metals*, Cleveland, 1956.
11. H. Yakowitz and J. R. Cuthill, "Annotated Bibliography on Soft X-Ray Spectroscopy," *Nat. Bur. Std.* (*U.S.*) *Monograph* **52** (1962).
12. *Conf. Applications X-Ray Spectry. to Solid State Problems.* Office of Naval Res. NP-4287 NAVEXOS, p-1033, ATI2O4728 (1950).
13. *Intern. Conf. Physics X-Ray Spectra, Cornell Univ., Ithaca, New York, 1965*; abstracts appear in *Bull. Am. Phys. Soc.* **10** (1965).
14. *Röntgenspectren und Chemische Bindung, Karl Marx Univ., Leipzig, 1965.* Leipzig, 1966.
15. *Advan. X-Ray Anal.* **9** (1966).
16. A. Fahlman, K. Hamrin, J. Hedman, R. Nordberg, C. Nordling and K. Siegbahn, *Nature* **210**, 4 (1966).
16a. V. H. Sanner, Dissertation, Uppsala Univ., Sweden (1941).
17. D. W. Fischer and W. L. Baun, *J. Appl. Phys.* **37**, 768 (1966).
18. D. W. Fischer and W. L. Baun, *J. Chem. Phys.* **43**, 2075 (1965).
19. J. E. Holliday, *in* "Handbook of X-Rays," Chap. 38. McGraw-Hill, New York (1968).
20. D. W. Fischer, *J. Chem. Phys.* **42**, 3814 (1965).
21. R. A. Mattson and R. C. Ehlert, *Advan. X-Ray Anal.* **9**, 471 (1966).
22. W. L. Baun and D. W. Fischer, The effect of chemical combination on K X-ray emission spectra from magnesium, aluminum, and silicon." Air Force Mater. Lab. Tech. Rept. 64–350, 1964.
23. D. W. Fischer and W. L. Baun, *Phys. Rev.* **145**, 555 (1966).
24. D. W. Fischer and W. L. Baun, *Advan. X-Ray Anal.* **10** (1967).
25. E. Schnell, *Monatsh. Chemie* **93**, 1383 (1962).
26. E. Schnell, *Monatsh. Chemie* **94**, 703 (1963).
27. D. W. Wilbur, Relationship between chemical bonding and the X-ray spectrum: studies with the sulfur atom. UCRL-14379, TID-4500 AEC Contract No. W-7405-eng.-48.
28. D. W. Fischer and W. L. Baun, *Anal. Chem.* **37**, 902 (1965).
29. L. Pauling, "The Nature of the Chemical Bond, " 3rd ed. Cornell Univ. Press, Ithaca, New York, 1960.
30. J. E. Holliday, *Advan. X-Ray Anal.* **9**, 365 (1966).
31. D. W. Fischer, *J. Appl. Phys.* **36**, 2048 (1965).
32. D. W. Fischer and W. L. Baun, unpublished results.
33. A. Appleton, *Contemp. Phys.* **6**, 50 (1964).
34. B. J. Thompson and P. K. Kellen, *Develop. Appl. Spectr.* **4**, 23–32 (1965).
35. E. W. White and R. Roy, *Solid State Commun.* **2**, 151 (1964).

Backscattered and Secondary Electron Emission as Ancillary Techniques in Electron Probe Analysis

J. W. COLBY

Bell Telephone Laboratories, Inc.
Allentown, Pennsylvania

I. Introduction

In the electron probe X-ray microanalyzer, the accelerated beam of electrons impinging on the surface of the specimen gives rise to a number of signals such as characteristic X rays, cathodoluminescence in the visible and near visible spectrum, backscattered and secondary electrons, and specimen current (absorbed electrons). Some or all of these signals may be used simultaneously to provide information concerning the sample. In this article, we shall be most concerned with the various electron signals since virtually all commercial electron probes have the capability to detect backscattered and transmitted electron signals, and some, to detect secondary electrons as well.

In fact, it has become rather common practice to obtain either backscattered electron (BSE) or specimen current (SC) images of the sample being analyzed as a reference. This is accomplished quite simply by causing the electron beam in the electron probe to scan in a raster over the specimen surface, and to use either the resultant BSE or SC signal after suitable amplification to modulate the intensity of a cathode ray tube whose electron beam is being swept in unison with the electron beam in the electron probe.

The images thus obtained are a function of mean atomic number and surface topography and so, may be analogous to optical micrographs.

Less understood, but perhaps more useful, are secondary electron (SE) signals. It has been difficult to separate true secondary electrons from backscattered electrons in the electron probe primarily because of the geometrical arrangement of lens pole pieces, beam scanning plates, etc., and only recently have SE images been obtained.

It has also been proposed that quantitative electron probe analyses using only specimen current measurements is possible, at least in some instruments. Such analyses are quite simple and can, under proper conditions, be quite accurate.

In this article, very little attention will be given to BSE or SC images, as these are already widely used by microprobe analysts, and are recently adequately discussed [1]. Rather, the distribution and properties of true secondary electrons and backscattered electrons will be discussed. Methods of detecting the two signals independently of each other and the uses of each will be covered in detail. Quantitative analysis by means of specimen current measurements will also be discussed with its advantages and limitations.

II. Qualitative Considerations

If the number of electrons having the energy E, emitted from a target during electron bombardment is plotted against the energy, the resulting curve [2] would appear as in Fig. 1. There is a group of electrons on the right-hand side of Fig. 1 whose energies are the same as that of the primary electrons. These are elastically reflected primaries. A second group of electrons having energies intermediate between 50 eV and the primary electron energies, are inelastically scattered, or rediffused primaries. A third group of electrons

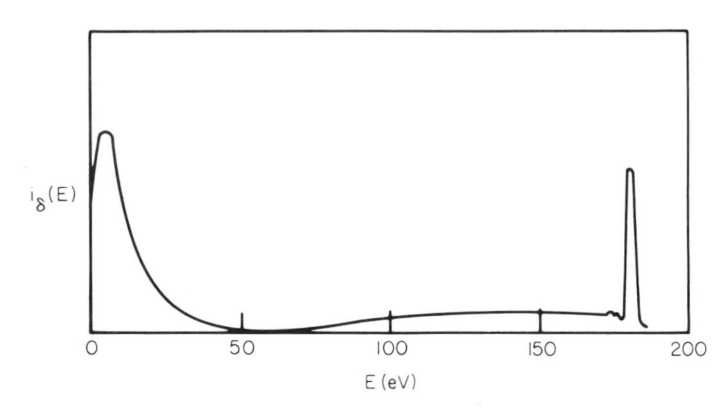

FIG. 1. General shape of the energy distribution of secondary and backscattered electrons.

appears at the left-hand side of Fig. 1, having energies less than 50 eV and characterized by a sharp maximum in the distribution at approximately 10 eV. These are the true secondary electrons. In principle, there may be rediffused primaries or inelastically scattered electrons having energies less than 50 eV, and true secondaries having energies greater than 50 eV. However, it has become common practice to separate the two groups at 50 eV.

The backscattered electron fraction is thus composed of elastically reflected primary electrons and inelastically scattered or multiply scattered electrons. The backscatter yield (number of electrons backscattered per incident primary electron) is related to the accelerating potential, and is proportional to the mean atomic number of the target. At low accelerating potential (i.e., less than 30 keV) the BSE yield increases with increasing primary energy for heavy elements [3], but decreases with increasing primary energy for light elements. Above 30 keV, the yield is virtually independent of accelerating potential. Archard [4] has shown that the predominant mechanism for backscattering of electrons in heavy elements is diffusion while for light elements large-angle single elastic scattering predominates. The backscattered electron yields at 5 and 30 keV as obtained by Bishop [5], are shown plotted in Fig. 2.

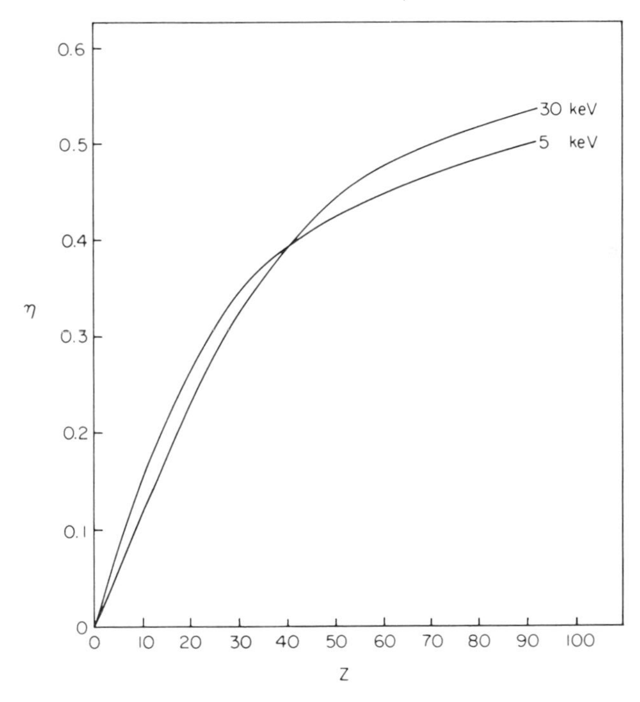

FIG. 2. Backscattered electron yield versus atomic number (after Bishop [5]).

Secondary electron emission is more complex than backscattered electron emission. The true secondary electron yield is related to the primary energy of the incident electrons, to the work function of the surface [6], to the number of outer shell electrons [7], and to the atomic radius [8]. Although several models have been proposed, no completely adequate theory of secondary electron emission exists.

The total secondary electron yield δ is defined as the ratio of the number of electrons emitted per incident primary electron, and contains both the backscattered electron yield η and the true secondary electron yield Δ. If δ is plotted as a function of the primary incident electron energy a characteristic yield curve is obtained as shown in Fig. 3. The shape of the curve is

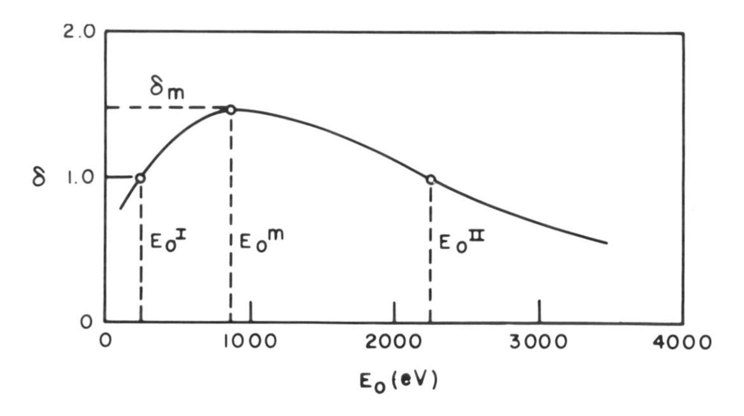

FIG. 3. General shape of the yield curve.

similar for all materials; however, the various parameters shown in the figure may vary considerably. In this figure, δ_m is the maximum yield, $E_0{}^m$ the incident energy at which the maximum yield occurs, and $E_0{}^{\text{I}}$ and E_0^{II} the incident energies for which the yield is unity. The maximum yields for metals are between 0.5 and 1.0 for the alkalis, alkaline earths, and other light metals ($Z < 23$), and from 1.1 to 1.7 for heavier metals.

Baroody [9] normalized the curves for different metals by plotting δ/δ_m as a function of $E_0/E_0{}^m$, and obtained the "universal" curve shown in Fig. 4. There have been numerous attempts to theoretically derive expressions for this curve, with limited success. The theoretical curve calculated by Lye and Dekker [10] is shown plotted in Fig. 4 as the solid line. There, it is seen that good agreement exists between experiment and theory up to $E_0/E_0{}^m \cong 4$. The expression of Lye and Dekker [10] was used to calculate the true secondary electron yield for gold at 5 keV. The value of 0.21 calculated [11] was in good agreement with the experimentally measured value of 0.19. However, at 20 keV, δ was found to be of the order of 0.01.

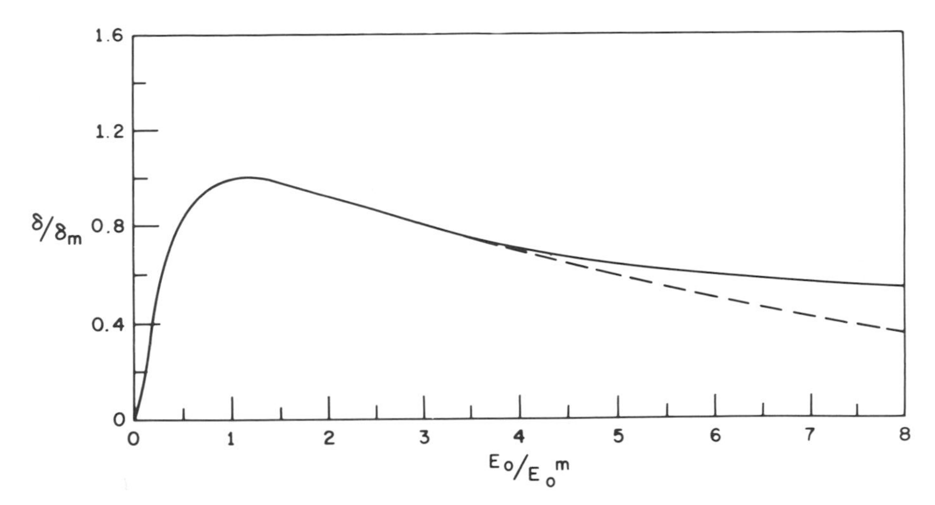

FIG. 4. Universal yield curves (theoretical, —; and experimental, — — —;).

Thus it might seem that in the electron probe microanalyzer, where analyses are commonly performed at accelerating potentials of from 20 to 30 keV, secondary electron emission is neither useful nor important. This is not the case, however. Due to the geometrical relationships which must be observed, electron probes are constructed in such a fashion that various parts of the instrument are in close proximity to the target. Thus electrons backscattered from the target may in turn strike various parts of the electron probe final lens specimen chamber. These backscattered electrons, having lost part of their energy, may produce a considerable number of secondaries, which subsequently may be collected by the specimen. Kanter [12] and more recently Seiler [8] have shown that backscattered electrons produce from 4 to 6 times more secondary electrons on the average than do primaries.

The complexity of the possible electron currents in the electron probe microanalyzer has made backscattered electron and secondary electron measurements somewhat less than attractive, and has led to large variations in reported backscatter yields. Three investigators [11, 13, 14] have reported attempting to measure secondary electron yields in the electron probe, and their results were widely separated. Some of these investigations and techniques and their results will now be discussed.

III. EXPERIMENTAL TECHNIQUES

In general, three different approaches have been taken in electron probe analysis to determine BSE yields. The first method was to simply measure the specimen current (the current flowing between the sample and ground) for various elements, and the beam current as collected in a Faraday cage [15]. The backscatter yield was then obtained from

$$(i_B - i_{SC})/i_B \tag{1}$$

where i_B is the current, and i_{SC} the specimen current.

The resulting backscatter yields when plotted as a function of atomic number yielded a smooth curve in fair agreement with published data. As will be seen this method is a gross oversimplification, and in some instances could cause serious errors.

A second method employed in the electron probe is the grid method [13, 16–18] in which a grid is introduced above the specimen and biased so as to retard secondary electron emission from the sample.

The third method employed is the composite target method [11, 19, 20] in which a small central target area is electrically isolated from the massive peripheral area. Both the target and the periphery can be independently biased and their respective currents measured.

The latter two methods will now be discussed in more detail. The first method described above will not be discussed as it should become apparent that the effects of low energy electrons neglected in the first method is considerable and consequently should not be ignored.

A. The Grid Method

Several investigators, notably Philibert and Weinryb [16, 17] Weinryb [13], and Burkhalter [18] have used the grid method. In this method, a fine mesh grid having approximately 80% electron transmission is inserted a few millimeters above the sample and between the sample and lens pole piece, as in Fig. 5. A negative bias of up to 450 V is applied to the grid. This has the effect of retarding secondary electron emission from the sample and from the lens pole piece (or screen). However, the primary electrons, and the backscattered electrons from the sample surface and from the lens, strike the grid, causing it to emit a large number of secondaries. This latter current, which can be appreciable, must be corrected for if meaningful results are to be obtained. However, Weinryb [13] assumes that at 30 keV primary energy, and with sufficient negative bias on the grid (−20 V), secondary emission from the sample and the screen are eliminated, and that the electrons emitted by the grid and collected on the sample are negligible, hence, the backscatter yield is measured directly. It is doubtful that this is true. If we write

$$i_{SC} = i_B - i_S - i_\eta + i_g \tag{2}$$

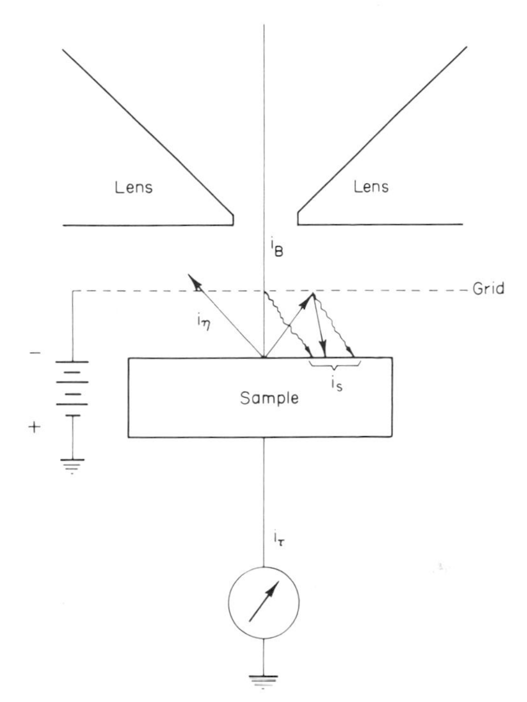

FIG. 5. Arrangement of specimen and grid, showing various currents.

where i_S is the secondary electron current from the specimen, i_η the backscattered electron current from the specimen, and i_g the electron current emitted from the grid and collected on the specimen, then

$$\delta = \Delta + \eta - i_g/i_B \tag{3}$$

where,

$$\delta = 1 - i_{SC}/i_B, \qquad \Delta = i_S/i_B, \qquad \eta = i_\eta/i_B.$$

With sufficient negative grid bias, secondary electron emission from the specimen is retarded, i.e., $\Delta \cong 0$, hence

$$\delta = \eta - i_g/i_B. \tag{4}$$

If the current from the grid were negligible, then indeed the backscattered electron yield would be simply given by δ. However, primary electrons striking the grid would cause some secondaries to be emitted (although probably not a significant number), and backscattered electrons from the sample striking the grid would cause secondary electrons to be emitted from the grid. These backscattered electrons having lost part of their energy would be much more effective in producing secondary electron emission from the grid [8, 12], and since the specimen is positive relative to the grid, they would be collected on it. In addition, there may be a small fraction of "doubly backscattered" electrons (backscattered first from the specimen, then from the grid back onto the specimen) collected on the specimen. Thus, it is not safe to assume the current emitted from the grid and collected on the specimen is negligible, so that the backscatter coefficient should be properly determined as

$$\eta = \delta + i_g/i_B. \quad (5)$$

The effect of neglecting the grid current is to lower the observed value of the backscatter yield. This is, in fact, supported by a comparison of yields obtained by the grid method with the backscattered electron yields measured by Bishop [5], who used a large scattering chamber, and biased the specimen to effectively eliminate all stray currents, and so measured the true backscatter yield. As noted by Bishop [5], the results of Weinryb [13], at higher energies are low by approximately 15%. However, the results obtained by Weinryb [13] at low primary energy (5 keV), to which he applied a grid correction, are in good agreement with those of Bishop [5].

Burkhalter [18], using a similar approach, but with a 30° inclined specimen obtained results which uncorrected for grid current agree quite well with those of Philibert and Weinryb [16]. After correction for grid current the results are more in agreement with those of Bishop [5]. Burkhalter [18] corrected his results for specimen tilt using the method outlined by Shimizu and Shinoda [21]. The backscattered electron yields obtained by Burkhalter [18] for specimens inclined 30° are higher than those obtained for normal beam incidence, in qualitative agreement with Weinryb [13]. The difference between yields obtained on normal and inclined specimens is greater for lower atomic number elements. These results may be explained simply by noting that in the case of the inclined specimen, the electron penetration normal to the surface is less than for the normal specimen. Consequently, the average electron for a similar number of scattering acts is closer to the surface at all times in the case of the inclined specimen and hence the probability of electrons being scattered out of the specimen is greater, especially in low atomic number elements where the large-angle single elastic scattering mechanism predominates [4]. In high atomic number elements where a diffusion mechanism predominates [4], complete diffusion of the primary electrons occurs rapidly; consequently, the probability of electrons emerging from the specimen is not appreciably affected by small changes in the angle of incidence.

Burkhalter [18] estimates an accuracy of 2% in his backscattered electron yields. However, this assumes a secondary electron yield for the carbon-coated grid of 0.1, and although this is not an unreasonable estimate for 25 keV primary electrons, it would be appreciably higher and more uncertain for lower energy electrons. Remembering that the backscattered electrons striking the grid may have lost appreciable energy in being scattered from the target, hence may have, on the average, energies considerably lower than 25 keV, the secondary electron yield of the grid could be as high as 1.0 (for 250 eV electrons) [2] introducing an error of 20 instead of 2%. While this is obviously not the case in the measurements of Burkhalter, it does serve to illustrate the magnitude of the errors that can occur in the grid method by uncertainties in estimating the current due to the grid.

The composite specimen method of Heinrich [19, 20] is simpler than the suppression grid method, allows a more complete understanding of the electron currents in the vicinity of the sample, and circumvents the errors inherent in the grid method.

B. The Composite Target

To aid in understanding why quantitative analysis by means of sample current measurements alone was possible, and to illustrate why discrepancies may exist in backscattered electron yield measurements made in the electron probe analyzer, Heinrich [19] devised the composite target shown in Fig. 6. The composite specimen has a small central target, electrically isolated from the massive periphery. Each portion may be independently biased so as to separate the various currents, also shown in Fig. 6. Later this device was

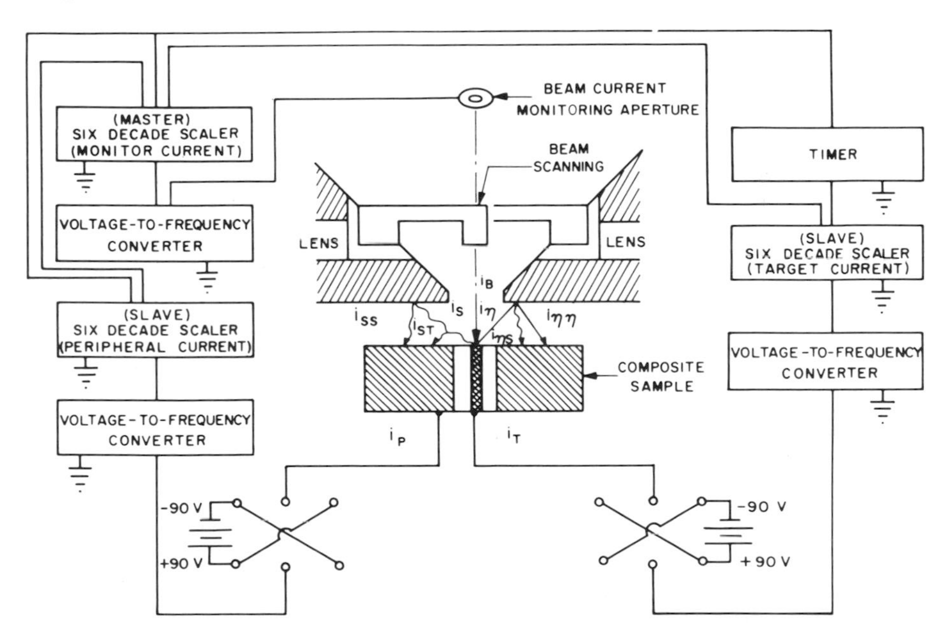

FIG. 6. Composite target showing possible currents and how they may be separated.

modified to contain 4 small central targets [20], Al, Cu, Ag, and Au. In this case, the specimen current is given by

$$i_{SC} = i_B - i_\eta - i_S + i_{\eta\eta} + i_{\eta S} + i_{ST} + i_{SS} \tag{6}$$

or

$$\delta = \eta + \Delta - [(i_{\eta\eta} + i_{\eta S} + i_{ST} + i_{SS})/i_B] \tag{7}$$

where $i_{\eta\eta}$ is the doubly backscattered electron current, $i_{\eta S}$ the secondary electron current from the lens due to backscattered electrons from the target, i_{ST} the secondary electron current due to secondaries from the target being refocused onto the target, and i_{SS} the secondary electron current due to secondaries released from the lens as a result of being struck by secondaries from the target.

In these experiments [19] the central specimen was biased from −80 V to +80 V with no bias applied to the periphery, and the peripheral current measured as a function of target bias. The experiment was conducted at an accelerating voltage of 30 keV. If secondary electrons are emitted from the central target, then at a high negative central bias (secondary electrons expelled from the target) the peripheral current should be high, whereas with a high positive central bias (secondary electrons prevented from leaving), the peripheral current should be quite a bit lower. However, it was found that the peripheral current was unchanged with central bias, and that i_{ST} and i_{SS} were negligible, i.e., there was virtually no secondary electron emission at 30 keV from the specimen.

Now if the central specimen is grounded and the periphery is biased from −80 V to +80 V, and again the peripheral current is measured, then under these conditions $i_{\eta\eta}$ should remain constant, while $i_{\eta S}$ should increase with increasing positive bias. At negative peripheral bias, the peripheral current was essentially 0, and it was concluded that $i_{\eta\eta}$ was also negligible [19].

Heinrich [20] later repeated the above experiments on a different instrument (of similar manufacture) and obtained essentially the same results. Colby *et al.* [11] using two similar instruments repeated Heinrich's [19, 20] measurements with essentially the same results, only small differences in the magnitude of the currents being noted. Wittry [14], however, using an instrument similar to those employed by Henrich [19, 20] and Colby *et al.* [11] found that peripheral current increased with negative central target bias to a degree which depended on whether the beam scanning plates were on or off. Neither Heinrich [20] nor Colby *et al.* [11] were able to confirm any effects due to the deflection plates being on or off, and although this is a significant difference, no explanation for the difference is known. The difference is important because the results of Wittry's [14] measurements indicate that there is an appreciable secondary electron yield from the sample at 30 keV, while the results obtained in the 4 instruments employed by Heinrich [19, 20] and Colby *et al.* [11] indicate negligible secondary electron yields from the sample at 30 keV.

From Henrich's [19, 20] results as verified by Colby *et al.* [11] it is evident that at 30 keV, $i_{\eta\eta}$, i_{ST}, and i_{SS} are essentially zero, the secondary electron yield of the target Δ is also zero, and therefore Eq. (7) may be reduced to

$$\delta = \eta - i_{\eta S}/i_B \tag{8}$$

It is also clear that $i_{\eta S}/i_B = i_P/i_B$ where i_P is the peripheral current measured in the composite specimen. The peripheral current as a function of peripheral bias for the composite specimen is shown in Fig. 7 [11], where it may be noted that this current is quite large, and that any direct measurement of backscatter yield without correction for this current would result in appreciable error. It should also be obvious that although the 30 keV primary

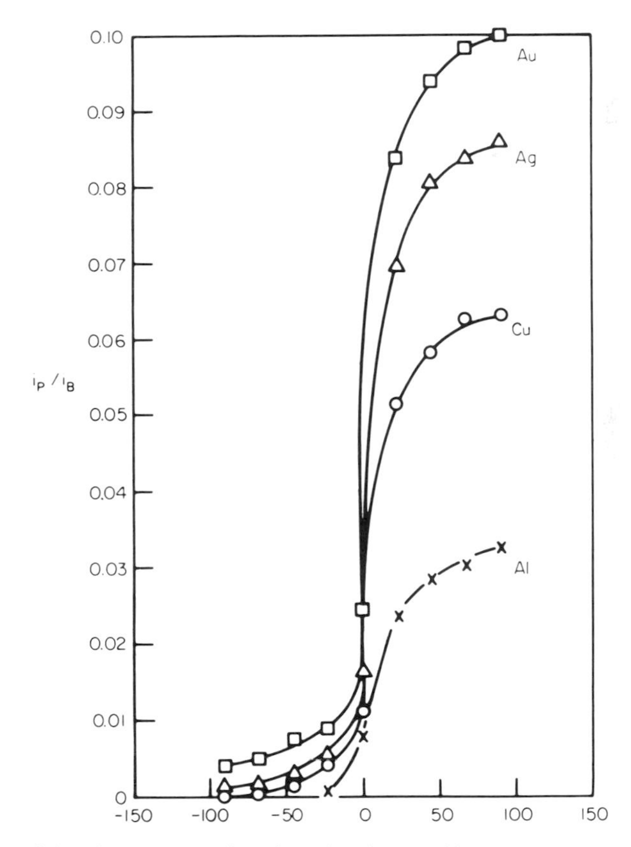

FIG. 7. Peripheral current as a function of peripheral bias (volts) at 30 keV (composite target).

electrons do not produce any appreciable secondaries from the sample, the backscattered electrons, many of them having lost considerable energy during scattering from the sample, are quite effective in producing secondaries from the lens. As a consequence, it would also be expected that a negatively biased grid positioned above the specimen would emit a considerable secondary electron current which would be captured by the sample, lowering the total yield, and hence the backscattered electron yield, unless a suitable correction is made.

Heinrich [20] noted that at 30 keV, peripheral current was linearly related to the backscatter yield of the central target. This was confirmed by Colby *et al.* [11] for 30 and 40 keV electrons. However, at lower primary energies the peripheral current deviated considerably from a linear relationship as shown in Fig. 8 [11]. This deviation is caused by enhanced secondary electron

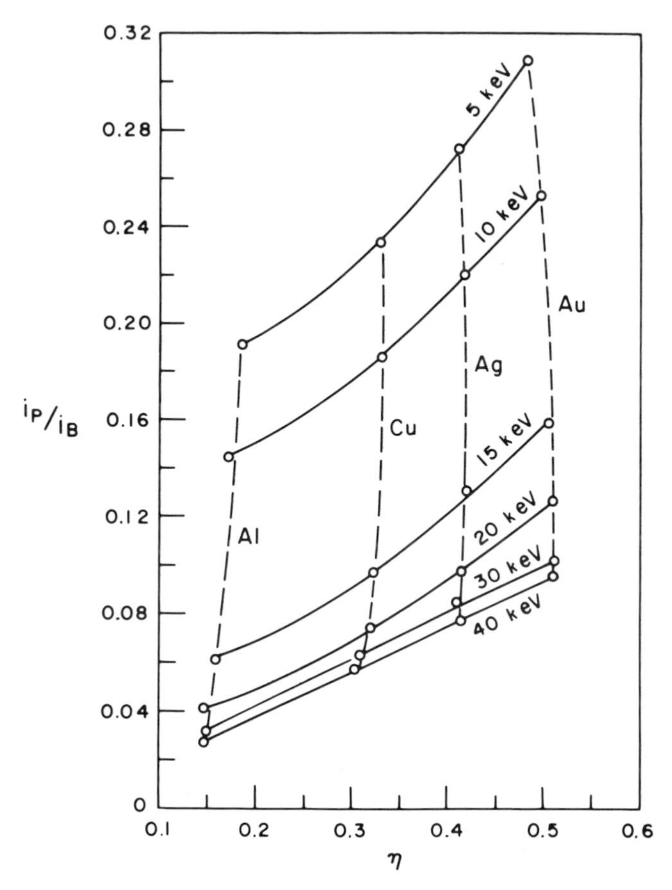

FIG. 8. Peripheral current as a function of backscattered yield and accelerating potential at a peripheral bias of +90 V.

emission from the lens due to the lower energies of the electrons backscattered from the target and striking the lens.

Colby *et al.* [11] used the composite specimen to determine secondary electron and backscattered electron yields by measuring the target current first with a negative peripheral bias, and then with a positive peripheral bias at primary electron energies of from 5 to 40 keV. The results are shown in Fig. 9 [11]. With a positive peripheral bias, secondary as well as backscattered

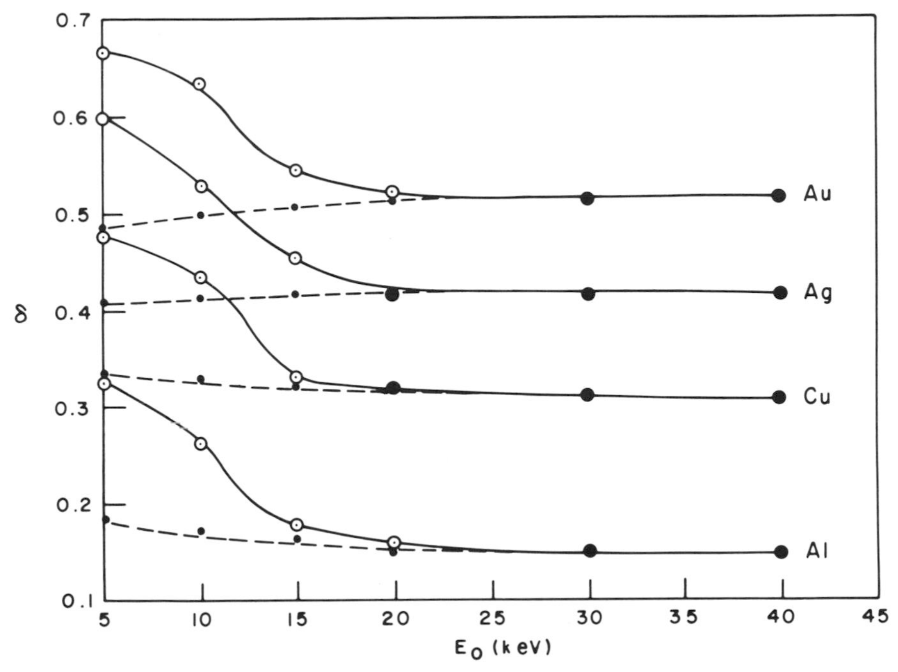

FIG. 9. Electron yields as a function of accelerating potential for the composite specimen (—, +90 V peripheral bias; — — —, −90 V peripheral bias).

electrons are emitted from the target, the former being collected on the periphery. Consequently, the solid curves in Fig. 8 [11] are the total yield (backscatter + secondary). With a negative peripheral bias, a field is set up so as to suppress the emission of secondaries from the target. Thus the dashed curves are the backscattered electron yields. Backscattered electron yields obtained in this way are compared with results obtained by other investigators in Table I [11]. At 30 keV, the results of Bishop [5], Henrich [20], and Colby *et al.* [11] are in excellent agreement, while those of Wittry [14] are consistently low. The results of Philibert and Weinryb [16, 17] and Weinryb [13] although not shown in the table are even lower than Wittry's [14]. At 10 keV the results obtained by Heinrich [20], Bishop [5], and Colby *et al.* [11] are in good

TABLE I
ELECTRON BACKSCATTER COEFFICIENTS

	Al	Cu	Ag	Au
5 keV				
Heinrich	—	—	—	—
Bishop	0.186	0.352	0.418	0.489
Wittry	0.168	0.298	0.361	0.432
Colby	0.185	0.328	0.407	0.482
10 keV				
Heinrich	0.148	0.299	0.428	0.502
Bishop	0.177	0.339	0.420	0.501
Wittry	—	—	—	—
Colby	0.172	0.329	0.416	0.496
30 keV				
Heinrich	0.148	0.306	0.412	0.505
Bishop	0.155	0.319	0.420	0.521
Wittry	0.135	0.291	0.388	0.481
Colby	0.150	0.311	0.411	0.513

agreement for the heavy elements, but the values of Heinrich are slightly lower for the lighter elements. At 5 keV, the results of Bishop [5] and Colby *et al.* [11] are still in good agreement, while those of Wittry [14] are again appreciably lower. As noted earlier, the results obtained by Weinryb [13] are in good agreement with those obtained by Bishop [5] and by Colby *et al.* [11] at 5 keV.

Secondary electron yields obtained at 5 keV are 0.19, 0.20, 0.15, and 0.14 for Au, Ag, Cu, and Al, respectively [11]. These were in good agreement with theoretical values calculated by means of an expression of Lye and Dekker [10, 11]. They are also in good agreement with those given by Seiler [8] for secondary yields due to 4 keV primaries.

Colby *et al.* [11] showed that there was an inherent error of $\sim 2\%$ in Wittry's [14] backscattered electron yield measurements at 30 keV, which would tend to bring his results into agreement with those of Bishop [5], Heinrich [20] and Colby *et al.* [11].

With a more complete understanding of the various electron-yields and currents in the electron probe it is possible to discuss the uses to which these phenomena may be applied.

IV. QUANTITATIVE ANALYSIS

Poole and Thomas [15] proposed that quantitative analysis by means of sample current measurements alone was possible, and suggested three relationships which might be applied. They tested these three equations using binary samples, and although their results were not conclusive, the equation proposed by Castaing [22] produced the best results. Philibert and Weinryb [16] concluded, however, that precise quantitative analysis was not possible using only sample current measurements.

Heinrich [19] repeated Poole and Thomas' [15] experiments and found, as they had, that the equation proposed by Castaing [22] gave the best results, and in most cases agreement between composition and sample current analyses was better than 1%. The method of sample current analysis was further substantiated by Colby [23] who analyzed six binary uranium alloys at various accelerating voltages and found that at 35 keV the sample current analyses agreed with the reported compositions to within 1%. At lower accelerating voltages however, the error increased due to the effects of low energy electrons.

Heinrich [19, 20] showed that since the peripheral current (secondary electron current from the lens) was linearly related to the backscattered electron yield of the sample (the number of electrons emitted from the lens is directly proportional to the number of electrons striking the lens) Castaing's [22] equation

$$\eta_{\text{alloy}} = \sum c_i \eta_i$$

could be reduced to

$$C_{\text{A}} = (i_{\text{B}} - i_{\text{alloy}})/(i_{\text{B}} - i_{\text{A}})$$

where i_{B} is the specimen current from pure B, i_{A} the specimen current from pure A, and i_{alloy} the specimen current from alloy.

Weinryb [13] contends, however, that the backscattered electron yield is not a linear function of mean atomic number, and that in the case of two elements widely separated in atomic number (e.g. U and Si), the backscatter would originate from two different mechanisms (large-angle single elastic scattering for Si and diffusion for U), and therefore it would not be possible to obtain a simple expression relating the backscatter yield to composition. Weinryb [13] made sample current measurements on several binary alloys and concluded that it was not possible to differentiate explicitly between Fe_3P and Fe_2P, UCu_5 and UCu_6, etc. His conclusions, however, are unjustifiable for several reasons. First, it should be possible to measure sample currents with a precision of 0.001 [11, 18–20]. Since there is a difference in sample currents (or backscatter yields) of $\sim$0.1 between Fe and P, it should be possible to detect compositional differences of as small as 1%. Second, the first 5 alloys used (Fe–Al) by Weinryb [13] do not have a clearly defined composition, but only a range, so it is difficult to see how he can justifiably say the sample current measurements do not agree with composition. Third, although Weinryb was unable to differentiate between Fe_3P and Fe_2P, from his own data, the specimen current measurements he reported give a mean atomic number differing by less than 1% (relative) from the theoretical mean atomic number (mass concentration). This is, in general, better than could be expected of X-ray measurements due to uncertainties in correction techniques. Fourth, in most of the other binaries analyzed by Weinryb [13], the mean atomic number deduced from sample current measurements is consistently lower by $\sim$4.5% than the theoretical mean atomic number calculated on a mass concentration basis. This suggests a systematic error in one of the measurements, either the yields for the binaries or the yields for the pure elements. Such an error might be caused by different sized sample mounts in the two cases. Finally, in the case of the uranium–copper binary (UCu_5) where agreement is admittedly poor, it has been shown [24] that UCu_5 may vary in composition from $UCu_{4.7}$ to $UCu_{5.3}$.

Consequently, Weinryb's [25] conclusion, that quantitative analysis by means of target currents is impractical, is unacceptable.

Direct proof of the linearity between sample composition and backscatter yield was given by Bishop [5] who measured the backscatter yield of several copper–gold alloys using a specially designed chamber to eliminate secondary emission from the specimen and from the chamber. He found that at 30 keV, the backscatter yield was quite linear with composition, but that considerable deviation from linearity occurred at lower accelerating potentials.

Colby *et al.* [11] demonstrated that at 30 keV specimen current measurements at the 1% level were indeed possible by analyzing a series of Cu–Au alloys. Concentrations of as little as 1% Au were easily detected, and it was clearly possible to distinguish between alloys containing 1, 2, and 3% Au.

These results [11] and the earlier results of Heinrich [19] and Colby [23] are summarized in Table II.

As pointed out by Heinrich [19], sample currents analyses are most useful when there is a large difference in atomic number between the two elements in the binary alloy. Since this is the condition for which X-ray analyses are most likely to be in error due to uncertainities in correction procedures for atomic number effects, the two methods should complement each other. In fact, if it is accepted that specimen current analyses are accurate, such analyses performed simultaneously with X-ray measurements would serve to ascertain whether differences between chemical composition and corrected X-ray measurements were due to errors in the chemical composition or to errors in the X-ray results.

V. Scanning Images

Scanning images obtained in the electron probe microanalyzer may serve as valuable references for subsequent analyses. Backscattered electron and

TABLE II
QUANTITATIVE ANALYSIS OF BINARY ALLOYS BY SPECIMEN CURRENT MEASUREMENTS

Alloy	Investigator	Mass concentration of heavier element	
		Theoretical	Measured
GaP	Heinrich (30 keV)	0.692	0.688
UB_4		0.846	0.853
UB_{12}		0.647	0.679
US_{13}		0.739	0.736
ZrB_2		0.808	0.807
ZrB_{12}		0.413	0.458
U_3Si	Colby (35 keV)	0.962	0.956
UP		0.885	0.884
UC_2		0.908	0.912
US		0.881	0.883
UN		0.944	0.936
U_6Fe		0.962	0.957
Au-Cu	Colby *et al.* (30 keV)	0.011	0.009
		0.020	0.019
		0.031	0.026
		0.050	0.056
		0.093	0.098
		0.524	0.519
		0.888	0.909
		0.951	0.963
		0.968	0.985
		0.979	0.989
		0.989	0.991

specimen current (absorbed electron) images are easily obtainable in most commercial electron probes; however, only a few are equipped at present to obtain secondary electron images. These images are obtained by electrostatistically or electromagnetically deflecting the focused electron beam in the electron probe so as to scan in a raster over the specimen surface. The same signal used to deflect the beam in the electron probe analyzer is applied to a cathode ray tube causing its electron beam to be deflected in synchronism with the electron probe. The intensity of the cathode ray tube is modulated by the backscattered electron, specimen current, or secondary electron signal, and the resultant image is related to the mean atomic number and/or the topography of the sample. The cathode ray tube is usually coated with two phosphors, one having a long decay time, and one having a short decay time.

The former facilitates visual observation of the image while photographic film is more sensitive to the latter.

A. Backscattered Electron Images

As noted earlier in Fig. 1, the electron yield consists of three fractions, elastically reflected primaries, inelastically reflected primaries, and secondaries. The elastically and inelastically reflected primaries constitute the backscattered electron yield of the specimen. As noted in Fig. 2, the backscattered electron yield is a function of mean atomic number, hence BSE images would also be related to the atomic number of the sample, and since the number of backscattered electrons seen by the detector is related to the angle between the sample surface and the detector, topographical features of the sample may also affect the BSE signal. In some instances, this double variation in backscattering may cause considerable confusion when viewing BSE images. This confusion may be minimized, however, by using multiple detectors as proposed by Kimoto and Hashimoto [26]. Subtraction of the signals seen by two diametrically opposed detectors accentuates topographical features and minimizes atomic number difference. Backscattered electron detectors may be unbiased scintillator-photomultiplier combinations, p–n junctions, biased scintillators-photomultiplier combinations (for low primary electron energies), or simply small beryllium plates.

B. Sample Current Images

Since the specimen current signal is inversely related to the backscattered electron yield of the specimen, SC images are complementary to those of the BSE. They are affected by both the topography of the specimen and by its mean atomic number. These images are obtained by isolating the specimen from ground and using the variation in current flow between sample and ground to modulate the intensity of the cathode ray tube.

In practice, the analyst would choose either BSE or SC image depending on the problem being investigated. For instance in studying inclusions in uranium, where virtually all inclusions have a lower mean atomic number than the matrix, SC images would most likely be used as the inclusions would appear merely as black holes in a BSE image. Where topographical features were of interest, a multiple collector BSE image would be preferred.

C. Secondary Electron Images

At high accelerating voltage (e.g., 30 keV), the backscattered electrons may originate from deep within the specimen (several microns), especially in the case of low atomic number elements where backscatter occurs primarily by a large-angle single elastic scattering mechanism [4]. Consequently, BSE and SC images have an inherent lack of resolution. At lower accelerating voltages, the number of backscattered electrons reaching the detector is reduced so that "grainy" images are likely to result.

However, as the accelerating potential is lowered, the secondary electron yield increases, as noted earlier. Seiler [8] has shown that the maximum exit depth of secondaries is 50 Å for metals and approximately 500 Å for insulators. Consequently, a significant improvement in resolution can be obtained by employing secondary electron images. The collection statistics are also improved because secondaries emitted over a large solid angle can be collected either electrostatically or electromagnetically. Care must be taken to properly position and shield the detector so as to collect only those secondaries which are emitted from the sample. Several investigators, notably Smith and Oatley [27], Smith [28], and Pease and Nixon [29], have obtained secondary electron images in the scanning electron microscope, by using a Mollenstadt-post accelerator and collection technique. Davidson and Neuhaus [30] recently applied such a detection system to a commercial microprobe and were able to achieve remarkable resolution and depth of focus. They employed a voltage biased ring collector ahead of a biased scintillator-photomultiplier combination, positioned below and to the side of the sample. They were able to obtain SE images with larger fields of view by lowering their specimen as much as 5.5 in. below the lower pole piece of the magnetic objective lens.

It is expected that SE images will become more useful and prominent as manufacturers incorporate suitable secondary electron detection equipment into their instruments.

References

1. K. F. J. Heinrich, Natl. Bur. Std. (US) Tech. Note 278 (1967).
2. O. Hachenberg and W. Brauer, *Advan. Electron. Electron Phys.* **11**, 413 (1959).
3. J. E. Holliday and E. J. Sternglass, *J. Appl. Phys.* **28**, 1189 (1957).
4. G. D. Archard, *J. Appl. Phys.* **32**, 1505 (1961).
5. H. E. Bishop, *in* "X-Ray Optics and Microanalysis," p. 153. Hermann, Paris, 1967.
6. K. G. McKay, *Advan. Electron. Electron. Phys.* **1**, 65 (1948).
7. E. J. Sternglass, *Phys. Rev.* **80**, 925 (1950).
8. H. Seiler, *Z. Angew. Phys.* **22**, 249 (1967).
9. E. M. Baroody, *Phys. Rev.* **78**, 780 (1950).
10. R. G. Lye and A. J. Dekker, *Phys. Rev.* **107**, 977 (1957).
11. J. W. Colby, W. N. Wise, and D. K. Conley, *Advan. X-Ray Anal.* **10**, 447 (1967).
12. H. Kanter, *Phys. Rev.* **121**, 681 (1961).
13. E. Weinryb, *Metaux* (*Corrosion Ind.*) **40**, 131 (1965).

14. D. B. Wittry, *in* "X-Ray Optics and Microanalysis," p. 168. Hermann, Paris, 1967.
15. D. M. Poole and P. M. Thomas, *J. Inst. Metals* **90**, 228 (1962).
16. J. Philibert and E. Weinryb, *in* "X-Ray Optics and X-Ray Microanalysis," (H. H. Pattee, V. E. Cosslett, and A. Engstrom, eds.), p. 451. Academic Press, New York, 1963.
17. E. Weinryb and J. Philibert, *Compt. Rend.* **258**, 4535 (1964).
18. P. G. Burkhalter, *US Bur. Mines Rept. Invest.* 6681 (1965).
19. K. F. J. Heinrich, *Advan. X-Ray Anal.* **7**, 325 (1964).
20. K. F. J. Heinrich, *in* "X-ray Optics and Microanalysis," p. 159. Hermann, Paris, 1967.
21. R. Shimiza and G. Shinoda, *in* "X-Ray Optics and X-Ray Microanalysis," (H. H. Pattee, U. E. Cosslett, and A. Engstrom, eds.), p. 419. Academic Press, New York, 1963.
22. R. Castaing, *Advan. Electron. Electron Phys.* **13**, 317 (1960).
23. J. W. Colby, *Advan. X-Ray Anal.* **8**, 352 (1965).
24. M. Beyeler and Y. Adda, *Compt. Rend.* **253**, 2967 (1961).
25. E. Weinryb, *Metaux* (*Corrosion Ind.*) **40**, 181 (1965).
26. S. Kimoto and H. Hashimoto, *in* "The Electron Microprobe," p. 480. Wiley, New York, 1966.
27. K. C. A. Smith and C. W. Oatley, *Brit. J. Appl. Phys.* **6**, 391 (1955).
28. K. C. A. Smith, *Proc. Eur. Regional Conf. Electron Microscopy* **1**, 177 (1960).
29. R. F. W. Pease and W. C. Nixon, *Brit. J. Sci. Inst.* **42**, 81 (1965).
30. E. Davidson and H. Neuhaus, *Natl. Conf. Electron Probe Microanalysis, 1st, 1966.*

The Influence of the Preparation of Metal Specimens on the Precision of Electron Probe Microanalysis*

G. HALLERMAN† and M. L. PICKLESIMER‡

Metals and Ceramics Division
Oak Ridge National Laboratory
Oak Ridge, Tennessee

I. Introduction

The precision of the local chemical analysis of a specimen by an electron microprobe is influenced by many factors such as secondary fluorescence, absorption of the characteristic X rays, the differences in the atomic numbers of the various elements in the specimen, the electron beam geometry and stability, the detector and crystal sensitivities, the take-off angle of the X-ray

* Research sponsored by the U.S. Atomic Energy Commission under contract with the Union Carbide Corporation.

† *Present address*: Inland Steel Research Laboratories, East Chicago, Indiana.
‡ *Present address*: Southern Research Institute, Birmingham, Alabama.

optical system, and the instrument and specimen X-ray backgrounds. Corrections must be made for these and many other factors if the recorded data are to have significant quantitative meaning. These problems are basic to the particular specimen composition and to the particular instrument being used. Independent of such factors, there is another of appreciable importance which is, to a great extent, under the control of the analyst. This is the preparation of the specimen surface for the analysis. There are glib answers to the problem this poses, one of which is "make the specimen surface as flat as possible." While such statements may be true, they tell nothing about (1) how flat the surface must be for the data to be unaffected by surface roughness; (2) how the specimen surface layer is affected by such preparation; (3) how such preparation influences the data obtained; and (4) some way that useful data can be obtained from a specimen with an unavoidable "rough" surface.

In this chapter, we will consider only the effects of the preparation of the surface of the specimen on the precision of the microprobe analysis. Factors we wish to consider are surface roughness, cold-working, smearing, etching, leaching and deposition of elements, and films deposited inadvertently or deliberately on the prepared surface.

Most microprobe analysts are aware of the standard procedures for cutting, mounting, grinding, polishing, and etching in the preparation of metal specimens for examination by optical microscopy. The details of these procedures are well presented in several books [1] and papers [2] and need not be described here. Rather, we wish to consider the effects of such procedures on the condition of the specimen surface, the faithfulness of the surface representation to the true structure, and the ways in which the condition of the surface affects the accuracy of the microprobe analysis.

II. Problems in Specimen Preparation

There are four general problem areas basic to the sectioning and preparation of a specimen for electron probe microanalysis. Each problem area seems quite obvious, but each deserves and requires separate and deliberate consideration before the specimen is prepared.

A. The Information Desired

The first, and possibly most important, problem area is that of determining exactly what information is desired from the specimen. Is it the composition or the constituents of a surface film? The composition range in a gradient structure? The composition gradient of a diffusion couple? The composition of a precipitate phase? Or is it the answer to a question such as:

what caused this specimen to crack (or corrode, or fail)? The answer determines most of the procedures and approaches that are used in both the preparation and the examination of the specimen. An incorrect answer can cause the waste of much time, effort, and money, and sometimes the loss of a valuable specimen.

B. Selection of the Section

The second problem area concerns the determination of the section of the specimen to be examined. The "section" is defined as that particular cross section of the specimen that is finally examined in the analysis. The section required for the analysis of a diffusion couple is different from that required for the analysis of a surface film, and both are different from that required for the determination of the composition of a precipitate particle. In the case of the diffusion couple, it seems obvious that the section desired is perpendicular to the original plane interface. Yet, suppose that the composition gradient is so steep that the entire diffusion zone is only three- to four-beam diameters deep. The section now most desirable is that of a plane surface cut at a relatively small angle to the original plane interface, so that the composition gradient is reasonably spread out (a "slant" section). The surface exposed must be at a known angle and location to the original interface to it if the analysis is to have quantitative meaning.

C. Specimen Preparation

The third problem area concerns the equipment, materials, and techniques used for cutting, mounting, grinding, and mechanically polishing the specimen. The specimen can easily be damaged, and the information it contains may be effectively lost by such things as clamping it too tightly in the vise of a cutoff machine, hacksawing, overheating during abrasive sawing, high pressure or high temperature during mounting, excessive pressure or either no lubricants or improper lubricants during grinding and polishing, or even the use of the wrong cloth or abrasive during final polishing. Specimens of alloys that can undergo various types of phase transformations or precipitation reactions at low temperatures can be damaged by mounting in thermosetting or thermoplastic resins. They can even be damaged by mounting in some of the "low temperature" epoxy resins that reach temperatures of 80 to 100°C during curing. Quite high temperatures, approaching the melting point of many alloys, can be reached in the first few thousandths of an inch below a surface being cut under pressure with abrasive saws. Severe cold working and distortion of the structure can occur in a layer many microns deep if too much pressure is used during grinding many of the softer metals and alloys. A layer several microns thick of a soft phase can be smeared over a harder phase during hand polishing on cloths if the specimen is pressed too hard against the wheel, or if the cloth runs dry of lubricant. Each of these possibilities, and others not discussed, must be considered and the procedures selected for each stage to produce the least damage to the specimen. If the specimen cannot be prepared in one of the steps without some damage, enough material must be left for that damage to be removed in the next step, so that the final surface to be examined is undamaged.

D. Etching and Polishing

The fourth and last problem area is that of the final surface preparation by chemical or electrolytic polishing, and chemical, electrolytic, or cathodic etching [3] to remove the last thin layer of damaged metal, which is unavoidably produced no matter how carefully the polishing was performed. This last layer may be only a small fraction of a micron thick, or it may be several tens of microns thick, depending on the material and the method of preparation. Enough material must be removed to be certain that the structure observed will not change in character with the removal of more material. Frequently, it is a difficult and tedious task to develop a solution or procedure that will remove the disturbed surface layer almost uniformly, without severe and preferential attack of one or more of the phases present. Yet, without this development, the specimen may be in a condition entirely unsatisfactory for electron probe microanalysis.

E. Example Problem

To illustrate the importance of each of the four problem areas and to show that they are not as obvious as they may seem, we present the following as a typical example of an analytical problem.

Pinhole leaks and/or small cracks observed in the neighborhood of a sealing fusion weld caused the rejection of tubular nuclear fuel elements having a nickel–base alloy as a cladding. The fusion weld was made between the fuel element tube and an end cap of the same alloy by the Heliarc process in a controlled atmosphere. The end of the weld pass overlapped the start by about 45°, and the welding power was steadily reduced after the overlap was begun. A photomacrograph of the cross section of the tube, end cap, and fusion weld is shown in Fig. 1. The problem presented was to find the cause of the pinholes and cracks so that the failures could be eliminated.

Conventional metallographic examination [4] showed that the pinholes and cracks were found only in the heat-affected zone of the base metal in the region of the overlap of the beginning and ending of the weld; that is, near or in the last part of the weld to freeze. The examination also showed that an

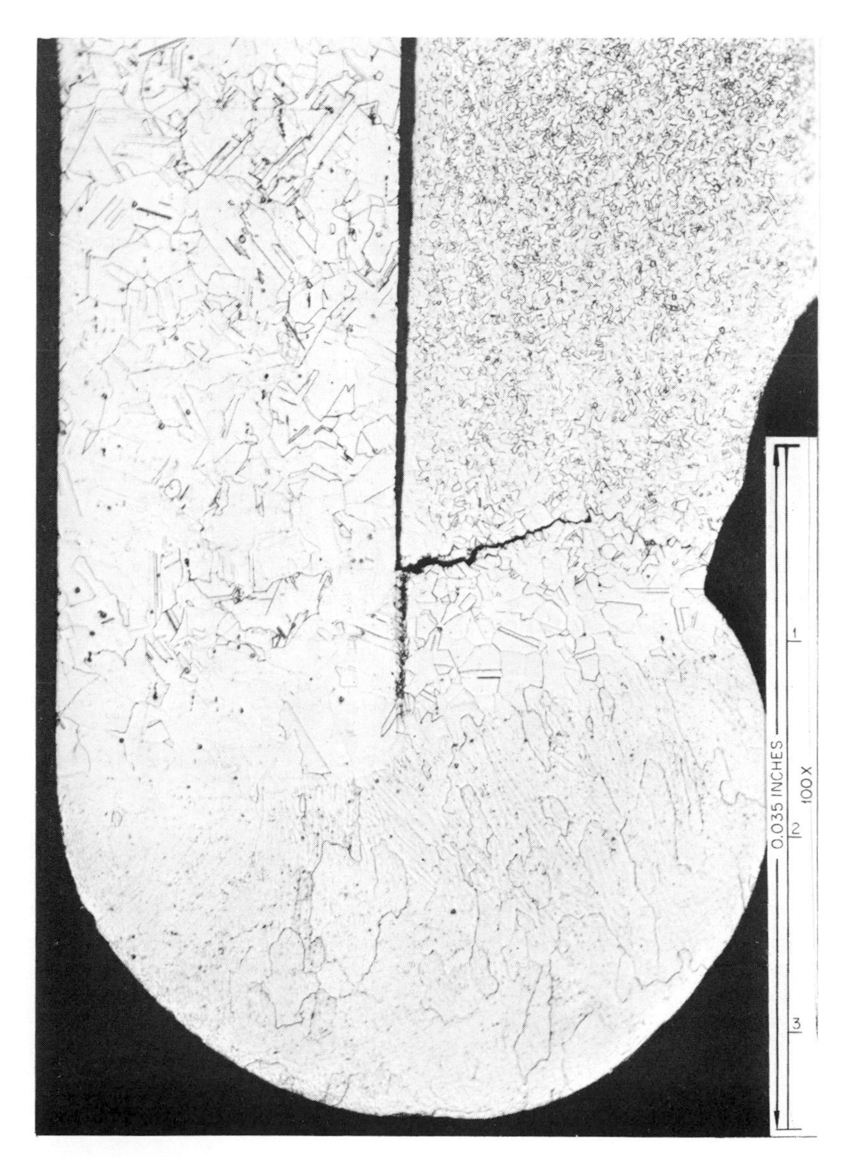

FIG. 1. Photomacrograph of the cross section of a tubular fuel element: fusion weld, end cap, and tube walls.

unknown second phase existed in the grain boundaries in the region containing the cracks, but in no other region, that all cracks occurred in grain boundaries containing the unknown phase, and that not all grain boundaries containing the unknown phase also contained cracks. The microstructure of the area concerned is shown in Fig. 2.

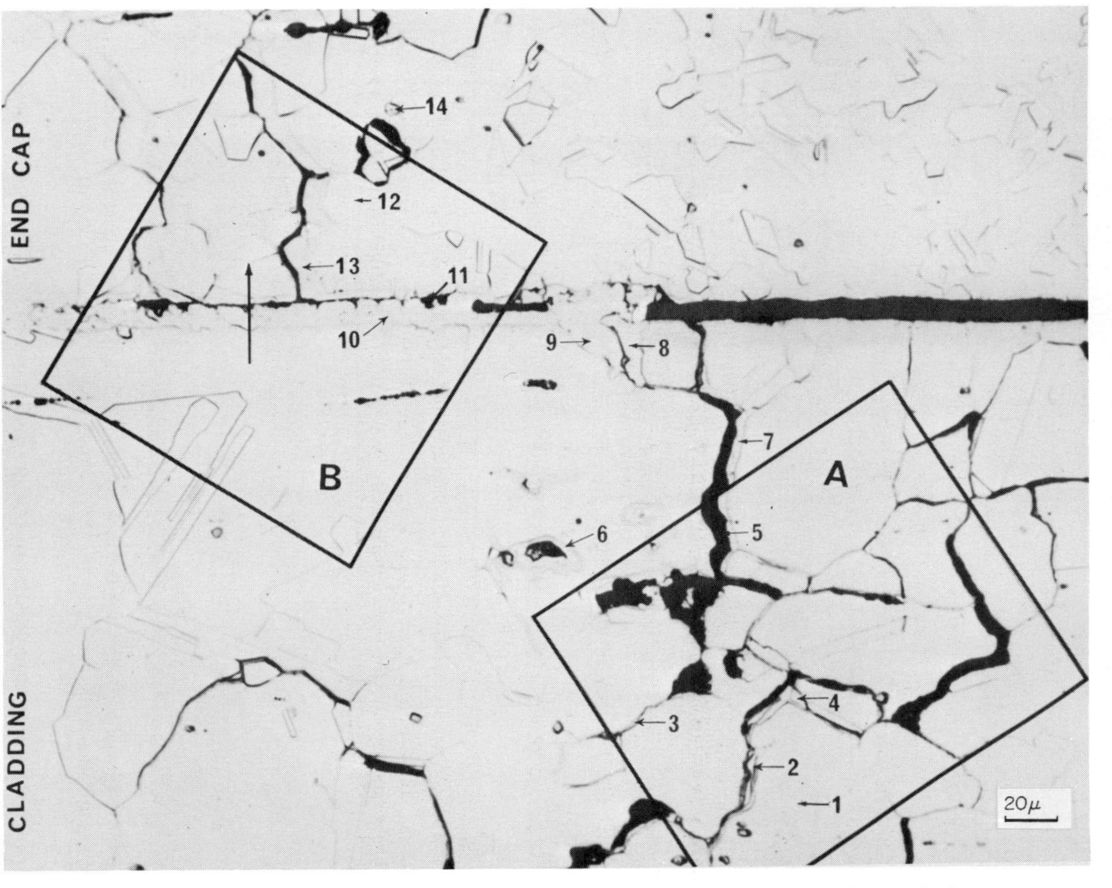

FIG. 2. Photomicrograph of areas examined by electron probe microanalysis.

The information needed for the solution of the problem was (1) What caused the cracks to occur? (2) What was the composition of the unknown phase or phases in the grain boundaries of the heat-affected zone? (3) From where could the "impurity" have come? (4) Was there more than one unknown phase? (5)Why did the unknown phase occur only in the region where it was observed? (6) What could be done to eliminate the unknown phase if it was the cause of the cracking? Electron probe microanalysis could produce only part of the information needed, but that part could lead to the final solution of the problem.

The microanalysis could conceivably determine answers to parts (2), (3), and (4). Thus, the information needed from the microprobe was determined. The examination by conventional metallography had established that the unknown phase and the cracks were found in only one region of the specimen. This established that the section of the specimen needed was one showing the unknown phase, the cracks, and portions of the weld metal and the unaffected base metal for comparison. The preparation and etching of the metallographic specimen showed that this preparation should be suitable for the specimen for microprobe analysis. The appearance of the microstructure was reasonable in the weld metal, the rest of the heat-affected zone, and in the base metal; that of the region containing the grain boundary phase and the cracks was not unreasonable. The crack edges remained sharp during the specimen preparation, and the unknown phase could be detected on both sides of the crack in some grain boundaries. The etchant used did not cause pitting, there was no appreciable etching relief at the phase boundaries nor was the surface of the specimen severely rumpled. Smearing of the microstructure was not apparent nor was there any evidence of cold working of the sample surface. Leaching was not evident nor was there any apparent deposition from the etchant. Both the section and the preparation of the specimen used for optical microscopy seemed quite suitable for the microprobe analysis specimen.

The microanalysis by the electron probe established that (1) the unknown phase in the grain boundaries differed from the parent metal only in that it contained about 6 wt % P; (2) the parent metal contained approximately 300 ppm P; (3) the weld metal contained less than 300 ppm P; and (4) there was no more than 300 ppm P in the center of grains that were completely surrounded by phosphorus-rich grain boundaries. Oscilloscope photographs of two area scans by specimen current and phosphorus K_α X rays are shown in Fig. 3.

This information provided the key to the rest of the answers required in the problem, with the exception of what could be done to eliminate the presence of the second phase. A knowledge of the hot-forging behavior of nickel-base alloys and the phase diagram of the nickel–phosphorus alloy

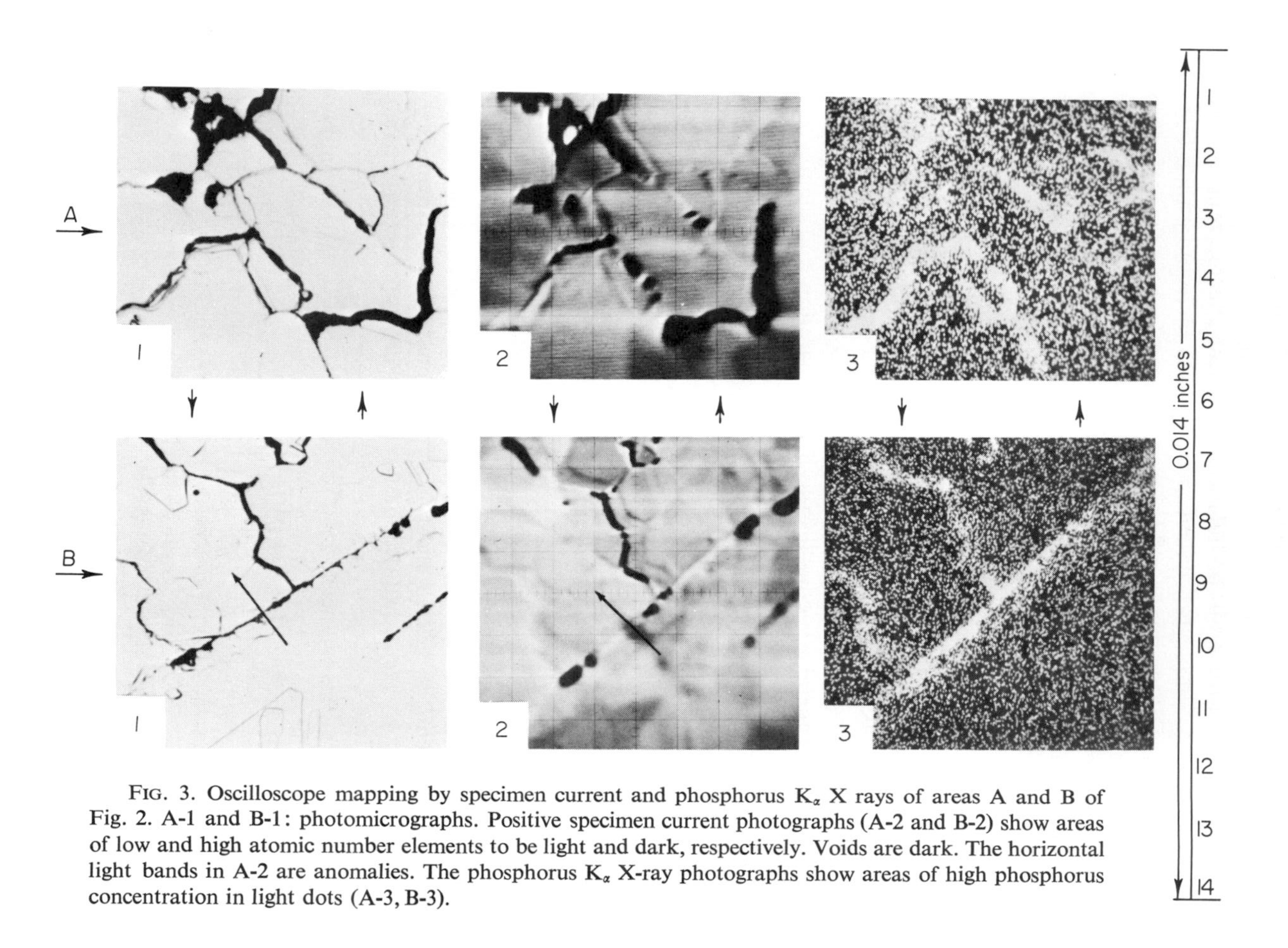

FIG. 3. Oscilloscope mapping by specimen current and phosphorus K_α X rays of areas A and B of Fig. 2. A-1 and B-1: photomicrographs. Positive specimen current photographs (A-2 and B-2) show areas of low and high atomic number elements to be light and dark, respectively. Voids are dark. The horizontal light bands in A-2 are anomalies. The phosphorus K_α X-ray photographs show areas of high phosphorus concentration in light dots (A-3, B-3).

system led to the conclusion that the following events occurred. Some of the phosphorus evaporated from the metal that was molten during welding condensed to form a liquid phase in the heat-affected zone in the gap between the tube and the end cap. The liquid phase reacted with the parent metal to form a grain boundary liquid rich in phosphorus. This liquid phase froze in place when the welding was stopped, being trapped in the heat-affected zone of the weld overlap. Nickel alloys containing moderate amounts of phosphorus are notoriously hot-short during forging and rolling, so that the presence of such compositions in the grain boundaries of the heat-affected zone could lead to a sensitivity to hot-cracking. Since the geometry and pertinent masses of the tube and end cap were appreciably different in the higher temperature regions around the weld bead, thermal stresses induced during cooling could be sufficiently high to lead to hot-cracking in at least some of the grain boundaries containing the phosphorus-rich phase.

The microprobe analysis did not provide all of the information necessary for the solution of this problem, but it did provide the key information. The selection of the wrong section for examination, improper preparation, or improper etching could each have caused the probe analysis to produce different information, leading to incorrect conclusions, or to confusion, and resulted in a failure to solve the problem.

III. Effects of Surface Roughness

Much of the literature on the applications of electron probe microanalysis makes passing reference to the effects of specimen roughness on the amount of X-ray absorption occurring in the specimen, but we have been unable to find any giving quantitative data such that a limit could be set on the scale of roughness that is tolerable. We have conducted experiments on specimens of known and characterized roughnesses produced by geometry, grinding, relief polishing, etching, and over-etching to establish such a limit. The results are presented below.

A. Geometrical Roughness

In order to eliminate any effects due to composition, preferential absorption, smeared, or leached surfaces, and to permit only the effects of surface topography to be seen on the precision of the analysis, it is necessary to use a pure metal specimen whose surface level varies in a known, and preferably regular, fashion. The effects we wish to examine are those of defocusing of the electron beam, the defocusing of the X-ray optical system, and the variation of

X-ray absorption by the specimen material, with all of the variations being produced by surface roughness alone. Thus, it would be best if the topographical variation occurred in a regular pattern and consisted of known abrupt changes of height between two level surfaces. To accomplish this, copper grids (200 mesh) normally used in electron microscopy were mounted on large copper bases in metallographic mounts. The grids were then ground and polished to various thicknesses to produce specimens of known step heights and geometries. The step heights were measured by interference microscopy for the smaller steps, and by microscope focusing at high magnifications for the larger steps.

Microprobe traverses were made over selected areas of grids having thicknesses of 40 to 50, 23 to 25, 11 to 12, 4 to 6, and 0.6 to 2 μ (all were tapered). Two electron probe machines were used, one having an actual take-off angle (ψ) for the X-ray optics of 35° (effective angle of 41°) and the other of 15.5°. This allowed a greater spread in the variables than was possible with only one take-off angle. A photomicrograph of the 25-μ grid is shown in Fig. 4. The geometry of the electron beam, the traverse, and the specimen grid for a take-off angle of 35° are shown in Fig. 5. Those for a take-off angle of 15.5° and an electron beam at normal incidence can be similarly drawn.

In Fig. 5, the traverse of the beam is from points 1, 2, 3, 6, and 7 to point 8.

FIG. 4. Photomicrograph of copper grid specimen 25 μ thick.

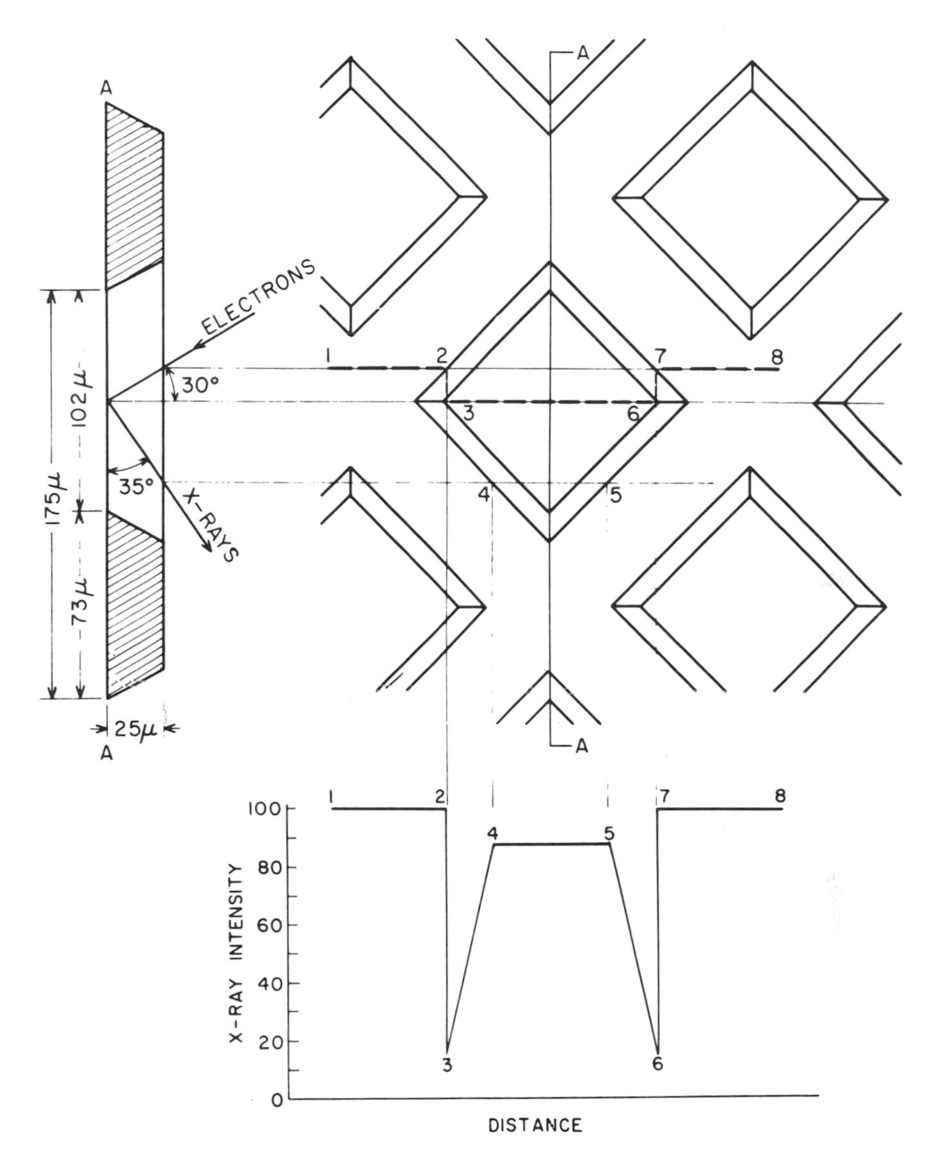

FIG. 5. Copper grid specimen (25 μ), beam and specimen geometry, and X-ray intensity for electron probe traverse at a take-off angle of 35°.

The beam was focused on the upper surface of the grid at point 1, and the focus was not changed throughout the traverse. When the traverse passed point 2, the beam abruptly dropped to point 3 on the copper base. At this point, the electron beam was out of focus on the surface encountered, the

X-ray optics were also out of focus, and the X rays generated by the electron beam were strongly absorbed by the specimen wall. As the traverse was continued, the fraction of X rays absorbed by the specimen wall decreased continuously until, at point 4, the X-ray beam no longer encountered the specimen wall. At this point, and up to point 5, the decrease in signal strength was due only to the lack of focus of the electron beam, the X-ray optical system, or both. At point 5, the X-ray beam again encountered the specimen wall and it was increasingly absorbed as the traverse continued to point 6. At this point, the electron beam abruptly encountered the upper surface again, at point 7, where both it and the X-ray optics were again in proper focus. If the X-ray intensity between points 1 and 2, and 7 and 8, is taken as 100%, then the difference in intensity between points 2 and 4 (and 5 and 7) shows the loss due to defocusing of the electron beam, the X-ray optical system, or both. The intensity loss between points 3 and 4 (and 5 and 6) is due to the absorption of the X rays in the specimen wall. The plot of X-ray intensity versus distance for the traverse shown in Fig. 5 is idealized, being drawn schematically from the rest of the illustration.

An actual traverse of the 25-μ-thick grid specimen made at a take-off angle of 35° is shown in Fig. 6. The fine structure observed in the intensity trace was smoothed out during drawing to simplify the figure. The intensity scale is given in arbitrary units. The fit with the schematic trace presented in Fig. 5 is generally good. It must be realized that the actual grid specimen was not completely smooth or regular, as can be seen in Fig. 4, and that the track of the beam was not exactly that shown schematically in Fig. 5. The sharp loss of X-ray intensity as the electron beam left the upper surface of the specimen, the increase as the beam continued its travel across the base, and the loss of intensity due to the lack of focusing of the system on the base are all apparent when Figs. 5 and 6 are compared.

A similar traverse made on the 11- to 12-μ-thick grid specimen is shown in Fig. 7. In this specimen, there is essentially no difference in intensities between the traces for the upper and lower surfaces of the specimen, indicating that there was little to no loss of intensity due to the lack of focus in the system. During a traverse of the 0.6- to 2-μ-thick grid, the step occurring between the upper and lower surfaces could not be detected.

Similar traverses were made on the same set of specimens at a take-off angle of 15.5°. The loss of X-ray intensity due to the lack of focus could not be made on these traverses, however, since the electron beam never traversed a region similar to that between points 4 and 5 of Fig. 5 (the X-ray beam was never both out of focus and unabsorbed by the walls at the same time). Only total loss measurements could be made.

The data collected for the traverses of all grids at both take-off angles (35 and 15.5°) are plotted in Fig. 8 and presented in Table I. The data show that there is no significant loss of X-ray intensity for step heights up to 2 μ when the

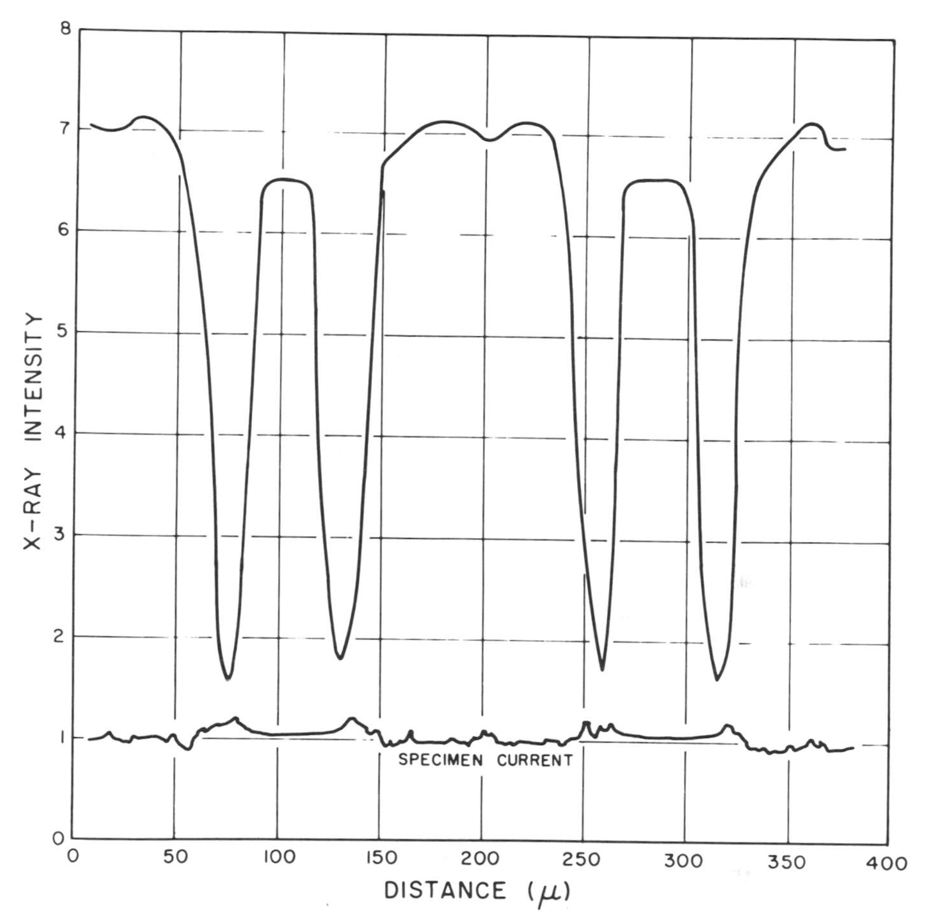

FIG. 6. Trace of the electron probe traverse across two grid openings of the 25-μ-thick copper grid. The fine wiggles in the actual trace were smoothed out to simplify drawing. Take-off angle is 35°.

take-off angle is 35°, and probably no large loss for a step height of 1 μ at a take-off angle of 15.5°.

As a further check of this conclusion, X-ray intensity data were collected for a take-off angle of 35° on a copper specimen in the as-polished condition and as-ground on 3/0 abrasive paper. The scratch depths and spacings produced by the abrasive measured an average of 1.5 and 3.5 μ, respectively, by interference microscopy. There were no detectable differences in the beam traverse data for the two conditions.

Both types of measurements lead to the conclusion that abrupt elevation differences, per se, do not cause an appreciable error in the analysis if they are approximately of the scale of the electron beam diameter.

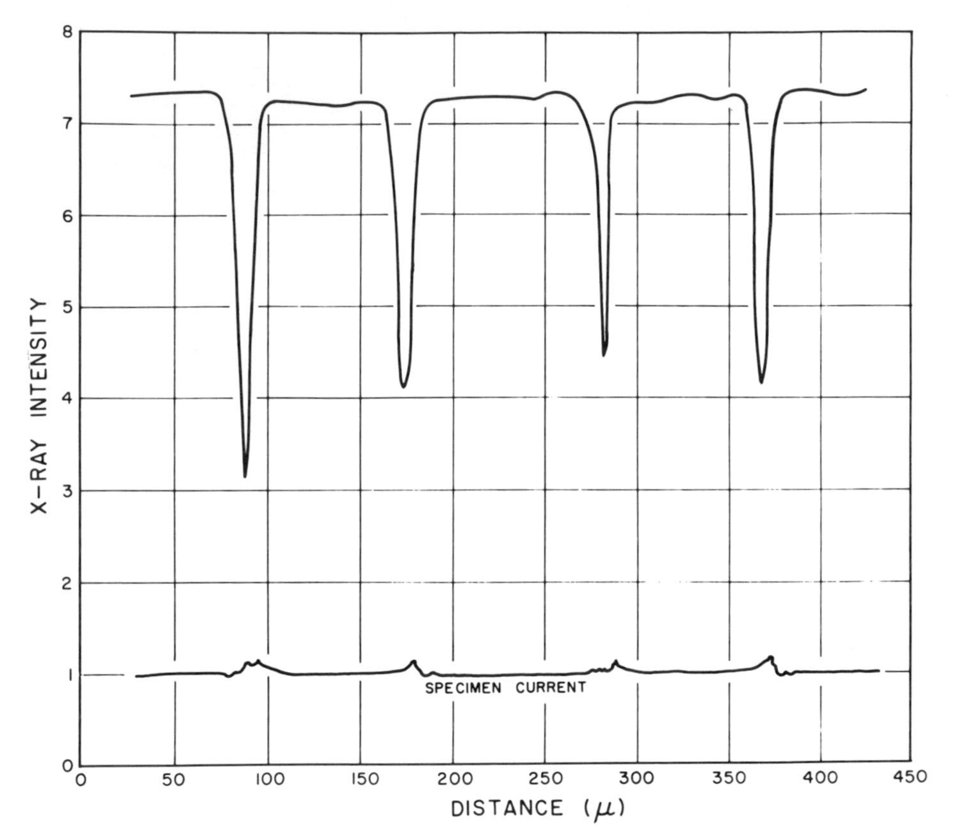

FIG. 7. Trace of electron probe traverse across two grid openings of 12-μ-thick copper grid. Fine wiggles in the actual trace were smoothed out to simplify drawing. Take-off angle is 35°.

TABLE I
INTENSITY MEASUREMENTS ON COPPER GRID SPECIMENS

Grid Thickness (μ)	Intensity Loss (%)			
	$\psi = 35°$			$\psi = 15.5°$
	Total	Defocusing	Absorption	Total
40–50	86			95
23–25	77	6	71	93
11–12	50	1	49	76
4–6	27	0	27	47
0.6–2.1	<2	0	<2	10

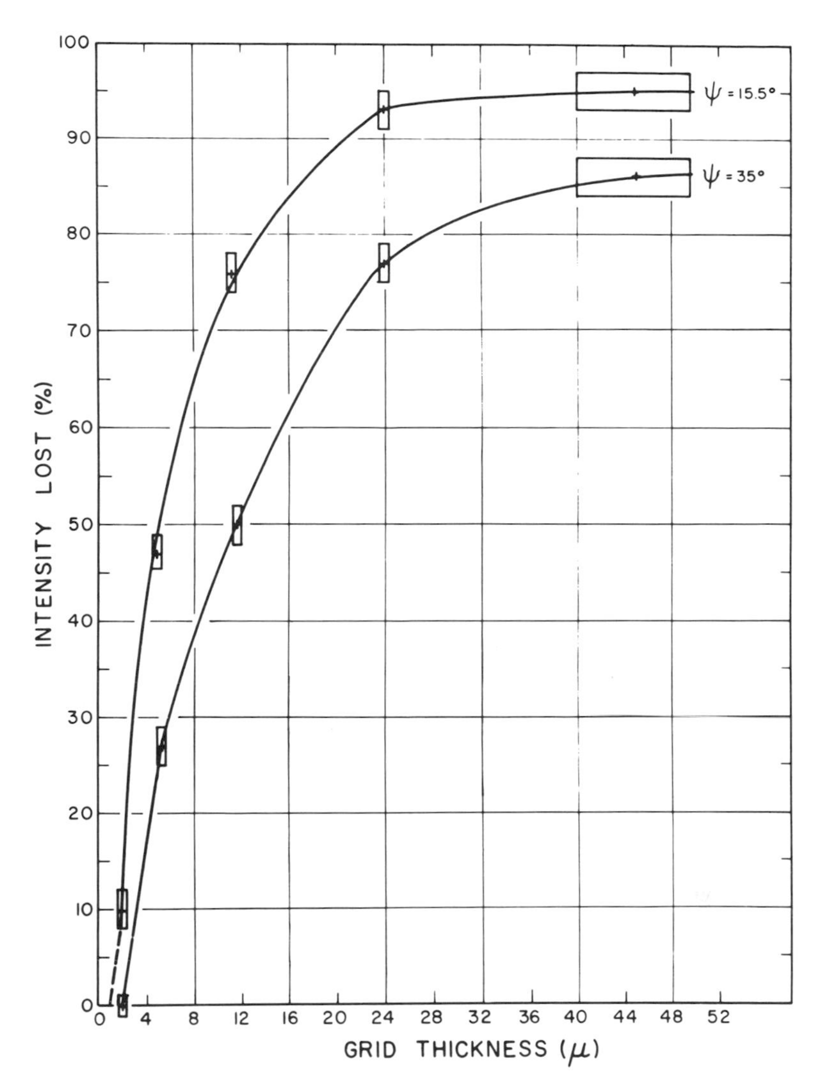

FIG. 8. Plot of the loss of X-ray intensity as a function of grid thickness. Copper grids. Take-off angles of 35 and 15.5°. Beam size less than 2 microns diameter.

B. Roughness Due to Scratches and Overetching

Specimens of two alloys were used to examine the effects of surface roughness produced by grinding on 3/0 abrasive paper, light etching, and heavy overetching in comparison to the same specimens in the as-polished condition. The two alloys were: (1) an as-cast, slowly cooled 50 Cu–50 Ni alloy showing

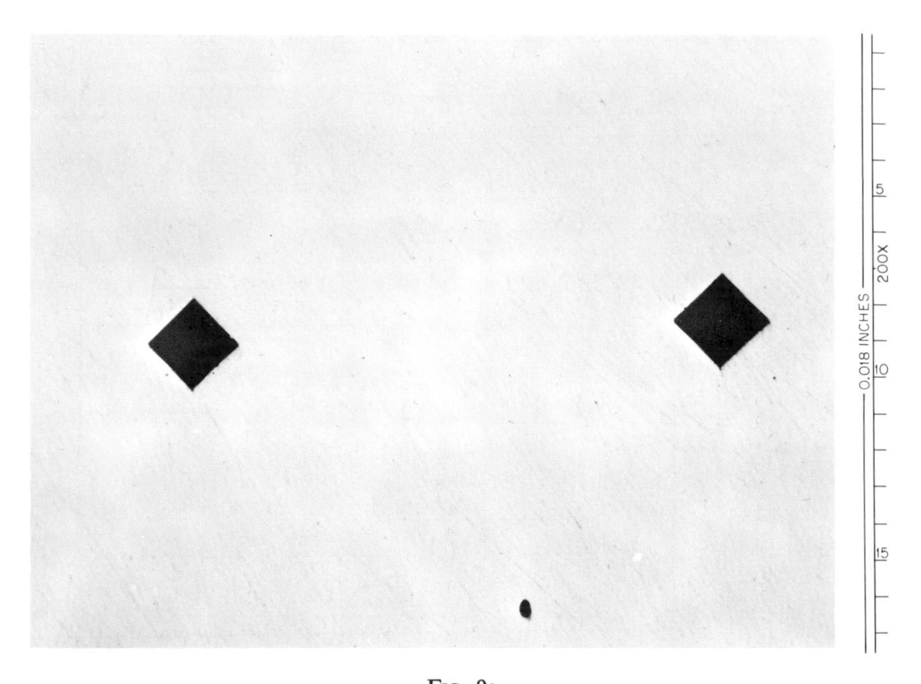

FIG. 9a.

FIG. 9. Microstructures in an as-cast, slowly cooled 50 Cu–50 Ni alloy. Severe, coarse dendritic segregation present. 200×. Bright field. Electron probe traverses were made between the hardness indentations. (a) As-polished conditions; (b) lightly etched in 70 ml H_2O, 20 ml H_2O_2, 10 ml H_2SO_4; (c) heavily overetched. Same etching solution.

severe and coarse dendritic segregation, and (2) an as-cast Al–40 wt % U alloy which had been annealed at 580°C for three weeks. Typical microstructures of the specimens are shown in Figs. 9 and 10. Electron probe microanalyses were made on both alloys in all conditions at a take-off angle of 35°, but only on the 50 Cu–50 Ni alloy at a take-off angle of 15.5°.

Intensity curves for nickel K_α X rays at a take-off angle of 35° versus distance of traverse between two hardness impressions on the copper–nickel alloy are shown in Fig. 11. Intensity curves for copper K_α X rays were determined over the same traverse and were found to be exact difference curves from those for nickel within 1%. Wherever the nickel content decreased at a given position in the traverse after a change of surface preparation, the copper content increased the same amount. Thus, the shift in composition at any point in the traverse was due more to a changing level of segregation (different section) in the specimen than due to the changes in surface preparation. However, for the purposes of determining the worst case, the composition changes at one position in the traverse were calculated and considered to be

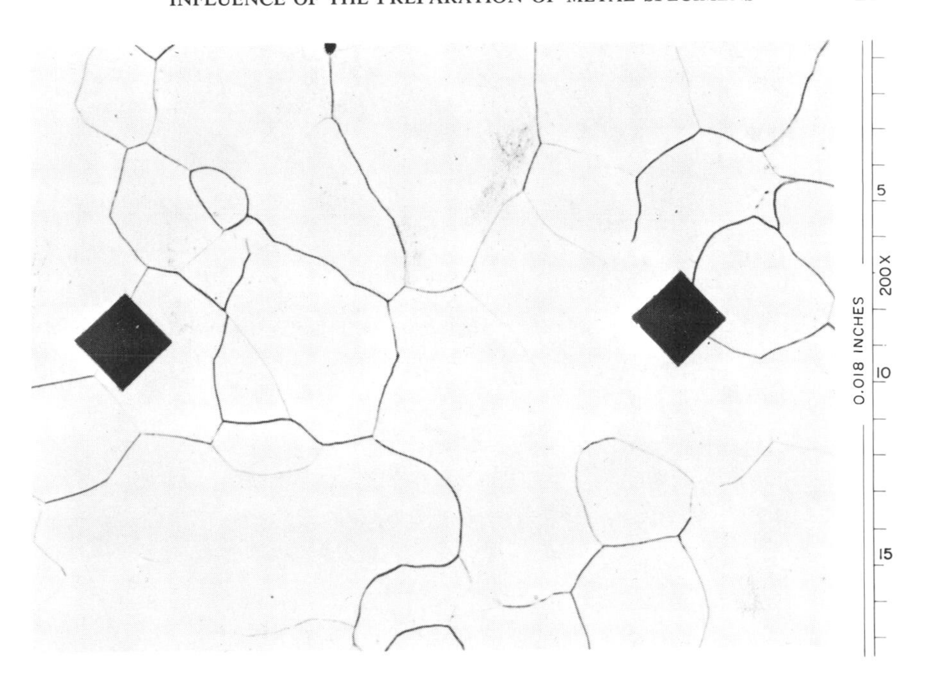

FIG. 9b.

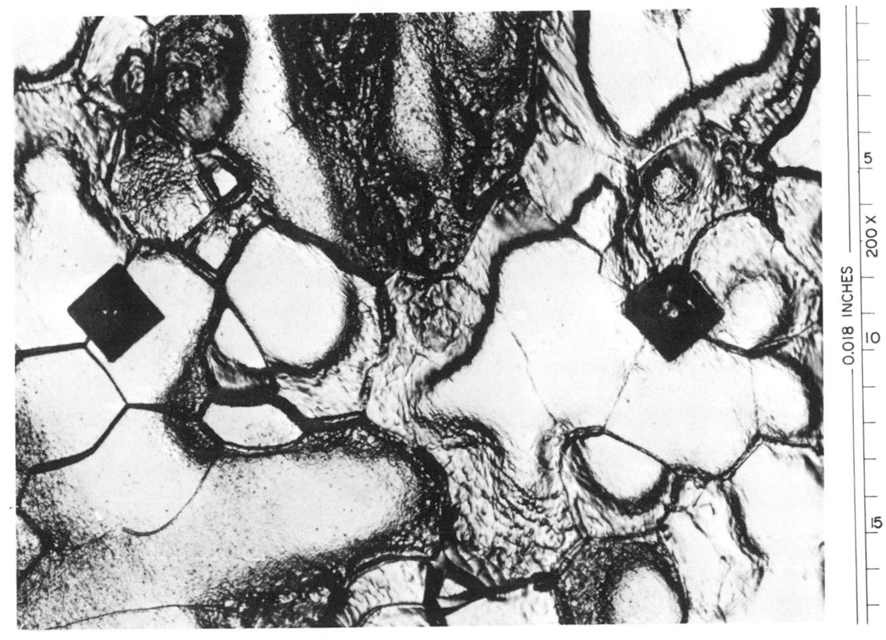

FIG. 9c.

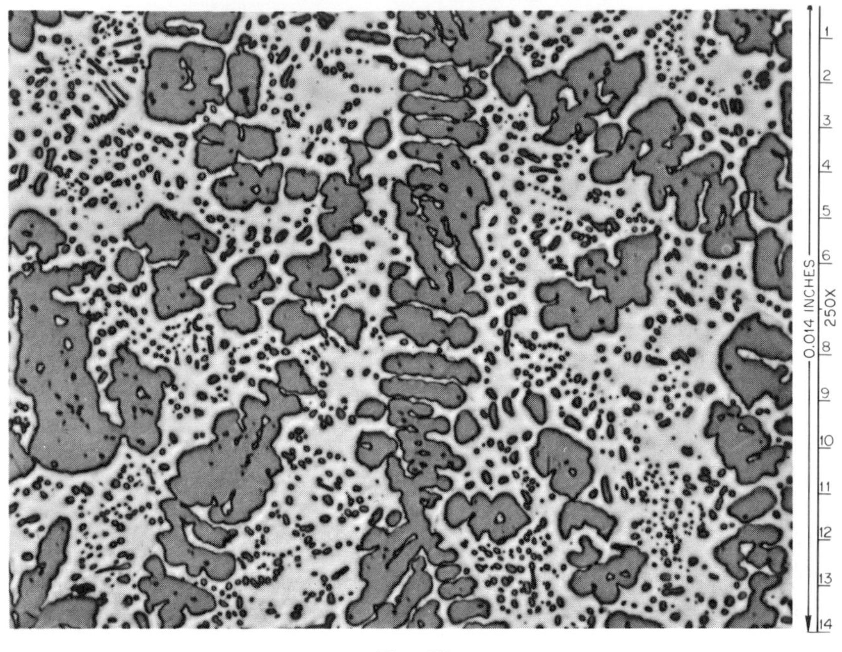

FIG. 10a.

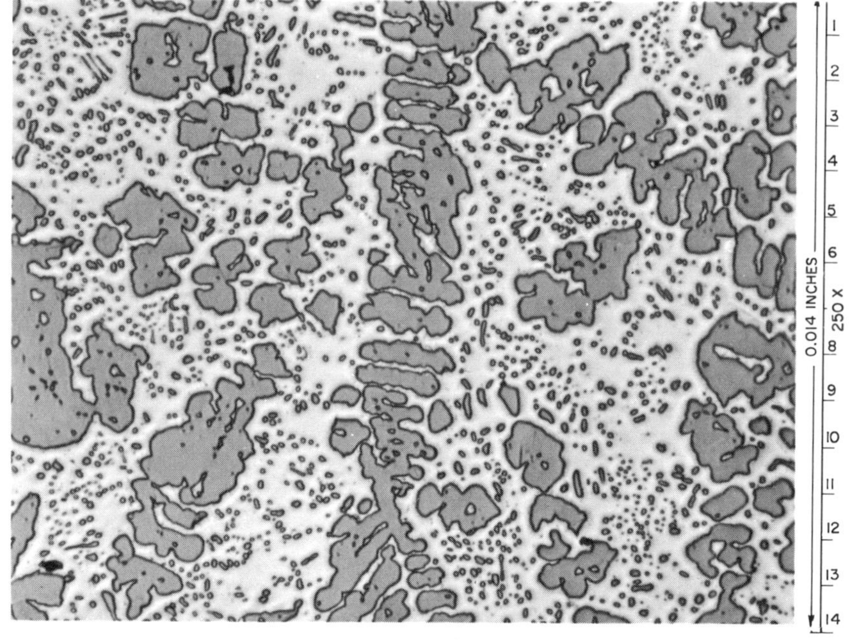

FIG. 10b.

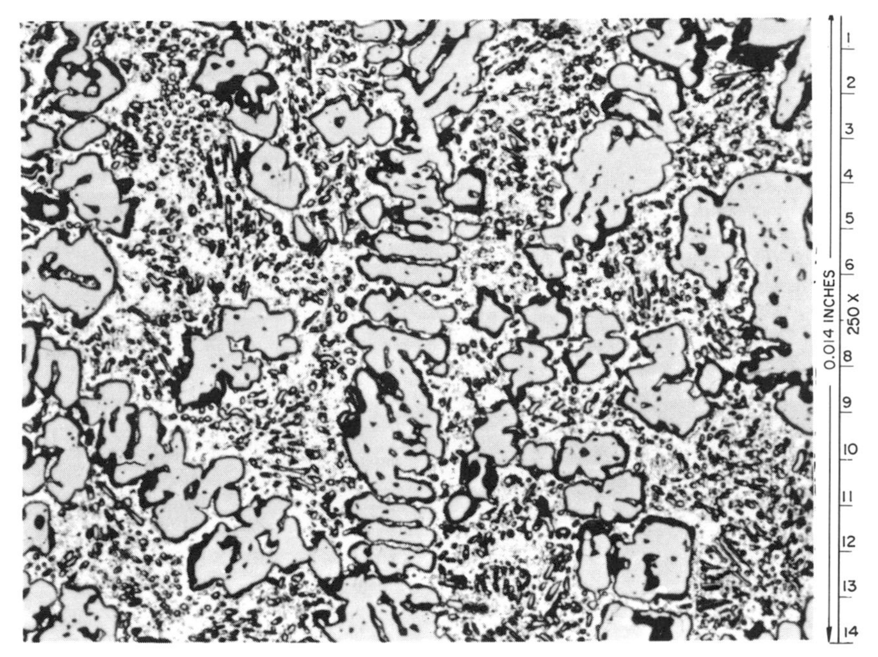

FIG. 10c.

FIG. 10. Microstructures of the Al–40 wt % U alloy. As-cast and annealed three weeks at 580°C. Primary UAl_4 plus spheroidized eutectic structure. 160×. Bright field. (a) As-polished. Polishing relief between intermetallic particles and matrix is between 0 and 0.3 μ. (b) Lightly etched; etching relief is 0.2 to 0.3 μ; etched in 10% NaOH in H_2O. (c) Heavily etched; etching relief varies between 2 and 3 μ; same etchant.

due entirely to differences caused by variation in surface preparation. A similar set of traverses were made at a take-off angle of 15.5°.

The data collected for all four conditions and two take-off angles are presented in Table II. These data show that there are no appreciable differences between the data for the four surface conditions at a take-off angle of 35°, and only small differences at a take-off angle of 15.5°. The maximum elevation differences found over the traversed line were: (1) less than 0.2 μ for the lightly etched condition, (2) 3–4 μ for the heavily etched condition, and (3) 1–1.5 μ for the ground condition. The specimen surface in the heavily etched condition was "rumpled" rather than scratched. There were no abrupt changes in surface elevation and the "wavelength" of the rumple was many times longer than the "amplitude."

Similar measurements were made on the Al–40 wt % U alloy specimen in the same four surface conditions at a take-off angle of 35° only. In the etched conditions, there were abrupt changes in elevation at the boundaries between

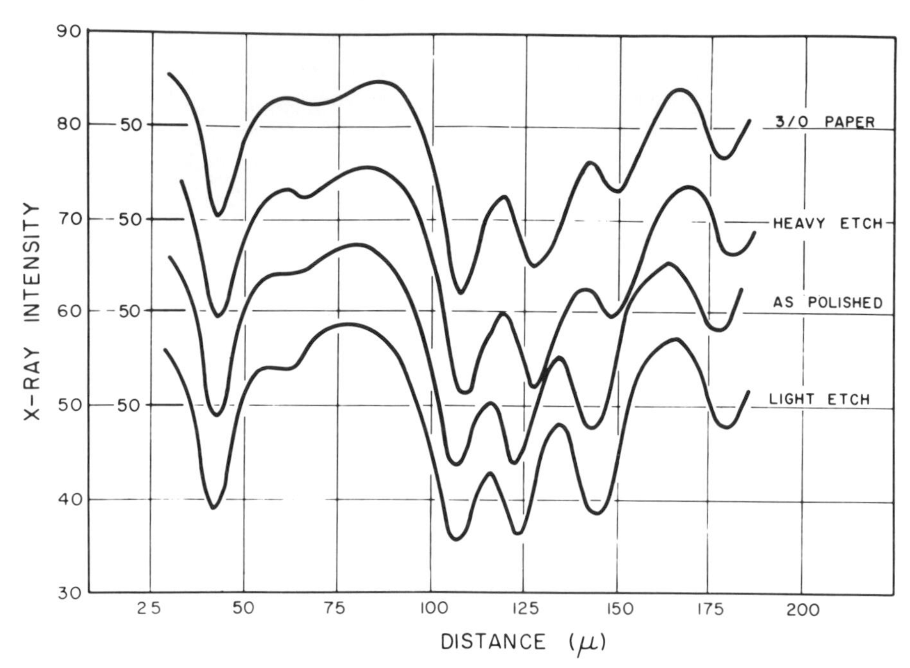

FIG. 11. Electron probe traverses of a 50 Cu–50 Ni alloy. Conditions: lightly etched; as-polished; heavily overetched; and ground on 3/0 abrasive paper. See Fig. 9 for relative microstructures. The traces are successively displaced relative to each other by a unit of 10 in X-ray intensity.

the intermetallic particles and the aluminum matrix. The elevation differences were: (1) less than 0.25 μ in the as-polished condition, (2) less than 0.25 μ in the lightly etched condition, (3) 1–2 μ in the heavily etched condition, and (4) scratch roughnesses of 1.5 to 2 μ in the ground condition (3/0 abrasive paper). No differences could be observed in the data for the four conditions within the normal recording and counting errors.

The results of these experiments show conclusively that surface roughness of a scale smaller than or equal to the diameter of the electron beam of the probe do not cause any appreciable errors in the microanalysis at the higher take-off angles, whether the elevation differences are abrupt, sharp steps, or are smoothed out over a distance of several beam diameters. In addition, the data for the heavily etched copper–nickel alloy show that changes in the slope of the specimen surface of a few degrees do not cause an appreciable error. Thus, specimen surfaces that are not too rough for examination by optical microscopy at high magnification are suitable for electron probe microanalysis at the higher take-off angles. There is negligible error if the local elevation differences are equal to or less than the diameter of the electron

TABLE II
COMPARISON OF ANALYSES AT TWO TAKE-OFF ANGLES (ψ) FOR FOUR CONDITIONS OF SLOW-COOLED 50 Cu–50 Ni ALLOYS[a]

	$\psi = 35°$				$\psi = 15.5°$			
			Difference				Difference	
Condition	% Ni	% Cu	%Ni	% Cu	% Ni	% Cu	% Ni	% Cu
As polished	37	63			36	61		
Light etch	37	63	0	0	34	62	2	2
Heavy etch	63	64	1	1	33	64	3	3
3/0 Paper	37	63	0	0	34	64	2	3

[a] Data corrected for background and dead time; not corrected for absorption or fluorescence.

beam of the probe. The results are not so conclusive for microanalyses at low take-off angles. However, it can be said that local elevation differences of the size of the electron beam do not cause appreciable error if they are not abrupt, but instead occur over a distance equal to 4 to 5 beam diameters.

IV. EFFECTS OF LEACHING, DEPOSITION, SMEARING, AND ETCHING

Etching of metallographically prepared specimens is used to remove the smeared, cold-worked surface after final polishing so that the underlying microstructure is revealed. The etched microstructure is then examined to determine the portion of the specimen to be covered in the electron probe microanalysis and to select the path of the electron beam during that analysis. Many analysts then mark the area and path by microhardness indentations, and repolish the specimen to a smooth unetched surface for the microanalysis, believing that this procedure produces better results.

Etching can cause one or more elements to be leached from the phases of the microstructure so that the composition of a region at the surface (or in a surface layer) is changed. The surface composition can also be changed by the deposition from the etchant by electrolysis of one of the elements removed from other regions, or by chemical displacement of some element from the etching solution which is foreign to the specimen. Etching must thus be used with caution on specimens for electron probe microanalysis. However, a layer of one phase may be smeared over another phase by the polishing procedure itself, so that a false analysis may result unless the smeared layer is removed by etching. A qualitative approximation of the severity of the problem can be gained by the following idealized analysis. In this analysis, it is assumed that the absorption properties of the layers and elements are the same. Also, since it is simpler and more general, this analysis is conducted in terms of volume or atom percent. The equations can be converted to weight percent at any time by identifying the elements A and B and making the proper substitutions in the equations for converting volume or atom percent to weight percent.

A. Smearing

Let us first consider the problems of the smearing of a layer of pure element A over a large mass of pure element B. Then we wish to determine how thick the smeared layer of A must be for it to be detected. There are two answers possible, depending on whether the analysis is being conducted for element A or for element B. In the analysis for element A, there is a certain detection limit which is usually in the range of 0.01 to 0.1% A in B. In the analysis for element B, there is also a limit to the difference in content of B than can be detected at high concentrations of B. This limit may well be in the range of 0.1 to 0.5% of element B.

Let the analysis be made for element A when A is smeared over pure B. Let the detection limit for A be L, the smeared layer thickness of A be x, the electron beam penetration into the specimen be in the shape of a cylinder of diameter D penetrating a distance h as a cylinder plus the distance $D/2$ as a hemispherical tip on the cylinder. Then the detection limit L, in terms of volume fraction, is equal to the ratio of the volume of the smeared layer V_x in the beam to the total volume of the specimen penetrated by the beam V_s. The equation becomes

$$L = \frac{V_x}{V_s} = \frac{(\pi/4)D^2 x}{(\pi/4)D^2 + \frac{1}{2}(\pi/6)D^3} = \frac{3x}{3h + D} \tag{1}$$

or

$$x = L\frac{D + 3h}{3}. \tag{2}$$

Now let $L = 0.01$ vol % A, $D = 1\ \mu$, $h = 1\ \mu$. Then $x = 0.0001(1 + 3)/3 = \frac{4}{3} \times 10^{-4}\ \mu$, or $\frac{4}{3}$ Å. This thickness is less than one atom layer, indicating that any smearing could be detected, even if the limit of detection L were 0.1 instead of 0.01%. Increasing the depth of penetration of the electron beam at a given thickness of smeared layer would cause the amount determined to decrease, or conversely, would require that the smeared layer be thicker for it to be detected. Thus, if smearing were suspected, it would be indicated by a decrease in the amount of A determined when the beam voltage was increased.

If the analysis of the smeared specimen is conducted for element B, the detection limit for a difference in concentration of B at high concentrations may well be 0.1–0.5% B. That is, an area containing 99.5% B can be distinguished from an area containing 99.6% B (or 99.0% B) but not from an area containing 99.55% B. The same equation, Eq. (2), applies in this case. The detection limit L is now equal to the minimum difference detectable between neighboring regions. If $L = 0.1\%$, $h = 1\ \mu$, and $D = 1\ \mu$, then $x = \frac{4}{3} \times 10^{-3}\ \mu$, or 13 Å. This is a layer of not more than 3 to 4 atoms in thickness, again indicating that smearing should be readily detectable.

Let us now consider a case of smearing wherein a layer of one phase α is smeared over another β. Both contain the same elements A and B but in different proportions. We are interested in how thick the smeared layer can be before it can be detected or can interfere with the precision of the analysis. The detection (or difference) limit L for an element will be equal to the ratio of the volume of the smeared layer of α in the beam to the total volume times the ratio of the concentration difference in the element between phases α and β to the concentration of the element in the deeper phase. That is, for the same beam geometry as before

$$L = \left(\frac{(\pi/4)D^2x}{(\pi/4)D^2h + \frac{1}{2}(\pi/6)D^3}\right)\left(\frac{c_2 - c_1}{c_2}\right) \tag{3}$$

where c_1 is the concentration of A in α, c_2 the concentration of A in β, and α is smeared over β. Then

$$x = L\left(\frac{3h + D}{3}\right)\left(\frac{c_2}{c_2 - c_1}\right). \tag{4}$$

If $L = 0.1\%$ A, $h = 1\ \mu$, $D = 1\ \mu$, $c_2 = 10.0\%$ A, and $c_1 = 9.0\%$ A, then $x = 133$ Å in thickness. If c_1 in this example were 5% A, then $x = 25$ Å in thickness.

From these approximations, it is seen that the smearing of one phase over another can be quite detrimental to the precision of the electron probe microanalysis. Thus, it would seem that such layers should be removed by etching or electropolishing the specimen surface for the microanalysis.

B. *Leaching, Deposition, and Etching*

Etching can cause changes in the surface condition of a specimen other than the removal of the disturbed layer. An element of the specimen can be leached from a thin layer at and near the surface, and foreign elements can be deposited from the etchant by the action of the electrolytic cell set up between phases of different composition. The maximum depth of the leached or deposited layer that can be tolerated can be approximated from the equations given previously for the smeared layer. The results are essentially the same, layer thicknesses ranging from approximately 1 to 100 Å, depending on the tolerance and beam geometries assumed. Leaching or deposition should also be detectable by a change in measured composition with different beam voltages, a decreasing content with increasing voltage indicating the presence of a leached or deposited layer on the surface of the specimen. Leaching should be a minimum when fast-attacking chemical or electrolytic etchants or polishes are used since there is little time for the required diffusion to take place relative to the time required for removal of the leached layer. Deposition can only be prevented by changing the composition of the etchant, or the amount can be determined and allowances made in the final calculations of composition.

The data presented in Table II for the copper–nickel alloy specimen and in Section III,B for the aluminum–uranium alloy specimen show that leaching is not a problem when the etching attack is rapid, even when the specimen is heavily overetched. In these specimens, a layer of metal several microns thick was removed in not more than 1 min of etching. It must be remembered while examining the data of Table II that the segregation in the specimen was three-dimensional. The differences in composition reported for the four surface conditions are actually more likely due to different cuts of that segregation than due to compositional changes in the surfaces produced by etching, leaching, and smearing.

C. Effects of Other Beam Geometries

Approximations of layer thicknesses required to interfere with the precision of the microanalysis can be made for other beam geometries and penetration depths. There is, however, no appreciable change in the thicknesses so approximated (all being of the order of 100 Å or less) if reasonable assumptions are made on detection limits, beam diameters, and beam penetration depths. Large differences between the atomic numbers of the elements of the specimen, or in their absorption coefficients, can affect the thickness required for a layer to interfere, but only by less than an order of magnitude. It can then be reasonably concluded that smeared, leached, or deposited layers of compositions different from the base material can cause detectable errors in the microanalysis when their thicknesses are of the order of 100 Å or less.

V. Effects of Anodic or Vapor Deposited Films

A technique very useful in optical metallography is that of anodizing the specimen to produce an oxide film which shows interference colors [5–7]. Under proper control, and in a given alloy system, the color developed on a given phase is a function only of the composition of that phase and the voltage applied to the specimen during anodizing. Phase boundaries can be accurately located by color differences, segregation gradients can be delineated by color, and phases can be identified through a series of specimens. The technique has been successfully used on alloy bases of Ti, Zr, Hf, Ta, Nb, W, and U.

A newer technique uses vapor deposited transparent films, such as TiO_2, thin enough to show interference colors [8]. Usually, a thickness of about 500 Å is sufficient. The difference in optical properties between metallic phases causes different phase shifts in the light reflected at the film-metal interface (the film is uniform in thickness). The color differences observed are mostly those of changes in shading within a color range, although differences in color can be produced when the average color is in the red-blue region of the color order. This technique has been used to discriminate between phases in alloys which cannot be anodized, such as steels, brasses, nickel-base alloys, etc.

Such color-producing films would be of considerable aid to the analyst in visual guidance of the electron beam over an area of interest, as well as in locating the area of interest. They should not, however, interfere with either beam operation or the precision of the analysis. An experimental check was made with the results reported below.

A. Anodized Films

The oxide film formed on alloys of Zr, Ti, Nb, Hf, Ta, W, and U by anodization is a stoichiometric (or nearly so) oxide of the metal on which it is formed. As such, it is an electrical insulator. Its thickness will vary from a few hundred to a few thousand angstroms, depending almost linearly on the voltage at which it is formed. Since the film is an insulator and its thickness can be appreciable, a static charge could be built up on it by the electron beam, causing the beam to be unstable. The addition of the oxygen to form the oxide also causes the base metal to be diluted. This could cause an error in much the same manner as leaching (previously discussed) unless it is somehow corrected.

To determine the effects of the anodized film alone, a high purity zirconium specimen was masked with tape into three regions. One region was left free of oxide, one was anodized at 30 V to form an oxide film about 450 Å thick, and the third was anodized at 100 V to produce an oxide film about 1500 Å thick. X-ray intensity measurements were made in the electron probe on each of the anodized films with the bare region serving as the standard. No deviation, deflection, or broadening of the electron beam was observed, indicating that the oxide films did not become sufficiently charged to interfere with the operation of the beam. The loss of X-ray intensity for each condition is shown in Table III. The oxide layer 450 Å thick caused a loss of only 0.5% of the

TABLE III
LOSS OF INTENSITY BY ANODIZATION

Intensity ratio	Zirconium (%) (uncorrected)	Difference in Zirconium (%) bare minus coated
Pure Zirconium		
$\frac{I_{Zr\ 450\ Å\ film}}{I_{Zr\ bare}} = 0.995$	99.5	0.5
$\frac{I_{Zr\ 1500\ Å\ film}}{I_{Zr\ bare}} = 0.943$	94.3	5.7
Nb–25% Zr Alloy		
$\frac{I_{alloy\ bare}}{I_{Zr\ bare}} = 0.252$	25.2	
$\frac{I_{alloy\ 450\ Å\ film}}{I_{Zr\ 450\ Å\ film}} = 0.249$	24.9	0.3
$\frac{I_{alloy\ 1500\ Å\ film}}{I_{Zr\ 1500\ Å\ film}} = 0.243$	24.3	0.9

X-ray intensity, while that 1500 Å thick caused a loss of 5.7%. If it is assumed that the zirconium content of the oxide film is reduced to 23 vol % (or atom percent), then the equations given previously for the "leached surface" layer can be used to predict that film thicknesses of 600 and 3000 Å would be required to produce the losses observed. Since the shape and penetration of the electron beam are not known with any precision, the agreement between calculated and actual oxide film thicknesses is good. Also, the predicted film thicknesses are of the order of magnitude of the known thicknesses, indicating that nothing other than dilution was likely present.

In much of the quantitative optical microscopy using anodization to produce color contrast, oxide film thicknesses of 1000 to 2000 Å are frequently required to obtain suitable color differentiation between phases. As comparison of bare zirconium with oxidized zirconium (1500 Å) showed an appreciable loss of X-ray intensity, it seemed desirable to compare results obtained during analysis of an alloy with a pure standard anodized to the same film thickness. A specimen of a Nb–25% Zr alloy was masked and anodized to produce the same colors (and the same film thickness) as those produced on the zirconium standard specimen (above). The data obtained are presented in Table III. In each case, the analysis of the alloy was based on the region of the zirconium standard having the same oxide film thickness. The film-free region of the specimen analyzed 25.2% Zr. The presence of the oxide film 450 Å thick on both the alloy, and the standard reduced it to 24.9% Zr, a loss of 0.3% Zr from that made on the bare specimen. The film 1500 Å thick on both specimen and standard resulted in an analysis of 24.3% Zr, a loss of 0.9% Zr from that made on the bare specimen. While it cannot be proved at present that the alloy was sufficiently homogeneous to insure that the loss measurements are accurate to 0.1% Zr, past experience has shown that for colors as uniform as these were, the alloy is homogeneous to at least 0.5% Zr by all other means of analysis, including microspark spectroscopy, microbeam X-ray fluorescence, and microdrilling for spark spectroscopy.

These results indicate that the anodized films can be used satisfactorily to aid the analysis provided that the standard is also anodized to the same film thickness. It is also desirable to use the thinnest films that will produce adequate color differentiation.

B. Vapor Deposited Films

It has been shown recently [8] that vapor deposited films of oxides having indices of refraction of about 2.0 or more can be used to reveal, by interference color, the variation of optical constants of the underlying metal phases, and, consequently, the variation and differences in composition of those phases. For example, TiO_2 can be evaporated onto nickel- or iron-base alloys having several phases or composition gradients, and the phases or gradients can be detected by color in white light optical microscopy. These films can then be used in the same way as the anodized films discussed above, provided that the standards are also coated to the same film thickness.

The effect of such vapor deposited films on the behavior of the electron beam and on the precision of the microanalysis was determined by vapor depositing approximately 500 Å of TiO_2 on a portion of the 50 Cu–50 Ni specimen used in determining the data reported in Section III,B. Pure copper and pure nickel standards were partially coated at the same time to the same film thickness. The segregation occurring in the sample (see Fig. 11) was observable in the white light optical microscope as a variation in color from a medium blue in the copper-rich regions to a very light blue-white in the nickel-rich regions. No sharp color boundaries were observed, just as no sharp composition boundaries were observed by the probe analysis (see Fig. 11). The copper and nickel K_α X-ray intensities were measured at 30 kV and 0.02 μA specimen current on both coated and uncoated samples and standards. The coating of TiO_2 did not cause an observable difference in the X-ray intensities nor did it interfere with the stability of the electron beam. About 0.4% Ti was detected at a beam voltage of 20 kV when the beam was focused on either the pure copper or pure nickel standard coated with TiO_2. This is proof of the detectability of very thin films on the specimens, lending further credence to the analysis made in Section IV on leaching, smearing, and deposition.

These results show that the thin vapor deposited oxide films do not interfere appreciably with the electron probe microanalysis, and that the errors produced by their presence can be allowed for should it be necessary to do so. The very great increase in optical contrast produced by the films should aid the analyst greatly in detecting areas of interest in the specimen, and in guiding the traverse of the beam across them.

VI. Conclusions

The effects of surface roughness and surface films on the precision of electron probe microanalysis of metal specimens have been examined. Surface roughness was produced by geometry, abrasion, and etching. Surface films were formed by anodization and by vapor deposition. The following conclusions may be drawn.

(1) Surface roughness, per se, does not affect the precision of the microanalysis at the higher take-off angles in the X-ray optical system if the elevation differences are of the size of the diameter of the electron beam or smaller. This is true whether they occur as abrupt steps or occur over a distance of several beam diameters. At low take-off angles, the elevation differences must be smaller, with an upper limit of about 1 μ.

(2) The elevation differences and slopes produced in an overetched specimen do not affect the precision of the microanalysis if they are of a size near that of the diameter of the electron beam or occur over a distance equivalent to several beam diameters.

(3) A layer of one phase or element smeared over another can seriously affect the precision of the microanalysis when the smeared layer is only a few tens to a few hundreds of angstroms thick, depending on the elements and compositions present.

(4) The leaching of one element from the specimen surface by the etchant, or the deposition of an element from it, can seriously affect the precision of the analysis when the layer thicknesses are of the same order as those for smearing.

(5) For a microanalysis of high precision, the mechanically polished specimens should be lightly etched or electropolished to remove the smeared and cold-worked layer always present on the surface. Attack polishing (polishing with a dilute etchant on the cloth) can decrease the severity of the problem. Etching with solutions that attack rapidly decreases the amount of leaching that can occur. Rapid chemical or electrolytic polishing seems best.

(6) Anodized and vapor deposited oxide films used for detecting compositional differences by color in segregation gradients and multiphase specimens do not interfere with electron probe operation. They can be used as an aid in the selection of the path of the electron beam over the specimen. The precision of the analysis will not be decreased appreciably if the analytical standards are coated to the same thickness with the film formed on the specimen.

Acknowledgments

The authors would like to acknowledge the contributions to this work made by T. J. Henson, Metals and Ceramics Division, in the metallographic preparation of all of the specimens, and by H. W. Dunn, Analytical Chemistry Division, in the operation of the electron probe for the collection of the data at the take-off angle of 15.5°.

References

1. G. L. Kehl, "The Principles of Metallographic Laboratory Practice." McGraw-Hill, New York, 1949.
2. E. L. Long, Jr. and R. J. Gray, *Metal Progr.* **74**, 145–48 (1958); P. Rothstein and R. F. Turner, *Symp. Methods of Metallographic Specimen Preparation.* ASTM-STP-285, pp. 90–102. Am. Soc. Testing Materials, Cleveland, 1960.
3. W. J. McG. Tegart, "The Electrolytic and Chemical Polishing of Metals," Pergamon Press, Oxford, 1959; D. M. McCutcheon and W. Pahl, *Metals Progr.* **56**, 674 (1949).
4. E. L. Long, Jr., Oak Ridge Nat. Lab., personal communication.
5. E. Ence and H. Margolin, *J. Metals* **6**, 346–48 (1954).
6. M. L. Picklesimer, Anodizing as a metallographic technique for zirconium base alloys. ORNL-2296 (1957).
7. M. L. Picklesimer, *Appl. Phys. Letters* **1**, 64–65 (1962).
8. J. O. Stiegler and R. J. Gray, Microstructural discrimination by deposition of surface films, *AEC Metallography Group Meeting, 20th, Denver, Colorado, 1966*; to be published by Dow Chem. Co., Rocky Flats Div., Golden, Colorado.

Electron Probe Microanalysis in Mineralogy*

CYNTHIA W. MEAD

U.S. Geological Survey
Washington D.C.

I. Introduction

The applications of electron probe microanalysis in the field of mineralogy are many and varied. They include the analysis of new minerals of limited quantity as well as other studies relating to minerals, such as identification of inclusions, intergrowth of phases and homogeneity within a single phase. Other studies have been made on both the metallic and nonmetallic constituents of meteorites, particles of cosmic and terrestrial origin, and cathodoluminescence as it occurs in minerals.† The advantage of this type of analysis is a nondestructive, in-place analysis of micron-size areas. Tedious, if not impossible, hand-picking of grains is not necessary, and it is possible to study minerals as they are found in nature when thin or polished sections are used.

The capabilities of the electron probe have increased greatly since the instrument was first used more than ten years ago. The instrumentation has been improved to such an extent that with an electron spot size now often

* Approved for publication by Director U.S. Geological Survey.

† The author would like to bring to the attention of the reader the review paper by Keil [1].

smaller than 1 μ, it is possible to analyze for all the elements from Be through U.

II. Sample Preparation

Sample preparation of minerals has been dealt with separately in this book in the chapter by Macres and Taylor [1a]; therefore suffice it to say that most mineralogical samples are prepared either as discrete grain mounts, as polished sections, or as polished thin sections. Because they vaporize in a vacuum, Canada balsam and Lakeside are not suitable for mounting thin sections which will be used in the electron probe. Clear epoxy resin is a suitable substitute. Regardless of the type of sample, all specimens should be polished to a smooth surface without any optically visible relief to avoid errors in quantitative analysis caused by X-rays emerging from the samples at varying angles. If this is not possible, as in the case in which loss of sample is feared, unpolished specimens may be used to obtain qualitative information on the major elements present. Electrically nonconducting specimens must be coated with an evaporated film of some conducting material, usually carbon or aluminum.

For the best quantitative results, it is helpful to get some preliminary qualitative information to determine what elements are present and in what relative amounts so that standards may be chosen and coated at the same time as the sample.

III. Problems Inherent in Mineral Analysis

Minerals are seldom simple systems chemically, and for this reason electron probe analysis in the field of mineralogy is not as easy and straightforward as might be expected. Minerals containing only two elements are rare, and minerals containing five or six elements are more frequently the case. Multicomponent samples with phases of varying hardness can cause difficulties in polishing. The conductive coating evaporated on nonconducting samples can enhance the phases of interest when viewed optically, but often it obscures them to such an extent that it is necessary to experiment to find the right type and thickness of coating which will give good conductance without obscuring the phase optically. Some nonmetallic and nonconducting minerals are adversely affected when the electron beam hits them and may actually decompose or have holes drilled in them. The latter problem can be alleviated by evaporating a heavier coating on the sample and/or lowering the specimen current, and in some cases defocusing the electron beam slightly.

Most electron probes have a maximum of three spectrometers for simultaneous analysis. In resetting the spectrometers for additional elements, a slight error may be introduced if the electron beam is moved to a different area because of contamination build-up. The error increases with the heterogeneity of the sample.

The excitation voltage for each series of X-ray lines increases with increasing atomic number, and when a sample is analyzed which has elements varying widely in atomic number, care must be exercised in the selection of operating voltage so that all the elements of interest can be detected. This may require using more than one voltage for a complete analysis because the optimum operating voltage for some of the lighter elements is lower than the excitation voltage for some of the heavier elements.

Many of the correction procedures for quantitative analysis were worked out for two component systems, and although some of them work well for complex minerals, they are not always satisfactory and in any case become more complicated.

IV. Qualitative Analysis of Minerals

Although it is undeniable that in some cases quantitative chemical analyses are essential, absolute chemical values are not required for many applications of the electron probe in mineralogy. Often it is sufficient to know what elements are present and how these elements are distributed within the sample. This type of electron probe microanalysis, when used in conjunction with other analytical methods, has been extremely helpful in providing evidence required to answer what otherwise might be an insoluble problem.

A number of electron probe procedures are used in the qualitative analysis of minerals. These include spectrometer traces and scans and the use of oscilloscope displays.

Spectrometer traces are made by attaching motors to the spectrometers in such a way that they will be driven in synchronization with strip charts. A spectrum is traced out on a chart which shows peaks for the elements which are present in the sample. If an electron probe has three spectrometers, it is possible to make three spectrometer traces simultaneously which, by careful choice of crystals, cover most, if not all, the elements detectable. Similar traces may be made on standard materials for comparison with the unknown, and in this way, it is possible to obtain qualitative information on the relative amounts of the element present. Unless the standard material is very similar to the sample, it is not valid to make a direct comparison for a given element on the basis of the spectrometer trace alone or to assign to it any more than a semiquantitative value.

As an example of this technique, Fig. 1 shows spectrometer traces for Ba and Ti in benitoite ($BaTiSi_3O_9$) and for Ti metal. The Ti content of benitoite is 11%.

Spectrometer scans are very useful when the material being studied is either zoned or not homogeneous. They are made by positioning one or more spectrometers for a given element, and then moving the sample under the beam at a constant speed. Variations in the element of interest across the sample are recorded on a strip chart.

Figure 2 shows spectrometer scans for Ni and S in a metallic inclusion in the Bonita Spring chondrite. The kamacite which contains about 7% Ni is surround by troilite which contains about 36% S. The remainder of both the kamacite and troilite is Fe.

One of the best means for a quick check for sample homogeneity is via an oscilloscope display of X-ray emission. In most instruments, this type of display of data is accomplished by causing the electron beam to sweep an

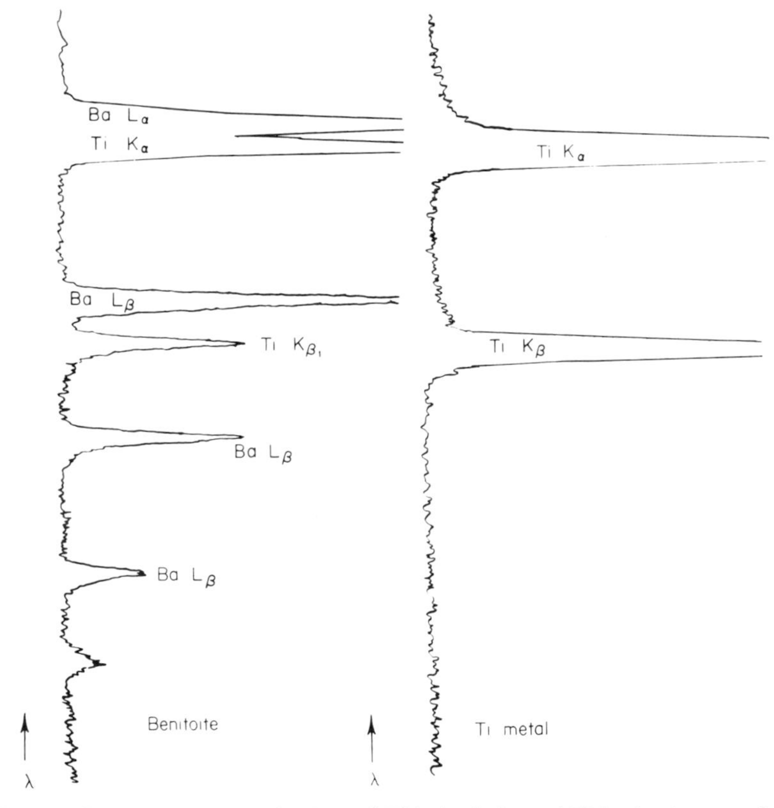

FIG. 1. Spectrometer traces for Ba and Ti in benitoite and Ti in titanium metal using a PET crystal for the analysis.

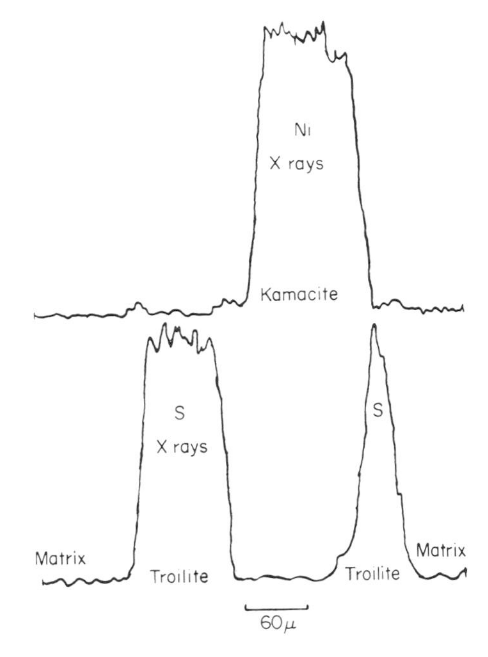

FIG. 2. Spectrometer scans for Ni and S in Bonita Spring chondrite using a LiF crystal for Ni and a PET crystal for S.

area of anywhere from 400 μ on a side down to an almost static spot while a spectrometer is peaked for the element of interest. As the electron beam sweeps over the sample, dots representing X-ray pulses for the element of interest are registered on the oscilloscope screen. In cases of low concentration of an element, it is helpful if these data can be accumulated on a storage oscilloscope over a long integration period. However, in some cases it is possible for continuous radiation, cosmic rays, and noise to also register dots on the screen, and if integrated for a long enough period, it can appear that some element which is not present is detectable. This possibility should be considered particularly when analyzing for elements of low concentration because it can lead to erroneous results. Long integration times can cause elements present in low concentrations to appear to be major constituents. Thus, to make a valid comparison of photographs taken of X-ray emission displays, all of the compared photographs should have been taken after the same length of integration on the screen. Figure 3 illustrates this possibility. The four pictures are all taken of the same particle, a kamacite inclusion in Bonita Spring chrondrite. The only difference among the pictures is the length of integration time. The scattered white dots represent background. Akin to this type of display are specimen current and backscattered electron images which show variation in atomic number across the sample. In addition, the backscattered electron images also show sample relief.

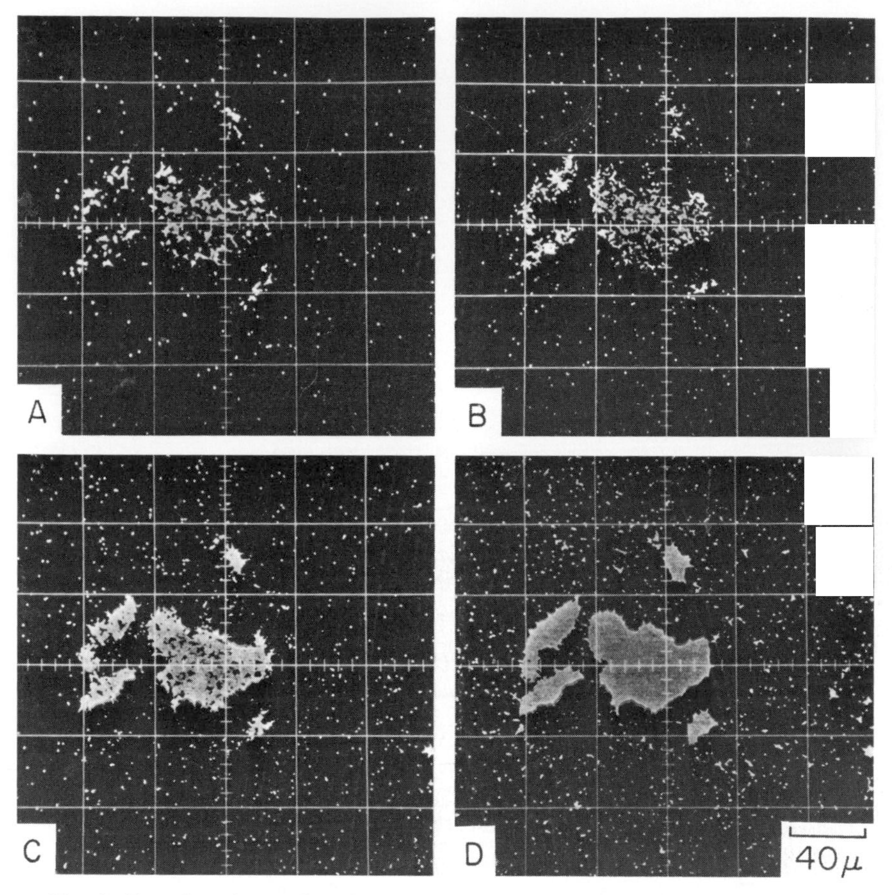

FIG. 3. Scanning pictures for Ni X-rays taken of the same kamacite inclusion in Bonita Spring chondrite but with the integration times (a) 30 sec, (b) 1 min, (c) 2 min, (d) 4 min.

V. Quantitative Analysis of Minerals

The methods of quantitative analysis have been described elsewhere in this volume by Brown [2] and by Macres and Wolfe [3]. However, it should be pointed out that in the case of complex minerals, even with the best operating conditions, the precision of a quantitative analysis is usually no better than ± 1–2% of the amount present. If the sample is to be analyzed by different people, and sometimes with different instruments, as is the case for "round robin" samples circulated by probe users groups, then a $\pm 5\%$ figure can be expected. In case of low concentration, a precision of $\pm 10\%$ is more probable. In nearly every case, this kind of precision is perfectly adequate for answering mineralogical problems, particularly when the information could not be obtained in any other way. Before reporting analytical results, geologists and mineralogists engaged in electron probe microanalysis would do well to learn something about the statistics of counting and the meaning of weight percent as used by the analytical chemist.

The easiest type of sample to analyze quantitatively is one for which a standard of similar composition is available. If a direct comparison of intensities can be made, the need for corrections is eliminated, and the most accurate results can be expected. Most mineralogical samples are not in this category. If samples are this well known, there is no need for quantitative electron probe microanalysis. However, this type of analysis can be useful for checking probe stability and for checking precision as well as for convincing the operator that he is using the proper electron probe techniques.

Another type of sample is one in which preliminary qualitative work by electron probe, spectrochemistry, or X-ray has made it possible to select standards not too different in composition from the sample. The corrections required are then kept to a minimum and very accurate analyses may be obtained.

The most difficult type of mineral analysis is the one in which the mineral is truly unknown and only a small amount of sample is available. All the information has been obtained by electron probe microanalysis, and no similar standard is available. In this case, pure element standards have to be relied upon, and because corrections are essential, assumptions based on the first approximation will have to be made as far as the composition of the unknown is concerned. If the concentration of an element in the unknown is low, the greater the correction needed and the greater sacrifice of accuracy. In order to arrive at a reliable analysis, several iterations will usually be necessary.

VI. Examples of Electron Probe Analysis in Mineralogy

A. Minerals

In the mineral group that includes the sulfides, arsenides, and sulfosalts are many of the most important ore minerals. Many are metallic and most of them are opaque. They are usually electrically conducting and are most frequently studied in polished sections. Some of them, including galena (PbS), pyrite (FeS_2), and pyrrhotite (FeS), make excellent mineral standards.

Many of these minerals occur as tiny inclusions or as intergrowths too fine to be identified positively by standard petrographic techniques and too small to be mechanically extracted. Qualitative electron probe analysis either by means of spectrometer traces or oscilloscope display is often sufficient to make an identification. An example of such identification is a study of copper arsenide minerals done at the U.S. Geological Survey [4] in which inclusions of niccolite, rammelsbergite, chalcocite, cuprite, maucherite, safflorite, tenorite, and also native silver and native copper were identified.

Some of these minerals show zoning, as in X-ray scanning pictures of bravoite (see Springer *et al.* [5]).

The oxides, multiple oxides, and hydroxides which have been studied with the electron probe include some of the main ore minerals of chromium, iron, and titanium, and several containing uranium or rare earth elements.

The sulfate, phosphate, and tungstate minerals are nonconductors, that, in some cases, are damaged by the electron beam; extra care must be used in their analysis. They are best studied in polished thin sections or as grain mounts.

Electron probe and X-ray diffraction studies of calcioferrite and montgomeryite which were done at the U.S. Geological Survey [6] made it possible to predict very accurately the formulas of these minerals.

A qualitative study for homogeneity and composition of zincian rockbridgeite made in this laboratory [6] showed that it is a mixture of zincian rockbridgeite and zincian lipscomite. Both of these minerals contain small, but approximately equal, amounts of zinc. X-ray work on the same material confirmed the identity of both minerals.

In the silicate group are some of the most important rock-forming minerals. They are best studied with the electron probe either in mounts of individual crystals or in polished thin sections. More electron probe work has been done on the pyroxenes and olivines than on any others in the group. Much of the study of these minerals has been done in conjunction with the study of meteorites.

It is increasingly common for descriptions of new minerals to include electron probe analyses. The recently described new minerals mackinawite [7], briartite [8], roquesite [9], hollingworthite [10], compregnacite [11], karelianite [12], markoite [13], mawsonite [14], and pabstite [15] have all been analyzed with the electron probe. Some of the minerals, exclusive of those in meteorites, which have been studied quantitatively with the electron probe are listed in Table I.

TABLE I

MINERALS ANALYZED QUANTITATIVELY WITH THE ELECTRON PROBE

Mineral	Formula	Authors and date	Reference
Native elements			
Maldonite	Au_2Bi	Boyer and Picot, 1963	[16]
Wairanite	CoFe	Challis and Long, 1964	[17]

TABLE I (continued)

Mineral	Formula	Authors and date	Reference
Sulfides, Arsenides, Sulfosalts			
Aikinite	$PbCuBiS_3$	Picot and Vernet, 1963	[18]
Algodonite	Cu_6As	Skinner *et al.*, 1962	[4]
Bravoite	$(Fe,Ni)S_2$	Springer *et al.*, 1964	[5]
Briartite	$Cu_2(Fe,Zn)GeS_4$	Francotte *et al.*, 1965	[8]
Chalcocite	Cu_2S	Mead, 1961	[19]
Chalcopyrite	$CuFeS_2$	Permingeat and Weinryb, 1960	[20]
		Birks *et al.*, 1959	[21]
Cubanite	$Cu_2S{\cdot}Fe_4S_5$	Birks *et al.*, 1959	[21]
		Oosterbosch *et al.*, 1964	[22]
		Chauris and Geffroy, 1966	[23]
Domeykite	Cu_3As	Skinner *et al.*, 1962	[4]
Dufrenoysite	$Pb_2As_2S_5$	Nowacki and Bahezre, 1963	[24]
Enargite	$3Cu_2S{\cdot}As_2S_5$	Geffroy and Lissillour, 1963	[25]
Galena	PbS	Mead, 1961	[19]
Germanite	$Cu_3(Fe,Ge)S_4$	Picot *et al.*, 1963	[26]
Gersdorffite	$NiS_2{\cdot}NiAs_2$	Geffroy and Lenfant, 1963	[27]
Hatchite	Tl–Pb–As sulfide	Nowacki and Bahezre, 1963	[24]
Hollingworthite	$(Rh,Pt,Pd)AsS$	Stumpfl and Clark, 1965	[10]
Hutchinsonite	Tl–Pb–Cu–Ag–As sulfide	Nowacki and Bahezre, 1963	[24]
Jordanite	$4PbS{\cdot}AsS_3$	Nowacki and Bahezre, 1963	[24]
Linneite	Co_3S_4 with Cu	Permingeat and Weinryb, 1963	[20]
Lengenbachite	$6PbS{\cdot}Ag_2S{\cdot}2As_2S_3$	Nowacki and Bahezre, 1963	[24]
Mackinawite	$(Fe_{0.96}Ni_{0.04})S$	Evans *et al.*, 1964	[7]
Mawsonite	$Cu_7Fe_2SnS_{10}$	Markham and Lawrence, 1965	[14]
Meneghinite	$4PbS{\cdot}Sb_2S_3$	Burnol *et al.*, 1965	[28]
		Fredriksson and Anderson, 1964	[29]
Pyrite	FeS_2	Mead, 1961	[19]
Pyrrhotite	FeS	Bizouard and Roering, 1958	[30]
Rathite	$3PbS{\cdot}2As_2S_3$	Nowacki and Bahezre, 1963	[24]
Rezbanyite	$Pb_3Cu_2Bi_{10}S_{13}$	Picot and Vernet, 1963	[18]
Roquesite	$CuInS_2$	Picot and Pierrot, 1963	[9]
Skleroclase	$PbS{\cdot}As_3S_3$	Nowacki and Bahezre, 1963	[24]

TABLE I (continued)

Mineral	Formula	Authors and date	Reference
Skutterudite	Fe–Co–Ni arsenide	Klemm, 1965	[31]
Smithite	$Ag_2S{\cdot}As_2S_3$	Nowacki and Bahezre, 1963	[24]
Sphalerite	(Zn,Fe)S	Bizouard and Roering, 1958	[30]
		Williams, 1967	[32]
Stannite	$Cu_2S{\cdot}FeS{\cdot}SnS_2$	Picot *et al.*, 1963	[33]
Tetradymite	$Bi_2(Te,S)_3$	Picot and Vernet, 1963	[18]
Trechmannite	$Ag_2S{\cdot}As_2S_3$	Nowacki and Bahezre, 1963	[24]
Oxides, Multiple Oxides, Hydroxides			
Compreignacite	$K_2U_6O_{19}{\cdot}11H_2O$	Protas, 1964	[11]
Crednerite	$CuO{\cdot}Mn_2O_3$	Gaudefroy *et al.*, 1963	[34]
Davidite	Rare earth oxide	Welin and Uytenbogaardt, 1963	[35]
Euxenite	$(Y,Ca,Ce,U,Th)(Nb,Ta,Ti)_2O_6$	Cruys *et al.*, 1964	[36]
Fergusonite	$(Y,Er,Ce,Fe)(Nb,Ta,Ti)O_4$	Cruys *et al.*, 1964	[36]
Franklinite	$(Fe, Zn, Mn)O{\cdot}(Fe, Mn)_2O_3$	Frondel and Klein, 1965	[37]
Guillimenite	$Ba(UO_2)_3(OH)_4(SeO_3)_2{\cdot}3H_2O$	Pierrot *et al.*, 1965	[38]
Hematite	Fe_2O_3	Bolfa *et al.*, 1961	[39]
		Agrell and Long, 1960	[40]
Hetaerolite	$ZnO{\cdot}Mn_2O_3$	Frondel and Klein, 1965	[37]
Ilmenite	$FeTiO_3$	Bolfa *et al.*, 1961	[39]
		Wright and Lovering, 1965	[41]
		Temple *et al.*, 1966	[42]
Karelianite	V,Fe,Cr,Mn oxide (V_2O_3)	Long *et al.*, 1963	[12]
Magnetite	Fe_3O_4	Babkine *et al.*, 1965	[43]
		Agrell and Long, 1960	[40]
Markoite	$CaMn_2O_4$	Gaudefroy *et al.*, 1963	[13]
Pseudobrookite	Fe,Ti oxide	Ottemann and Frenzel, 1965	[44]
Rutile	TiO_2	Niggli, 1965	[45]
Spinel	$MgAl_2O_4$	Babkine *et al.*, 1965	[43]
		Agrell and Long, 1960	[40]
		Long *et al.*, 1963	[12]
Uraninite	UO_2	Welin, 1961	[46]
Titanomagnetite	Fe_3O_4 with Ti	Babkine *et al.*, 1965	[43]
		Wright and Lovering, 1965	[41]
Sulfates, Phosphates, Tungstates			
Calcioferrite	$Ca_2Fe_2(PO_4)_3(OH){\cdot}7H_2O$	Mead and Mrose, 1968	[6]
Ferritungstate	$Fe_2O_3{\cdot}WO_3{\cdot}6H_2O$	Burnol *et al.*, 1964	[47]
Graftonite	$(Fe,Mn,Ca)_3P_2O_8$	Hurlbut, 1965	[48]

TABLE I (continued)

Mineral	Formula	Authors and date	Reference
Jarosite	$KFe_3(SO_4)_2(OH)_6$	Braitsch and Keil, 1965	[49]
Montgomeryite	$Ca_2Al_2(PO_4)_3(OH)\cdot 7H_2O$	Mead and Mrose, 1968	[6]
Sarcopside	$(Fe,Mn,Mg)_3(PO_4)_2$	Hurlbut, 1965	[48]
Vivianite	$Fe_2P_3O_8\cdot 8H_2O$	Hurlbut, 1965	[48]
Silicates			
Bazzite	Sc, rare earth silicate	Nowacki and Phan, 1964	[50]
Biotite	$K(Mg,Fe,Al)_3(Si,Al)_4O_{10}(OH)_2$	Bahezre *et al.*, 1965	[51]
Clino-ferrosilite	$(Fe_{0.95}Mn_{0.05})SiO_3$	Bown, 1965	[52]
Clinopyroxene	$RSiO_3$[a]	Muir and Long, 1965	[53]
		Smith, 1966	[54]
Cordierite	$(Mg,Fe,Mn)_2(Al,Fe)_4Si_5O_{18}$	Agrell and Long, 1960	[40]
Feldspar			
Alkali	$KAlSi_3O_8$–$NaAlSi_3O_8$	Smith and Ribbe, 1966	[55]
Plagioclase	$NaAlSi_3O_8$–$CaAl_2Si_2O_8$	Ribbe and Smith, 1966	[56]
Olivine	$(Mg,Fe)_2SiO_4$	Smith and Stenstrom, 1965	[57]
Orthopyroxene	$(Mg,Fe)SiO_3$	Green, 1963	[58]
		Muir and Long, 1965	[53]
		Howie and Smith, 1966	[59]
Pabstite	$Ba(Sn_{0.77}Ti_{0.23})Si_3O_9$	Gross *et al.*, 1965	[15]
Phlogopite	$H_2KMg_3Al(SiO_4)_3$	Metais *et al.*, 1962	[60]
Rustumite	$Ca_4Si_2O_7(OH)_2$	Agrell, 1965	[61]
Thorite	$ThSiO_4$	Welin and Uytenbogaardt, 1963	[35]
Thortveitite	$(Sc,Y)Si_2O_7$	Phan, 1965	[62]
Tremolite	$Ca_2(Mg,Fe)_5Si_8O_{22}(OH)_2$	Welin, 1961	[48]
Zircon	$ZrSiO_4$	Pigorini and Veniale, 1966	[63]

[a] R = Ca, Mg, Fe″, Fe‴, Mn″, Zn, K, Na, Al, Ti.

B. Meteorites

Partially as a direct result of the space program, an increased interest in objects of outer space has resulted in the electron probe study of many meteorites. The studies have sought to learn something about the origin, composition, structure, and thermal history of all types of meteorites. Valuable additions to the phase diagrams of some pertinent systems have been made. In the iron meteorites the minerals which have been studied most thoroughly include the Ni/Fe phases, kamacite, taenite, and plessite [64–67], the phosphides, schreibersite, and rhabdite [68–70], the sulfide, troilite [65, 69] and the phosphates, sarcopside, and graftonite [71]. The olivines and pyroxenes in stony meteorites or chondrites have also been studied [71–74].

Mineralogical studies in which a number of minerals in iron meteorites were analyzed with the electron probe include studies by El Goresy [75], by Marshall and Keil [76], and by Park *et al.* [77]. Similar studies of chrondrites have been made by Keil and Anderson [78], by Jobbins *et al.* [79] and by Bostrom and Fredriksson [80]; Keil and Fredriksson [81] have also studied a number of minerals in an anchondrite.

Several new minerals have been discovered during electron probe microanalysis of meteorites. Some of these are sinoite described by Keil and Mason [82], roedderite described by Fuchs *et al.* [83], djerfisherite described by Fuchs [84], merrihueite described by Dodd *et al.* [85], and ureyite described by Frondel and Klein [86]. Two others, still unnamed, were described by El Goresy [75].

Table II lists some of the minerals in meteorites which have been studied with the electron probe.

TABLE II
MINERALS IN METEORITES WHICH HAVE BEEN ANALYZED WITH THE ELECTRON PROBE

Native Elements	*Phosphides and Nitrides*	*Silicates*
Cliftonite	Rhabdite	Albite
Copper	Schreibersite	Anorthite
Graphite	Sinoite	Merrihueite
Kamacite	*Oxides*	Olivine
Pentlandite	Brucite	Fayalite
Plessite	Chromite	Forsterite
Taenite	Goethite	Orthoclase
	Ilmenite	Oligoclase
Carbides	Maghemite	Roedderite
Cohenite	Magnetite	Pyroxene
Moissanite	Quartz	Augite
	Rutile	Bronzite
Sulfides		Chromium clinopyroxene
Chalcopyrrhotite	*Phosphates and Carbonates*	Clinoenstatite
Daubreelite	Apatite	Diopside
Djerfisherite	Breunnerite	Enstatite
Ferro-alabandite	Calcite	Hypersthene
Mackinawite	Chlorapatite	Pigeonite
Oldhamite	Farringtonite	Ureyite
Sphalerite	Graftonite	
Troilite	Sarcopside	

C. Dust Particles

For many years, scientists have been making estimates of how much cosmic dust falls on the earth each year. The variations in the estimates have been enormous because there has been until recently no certain way of differentiating the "cosmic" dust from any other kind of dust. For this reason, extensive studies have been undertaken in order to find methods of distinguishing "cosmic," or extraterrestrial dust, from terrestrial dust. However, it was not until the electron probe became available that individual micron-sized dust particles could be characterized chemically in a nondestructive way. One of the first published electron probe studies [87] involved the analysis of deep-sea spherules collected by the Swedish Deep Sea Expedition in 1947–1948.

For comparative dust studies with the electron probe, materials from various sources have been examined. Particles found in soil around craters which are of obvious meteoritic origin have been compared with particles collected from the soil around craters believed to be of meteoritic origin and with the metallic particles found in tektite glass. Particles collected from the atmosphere, from deep-sea sediments, from polar ice, and from other places thought to be free from terrestrial contamination have been compared with particles of known terrestrial origin such as from volcanic and industrial sources.

The diameter of the particles studied ranges from less than 1 μ to about 200 μ. Their shapes vary from perfect spheres to irregular particles. Some are metallic, others are nonmetallic or even glassy. In most cases the presence of Ni, Co, and sometimes Si is used as a positive indicator for possible meteoritic origin. For example, the metallic spherules, extracted from tektite glass from the Philippine Islands, South Viet Nam, and Indonesia and described by Chao *et al.* [88, 89], have an Ni content very similar to that of the metallic spherules from Bosumtwi Crater glass described by El Goresy [90] and to the metallic spherules from Meteor Crater, Arizona, described by Mead *et al.* [91].

In what is probably the most comprehensive published series of articles on the nature of dust particles [92–98] more than 500 individual dust particles have been examined with the electron probe. The particles studied came from a variety of sources including polar ice, soil from around the Sikhote–Alin meteorite shower, ice caves, the atmosphere, sediments, industrial sources, volcanoes, and even from spherules produced artificially by melting pieces of an iron and a stony-iron meteorite with an electric arc welder and an acetylene torch.

Several benefits have already accrued from the electron probe analysis of dusts. Methods for handling individual micron-sized particles have been developed. It has become possible to distinguish at least some kinds of "cosmic" dust from dust of terrestrial origin. The dust of "cosmic" origin has been found to constitute only a relatively small portion of the dust studied, and therefore the figures on the accumulation of cosmic dust on the earth's surface probably favor the lower estimates.

D. *Cathodoluminescence*

Certain minerals fluoresce when placed under the electron beam. This fluorescence, known as cathodoluminescence, can serve several useful purposes. It can be used to identify or differentiate minerals in thin or polished sections, and it can be an indicator of minor impurities which may be present in concentrations less than can otherwise be detectable by the probe. It is also possible that the intensity of the luminescence may be an indicator of the temperature of formation of the mineral. The bright blue cathodoluminescence of benitoite, a barium titano-silicate, is routinely used as a means for checking the focus of the electron beam and column alignment.

Any probe user who has analyzed minerals is aware of the phenomena of cathodoluminescence, but surprisingly little has been written about it as it occurs in minerals. Long and Agrell [99] describe cathodoluminescence in thin sections of minerals, and Keil [100] describes how cathodoluminescence can be used in modal analysis of minerals.

Table III presents a list of a few of the minerals exhibiting cathodoluminescence.

TABLE III

LIST OF MINERALS SHOWING CATHODOLUMINESCENCE

Mineral	Formula	Color of luminescence
Alumina	Al_2O_3	Blue
Apatite	$Ca_4(Ca,F)(PO_4)_3$	Light blue
Barite	$BaSO_4$	Green
Benitoite	$BaTiSi_3O_9$	Bright purplish-blue
Cadmium sulfide	CdS	Bright green
Calcite	$CaCO_3$	Light yellow to orange red
Celestite	$SrSO_4$	Blue
Dolomite	$CaCO_3 \cdot MgCO_3$	Bright red-violet to blue
Enstatite[a]	$MgSiO_3$	Blue and/or red
Fluorite	CaF_2	Blue or red-violet
Hinsdalite	$(Pb,Sr)Al_3(PO_4)(SO_4)(OH)_6$	Blue
Magnesite	$MgCO_3$	Blue
Oldhamite[a]	CaS	Yellow
Orthoclase	$(K,Na)AlSi_3O_8$	Blue
Quartz	SiO_2	Orange, gray
Scheelite	$CaWO_4$	Bright blue
Sinoite[a]	Si_2N_2O	Green

[a] From Keil [82]. All others observed in the author's laboratory.

VII. Future Trends

Further developments in the electron probe analysis of minerals will probably take place in two areas: (1) the continual upgrading of presently available equipment and (2) greater use of multipurpose instruments.

Included in the first category would be greater efficiency of detection elements Be to Na by means of improved detectors, more highly reflective crystals, and built-in decontamination devices. Improvements in resolution of the electron beam via a smaller spot size should increase the capabilities for small particle analysis. Greater use of programmed spectrometers and/or multichannel analyzers can shorten the time required to make spectrometer traces and to do some routine analyses.

Greater use of computers can facilitate quantitative analysis by reducing personnel time and human error. Their use also lessens the need for obtaining standards similar in composition to the unknown in the hope of avoiding doing the calculations with a slide rule or desk calculator.

In the second category, greater use will be made of multipurpose instruments which combine electron microscopy, electron diffraction, and electron probe analysis. Such an instrument, known as EMMA, was described by Duncumb [101]. This type of instrument would allow correlation of the structure and morphology with the chemical composition of particles as small as 0.1 μ in diameter.

VIII. Conclusions

In the brief span of ten years, the electron probe has proved to be one of the most valuable methods of analysis which the mineralogist has at his disposal. However, analysis of the light elements is still not a routine procedure. Some of the factors used when making absorption, fluorescence, or atomic number corrections are still incompletely known, and some of the procedures in general use are inadequate for complex minerals. Nevertheless, in the future the electron probe will continue to be an invaluable tool in solving problems which cannot be solved by other methods alone. But let no mistake be made, the electron probe is not a panacea for all mineralogical problems nor it is an instrument which can, nor indeed should, stand alone. To obtain the greatest benefits it can afford, it must be used in conjunction with other types of analysis and preferably after the other methods have been carried to their utmost capability.

Acknowledgments

The author would like to thank H. J. Rose, Jr., and E. J. Dwornik of the U.S. Geological Survey for a critical reading of the manuscript and for their valuable suggestions.

References

1. K. Keil, *Fortschr. Miner.* **44**, 4 (1967).

1a. V. G. Macres and C. M. Taylor, this volume.

2. J. D. Brown, this volume.
3. V. G. Macres and R. C. Wolfe, this volume.
4. B. J. Skinner, I. Adler, and C. W. Mead, US Geol. Surv., unpublished (1962).
5. G. Springer, D. Schachner-Korn, and J. V. P. Long, *Econ. Geol.* **59**, 475 (1964).
6. C. W. Mead and M. E. Mrose, *US Geol. Surv. Profess. Papers* **600-D**, D-204 (1968).
7. H. T. Evans, Jr., C. Milton, E. C. T. Chao, I. Adler, C. Mead, B. Ingram, and R. A. Berner, *US Geol. Surv. Profess. Papers* **475-D**, D-64 (1964).
8. J. Francotte, J. Moreau, R. Ottenburgs, and C. Levy, *Bull. Soc. Franc. Mineral. Crist.* **88**, 432 (1965).
9. P. Picot and R. Pierrot, *Bull. Soc. Franc. Mineral Crist.* **86**, 7 (1963).
10. E. F. Stumpfl and A. M. Clark, *Am. Mineralogist* **50**, 1068 (1965).
11. J. Protas, *Bull. Soc. Franc. Mineral. Crist.* **87**, 365 (1964).
12. J. V. P. Long, O. Kuovo, and J. Vuorelainen, *Am. Mineralogist* **48**, 33 (1963)
13. C. Gaudefroy, G. Jouravsky, and F. Permingeat, *Bull. Soc. Franc. Mineral. Crist.* **86** 359 (1963)
14. N. L. Markham and L. J. Lawrence, *Am. Mineralogist* **50**, 900 (1965).
15. E. B. Gross, J. E. N. Wainwright, and B. W. Evans, *Am. Mineralogist* **50**, 1164 (1965).
16. F. Boyer and P. Picot, *Bull. Soc. Franc. Mineral. Crist.* **86**, 429 (1963).
17. G. A. Challis, and J. V. P. Long, *Mineral. Mag.* **33**, 942 (1964).
18. P. Picot and J. Vernet, *Bull. Soc. Franc. Mineral Crist.* **86**, 87 (1963).
19. C. W. Mead, US Geol. Surv., unpublished (1961).
20. F. Permingeat and E. Weinryb, *Bull. Soc. Franc. Mineral. Crist.* **83**, 65 (1960).
21. L. S. Birks, E. J. Brooks, I. Adler, and C. Milton, *Am. Mineralogist* **44**, 974 (1959).
22. R. Oosterbosch, P. Picot, and R. Pierrot, *Bull. Soc. Franc. Mineral. Crist.* **87**, 613 (1964).
23. L. Chauris and J. Geffroy, *Bull. Soc. Franc. Mineral. Crist.* **89**, 137 (1966).
24. W. Nowacki and C. Bahezre, *Schweiz. Mineral. Petrog. Mitt.* **43**, 407 (1963).
25. J. Geffroy and J. Lissillour, *Bull. Soc. Franc. Mineral. Crist.* **86**, 14 (1963).
26. P. Picot, P. Sainfeld, and J. Vernet, *Bull. Soc. Franc. Mineral. Crist.* **86**, 299 (1963).
27. J. Geffroy and M. Lenfant, *Bull. Soc. Franc. Mineral. Crist.* **86**, 201 (1963).
28. L. Burnol, P. Picot, and R. Pierrot, *Bull. Soc. Franc. Mineral. Crist.* **88**, 290 (1965).
29. K. Fredriksson and C. A. Anderson, *Am. Mineralogist* **49**, 1467 (1964).
30. H. Bizouard and C. Roering, *Geol. Foren. Stockholm Forh.* **80**, 309 (1958).
31. D. D. Klemm, *Beitr. Mineral. Petrog.* **11**, 323 (1965).
32. K. L. Williams, *Am. Mineralogist* **52**, 475 (1967).
33. P. Picot, G. Troly, and H. Vincienne, *Bull. Soc. Franc. Mineral. Crist.* **86**, 373 (1963).
34. C. Gaudefroy, J. Dietrich, F. Permingeat, and P. Picot, *Bull. Soc. Franc. Mineral. Crist.* **89**, 80 (1966).
35. E. Welin and W. Uytenbogaardt, *Arkiv Mineral. Geol.* **3**, 277 (1963).
36. A. Cruys, A. Parfenoff, and D. Fauquier, *Bull. Soc. Franc. Mineral. Crist.* **87**, 625 (1964).
37. C. Frondel and C. Klein, Jr., *Am. Mineralogist* **50**, 1670 (1965).
38. R. Pierrot, J. Toussaint, and T. Verbeek, *Bull. Soc. Franc. Mineral. Crist.* **88**, 132 (1965).
39. J. Bolfa, H. de la Roche, R. Kern, M. Capitant, and K. D. Phan, *Bull. Soc. Franc. Mineral. Crist.* **84**, 400 (1961).

40. S. O. Agrell and J. V. P. Long, *in* "X-Ray Microscopy and X-Ray Microanalysis" (A. Engstrom, V. Cosslett, and H. Pattee, eds.), p. 391. Elsevier, Amsterdam, 1960.
41. J. B. Wright and J. F. Lovering, *Mineral. Mag.* **34**, 604 (1965).
42. A. K. Temple, K. F. J. Heinrich, J. F. Ficca, Jr., *in* "The Electron Microprobe" (T. D. McKinley, K. F. J. Heinrich, and D. B. Wittry, eds.), p. 784, Wiley, New York, 1966.
43. J. Babkine, F. Conquere, J.-C. Vilminot, and K. D. Phan, *Bull. Soc. Franc. Mineral. Crist.* **88**, 447 (1965).
44. J. Ottemann and G. Frenzel, *Schweiz. Mineral. Petrog. Mitt.* **45**, 819 (1965).
45. C. R. Niggli, *Schweiz. Mineral. Petrog. Mitt.* **45**, 807 (1965).
46. E. Welin, *Geol. Foren. Stockholm Forh.* **83**, 129 (1961).
47. L. Burnol, Y. Laurent, and R. Pierrot, *Bull Soc. Franc. Mineral. Crist.* **87**, 374 (1964).
48. C. S. Hurlbut, Jr., *Am. Mineralogist* **50**, 1698 (1965).
49. O. Braitsch and K. Keil, *Beitr. Mineral. Petrog.* **11**, 247 (1965).
50. W. Nowacki and K. D. Phan, *Bull. Soc. Franc. Mineral. Crist.* **87**, 453 (1964).
51. D. Bahezre, R. Michel, and P. Vialon, *Bull. Soc. Franc. Mineral. Crist.* **88**, 267 (1965).
52. M. G. Bown, *Mineral. Mag.* **34**, 66 (1965).
53. I. D. Muir and J. V. P. Long, *Mineral. Mag* **34**, 358 (1965).
54. J. V. Smith, *J. Geol.* **74**, 463 (1966).
55. J. V. Smith and P. H. Ribbe, *J. Geol.* **74**, 197 (1966).
56. P. H. Ribbe and J. V. Smith, *J. Geol.* **74**, 217 (1966).
57. J. V. Smith and R. C. Stenstrom, *Mineral. Mag.* **34**, 436 (1965).
58. D. H. Green, *Bull. Geol. Soc. Am.* **74**, 1397 (1963).
59. R. A. Howie and J. V. Smith, *J. Geol.* **74**, 443 (1966).
60. D. Metais, J. Ravier, and K. D. Phan, *Bull. Soc. Franc. Mineral. Crist.* **85**, 321 (1962).
61. S. O. Agrell, *Mineral. Mag.* **34**, 1 (1965).
62. K. D. Phan, *Bull. Soc. Franc. Mineral. Crist.* **88**, 97 (1965).
63. B. Pigorini and F. Veniale, *Atti. Soc. Ital. Sci. Nat. Museo Civico Storia Nat. Milano*, **105**, 207 (1966).
64. S. J. B. Reed, *Geochim. Cosmochim. Acta* **29**, 535 (1965).
65. K. Fredriksson and K. Keil, *Geochim. Cosmochim. Acta* **27**, 717 (1963).
66. J. I. Goldstein, *J. Geophys. Res.* **70**, 6223 (1965).
67. J. M. Short and C. A. Anderson, *J. Geophys. Res.* **70**, 3745 (1965).
68. S. J. B. Reed, *Geochim. Cosmochim. Acta.* **29**, 513 (1965).
69. J. I. Goldstein and R. E. Ogilvie, *Geochim. Cosmochim. Acta* **27**, 623 (1963).
70. I. Adler and E. J. Dwornik, *US Geol. Surv. Profess. Papers* **424-B**, B-263 (1961).
71. E. Olsen and K. Fredriksson, *Geochim. Cosmochim. Acta* **30**, 459 (1966).
72. K. Keil and K. Fredriksson, *J. Geophys. Res.* **69**, 3487 (1964).
73. K. Fredriksson and K. Keil, *Meteoritics* **2**, 201 (1964).
74. K. Keil, B. Mason, H. B. Wiik, and K. Fredriksson, *Am. Museum Novitates* **2173**, 1 (1964).
75. A. El Goresy, *Geochim. Cosmochim. Acta* **29**, 1131 (1965).
76. R. R. Marshall and K. Keil, *Icarus* **4**, 461 (1965).
77. F. R. Park, T. E. Bunch, and T. B. Massalski, *Geochim. Cosmochim. Acta* **30**, 399 (1966).
78. K. Keil and C. A. Anderson, *Geochim. Cosmochim. Acta* **29**, 621 (1965).
79. E. A. Jobbins, F. G. Dimes, R. A. Binns, M. H. Hey, and S. J. B. Reed, *Mineral. Mag.* **35**, 881 (1966).
80. K. Bostrom and K. Fredriksson, *Smithsonian Inst. Misc. Collections* **151**(3), 1 (1966).
81. K. Keil and K. Fredriksson, *Geochim. Cosmochim. Acta* **27**, 939 (1965).
82. K. Keil and B. Mason, *Science* **146**, 256 (1964).

83. L. H. Fuchs, C. Frondel, and C. Klein, Jr., *Am. Mineralogist* **51**, 949 (1966).
84. L. H. Fuchs, *Science* **153**, 166 (1966).
85. R. T. Dodd, W. R. van Schmus, and U. B. Marvin, *Science* **149**, 972 (1965).
86. C. Frondel and C. Klein, Jr., *Science* **149**, 742 (1965).
87. R. Castaing and K. Fredriksson, *Geochim. Cosmochim. Acta* **14**, 114 (1958).
88. E. C. T. Chao, E. J. Dwornik, and J. Littler, *Geochim. Cosmochim. Acta* **28**, 971 (1964).
89. E. C. T. Chao, I. Adler, E. J. Dwornik, and J. Littler, *Science* **135**, 97 (1962).
90. A. El Goresy, *Earth Planetary Sci. Letters* **1**, 23 (1966).
91. C. W. Mead, E. C. T. Chao, and J. Littler, *Am. Mineralogist* **50**, 667 (1965).
92. P. W. Hodge and F. W. Wright, *J. Geophys. Res.* **69**, 2449 (1964).
93. P. W. Hodge, F. W. Wright, and C. C. Langway, Jr., *J. Geophys. Res.* **69**, 2919 (1964).
94. F. W. Wright and P. W. Hodge, *J. Geophys. Res.* **70**, 3889 (1965).
95. P. W. Hodge, F. W. Wright, and C. C. Langway, Jr., *J. Geophys. Res.* **72**, 1404 (1967).
96. F. W. Wright, P. W. Hodge, and R. V. Allen, *Smithsonian Astrophys. Obs. Spec. Rept.* 228 (1966).
97. F. W. Wright and P. W. Hodge, *Res. Space Sci. Spec. Rept.* No. 192 (1965).
98. F. W. Wright and P. W. Hodge, *Smithsonian Astrophys. Obs. Spec. Rept.* 172 (1965).
99. J. V. P. Long and S. O. Agrell, *Mineral. Mag.* **34**, 318 (1965).
100. K. Keil, *Am. Mineralogist* **50**, 2089 (1956).
101. P. Duncumb. *Proc. Intern. Symp. X-Ray Microscopy X-Ray Microanal. 3rd, Stanford Univ. 1962*, p. 341. Academic Press, New York, 1963.

Electron Probe Analysis in Metallurgy

JOSEPH I. GOLDSTEIN*

Goddard Space Flight Center
National Aeronautics and Space Administration
Greenbelt, Maryland

I. Introduction

Since the pioneering work of Castaing [1], the electron probe has been utilized in an increasing number of ways to investigate metallurgical problems. This instrument has several capabilities which make it unique and useful for metallurgical studies. The region analyzed is usually between 1 to 10 μ in diameter, and 10^{-14} to 10^{-11} g of material are excited by the electron beam which in most cases is nondestructive to the specimen. Elements of $Z = 5$ (boron) and above can be detected by dispersive crystal methods. The amount of any elemental constituent can be determined to the degree of accuracy desired if appropriate standards are available. To obtain the distribution of an element over a selected area of interest larger than the size of the electron beam, a scanning technique is employed. Backscattered electrons, sample current, or X-ray intensity produced within the selected area is then measured. Precision of analysis is, however, sacrificed in this case since a large area is

* *Present address*: Department of Metallurgy and Materials Sciences, Lehigh University, Bethlehem, Pennsylvania.

analyzed at one time. Most of these capabilities are now available on commercial instruments.

The electron probe has been most useful to the metallurgist in the investigation of metallographic structures and kinetic processes. The composition and distribution of inclusions and major phases in a sample can be determined. Material transport processes such as fractional crystallization, oxidation, and the precipitation and growth of phases can be studied in detail, and both interdiffusion coefficients and phase diagram boundaries can be determined by analysis of kinetic data. The use of the electron probe to solve metallurgical problems will be discussed in some detail in this article. Actual applications will be cited only to illustrate the technique of electron probe microanalysis. No attempt will be made to list a complete bibliography since this is readily available [2, 3]

II. Applications in Metallography

A. *Inclusion and Phase Analysis*

1. *Introduction*

The objective of inclusion analysis is usually the identification of areas clearly visible under the optical microscope. The analysis with the electron probe, being nondestructive and accepting metallographic mounts, allows easy transfer of specimens from metallograph to electron probe.

2. *X-ray Spectroscopy*

Useful information can be obtained by positioning the electron beam on the inclusion and measuring the characteristic X-ray spectrum emitted from the area excited. The X-ray spectrometer angle is varied and the resultant intensity is measured on a chart recorder run synchronously with the spectrometer. In most cases, it is possible to detect elements present in amounts greater than 0.5 wt % by varying the spectrometer angle.

Figure 1 shows a spectral scan of inclusions in a stainless steel [4]. In the specimens studied, alloy carbide inclusions rich in Ti were formed either due to excessive carbon content or abnormal heat treatment. This data was used in a study which attempted to identify such inclusions. The spectral patterns shown were obtained first on a 5 μ carbide inclusion and then on the matrix. The carbide is rich in Ti, Mn, and Cr while the matrix exhibits normal Fe, Cr, Ni, and small amounts of Co and Mn (Fig. 1).

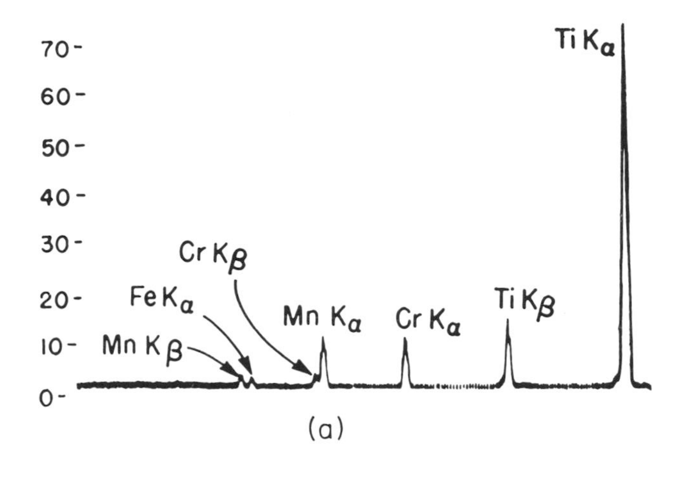

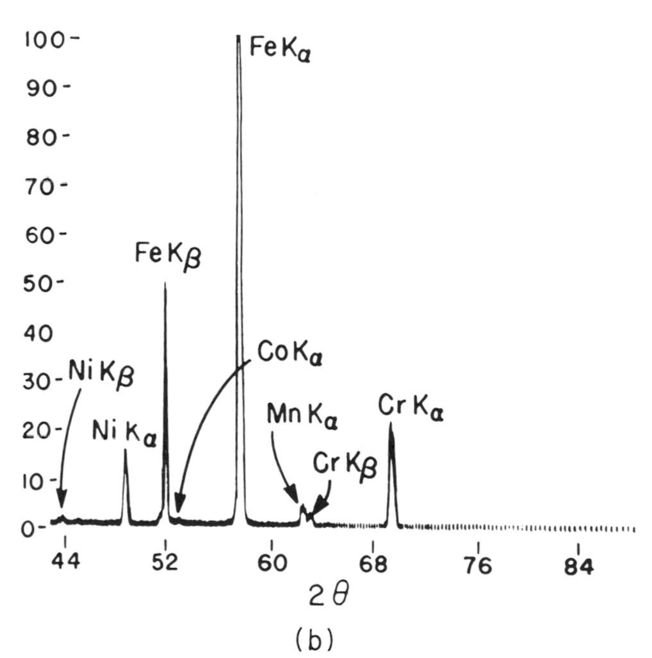

FIG. 1. Spectral scan of a stainless steel (a) inclusion and (b) matrix [4]. Intensity versus spectrometer angle (2θ).

In many applications, all that may be necessary is to identify the elements present in an inclusion or phase. However, in many cases the elemental analysis of the area is desired. Several calculation techniques can be used to obtain a first approximation of the composition of the area of interest. One method is to compare the measured characteristic X-ray intensities with those measured for either pure elements or compounds. As a first approximation, a linear ratio between intensity and composition can be used

$$C_A = (I/I_s)C_s \tag{1}$$

where C_A is the amount of element A present in the sample, C_s the amount of element A present in the standard, I/I_s the relative intensity ratio of sample to standard for element A.

Another method is to compare the measured elemental intensities with those measured for well-characterized samples containing the elements which are found in the unknown. A ratio of X-ray intensities is also used to determine the composition. These methods will usually yield accurate enough analyses for most inclusion studies. In cases in which more accurate analyses are desired ($\pm 2\%$ of the amount present), either the correction procedures which are discussed in other chapters of this volume must be used or standards near the actual composition of the unknown must be used.

It should be emphasized that accurate inclusion analyses are subject to several subtle sources of error. When an inclusion is analyzed, we take great pains to analyze particles larger than the excitation volume produced by the electron beam. However, because the electron beam penetrates beneath the specimen surface, it is possible for part of the beam to penetrate to the matrix (Fig. 2a). It is also possible that inclusion particles lying just below the sample surface may contribute to the analysis of the matrix (Fig. 2b). Consequently, it is necessary to repeat measurements on several inclusions to be assured that these effects are minimized. Another subtle source of error can be caused by X-ray fluorescence. Even if the inclusion is larger than the X-ray excitation volume, the characteristic and continuous spectrum pro-

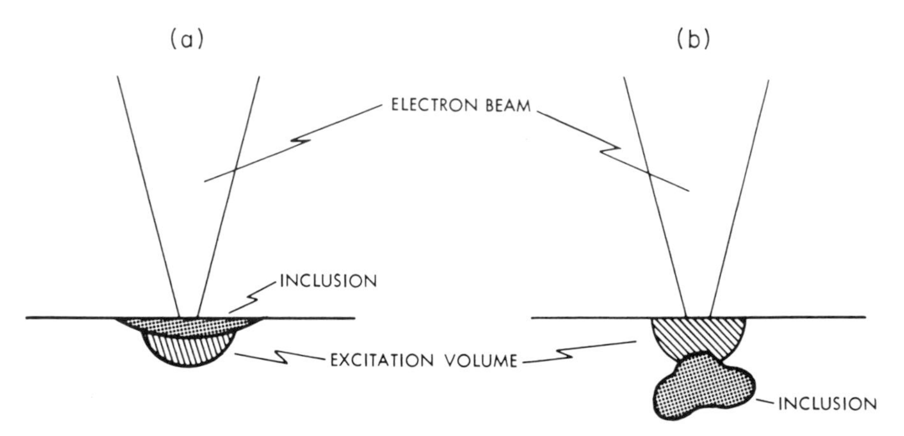

FIG. 2. Effects (a) matrix and (b) inclusions: Direct excitation by the electron beam.

duced in the inclusion may penetrate into the matrix. In certain cases fluorescence of the matrix by characteristic or continuum radiation may then occur. This radiation will be measured, in addition to the X-ray intensity from the inclusion if sample absorption is low and the area fluoresced is still on the X-ray focusing circle.

A practical consequence of such an effect was discussed by Duke and Brett [5] in their study of metallic Cu inclusions in meteorites. Before their work, Fe was reported in amounts of up to 1.2 wt % in the copper inclusions. However, the solubility of Fe in Cu has been reported as less than 0.35 wt % at 700°C [6]. At lower temperatures, the solubility limit is even less than 0.35 wt %. The analyzed Cu grains ranged from 10 to 20 μ in diameter, which is larger than the excitation diameter of the electron beam (2 μ). The principle uncertainty lay in the possibility that Cu K_α radiation from the analyzed Cu grain excited Fe K_α radiation in the surrounding Fe-rich matrix. To illustrate the possible effect of X-ray fluorescence, a pure Cu–Fe boundary was examined with an ARL electron probe ($\theta = 52.5°$) at an excitation potential of 25 kV. The intensity of Fe K_α radiation arising from points within the copper was determined at measured distances from the Cu–Fe boundary. The results are shown in Fig. 3. There is an observable fluorescence effect at distances of as much as 30 μ from the boundary. For points within 20 μ of the boundary, the effect is extremely pronounced. Therefore, the measurement of the Fe content in a Cu inclusion surrounded by an Fe-rich matrix can only be made on inclusions which are larger than 60 μ in diameter. Similar effects can occur in any type of inclusion analysis, and such a possibility must always be investigated. In analyzing the same type of Cu inclusions, the author measured both the Fe K_α (1.937 Å) and the Fe L_α (17.59 Å) radiation from the inclusion. The Fe L_α line is not noticeably fluoresced by Cu K_α. The Fe K_α intensity indicated an apparent Fe content of 3.4 wt %. As expected, the Fe L_α indicated <0.3 wt % Fe present in the inclusion.

In recent years, it has become possible to analyze for the X-radiation of light elements. The use of stearate crystals and thin-window proportional counters has enabled accurate intensity measurements to be made of the emission from the light elements down to boron (K-emission, 67 Å), even for concentrations below 1 %. Reliable quantitative results are best obtained by comparison with standards of known composition close to that of the specimen. When these standards are not available, analyses are difficult because the correction models must be used in regions in which they are probably not applicable, and several of the input parameters are not well known. Complications in light element analysis also arise because of the presence of the L spectra from heavier metals. For example, in the analysis of MnO the Mn L_α at 19.5 Å is well separated from O K_α at 23.7 Å. However, the Mn L_1 at 22.3 Å cannot be completely resolved from O K_α [7].

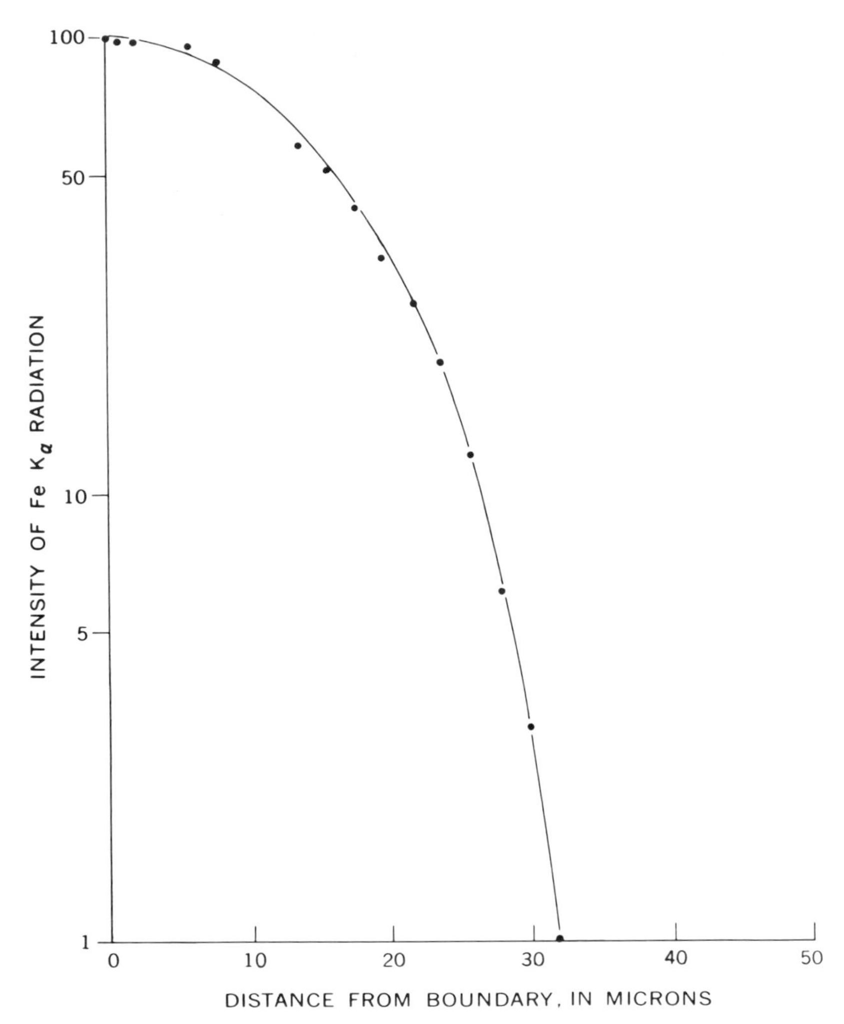

FIG. 3. Intensity of Fe K_α radiation measured in pure Cu as a function of distance from the Cu/Fe interface [5].

Duncumb and Melford [7] have shown that even with the above difficulties, a qualitative analysis can be obtained. For example, both nitride and carbide phases are formed in a specimen of graphitic cast iron. Figure 4 shows a comparison of the titanium L spectra obtained from these phases together with that from pure Ti. The phase referred to as TiNC gave an emission more intense than that received from pure Ti at the Ti L_1 wavelength. This peak consists mainly of Ti L_1 at 31.4 Å (18.3° θ) together with a small amount of N K_α emission indistinguishable from it at 31.6 Å (18.5° θ)

(Fig. 4). The Ti L_α line at 27.4 Å (16.0° θ), being heavily absorbed by nitrogen, is about one-third as intense as that from pure Ti. The Ti L_1 emission, however, is only slightly absorbed by nitrogen.

The analysis of this type of inclusion may appear to be impossible, but Duncumb and Melford analyzed the Ti content by using the K_α radiation and analyzed the C content by use of the TiC standard (Fig. 4). The results indicate about 80 wt % Ti and 4 wt % C. The analysis for nitrogen was not possible for reasons already stated and was obtained by difference from 100 %.

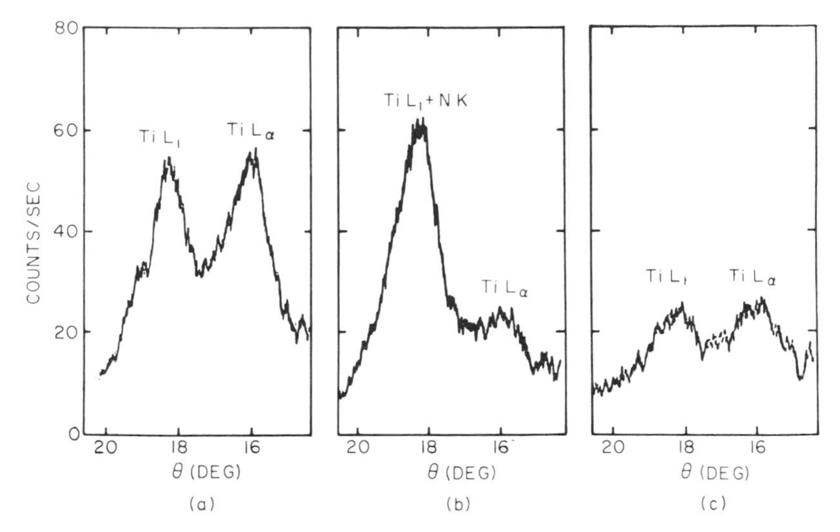

FIG. 4. X-ray spectra from (a) pure Ti (b) TiN, and (c) TiC at 10 kV operating potential (Duncomb and Melford [17]).

The technique of light element analysis looks promising, especially as a qualitative method. Its applicability as a quantitative tool is in doubt, however, because the structure of long wavelength X-ray bands ($\lambda > 18$ Å) that are used for the light elements depend strongly on the state of chemical combination. A detailed analysis of this problem has been presented by Holiday [8, 9] and by Baum in this volume. The K emission spectra of the light elements are relatively broad bands in comparison with the sharp lines of the heavy elements. Not only does the intensity of the emission bands vary with composition but also the shape of the bands. Because of peak shifts, the measurement of peak intensity is difficult to make. Therefore, we are not certain that the measurements of peak intensity for the light elements is the correct method to use [8]. More work is necessary before quantitative analysis using the light elements can be routinely performed.

3. *Electron Beam Scanning*

In many problems, the metallurgist not only wants to know what elements are present in the various phases or inclusions but also how a selected element is distributed between them. To obtain an area distribution, the electron beam is scanned over the specimen by electrostatic or electromagnetic beam deflection or by mechanical displacement of the specimen. The resultant X-ray emission, electron backscatter or specimen current image is measured and used to trigger the cathode beam of a display tube scanned in synchronism with the specimen. The resultant area display is recorded photographically. The electron backscatter and specimen current pictures indicate areas of differing average atomic number and topology, while the X-ray pictures show the elemental distribution throughout the area. The attraction of this form of data gathering is that detailed microcompositional information about a specimen is presented in a way that greatly facilitates correlation with optical metallography.

An example of the type of information that can be gathered by scanning techniques is shown in Figs. 5 and 6. Figure 5 shows the comparison between the optical and specimen current picture obtained for a complicated metal alloy composed of Ag–Hg–Sn. The lighter areas represent phases with lower average atomic numbers. The greater depth of focus available with the electron beam enables one to obtain information on a sample which is not optically flat. Figure 6 shows the distribution of Ag, Hg, and Sn between the various phases. The areas scanned are 275×220 μ. It is obvious that a point to point analysis could never show the type of variations seen in a sample in which severe segregation occurs.

It should be noted, however, that the information gathered by this technique is qualitative in nature. A sacrifice in terms of compositional information in each micron square area must be made in order to obtain qualitative compositional information over a large area. The general procedure for an analysis of a specimen would then be to use the scanning technique to determine the distribution of major elements within the various phases and inclusions both by X ray and electron images. After this, the composition of any particular inclusion can be determined qualitatively with the static probe, using a spectrometer scan and linear ratios to elemental standards. If necessary, quantitative measurements can be made using calculation models or standards.

A more quantitative approach to scanning can be taken by using a device that produces abrupt changes of the oscilloscope beam brightness at predetermined ratemeter signal levels. The contrast produced by a signal which is only slightly higher than the background can be enhanced so that only signal levels above background are recorded. Melford [10] demonstrated such an

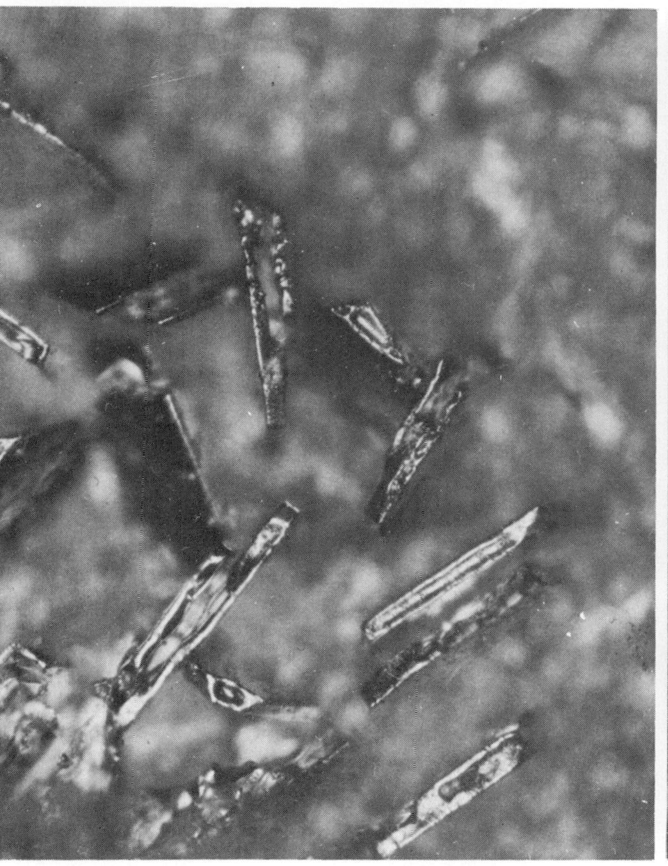

FIG. 5. Recrystallized dental alloy. Optical photomicrograph (X250) left and target current image (right). Instrumental conditions: 20 kV, 5×10^{-8} A. Scanned area: 550 μ $\times$ 440 μ (Heinrich [11]).

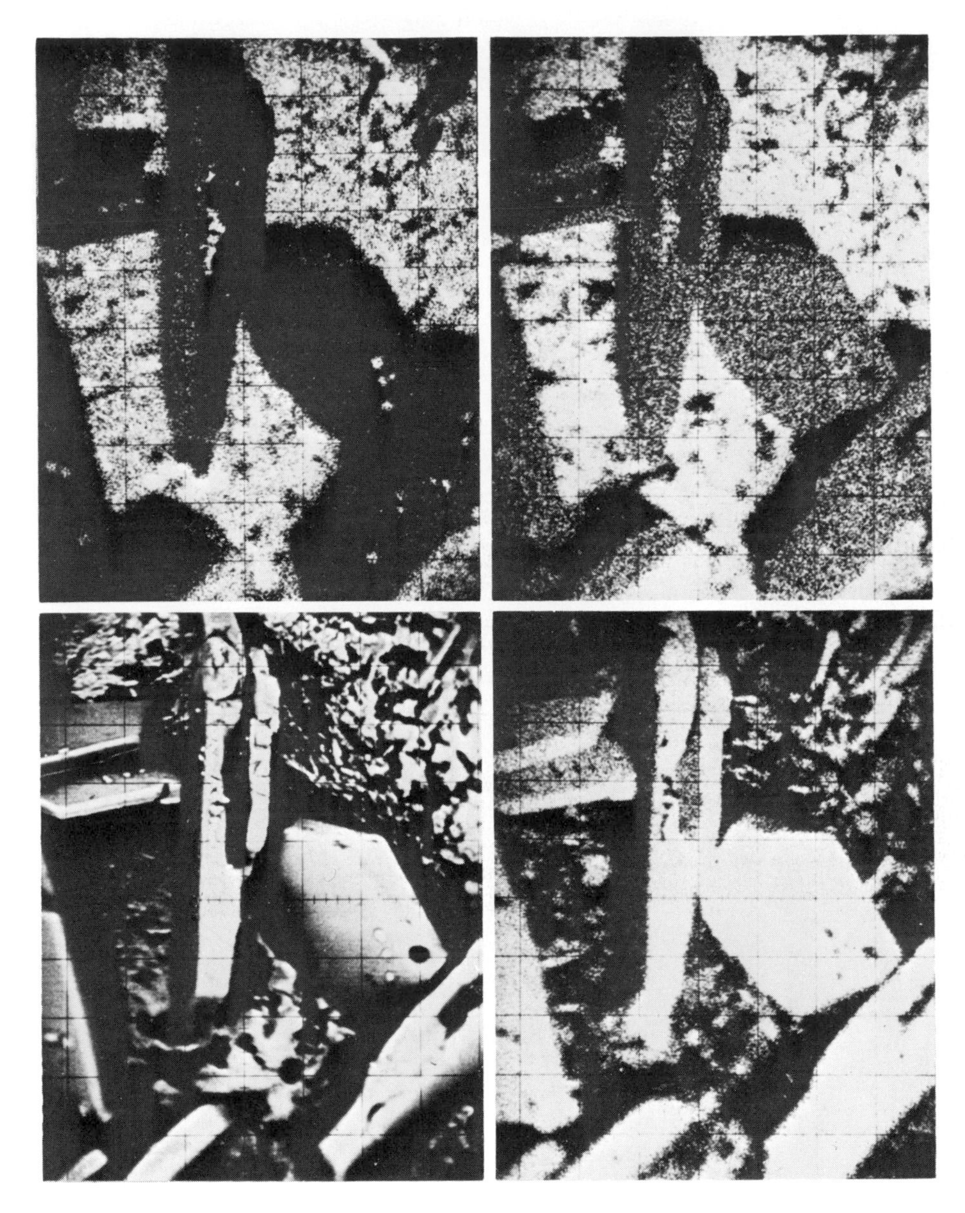

FIG. 6. Recrystallized dental alloy unprepared surface. Same focus and conditions as Fig. 5. Upper left: Ag L_α, X-ray scan. Upper right: Hg M_α X-ray scan. Lower left: target current scan. Lower right: Sn L_α, X-ray scan. Area scanned: 275 μ $\times$220 μ (Heinrich [11]).

"expanded contrast method" in which brightness modulation by the ratemeter signal has been employed for image formation. This method can also be used to enhance differences at high concentration levels.

Another technique, developed by Heinrich [11], is called concentration mapping. In principle, this method shows the location of defined concentration ranges. To obtain this information, all reference to concentration differences within each range must be suppressed. Figure 7 shows an illustration of the type of images that have been obtained from the Tazewell iron meteorite. Two Fe–Ni phases, kamacite >90% wt % Fe and taenite 50–85 wt % Fe, predominate. The upper left-hand area of Fig. 7 is a specimen current picture. The light phase having the lowest average atomic number is taenite, and superimposed on it is an Fe K_α line scan across the phases marked by the straight line L. The zero X-ray intensity level is marked by 0. The upper right-hand portion of this figure shows the Fe K_α scanning presentation usually obtained with the electron probe where it is difficult to see differences in the Fe concentration. The lower left-hand area shows an expanded contrast image obtained using the technique of Melford, while in the lower right-hand area, a four-level concentration map of the same region is shown. By this method, concentration ranges are defined: white-kamacite > 90% Fe, dark gray-taenite 50–65% Fe, black-taenite 65–80% Fe, and light gray 80–85% Fe. The use of the newly developed contrast technique may bridge the gap between the normal qualitative scanning procedure and the normal point to point analysis.

4. *Combined Electron Probe–Electron Microscopy*

In many practical situations, precipitates and inclusions are submicron in size. Since the X-ray emission volume from the electron probe is larger than this, a positive compositional measurement is difficult if not impossible. Usually, submicron particles are identified after observation in the electron microscope by electron diffraction. However, electron diffraction yields precise and unambiguous results only under ideal conditions. The diffraction rings quite often consist of diffuse spots which are of low intensity compared with the background. In these cases, approximate analysis of the precipitate, extracted on a carbon replica by the electron probe, can distinguish between the various possible phases suggested by electron diffraction and thus provide a specific identification. Therefore, combined electron microscopy–electron probe analysis is usually necessary for submicron particles.

A study of precipitates smaller than 0.1 μ in maraging steels of complex composition has been performed by Fleetwood *et al.* [12]. These steels were substantially carbon-free alloys hardened to a high strength by precipitation during aging. Positive identification of the age hardening precipitate was

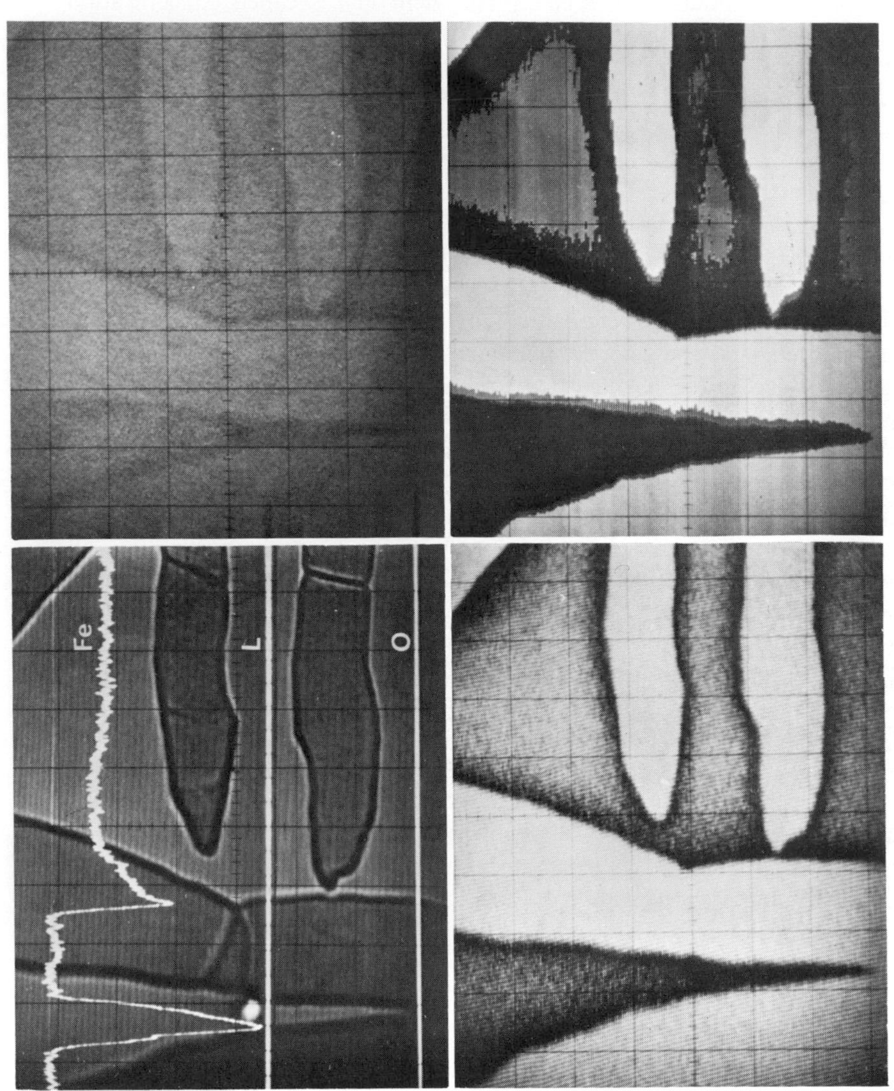

FIG. 7. Tazewell meterorite: representations of the FeK signal. Upper left: X-ray line scan of FeK_{α}, and zero intensity level 0, superimposed upon a target current area scan. The line L marks the locus of the line scan, with respect to the target current area scan. Upper right: pulse registration X-ray area scan. Lower left: expanded contrast image. Lower right: four-level concentration map. Scanned area: $138\,\mu \times 110\,\mu$ (Heinrich [11]).

achieved by use of a combined technique. Carbon replicas were stripped from samples of the overaged steels and were examined first in the electron microscope from which electron diffraction patterns of the precipitates were taken. These patterns of the extracted precipitates were of the "spotted" ring type, permitting measurement of the rings and calculation of the *d* spacings to within about ±2%. Subsequently, electron probe analysis of the same precipitates for Ni, Fe, Mo, Ti, and Co were made. Since many electrons capable of further ionization pass completely through the layer of precipitate, the above authors found that the total of the concentrations in each set of measurements falls below 100%. Consequently, the results must be regarded only as a measure of the relative proportions of the elements present in the precipitates.

The type of analysis just discussed is difficult because of the practical difficulties of transferring the replica from one instrument to the other and of relocating the precise field of view. Furthermore, it is not usually possible to plan the investigation sufficiently well in advance to avoid more than one interchange; information obtained during examination in one instrument frequently raises questions which demand immediate resource to the other.

A combined electron miscroscope–electron probe (EMMA) has been designed by Duncumb [13]. In this instrument the specimen can be studied by electron microscopy, electron diffraction, and either point or scanning electron probe analysis without altering the field of view. A cross section of the electron-optical system and X-ray spectrometer of EMMA is shown in Fig. 8. The electron beam can be focused down to a spot of about 0.1 μ. The maximum magnification from the objective and projector lenses is about 12,000×. To obtain structural information, the intermediate lens is adjusted to image the diffraction pattern from the particle; for chemical information, the characteristic X-ray emission is analyzed by means of a crystal spectrometer and proportional counter. The chief requirement of the X-ray spectrometer is that it should have as high a collection efficiency as possible since the X-ray intensity emitted from the submicron particles is very weak. In certain applications, as the X rays are allowed to pass directly into the proportional counter, the counting efficiency increases. The simultaneous decrease in signal to noise ratio is accompanied by an increased uncertainty in element identification. Because of the rapid decrease of counting rate with particle size and contamination difficulties, the smallest sized particle that can be analyzed using a focusing spectrometer appears to be a particle of about 0.1 μ in size. A similar type of instrument has recently been developed in the United States [14]. Also X-ray spectrometers are now being used to obtain compositional information right in the electron microscope [15].

Duncumb [13] has studied grain boundary precipitation in an 18 : 8 stainless steel using EMMA. After a heat treatment of 5 min at 700°C, mixed

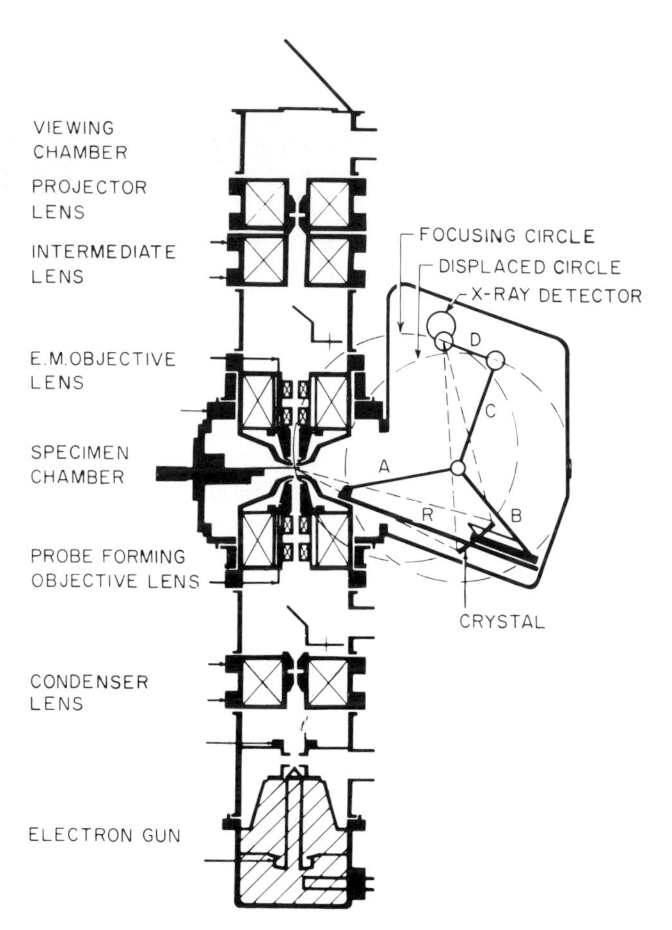

FIG. 8. Cross section of EMMA (electron microscope–electron microanalyzer) (Duncumb [13]).

Fe–Cr carbides are formed at the grain boundaries. A number of these precipitates are shown in transmission in Fig. 9. The maximum particle size was about 0.3 μ. The diffraction pattern of one particle selected for analysis is also shown together with the spectra for chromium and iron. The diffraction pattern indicates that the particle is of the $M_{23}C_6$ type. Sufficient counts were available to permit a measurement of the iron to chromium ratio so that a qualitative composition can be obtained from the thin specimen.

B. Quantitative Metallography

In many practical cases, it is of interest to know the size and size ranges as well as the distribution of various types of inclusions in a sample. The microanalyzer used in the scanning mode is ideally suited for the task. Melford and

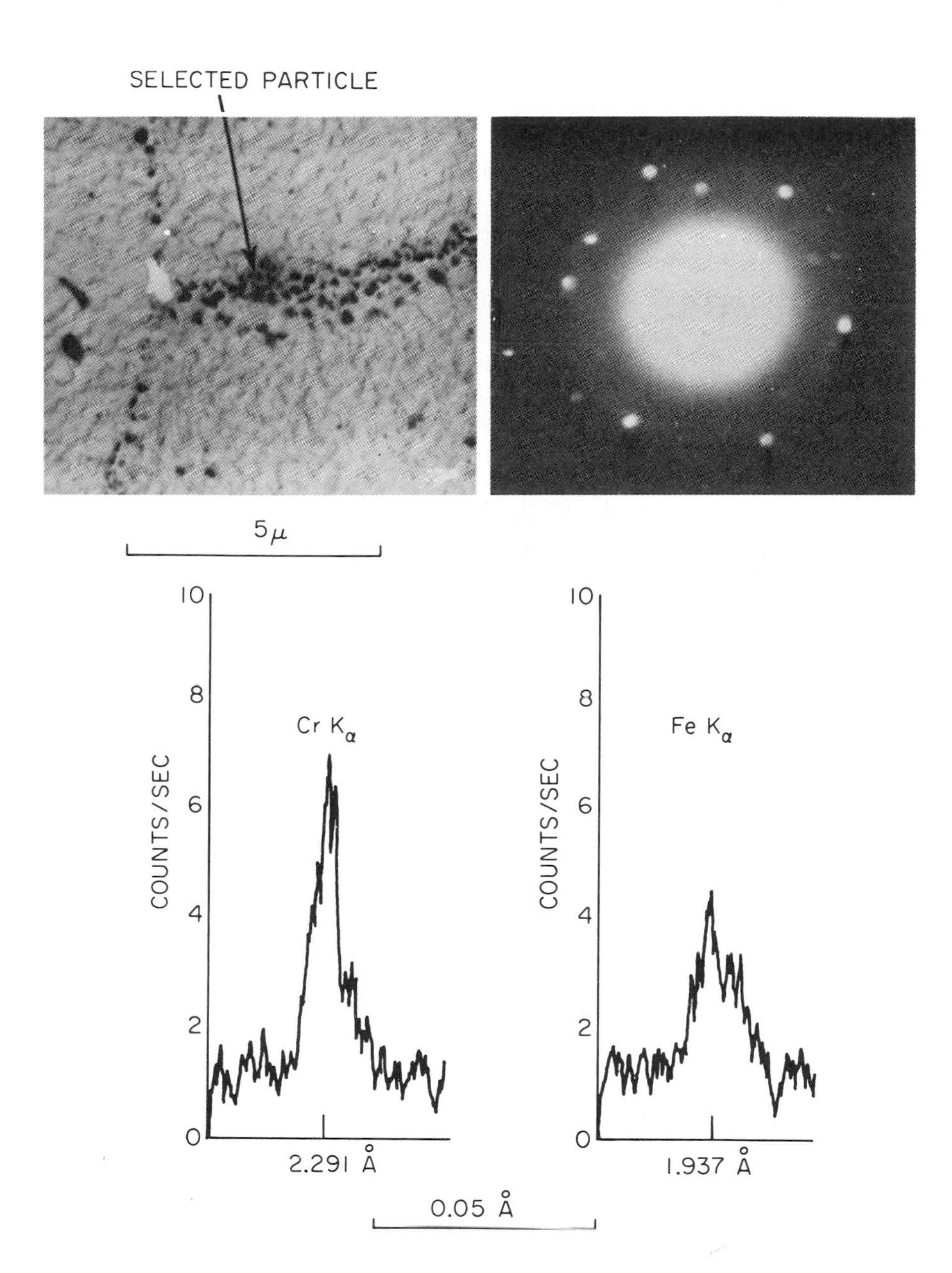

FIG. 9. Study of mixed Fe–Cr carbides with EMMA. Upper left: transmission electron microscope picture of carbides at grain boundaries. Upper right: diffraction pattern from particle. Bottom: spectrometer traces of carbides (Duncumb [13]).

Whittington [16] have developed a special purpose computer to handle the scanning probe data. For example, the Al K_α, S K_α, and Mn K_α intensities are obtained from a line scan through inclusions of MnS, MnO, and Al_2O_3. Using the two X-ray signals, logic circuitry compares them in such a way that the inclusions can be recognized. An oxide particle might be recognized by the fact that it contained either (a) Al, (b) Mn + Al, or (c) Mn but not S. The size of the inclusion can also be determined from the line scan and the information stored. To avoid double counting due to the beam crossing the same inclusion on more than one successive line, the signal obtained along one line is stored for comparison with the signal obtained along the subsequent line. Signals are counted only if they are not inhibited by a similar signal in the previous line. The inclusion type, size range, and volume percent can be determined.

Dorfler and Plockinger [17] have developed a technique for quantitative metallography of phases containing the same elements but in varying concentrations. The various phases are delineated by measuring the voltage output of the ratemeter. A voltage band (corresponding to a certain concentration range of a characteristic element for the one phase) is selected by two variable thresholds. One phase is identified by being within the voltage band, and the other phase is identified as located outside the band. The ratio of the measured time for one phase to that of the entire analysis gives the fraction of that particular phase. The phase integrator has been applied to the quantitative determination of the volume percentage of ferrite and austenite in Cr–Ni steels [17].

Both these techniques involve the use of line scans for phase identification. Another technique called computer metallography can be used with a scanning area image. Accurate quantitative metallography is made economically practical by employing a digital computer which both accepts suitable photomicrographs and prints out the desired information, directly [18]. The number of inclusions or separate phases can be determined as well as the size and distribution. The photomicrographs to be scanned must be truly representative of the sample at the magnification shown. In many cases, it is often hard to prepare metallographic specimens in which the inclusion or phase of interest is suitable for computer analysis. One ideal way to prepare suitable area pictures is to use the X-ray area scan. One element which characterizes the inclusion or phase can be used.

C. Characterization of Materials: Homogeneity, Composition Differences, and Trace Analysis

Much of the effort in electron probe work is centered around characterizing materials. What is generally desired is a determination of homogeneity, a measurement of compositional differences between two phases, or the determination of the presence or absence of a particular element. Problems of this type are largely statistical in nature. At this time, there is a tendency to overrate the ability of the electron probe technique to study these problems. The statistical basis and practical limits for many of these types of analyses are described in the following section.

Under ideal conditions, the individual X-ray counts must lie upon the unique Gaussian curve for which the standard deviation is the square root of the mean ($\sigma_c = \sqrt{\bar{N}}$). Here, $\bar{N}$ is considered to be the most probable value of N, the total number of counts for a given time t. In as much as σ_c results from fluctuations that cannot be eliminated as long as quanta are counted, this standard deviation is the irreducible minimum for X-ray emission spectrography. For example, to obtain a number with a minimum of a 1% deviation in N, at least 10,000 counts must be accumulated.

However, as Liebhafsky [19] pointed out, the actual standard deviation S_c, given by

$$S_c = \left[\sum_{i=1}^{n} (N_i - \bar{N}_i)^2/(n-1) \right]^{1/2} \tag{2}$$

where n is the number of determinations of i, equals σ_c only when operating conditions are ideal. In most electron probes, instrument drift and specimen positioning create operating conditions which are not necessarily ideal. The high voltage-filament supply, the lens supplies, and other associated electron equipment may drift with time. After a specimen is repositioned under the electron beam, a change in measured X-ray intensity may occur if the effective "depth of focus" for the X-ray spectrometers is smaller than the "depth of focus" of the light optical system. In actual practice, for the usual counting times of 10 to 100 sec/point, S_c is about twice σ_c. If larger counting times are used, S_c/σ_c increases due to instrument drift. Only when counting times are short and the instrument has stabilized does S_c approach σ_c. Besides the sample signal, sources of variation may also occur if data from reference standards and/or background are required [20]. These as well as faulty specimen preparation may also effect the precision of an analysis. Therefore, both instrumental factors and signal variations must be considered when the precision of an analysis is given. Very rarely will the percent coefficient of variation of an analysis ($S_c/\bar{N}$ (100)) approach the theoretical limit ($\sigma_c/\bar{N}$ (100)).

Several studies have been made characterizing standard materials for use as electron probe standards [21, 22]. The usual procedure for checking these standards is first to investigate inclusions and secondary phases if present. After this is done, a preliminary check on homogeneity of the matrix is made using a set of mechanical line scans. These line scans will point out any gross inhomogeneities (>10% of the amount present) on the 1 to 100 μ level. The possibility of gross inhomogeneities on the 1 mm to 1 cm level should also be investigated by conventional means such as X-ray fluorescence. To check for inhomogeneities of less than 10% of the amount present, a static probe is used and X-ray quanta are accumulated at each point. The procedure normally used is to take data at many points, usually between 10 and 100 spread across the sample. The criterion used for homogeneity is [21–23] that all points fall within the $\bar{N} \pm 3\sqrt{\bar{N}}$ limits. If 100,000 counts are accumulated at each point, a variation of more than 1% of the amount present can be detected. Therefore, a homogeneity level of within ±1% of the amount present can be insured.

A more exacting criterion for homogeneity would include (a) the use of the actual standard deviation S_c which accounts for instrument drifts and refocusing errors, and (b) the use of a confidence level for the determination of $\bar{N}$. The confidence level usually chosen is 95 or 99%. This means that we would expect, on the average, only 5% (or 1%) of repeated random points to be outside the limits

$$\bar{N} \pm t_{1-a,n-1}[(n+1)/n]^{1/2} S_c = \bar{N} \pm k_a S_c \quad (3)$$

where $t_{1-a,n-1}$ is the Student-t value for $n-1$ degrees of freedom which is exceeded in absolute value with probability $1-a$ [24]. A short table of values of k_a for $a = 95$ and 99% is given in Table I. Using this table combined with

TABLE I
FACTORS k_a FOR TOLERANCE INTERVALS[a]

Sample size n	k_{95}	k_{99}
2	15.56	77.96
3	4.97	11.46
4	3.56	6.53
8	2.51	3.71
16	2.20	3.04
30	2.08	2.80
120	1.99	2.63
∞	1.96	2.58

[a] Such that $\bar{N} \pm k_a S_c$ will include a proportion a of the population on the average.

values of S_c, the homogeneity limits in terms of percentage of the amount present $[(k_a S_c/\bar{N})(100)]$ can be determined for any sample size n and confidence level a. If $S_c \simeq \sigma_c$ and a tolerance interval of 95 or 99% is used,

$$k_a S_c \simeq 3\sigma_c \quad (4)$$

The criterion of $\pm 3\sigma_c$ is useful as long as operating conditions are ideal.

Ideally, one would like to predict the degree of homogeneity

$$\pm \left[\frac{k_a S_c}{\bar{N}}(100)\right] \tag{5}$$

as a function of N, n, and a that would be measured under typical operating conditions if the sample was homogeneous. The expected degree of homogeneity for a homogeneous sample would then serve as a baseline for comparison purposes. The measured degree of homogeneity would be expected to be the same or larger than this baseline value.

An expected degree of homogeneity for a homogeneous sample can be obtained if S_c is determined before the analysis. Repeated analyses in a small area of a homogeneous sample, refocusing at each point for counting times of from 10 to 100 sec, show that S_c is usually about twice σ_c. Assuming that this is the case for most analyses, the expected degree of homogeneity can be calculated, as follows:

$$\pm \left[\frac{k_a S_c}{\bar{N}}(100)\right] \simeq \pm \left[\frac{k_a 2\sqrt{\bar{N}}}{\bar{N}}(100)\right] \simeq \pm \left[\frac{k_a 2}{\sqrt{\bar{N}}}(100)\right]. \tag{6}$$

Graphs can be constructed for the expected degree of homogeneity versus accumulated counts per point N, the number of points analyzed n, and the confidence limit a. One graph for $a = 95\%$ is given in Fig. 10. The graph shows that to measure homogeneity on the 1% scale, about 200,000 counts must be accumulated per point. It also appears that little improvement in sensitivity can be gained by analyzing more than 16 points per sample. For example, by accumulating 100,000 counts for 32 points rather than 16 points, the degree of homogenization that can be detected improves only from 1.4 to 1.3%. However, by using the same amount of time as for the 32 points but now accumulating 200,000 counts for 16 points, the degree of homogenization that can be detected improves from 1.4 to 1.0%. Therefore, to improve sensitivity, it is best to increase the number of counts accumulated per point. In many samples, however, where precipitates or inclusions are present, increasing the number of points studied is an overriding consideration.

Analytical sensitivity usually indicates the ability to distinguish between two compositions that are close together. If we determine two compositions C and C^1 by n repetitions of each measurement taken for the same fixed time interval τ, then we can say that these two values are significantly different with a certain confidence a, if

$$C - C^1 \geq \frac{\sqrt{2}S_c t_{1-a}}{\sqrt{n}}\left(\frac{C}{\bar{N} - \bar{N}_B}\right) \tag{7}$$

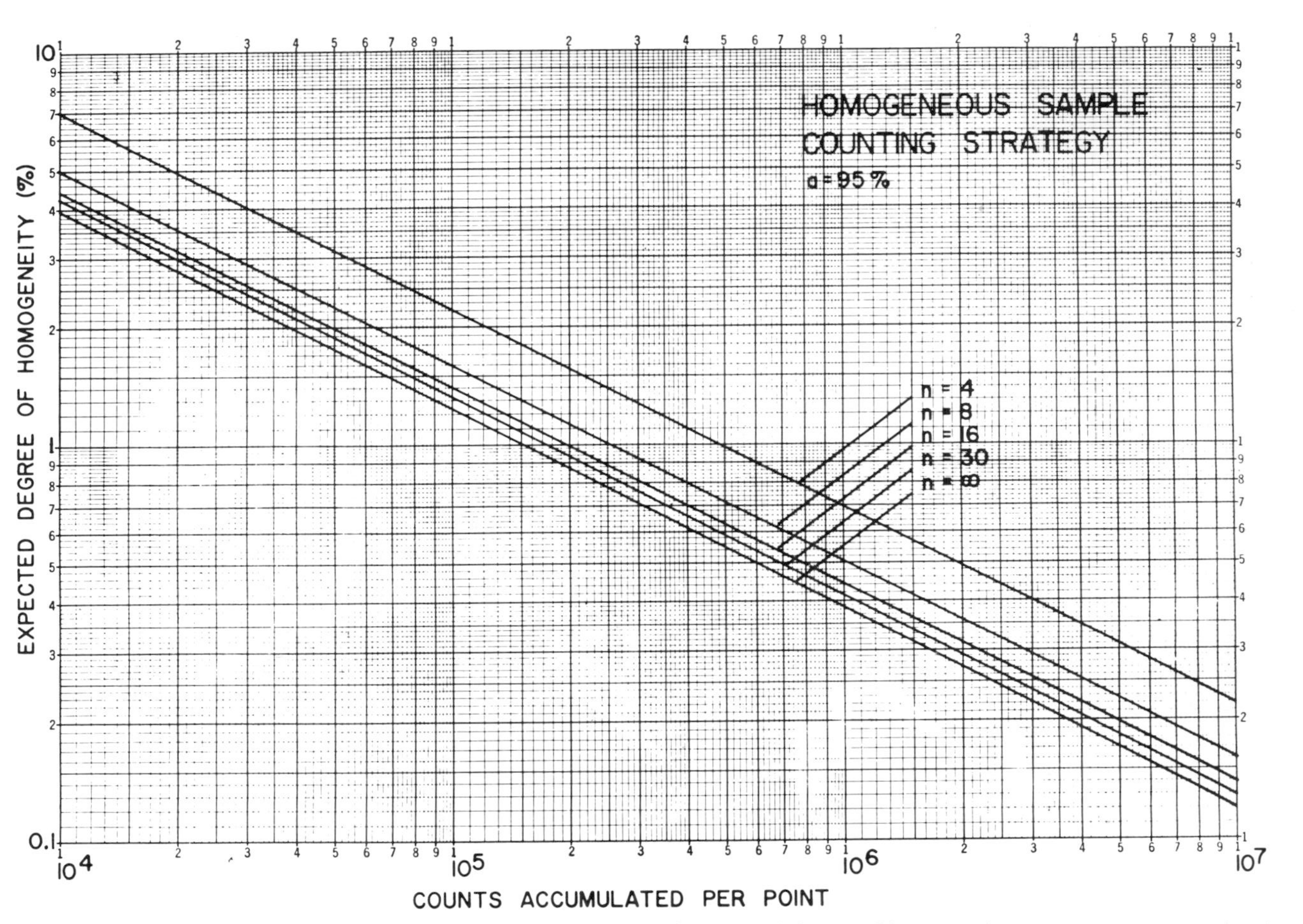

FIG. 10. Counting strategy for a homogeneous sample, $a = 95\%$. Expected degree of homogeneity versus counts accumulated per point.

in which C is the composition of sample; $\bar{N} - \bar{N}_B$ is the number of X-ray counts for the sample minus its background; t_{1-a} is the "Student's" factor dependent on the specified confidence level a, and n is the number of repetitions [25]. In this case signals from each sample have a statistical variation. Ziebold [25] has shown that the analytical sensitivity for a specified confidence level of 95% is

$$(C - C^1)_{\min} \geq \frac{2.33\sigma_c}{\sqrt{n}}\left(\frac{C}{\bar{N} - \bar{N}_B}\right). \tag{8}$$

He also calculated the term

$$\frac{2.33\sigma_c}{\sqrt{n}} = \frac{2.33}{(n\tau I_0 R)^{1/2}}\left(\frac{a-(a-1)C}{C}\right)(CR[a-(a-2)C] + 2[a-(a-1)C]^2)^{1/2} \tag{9}$$

where I_0 is pure element counting rate, R is peak/background ratio for pure element and a relates composition and intensity through the Ziebold and Ogilvie [26] empirical relation

$$(1 - I/I_0)/(I/I_0) = a[(1 - C_1)/C_1] \tag{10}$$

where C_1 is the weight fraction of element 1 in the mixture. Therefore, the nonlinear relationship between I/I_0 and C will effect the analytical sensitivity. The relation for $(C - C^1)_{\min}$ represents an estimate of the maximum sensitivity that can be achieved when signals from both compositions have their own errors but instrumental errors are disregarded. Since the actual standard deviation in the individual measurements will be about 2 times larger than the standard deviation due solely to X-ray statistics $(C - C')_{\min}$ is approximately 2 times that given by Eqs. (8) and (9). Below about $C = 20\%$, $C - C'_{\min}$ increases rapidly with decreasing C. The ability to detect differences between two compositions at low values of composition poses a difficult problem to the analyst.

For analysis in the ppm range (trace analysis) the X-ray calibration curves may be taken as a simple linear function. For trace analysis C_x, the unknown composition can be related to $\bar{N}_x$, the mean counts for the unknown, by the equation

$$C_x = \left(\frac{\bar{N}_x - \bar{N}_B}{\bar{N}_s - \bar{N}_{BS}}\right)C_c \tag{11}$$

where $\bar{N}_B$ is the mean count determined for sample background; $\bar{N}_S$ is the mean count determined for sample standard; and $\bar{N}_{BS}$ is the mean count determined for sample standard background.

Usually the pure element is used as the standard and $\overline{N}_{BS} \simeq \overline{N}_B$ so that Eq. (11) can be written

$$C_x = (\overline{N}_x - \overline{N}_B)/\overline{N}_S. \tag{12}$$

From this relation, it is apparent that the detectability limit is governed by the minimum value of the difference $\overline{N}_x - \overline{N}_B$ which can be measured with statistical significance.

Trace analysis is therefore dependent on the ability to distinguish two compositions that are close together. Therefore, Eq. (8) used for determining composition differences is also applicable. Letting the composition go to zero in Eq. (9) gives us a measure of the detectability limit, that is, the sensitivity at trace levels. Ziebold [25] has shown the sensitivity to be

$$C_{DL} \geq 3.29a/(n\tau I_0 R)^{1/2} \tag{13}$$

where the terms have been defined previously. To illustrate the use of this relation, the following values were used for calculating the detectability limit for Ge in iron meteorites [27]. The operating conditions were:

high voltage, 35 kV	$\tau = 100$ sec
specimen current, 0.2 μA	$n = 16$
$I_0 = 150{,}000$ counts/sec	$a \simeq 1.$
$R = 200$	

Using these numbers, $C_{DL} \geq 15$ ppm. The actual detectability limit after calculation of S_c and Eq. (8) was 20 ppm. Eqs. (9) and (13) which were developed by Ziebold, are very useful for determining the operating conditions for trace analysis before the actual data are taken.

Trace analysis below 10 ppm can be done only under carefully controlled conditions. In most cases long counting times are required, necessitating a very stable probe. Very few studies investigating trace element solubility have been made to date, but trace element analysis does prove to be an interesting new application for the probe.

III. Kinetic Processes

A. Diffusion Analyses

Mass transport through materials has been experimentally demonstrated to depend on a potential gradient in free energy, temperature, composition, etc. The rate of mass transport is usually expressed by an equation of the form

$$J_i = \sum_j D_{ij}(\partial \mu_j/\partial x) \tag{14}$$

where J_i is the flux of component i; $\partial u_j/\partial x$ is the gradient of free energy, temperature, composition, etc.; and D_{ij} is the coefficient which relates the flux and the potential gradient. Mass transport can occur through the material, along the grain boundaries, or on the surface. Each of these modes of mass transport are caused by a potential gradient. However, D_{ij} varies for each mode due to the different paths available for mass transport.

In most metallurgical systems, D_{ij} is the diffusion coefficient which relates the flux of component i to the potential free energy gradient of components $j = i, \ldots, n$. In many cases, the compositional gradient rather than the activity gradient is used. The diffusion coefficient varies with composition, temperature, and pressure so that if D_{ij} is to be measured, the experimental conditions must be well controlled.

For the case of binary diffusion, Eq. (14) has the form

$$J_A = -\tilde{D}(\partial C_A/\partial x) \tag{15}$$

where J_A is the flux of component A and $\tilde{D}$ is the interdiffusion coefficient. for the binary A–B. This equation is normally referred to as Fick's first law.

Since $\tilde{D}$ is usually a function of composition, the flux J_A will vary along the compositional gradient. This variation can be expressed by Fick's second law

$$\frac{\partial C_A}{\partial t} = \frac{\partial}{\partial x}\left(\tilde{D}\,\frac{\partial C_A}{\partial x}\right). \tag{16}$$

If $C = f(\lambda)$ where λ is given by the Boltzman [28] relation $\lambda = x/\sqrt{t}$ and the partial molar volumes of components A and B are not a function of composition, Fick's second law can be solved for $\tilde{D}$. The solution for $\tilde{D} = f(c)$, as derived by Matano [29] is

$$\tilde{D}(C') = \frac{-1}{2t}\,(dx/dc)_{c'} \int_{C_i}^{C'} x\,dc \tag{17}$$

for constant time t, pressure P, temperature T, and for the boundary conditions

$$\begin{aligned} C &= C_0, \quad \text{for} \quad x < 0, \quad \text{at} \quad t = 0 \\ C &= C_i, \quad \text{for} \quad x > 0, \quad \text{at} \quad t = 0 \\ dc/dx &= 0 \quad \text{at} \quad C = C_0, \end{aligned} \tag{18}$$

and where the origin of coordinates, the Matano interface, satisfies the relation

$$\int_{C_0}^{C_i} x\,dc = 0. \tag{19}$$

The value of $\tilde{D}(C')$ may be determined from the concentration gradient by measuring the slope dx/dc and the area $\int_{c_i}^{c'} x\,dc$, at c'. Since $\tilde{D}$ varies with temperature, $\tilde{D}(c') = D_0 \exp(-Q(c')/RT)$, the activation energy $Q(c')$, and D_0 can be obtained by measuring $\tilde{D}(c')$ at various temperatures. The measurement of Q and D_0 allows one to calculate $\tilde{D}$ at low temperature where it cannot be measured.

Before the electron probe was available, chemical composition was measured conventionally or by an X-ray absorption technique [30]. Both of these techniques required long diffusion gradients ($\gg 1$ mm) and were very tedious. It is, therefore, not surprising that diffusion couples were among the first specimens studied with this instrument. Castaing, in his thesis [1], chose the copper–zinc system to demonstrate the potentialities of this technique. Many of the initial binary diffusion studies were done by Adda and Philibert [31] and Adda *et al.* [32] on uranium base alloys. Since that time, a great number of binary systems have been studied [2, 3]. Since the Matano analysis works equally well for binary systems with two-phase discontinuities and intermediate phases, measurements in systems of this type have also been made.

Interdiffusion coefficients have also been measured at high pressures [33, 34]. The application of pressure lowers the diffusion coefficients because the equilibrium number of vacancies present is decreased and the energy required for an atom to move from one lattice site to another is increased [35]. The relation between $\tilde{D}$ at 1 atmosphere (D_1) and $\tilde{D}$ at a certain pressure D_p is

$$D_p = D_1 \exp(-P\,\Delta v/RT) \tag{20}$$

where Δv is the activation volume for the process. The value of the activation volume can, therefore, be easily obtained. The ability to measure mass transport processes under pressure adds a new dimension to our understanding of metallurgical processes.

Diffusion coefficients in ternary systems have also been measured [36, 37]. The solution of the equation for the generalized mass transport Eq. (14) is more difficult, and the number of diffusion coefficients rises from 1 to as many as 6. The measurement of composition using the probe is more difficult, and often necessitates the use of ternary standards.

Many experiments have shown that at low temperatures the diffusion rates in the grain boundaries may be several orders of magnitude greater than for lattice diffusion. It is now well established that the mean jump frequency of an atom in these areas is much higher than that of an atom in the lattice. Some studies have indicated a variation of grain boundary diffusion with the degree of misalignment of the grains and also with the orientation of the diffusion direction. By measuring the diffusion coefficient in various types of grain boundaries, it is possible to learn more about the structure of these paths and about how the atoms move in them.

The present mathematical solutions [38, 39] of grain boundary diffusion are based on a model of a grain boundary as a narrow region having a high diffusion rate relative to the adjacent grain. The lattice and grain boundary diffusion rates are assumed to be independent of concentration. The idealized grain boundary model is shown in Fig. 11; Y is the diffusion direction, $\tilde{D}'$

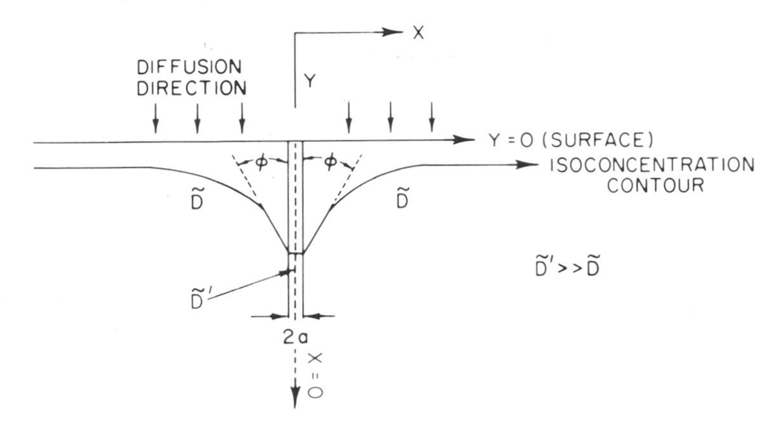

FIG. 11. Idealized grain boundary model, showing isoconcentration lines.

is the diffusion coefficient in the grain boundary and $2a$ represents the width of the grain boundary. The solutions of these idealized models predict concentration contours resulting from lattice diffusion away from the boundary. A hypothetical isoconcentration line is shown in Fig. 11 where ϕ is the angle included between the boundary on an isoconcentration line. Such concentration contours have been shown qualitatively by differential etching and autoradiography.

The electron probe provides an excellent means of determining such concentration contours and, therefore, an excellent means of measuring $\tilde{D}'$. Austin and Richard [40] have studied Ni diffusion in Cu bicrystals. They measured lateral concentration gradients and the lattice diffusion coefficients $\tilde{D}$ as inputs for the approximate solution of Fisher [38] and the steepest-descent approximation of Whipple [39]. They found, interestingly enough, that $\tilde{D}'$ was concentration dependent. Koffmann [41] measured Zn diffusion in low angle (7.5 and 16°) Ag bicrystals. He found that it was necessary to apply the exact solution of Whipple [39]. An example of some of the isoconcentration curves measured as a function of X, Y, and composition are given in Fig. 12. He found that for chemical diffusion of zinc in silver, the dislocations also act as "pipes" for the diffusion of zinc atoms.

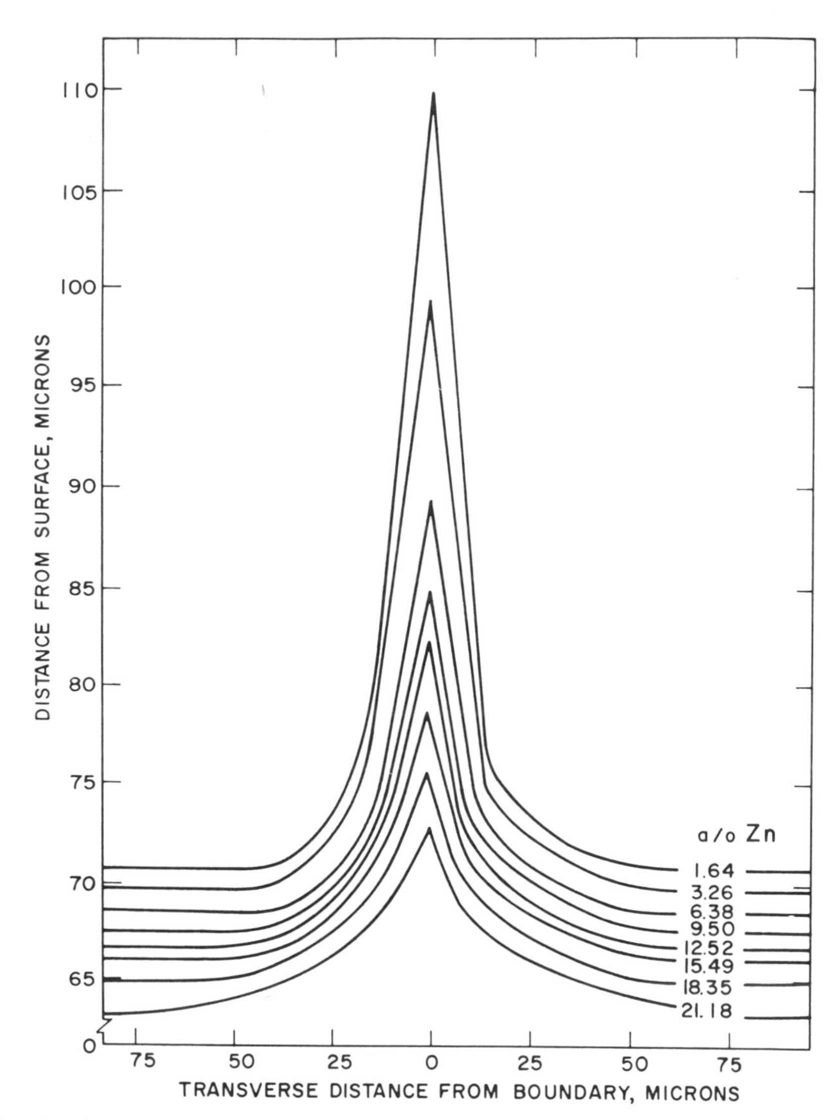

FIG. 12. Isoconcentration lines of Zn in Ag bicrystal diffused at 399°C (Koffmann [41]).

The use of the electron probe for diffusion analysis may appear to be rather simple at first, and in most cases this is true; however, certain precautions must be taken to insure an accurate probe analysis. To obtain accurate data, the absorption path for the X rays which are measured must have the same composition as the excited volume which is giving rise to the X rays. In order for this condition to be satisfied, the specimen must be oriented in such a way that the X rays produced within the specimen pass out through material of the same composition. For the diffusion sample this would mean orientation of the specimen, such that the excited X rays seen by the X-ray spectrometer lie in a plane of constant composition, i.e., perpendicular to the diffusion gradient. For materials in which the absorption parameter is small and does not vary with composition, it is not necessary to orient the sample.

The assumption which allows one to use the Matano analysis requires that the controlling diffusion mechanism be that of lattice diffusion. Therefore, if grain boundaries are present within the measured diffusion zone, the coefficient determined will be one for both lattice and grain boundary diffusion. To avoid this effect, the experimenter must prepare his diffusion specimens with a grain size of 1 mm or larger and avoid recrystallizing the sample and forming small grains when bringing the diffusion couple to diffusion temperatures. Speich *et al.* [42] observed that in diffusion of Cu in γ-iron, they had considerable difficulty in reproducing concentration-penetration curves because of grain boundary diffusion that occurred during one of their diffusion runs. To avoid the grain boundary zones, they scanned the diffusion specimens beforehand and selected large grains where the concentration-penetration curves taken at their centers would be truly representative of only volume diffusion. Repeated measurements on several areas help to ensure that the measurements are representative.

The accurate conversion of microprobe data to composition is not easily accomplished. In diffusion analysis, the investigator wishes to have both his precision and accuracy within 1% of the amount present except near the ends of the couple where C approaches C_0 or C_i. There are only a few binary or ternary systems in which the correction procedures and their input parameters are accurate enough to measure compositions to 1%. Large atomic number differences between the elements or a high mass absorption coefficient help to increase the uncertainties. In most cases, intermediate standards between the pure end members are necessary. These standards are unfortunately not easy to obtain. As discussed earlier, to obtain a precision of 1% of the amount present, at least 100,000 to 200,000 counts per point should be accumulated. When many points are to be counted across the diffusion gradient, this procedure may be rather time-consuming. Therefore, it is important to obtain high counting rates without the loss of X-ray spacial resolution.

During the analysis of a point on a diffusion gradient, both the characteristic and continuum radiation produced may excite radiation from the surrounding material. This effect has been discussed recently for undiffused diffusion couples by Maurice *et al.* [43], and is similar to that discussed previously in the section on inclusion analysis. However, since most data are taken within single phase areas not close to composition discontinuities, the effect of secondary excitation is minimal.

Although the electron probe gives a local analysis, the X-ray intensity obtained depends on the shape of concentration-penetration curve and the beam diameter for X-ray excitation. A correction for the effect of a finite beam size on a continuous diffusion gradient has been developed by Ziebold [44]. The correction is based on the assumption that the electron beam current density distribution is Gaussian and radially symmetrical about the beam axis and that the X-ray intensity is proportional to the beam current density. The effect of the finite beam size can be minimized if the diffusion experiment is designed such that the length of the gradient is at least 10 times the size of the effective beam diameter.

To obtain accurate values of $\tilde{D}(C')$, the Matano interface, dx/dc and the $\int_{C_i}^{C'} x\,dc$ must be accurately measured. To date, most analyses have been done by hand. To determine the Matano interface and $\int_{C_i}^{C'} x\,dc$, one counts squares under the concentration-penetration curve. The error in this is only about 1%. The largest error, however, occurs in determining the slope dx/dc. A subjective judgment must be made as to the best curve through the data points and the best slope of the curve. The variance in measured slope is not more than 5% if the slope is not taken near the ends of the concentration curves. The sum of the errors in obtaining $\tilde{D}(C')$ is usually about $\pm 10\%$.

A computer program has been developed for the direct conversion of electron probe data to diffusion coefficients [45]. The electron probe data are first converted to composition by a predetermined calibration curve. The Matano analysis is then completed by the computer and $\tilde{D}$ is reported as a function of composition. The major error, that of obtaining the slope dx/dc, is minimized by using several points on either side of the point selected, and fitting these points with a suitable function so that the slope can be obtained analytically. The diffusion couple program should help greatly in decreasing the amount of tedious work necessary to determine diffusion coefficients.

B. *Equilibrium Phase Diagrams*

It might seem unusual at first to discuss the determination of an equilibrium property, that of phase boundary compositions, in a section devoted to kinetic effects. The electron probe microanalyzer, however, can be used to measure phase equilibria in multicomponent systems, even if the various phases analyzed are not homogeneous and do not have their equilibrium compositions. This is because the probe can be used to measure the interface compositions of two coexisting phases. Since equilibrium conditions must exist at phase interfaces [46], the solubility limits of the coexisting phases can be determined. The following discussion parallels that of a paper on the subject by Goldstein and Ogilvie [47].

Two general techniques can be used to prepare suitable samples for probe analysis: the diffusion couple technique and the quench and anneal technique.

1. *Diffusion Couple Technique*

The diffusion couple technique involves the bonding of two pure metals or alloys, annealing this sample at a temperature T_0, and measuring the resulting concentration gradient. The relation of the composition-penetration curve for a binary alloy diffusion couple to the corresponding binary phase diagram is simple. The curve must vary continuously through solid solutions and exhibit sharp discontinuities at any two-phase regions (see Fig. 13).

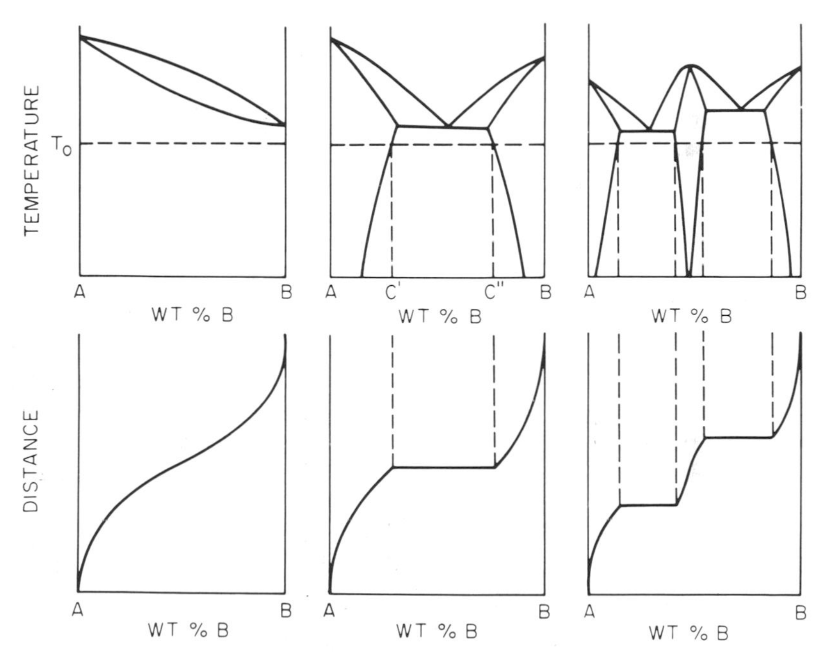

FIG. 13. Composition gradients obtained from binary diffusion couples. Single phase, intermediate phase and two intermediate phases.

The compositions at the terminal points of the diffusion couple must be the compositions in equilibrium at the two-phase gap of the phase diagram at the diffusion temperature. If the electron probe has sufficient resolution to define the interface compositions and solid solution zones, then the entire binary constitution may be obtained from a single diffusion couple. This technique has been applied extensively to several binary systems [48, 49].

Diffusion couples can also be used to define ternary alloy phase equilibria. In this case, the relation between the composition-penetration curve and the phase diagram is not so simple. The composition path does, however, represent an equilibrium configuration. If the multiphase regions are present, composition paths which cross these regions of the phase diagram exhibit the equilibrium compositions for coexisting phases.

It is possible that an intermediate phase may not form when using the diffusion couple technique. This is because the phase may not nucleate or the diffusion coefficient of the phase is so small that no apparent discontinuity in the concentration gradient appears. For example, Adda *et al.* [48, 50], when studying the U/Ni system, found only four phases at 600°C using a U/Ni couple; but when couples between U and UNi_5 were used, all seven phases predicted from the diagram were present. It is quite possible, therefore, that the diffusion couple technique may not be conclusive.

2. *Quench and Anneal Technique*

In the quench and anneal technique, an alloy is annealed in a two or three phase region of the phase diagram. It is then quenched and the interface compositions between two coexisting phases are measured. If a sufficient number of alloys are prepared at several temperatures, the phase diagram can be determined accurately. Up to the present, most studies that have been made using this technique have directly determined the composition of each phase in equilibrated two-phase alloys [51, 52]. The state of equilibration in these alloys was ascertained by noting no composition variations in either phase of the two-phase alloys. As discussed earlier, in many systems, it is not possible to obtain equilibrated samples because of low diffusion coefficients. The quench and anneal technique is most powerful for these unequilibrated samples since only the interface compositions between two coexisting phases need be measured.

Three methods can be chosen to obtain the equilibrium phases. These are:

(a) Quench a homogeneous alloy into a two-phase field and anneal to form the equilibrium phases.

(b) Quench from the liquid to room temperature, cold work, and reheat to the temperature of interest.

(c) If a martensitic phase forms, quench a homogeneous alloy to below the martensite start temperature and anneal to form the equilibrium phases.

The choice of which method to use for a particular set of phases is dependent on the growth rates of the equilibrium phases. The same type of annealing treatments can be used for ternary alloys. In this case, a few strategically located alloys may serve to determine an entire isothermal section.

3. *Accuracy of Phase Diagram Determination*

The ultimate accuracy of phase diagram determination by electron probe microanalysis is limited by the uncertainties involved in the conversion of X-ray intensity ratios to composition. In phase diagram determination as in other probe work, the ability to prepare standard alloys is of great importance.

Second, the accuracy is limited by three main factors: X-ray beam size, absorption effects, and concentration gradients at interfaces. Since the volume in which primary X rays are produced is much larger than the size of the electron beam, the direct measurement of the phase boundary composition is impossible within one beam diameter of the phase interface. However, operating conditions should be set so that a minimal excitation area is obtained, even at the loss of X-ray intensity. Possible excitation of characteristic X rays by the continuum and the characteristic spectrum of other lines at a considerable distance away from the electron beam can also occur. This effect has already been discussed in this review. In such a situation, it is best to determine how large the effect is [5], and to correct the data.

The absorption effect occurs when X rays produced at one point within the specimen travel through material of different composition, and hence different mass absorption coefficients on their way to the spectrometer. This effect has been discussed in detail by Reed and Long [53]. To avoid the absorption effect, it is necessary, as in diffusion analysis, to orient the interface perpendicular to the surface of the specimen and the interface parallel to the X-ray path to the spectrometers.

If the concentration gradients near phase boundaries are too steep, meaningful extrapolations to these phase boundaries are impossible. If enough time is available experimentally, the gradients can generally be made small enough at the growth or diffusion temperatures so that accurate extrapolations can be made. However, it should be noted that a minimum extrapolation equivalent to one beam diameter is always necessary. In most cases, the necessity for extrapolation, brought about by the presence of gradients near interfaces, is the most serious limitation of the electron probe method.

Both the diffusion technique and the quench and anneal technique were used in the redetermination of the Fe–Ni phase diagram [54]. Figure 14 shows the composition versus distance curves determined from the diffusion couple technique and the quench and anneal technique at 700°C. Diffusion in the phase was so slow that no value of $C_\gamma/C_\alpha + C_\gamma$ could be obtained using the diffusion couple technique. The compatibility of the two methods is shown by the excellent agreement of the values of $C_\alpha/C_\alpha + C_\gamma$.

The sources of error caused by the finite size of the beam and the absorption effect were negligible. For the maximum discontinuity of 25% Ni, the

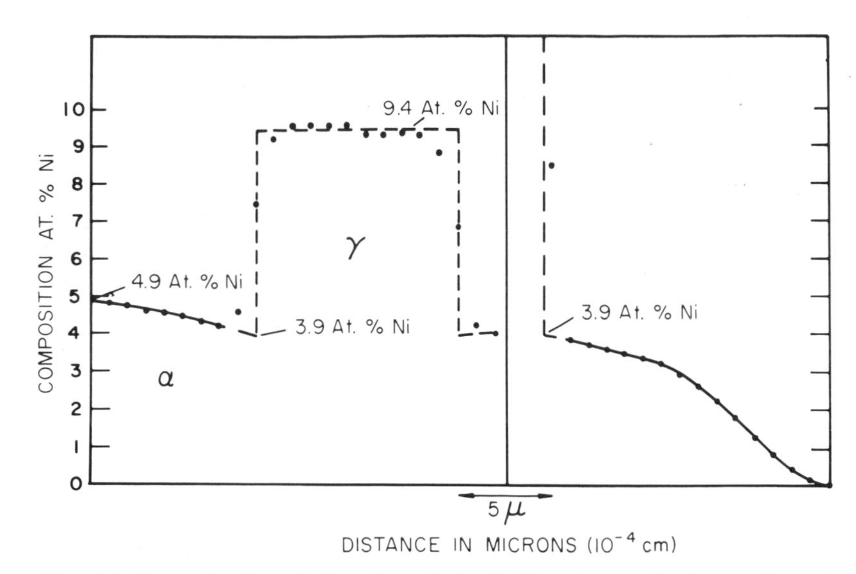

FIG. 14. Determination of phase boundaries in Fe–Ni at 700°C (●, probe values), two techniques: (left) quench and anneal technique, alloy 4.9 at % Ni; and (right) diffusion couple technique, 0–14.5 at % Ni couple (Goldstein and Ogilvie [54]).

contribution of the finite beam size effect was less than 1% at a distance of one X-ray beam diameter, 2.5 μ, or more from the interface. The absorption effect is minimal because of the high take-off angle of the probe ($\theta = 52.5°$). The concentration gradients measured near the α/γ interfaces are relatively small (Fig. 14). Because of the relatively flat gradients, interface extrapolations were never larger than 0.3% Ni (absolute). These extrapolation procedures contributed practically all of the uncertainties found in the phase boundary determinations.

C. *Mass Transport Processes*

The electron probe is ideally suited for investigation of problems which involve mass transport. Several areas in metallurgy such as solidification, oxidation and phase growth have been particularly aided by the use of this instrument. In the following section, several of these areas will be discussed, and the use of the probe will be described.

1. *Solidification*

In most cases when a liquid solution, initially of uniform composition, is solidified progressively, the composition of the solid is not uniform; the distribution of solute in the solid, when solidification is complete, is different from that in the liquid although the total amount of solute is unchanged [55]. Several limiting cases can be discussed for solute redistribution in liquid and solid.

(a) *Equilibrium maintained at all times.* In this case, the interface advances so slowly that the rejected solute is uniformly mixed into the whole of the liquid at all times. The diffusion of solute in the solid is sufficient to keep the whole of the solid at the composition that is in equilibrium with the liquid. This can happen only if the rate of advance of the interface is slow compared with the diffusion rate of the solute in the solid, and if the required diffusion distance is small. These conditions are never completely satisfied.

(b) *Mixing in the liquid by diffusion only; no diffusion in the solid.* The amount of solute taking part in the diffusion process is constant and does not change as the interface moves. In this type of solidification, a steady state condition exists where a solid with the same composition as the initial liquid is formed. This is a steady state process in which the amount of solute rejected from the solid-liquid interface is just balanced by the amount that diffuses away from it. Good experimental evidence was presented by Kohn and Philibert [56] for the existence of the enriched layer in contact with an advancing solid-liquid interface (Fig. 15). They rapidly quenched an Al–1.8% Cu alloy after partial solidification in a slow cooling (6°C/min) experiment and examined solute composition around the solid-liquid interface with the electron probe. They found that the liquid in the neighborhood of the solid-liquid boundary was richer in alloying elements. However, the solid did not

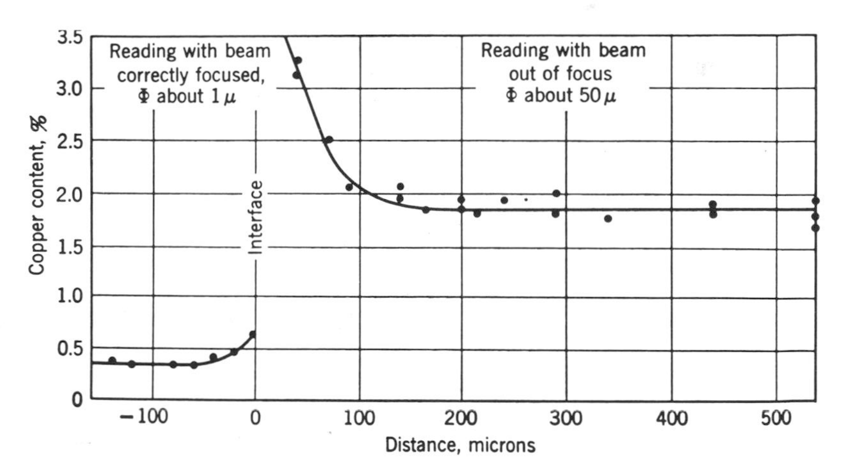

FIG. 15. Distribution of copper in an aluminum–copper alloy quenched during solidification. An enriched layer in contact with an advancing solid-liquid interface is present (Kohn and Philibert [56]).

have the same composition as the initial composition of the liquid. This probably occurred because the solid-liquid interface was not planar, and therefore linear solute transport did not occur.

(c) *Complete mixing in the liquid; diffusion in the solid.* In this case, the liquid is mixed at all times except in the immediate vicinity of the solid-liquid interface. The liquidus temperature of the liquid in contact with the interface is lower than that of the liquid at a greater distance from the interface. This situation gives rise to constitutional supercooling [55]. In the planar case, a gradient of supercooling can be resolved by the development of a transverse periodicity in the solidification process, the cellular substructure. The cell surface is convex towards the liquid; the concentration of solute is higher at the cell walls than in the cell centers and is a maximum at the cell corners. The development of cells rather than plane front growth is dependent on a high speed of growth, a large solute content, and a small temperature gradient in the liquid. In cellular solidification, it is of interest to determine the amount of segregation occurring.

In a study of segregation produced during cellular solidification of "pure" aluminum [57], the effect of residual impurities in creating the cellular substructure during controlled growth was studied. Two types of pure Al were chosen, one with less than 10 ppm impurities, and the other with less than 100 ppm impurities. Crystals were grown at three rates: 1, 10, and 60 cm/hr, and the temperature gradient was in the range of 5 to 10°C/cm. The iron impurity was calculated to have the greatest tendency to segregate. The impurity atoms, as expected, were found at the nodes between various cells. The various nodes were examined with the electron probe for Fe, Cu, and Si. The detection of these impurities was difficult because not all the nodes contained the impurities, and the nodes were of different sizes. The authors, therefore, chose to use two beam diameters, 5 and 20 μ, and claimed a detection limit of about 10 ppm for iron. For only 2 ppm of iron overall, the enrichment at the nodes was as high as 500 ppm with an average value of 170 ppm; for 20 ppm iron the enrichment was as high as 1790 ppm with an average value of 510 ppm.

The authors conclude that, since each node did not contain iron, the iron-rich solid exists as separated segments. The solubility limit of iron in aluminum falls below 500 ppm at temperatures below the eutectic. Therefore, it is improbable that the segregation of iron can be made to "anneal out" under any treatment. Several characteristics of "pure" aluminum, such as its behavior during corrosive attack, may not be attributed to the aluminum, but rather to its segregated state.

Dendritic growth occurs by conduction of latent heat outward from the growing crystal into a supercooled melt. In dendritic growth, the rate of growth is determined by the temperature and composition. The shape of a crystal growing dendritically is determined by local conditions of growth and not by any externally imposed temperature gradient. Solute redistribution occurs during solidification. Several studies have been carried out in which the amount of segregation in dendrites has been measured with the electron probe [58].

The classical nonequilibrium solidification equation,

$$C_s^* = kC_0(1 - f_s)^{k-1} \tag{21}$$

where C_s^* is the interface composition of the solid when the weight-fraction solid is $f_s(wt\ \%)$, k is the equilibrium partition coefficient between solid and liquid, and C_0 is the initial alloy composition (wt %), accurately describes solute redistribution between dendrite arms, provided diffusion in the solid is negligible. However, the assumption of no diffusion in the solid is not always valid. Brody and Flemings [59] have shown that

$$\alpha k = (D_s \theta_f / L^2)k \tag{22}$$

(where D_s is the volume diffusion coefficient of solute, θ_f is the local solidification time, and L is one-half the dendrite arm spacing) determines the extent of diffusion in the solid phase. For $\alpha k \ll 1$, microsegregation approaches the maximum predicted by the classical nonequilibrium solidification equation, Eq. (21). For $\alpha k \gg 1$, the composition of the primary solid phase approaches uniformity. Since solid diffusion does occur during solidification, these authors calculated its effect using a computer analysis for a platelike geometry. The analyses predict the "severity of microsegregation" in a casting or ingot after solidification and cooling to room temperature. This analysis was applied to an Al–4.5% Cu alloy, and compared with experimental measurements [60]. The dendrites in this alloy had a platelike morphology so that the computer analysis could be used. The effect of diffusion in the solid during solidification and cooling to room temperature is that it (a) decreases the amount of nonequilibrium second phase, (b) results in continuously increasing solute content of the primary solid phase during solidification, and (c) results in a higher minimum solute content within a dendrite arm than given by normal nonequilibrium considerations.

The minimum solute content (at the centers of secondary or primary dendrite arms) can be measured with the electron probe. Figure 16 shows a microprobe trace across a columnar dendrite, 7 in. from the chill and perpendicular to the heat flow direction in a water cooled ingot. Several minima are observed; the lowest corresponds to a location at the center of a primary dendrite arm. Measurements of minima at the centers of the primary dendrite arms were made for Al–Cu samples with solidification times from approximately 10^3 to 10^4 sec. The calculated and experimental values agree within

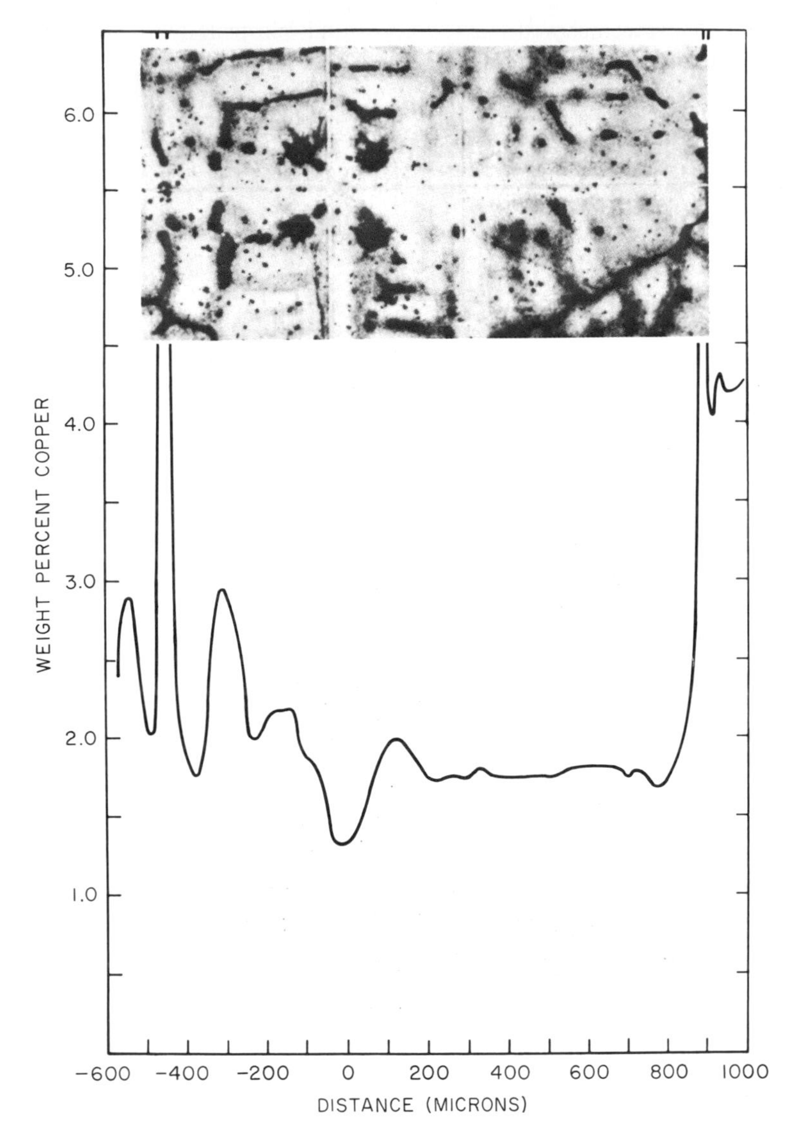

FIG. 16. Composite figure showing microprobe trace and photograph of microstructure of water-chilled ingot at 7 in. from the chill. Both graph and photograph have same scale in the horizontal direction Al–4.5 wt % Cu (Brody and Flemings [59]).

±0.2% in alloy content, and the minima ranged from 1.3 to 2.2% Cu, depending on local solidification time and dendrite arm spacing. The minimum determined from the phase diagram on the basis of no diffusion in the solid was 0.6%. Measurements of this type, using the electron probe, will certainly prove valuable in understanding the mechanism of solidification.

2. *Oxidation*

One of the most important properties of metals is their corrosion and/or oxidation resistance. The rate of oxidation of a metal surface exposed to air depends on the temperature and the rate of access of metal and oxygen atoms to each other. This in turn, depends on the structure of the oxidation film that has already formed. For pure metals, the growth of an oxide film is usually controlled by the rate of diffusion of the metal ion through the oxide film. However, if the oxide film is not adherent and spalls off, continuous oxidation occurs. In metal alloys, alloying elements are added to help form a protective oxide film which in turn helps decrease the oxidation rate. Internal oxidation may also occur in which oxide particles are formed with elements originally in solid solution. Because mass transport in the alloy controls most of the oxidation processes, the electron probe is an ideal instrument to study oxidation.

One of the first uses of the microprobe was to study the oxidation of ferrous alloys. In these studies, attention was focused upon the composition of both the oxide and the adjacent metal. For example, when billets of steel are heated in an oil-fired furnace, they are coated with complex scales. With the microanalyzer, Philibert and Crussard [61] studied the scale formed on a typical open-hearth billet (0.4% C, 0.6% Mn, 0.31% Si, 0.22% Ni, 0.14% Cu). They found several inclusions of the approximate compositions FeO, FeS, and a dark gray silicate. Considerable enrichment of nickel was observed in the surrounding metal, whereas the inclusions contained no trace of nickel. The rate of oxidation is probably controlled by the bulk movement of material out of the area where the oxide forms.

As we have discussed before, it is now possible to completely identify an inclusion by measuring radiation from elements $Z = 4$ and above. In one of the first applications of light element analysis, Lovering and Anderson [62] investigated the composition of an unknown phase in the Santa Catherina iron meterorite. This meteorite contained two co-existing iron–nickel-rich phases which had never been found in other iron meteorites or man-made alloys. From the scanning photographs of Ni, Fe, and O (Fig. 17), they found that one of the phases was a Ni-rich oxide. The oxygen-rich area had apparently been selectively oxidized by terrestrial oxidation along cracks in the meteorite. The other phase or parent phase had remained unoxidized. The use of low atomic number X-ray analysis is a great help in studying any problem involving oxidation.

To make quantitative studies of alloy oxidation, the simplest case, that

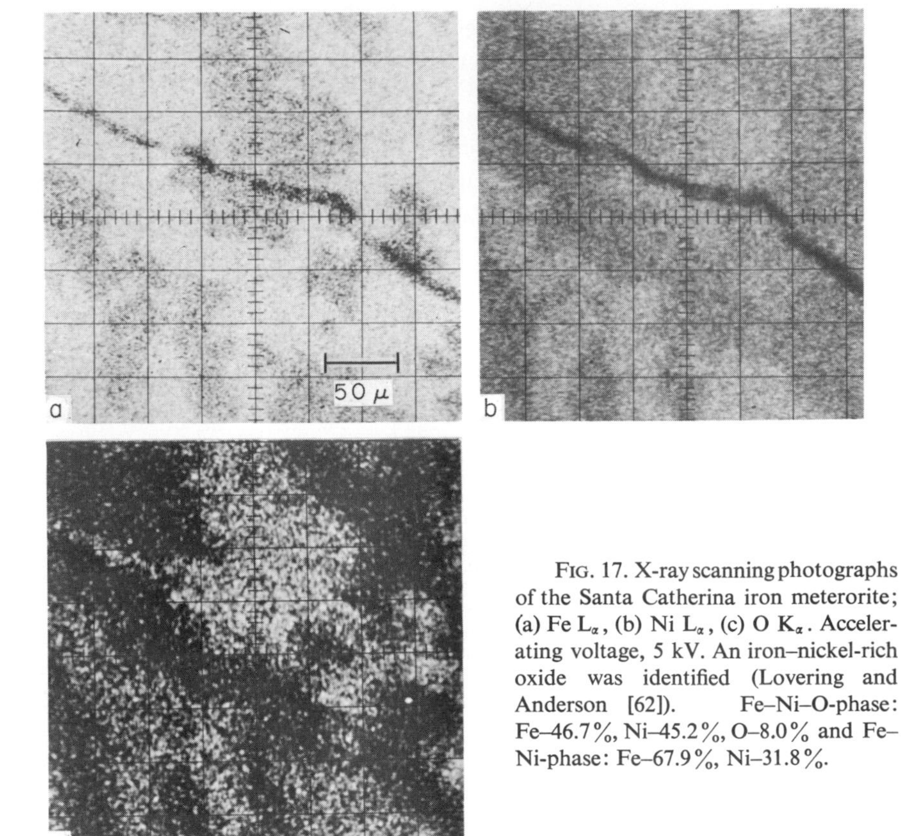

FIG. 17. X-ray scanning photographs of the Santa Catherina iron meterorite; (a) Fe L_α, (b) Ni L_α, (c) O K_α. Accelerating voltage, 5 kV. An iron–nickel-rich oxide was identified (Lovering and Anderson [62]). Fe–Ni–O-phase: Fe–46.7%, Ni–45.2%, O–8.0% and Fe–Ni-phase: Fe–67.9%, Ni–31.8%.

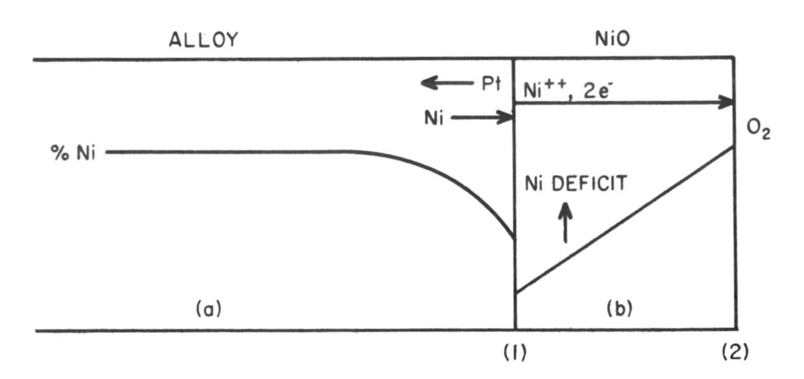

FIG. 18. Schematic representation of the oxidation of Ni–Pt alloys. The rate of oxidation is controlled by diffusion in the alloy; (a) alloy; (b) NiO. At (1) $Ni \rightarrow Ni^{++} + 2e^-$ and at (2) $Ni^{++} + 2e^- + O \rightarrow NiO$.

of binary alloys in which one element is a noble metal, has been used. In the case of such a binary alloy, only the reactive element is oxidized. For example, in Ni–Pt, only NiO forms. The oxidation process is shown schematically in Fig. 18. The migration rate of nickel ions is essentially proportional to the gradient of the concentration of nickel ion vacancies. Oxidation then actually occurs at the oxide-oxygen interface. The nickel is supplied to the metal oxide interface by the outward diffusion of Ni and the corresponding enrichment of platinum. The oxidation rate for the specialized case of noble metal binary alloy has been calculated on a theoretical basis by Wagner [63]. He assumed that there was no internal oxidation and that the diffusion coefficient $\tilde{D}$ of Ni–Pt was not a function of composition. The analysis takes into account the oxidation rate as a function of alloy composition, temperature, thickness of oxide layer, and oxygen pressure. The analysis shows that for oxidation, where the diffusion of metal to the interface through the alloy is controlling, the displacement of the metal surface per unit time $(d\,\Delta x_{\text{metal}}/dt)$ is proportional to the gradient of Ni at the oxide-metal interface $(\partial N_A/\partial x)_{X=\Delta x_{\text{metal}}}$ by

$$(1 - N_{A(i)})\left(\frac{d\,\Delta x_{\text{metal}}}{dt}\right) = \tilde{D}(\partial N_A/\partial x)_{X=\Delta x_{\text{metal}}} \tag{23}$$

Here, $N_{A(i)}$ is the mole fraction of Ni at the metal oxide interface and $\tilde{D}$ is the interdiffusion coefficient of Ni–Pt. Therefore, the rate of oxidation can be calculated if $\tilde{D}$ is known. Oxidation in the Ni–Pt system was studied recently by Koopman [64] and the Wagner analysis was checked. The electron probe was used to measure the Ni concentration gradients so that $\partial N_A/\partial x)_{X=\Delta x}$ could be obtained. Koopman measured $(d\,\Delta x_{\text{metal}}/dt)$ by oxidation experiments conducted at two temperatures for several times and several alloy compositions. The value of $\tilde{D}$ was then determined from Eq. (23). For comparison purposes, measurements of $\tilde{D}$ as a function of composition were taken at the two temperatures. Koopman found that Wagner's theoretical analysis was substantially correct. Deviations did occur in nickel-rich alloys, but these were due to the subscale formation (internal oxidation) which is not accounted for in Wagner's analysis. The oxygen in the internal oxide comes from the decomposition of NiO at the oxide-suboxide interface. This added reaction tends to decrease the rate at which the oxide layer grows.

Although an analysis with the probe of Ni in a Pt matrix might appear to present no difficulties, this is far from the truth. In the Ni–Pt system, Pt L_α (1.31 Å) is shorter than the absorption edge of Ni K_α (1.49 Å). The effect of Pt fluorescence and an atomic number correction presented no difficulty since well-analyzed standards were available. However, in oxidized Ni–Pt alloys two difficulties are encountered (Fig. 19). As the metal-NiO interface is approached, even though the actual Ni concentration drops to very low values, secondary fluorescence of Ni K_α by Pt K_α and the continuum

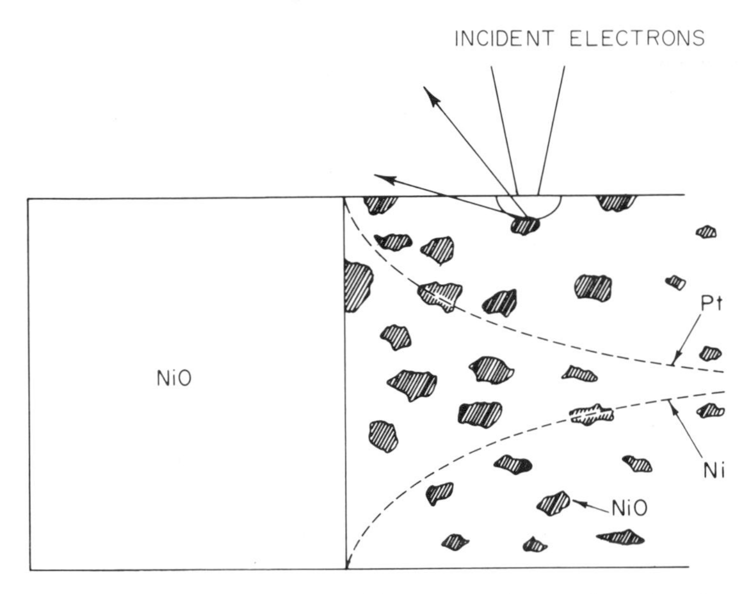

FIG. 19. Oxidation of Ni–Pt alloys with internal oxidation. Secondary fluorescence presents difficulties in obtaining an accurate analysis.

spectrum in both the NiO and in the Ni gradient up to 10 μ from the metal-oxide interface occurs. The effect of Pt on the Ni gradient can be approximately corrected for by measuring the Pt concentration. However, oxide particles which are very fine as positioned just below the surface tend to increase the measured nickel concentration at that point and decrease the measured Pt concentration. There is no way to correct for this.

3. *Microsegration*

In many cases of practical interest, segregation occurs at the grain boundaries of materials. A classical example of such a case is the grain boundary corrosion of stainless steel described by Philibert [58]. To explain grain boundary corrosion of stainless steels, most metallurgists have assumed the presence of a chromium-depleted zone near grain boundaries, caused by carbide precipitation. However, recent attempts with the electron probe have failed to demonstrate the existence of such a depleted zone in an 18 : 8 stainless steel. Since the grain boundary carbides are thin and the depleted zone probably narower than 1 μ, this failure is really not unexpected. Investigations with a 36 % Ni, 11 % Cr stainless steel were successful, however. The results of grain boundary investigation with the probe did demonstrate that the zone was chromium depleted and nickel enriched. During the growth of the carbides, the carbide obtains chromium from the neighboring austenite and rejects nickel into the austenite.

4. *Growth Kinetics*

Understanding the mechanisms through which alloying elements influence the kinetics of decomposition reaction that proceed by nucleation and diffusion growth has been a major research objective of physical metallurgy for almost 100 years. The mass transport of alloying elements from the parent phase to the decomposition phase may, in many cases, control the rate of growth of the new phase. With the electron probe, one can now study the method of mass transport occurring in various materials and ultimately understand the mechanism by which this occurs.

One study of this type has recently been performed by Aaronson and Domain [65] in which they studied the partition of alloying elements between the parent austenite and proeutectoid ferrite or bainite in high purity Fe–C–X alloys. The partition was studied with steels, usually of two levels of carbon content for various times at temperatures between the Ms (martensite start temperature) and the Ae_3 (the temperature at which the proeutectoid reaction $\gamma \rightarrow \alpha + \gamma$ occurs).

After the appropriate heat treatments, the ferrite that formed was located with a light etch and the composition of ferrite and austenite measured with the probe. The known amount of X in austenite served as a built in calibration for composition determination. Measurements of the composition of ferrite were taken in at least three different crystals. A Student's t test similar to that discussed earlier was used to determine the significance of the difference between the austenite and ferrite averages for each specimen.

At both levels of carbon content, no partitioning of alloying element was measured for Si, Co, Mo, Al, Cr, and Cu steels. In the manganese, nickel, and platinum steels, the concentration of X in ferrite was found to be significantly less than in austenite at high reaction temperatures. Figure 20b illustrates the composition profile taken through a ferrite crystal which underwent partition in a nickel steel. The crystal is shown in Fig. 20a crossed by arrow 1. The concentration profiles in the austenite adjacent to the opposite sides of the ferrite crystal are due to the nature of the austenite (γ)/ferrite (α) interface. The planar facets in the left-hand γ/α boundary are considered to have a predominantly dislocation structure, the nature of which substantially impedes the migration of the boundary. The substantial buildup of nickel in austenite at the left-hand side of the ferrite region of the concentration profile in Fig. 20b took place at this boundary. The smoothly curved boundary on the right-hand side of this crystal can be expected to have a disordered structure, which permits unimpeded movement of the boundary. The austenite adjacent to this boundary is seen at the right-hand side of Fig. 20b and appears to have been only slightly enriched in nickel. Figure 20c shows the concentration profile taken through the ferrite crystal traversed by arrow 2 in Fig.

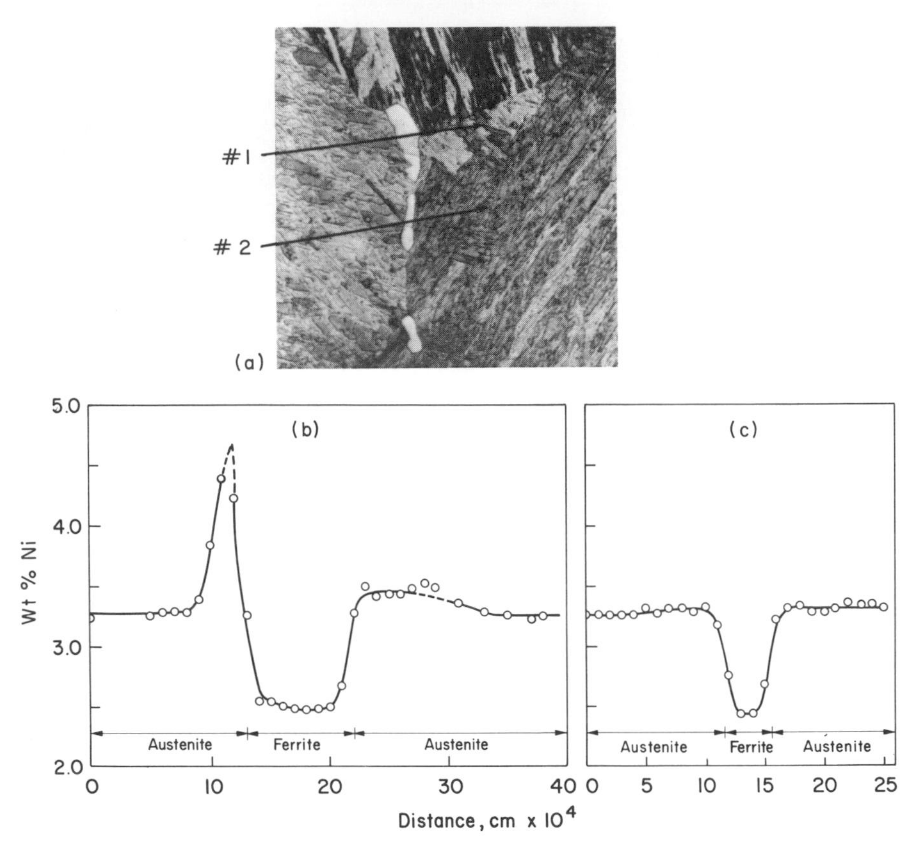

FIG. 20a. Ferrite crystals whose concentration profiles are given in Fig. 20b and c. Etched in 2% nital, 250×. Partition of Ni between parent austenite and proeutectoid ferrite (Aaronson and Domain [65]). (b) and (c). Concentration profiles through (b) ferrite crystal #1, and (c) ferrite crystal #2 of Fig. 20a, formed by isothermal reaction of 0.11% C, 3.28 % Ni steel held for 811,900 sec at 734°C.

20a. The smoother and more symmetrical shape of this crystal suggests that its α/γ boundaries have a largely disordered structure on both sides of the crystal. Accordingly, no detectable buildup of nickel is to be expected in the austenite on either side of the crystal. The investigation of the mechanism of growth as controlled by alloy element partition is greatly aided by the use of the electron probe.

In a more exotic application of the probe, the growth of the Widmanstatten pattern of iron meteorites has been studied [66]. Metallic meterorites are Fe–Ni alloys which often exhibit a two phase structure of co-existing Ni-poor alpha iron (kamacite), and Ni-rich gamma iron (taenite), the kamacite appearing in a Widmanstatten pattern (Fig. 21). Iron meteorites

FIG. 21. Precipitation pattern in the Butler meteorite. A well-defined Widmanstatten pattern developed late in the meteorite's cooling history.

are formed by the segregation of molten iron–nickel in an asteroidal-sized body in outer space. As the large body cools, the metal solidifies and the kamacite precipitates from the parent taenite phase in the form of a Widmanstatten pattern and grows. The growth of the kamacite is controlled by the rate at which Ni can be removed from the kamacite across the interface into the taenite. Even though growth may occur over 100 million years, segregation of Ni is not completed and a concentration gradient of Ni–Fe and impurity elements remain in both kamacite and taenite. A typical Ni and Ge concentration gradient for the Butler meteorite is shown in Fig. 22. The electron probe is used here as an aid in determining the cooling rate for meteorites at the time when the Widmanstatten pattern developed. This was accomplished by using the probe to measure the concentration gradients in

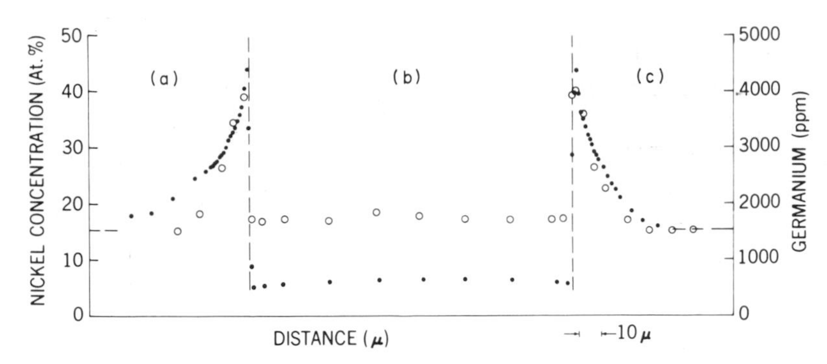

FIG. 22. The Ni–Ge distribution in the Widmanstatten pattern of the Butler meteorite. The meteorite has 16 wt % Ni and 2000 ppm Ge; ○, Ge; ●, Ni data. (a) Taenite; (b) kamacite; (c) taenite (Goldstein [58]).

several kamacite-taenite areas. Comparisons were made with gradients calculated by a theoretical growth analysis for the Widmanstatten pattern [66, 67]. The calculated gradients vary with the cooling rate assumed. The cooling rate is determined when a fit is obtained between the measured and calculated gradients. Cooling rates from one meteorite to another vary from 0.5 to 500°C per million years. The Butler meteorite cooled at 0.5°C per million years [67].

IV. CONCLUSION

The electron microprobe has already been used to solve a large number of metallurgical problems. Industrial firms have made by far the greatest use of the instrument, mostly for the characterization of various kinds of materials and material problems of an industrial nature. Although the electron probe has been used by a small number of research metallurgists for many years, it is only recently that uses of such a versatile instrument have been realized. The electron microprobe is now attaining the same importance in the metallurgical laboratory that the electron microscope achieved some years ago.

REFERENCES

1. R. Castaing, Thesis, Univ. of Paris. ONERA No. 55, 1951.
2. K. F. J. Heinrich, "The Electron Microprobe" (T. D. McKinley, K. F. J. Heinrich and D. B. Wittry, eds., p. 841. Wiley, New York, 1966.
3. K. F. J. Heinrich, ASTM Spec. Tech. Publ. No. 349 (1963).
4. S. H. Moll, *Norelco Rept.* **11**, 55 (1964).
5. M. B. Duke and R. Brett, Geol. Surv. Res., B101–103 (1965).
6. M. Hanson, "Constitution of Binary Alloys." McGraw-Hill, New York, 1958.

7. P. Duncumb and D. A. Melford, Quantitative applications of ultra-soft X-ray microanalysis in metallurgical problems. *Actes Congr. Intern. Optique Rayons X Micro-Analyse, 4th, 1967.* Hermann, Paris.
8. J. E. Holiday, Investigation of the carbon K emission band for stoichiometric and nonstoichiometric carbides. "Applications of X-ray Analysis," Vol. 10. Plenum Press, 1967.
9. J. E. Holiday, "The Electron Microprobe" (T. D. McKinley, K. F. J. Heinrich, and D. B. Wittry, eds., p. 3. Wiley, New York, 1966.
10. D. A. Melford, *J. Inst. Metals* **90**, 217 (1962).
11. K. F. J. Heinrich, Scanning electron probe microanalysis, to be published in an ASTM Spec. Tech. Publ. (1966).
12. M. J. Fleetwood, G. M. Higginson and G. P. Miller, *Brit. J. Appl. Phys.* **16**, 645 (1965).
13. P. Duncumb, "The Electron Microprobe" (T. D. McKinley, K. F. J. Heinrich and D. B. Wittry, eds.), p. 490. Wiley, New York, 1966.
14. M. A. Schippert and R. E. Ogilvie, A combined electron microscope-electron microprobe, Advanced Metals Res. Corporation Rept., 1967.
15. V. E. Fuchs, *Rev. Sci. Instr.* **37**, 623 (1966).
16. D. A. Melford and K. R. Whittington, The application of the scanning microanalyzer to particle counting and identification. *Actes Congr. Intern. Optique Rayons X Micro-Analyse, 4th, 1967.* Hermann, Paris.
17. G. Dorfler and E. Plockinger, A new apparatus for the determination of phase-contents of metallic and nonmetallic samples by electron microprobe analysis. *Actes Congr. Intern. Optique Rayons X Micro-Analyse, 4th, 1967.* Hermann, Paris.
18. G. A. Moore, L. L. Wyman, and H. M. Joseph, *in* "Quantitative Metallography" (F. N. Rhines, ed.), Chap. 15. McGraw Hill, New York, 1964.
19. H. A. Liebhafsky, H. G. Pfeiffer, and P. D. Zemany, *Anal. Chem.* **27**, 1257 (1955).
20. T. O. Ziebold, M.I.T. Summer Course Notes (T. O. Ziebold and R. E. Ogilvie, ed.), 1965.
21. R. E. Michaelis, H. Yakowitz, and G. A. Moore, *J. Natl. Bur. Std. A* **68**, 343 (1964).
22. H. Yakowitz, D. L. Vieth, K. F. J. Heinrich, and R. E. Michaelis, *Natl. Bur. Std. (U.S.), Misc. Publ.* 260–10 (1965).
23. J. I. Goldstein, F. J. Majeske and H. Yakowitz, Preparation of electron probe microanalyzer standards using a rapid quench method *in* "Applications of X-ray Analysis," Vol. 10, p. 431. Plenum Press, 1967.
24. F. Proschan, *J. Amer. Statist. Assoc.* **48**, 550 (1953).
25. T. O. Ziebold, M.I.T. Summer Course Notes (T. O. Ziebold and R. E. Ogilvie, eds.), 1966.
26. T. O. Ziebold and R. E. Ogilvie, *Anal. Chem.* **36**, 322 (1964).
27. J. I. Goldstein, and F. Wood, Experimental procedure for trace determination with the electron probe. *Natl. Conf. Electron Probe Microanalysis, 1st, College Park, Maryland, 1966.*
28. L. Boltzmann, *Ann. Physik* **53**, 960 (1894).
29. C. Matano, *Japan. Phys.* **8**, 109 (1933).
30. R. E. Ogilvie, ScD Thesis. Mass. Inst. Technol. (1955).
31. Y. Adda and J. Philibert, *Compt. Rend.* **742**, 3081 (1956).
32. Y. Adda, J. Philibert, and C. Mairy, *Compt. Rend.* **243**, 1115 (1956).
33. R. E. Hanneman, R. E. Ogilvie, and H. C. Gatos, *Trans. Met. Soc. AIME (Amer. Inst. Mining. Met. Petrol. Eng.)* **233**, 691 (1965).
34. J. I. Goldstein, R. E. Hanneman, and R. E. Ogilvie, *Trans. Met. Soc. AIME (Amer. Inst. Mining, Met. Petrol. Eng.)* **233**, 812 (1965).

35. P. G. Shewmon, "Diffusion in Solids." McGraw Hill, New York, 1963.
36. J. S. Kirkaldy and L. C. Brown, *Can. Met. Quart.* **2**, 89 (1963).
37. T. O. Ziebold, PhD Thesis. Mass. Inst. Technol., 1965.
38. J. C. Fisher, *J. Appl. Phys.* **22**, 74 (1951).
39. R. T. P. Whipple, *Phil. Mag.* **45**, 1225 (1954).
40. A. E. Austin and N. A. Richard, *J. Appl. Phys.* **32**, 1462 (1961).
41. D. M. Koffman, ScD Thesis. Mass. Inst. Technol., 1964.
42. G. R. Speich, J. A. Gula, and R. M. Fisher, "The Electron Microprobe" (T. D. McKinley, K. F. J. Heinrich, and D. B. Wittry, eds.), p. 525, New York, 1966.
43. F. Maurice, R. Seguin and J. Henoc, Phenomena of fluorescence in diffusion couples. *Actes Congr. Intern. Optique Rayons X Micro-Analyse, 4th, 1967*. Hermann, Paris.
44. T. O. Ziebold, M.S. Thesis. Mass. Inst. Technol., 1963.
45. R. E. Hanneman, Gen. Elect. Res. Develop. Center, private communication.
46. W. Jost, "Diffusion in Solids, Liquids and Gases." Academic Press, New York, 1952.
47. J. I. Goldstein and R. E. Ogilvie, Metallurgical considerations for the determination of phase diagrams with the electron probe microanalyzer. *Actes Congr. Intern. Optique Rayons X Micro-Analyse, 4th, 1967*. Hermann, Paris.
48. Y. Adda, A. Kirianenko, M. Beyeler, F. Maurice, *Mem. Sci. Rev. Met.* **58**, 716 (1961).
49. N. L. Peterson and R. E. Ogilvie, *Trans. AIME* **218**, 439 (1960).
50. Y. Adda, M. Beyeler, and A. Kirianenko, *Compt. Rend.* **250**, 115 (1960).
51. N. Swindells, *J. Inst. Metals* **90**, 167 (1962).
52. E. J. Rapperport and M. F. Smith, *Trans. AIME* **230**, 6 (1964).
53. S. J. B. Reed and J. V. P. Long, "X-ray Optics and X-ray Microanalysis," p. 317. Academic Press, New York, 1963.
54. J. I. Goldstein and R. E. Ogilvie, *Trans. AIME* **233**, 2083 (1965).
55. B. Chalmers, "Principles of Solidification." Wiley, New York, 1964.
56. A. Kohn and J. Philibert, *Mem. Sci. Rev. Met.* **57**, 291 (1960); *Metal Treat. Drop Forging* **27**, 327 (1960).
57. H. Biloni, G. F. Bolling, and H. A. Domian, *Trans. AIME* **233**, 1926 (1965).
58. J. Philibert, *J. Inst. Metals* **90**, 241 (1962).
59. H. D. Brody and M. C. Flemings, *Trans. AIME* **236**, 615 (1966).
60. T. F. Bower, H. D. Brody and M. C. Flemings, *Trans. AIME* **236**, 624 (1966).
61. J. Philibert and C. Crussard, *J. Iron Steel Inst.* (*London*) **183**, 42 (1956).
62. J. F. Lovering and C. A. Anderson, *Science* **147**, 734 (1965).
63. C. Wagner, *J. Electrochem. Soc.* **99**, 369 (1952).
64. N. G. Koopman, M.S. Thesis. Mass. Inst. Technol., 1963.
65. H. Aaronson and H. A. Domain, *Trans. AIME* **236**, 781 (1966).
66. J. I. Goldstein and R. E. Ogilvie, *Geochim. Cosmochim.* Acta **29**, 893 (1965).
67. J. I. Goldstein and J. M. Short, Cooling rates of 27 iron and stony-iron meteorites. *Geochim. Cosmochim. Acta* **31**, 1001 (1967).
68. J. I. Goldstein, *Science* **153**, 975 (1966).

Scanning Electron Probe Measurement of Magnetic Fields

JOHN R. DORSEY*

Department of Defense
Fort George G. Meade
Maryland

I. Introduction

The electron probe is an extremely versatile instrument. Not only has the probe been used to study the chemical and structural properties of materials by measuring the energy and scattering pattern of the primary X rays generated by the beam but also it has been used as a voltage probe to measure electrical potentials in microelectronic circuits and solid state devices.

The measurement of magnetic fields in the scanning electron probe is, therefore, but one more in a growing list of applications of this valuable instrument. Described in this paper is a means of measuring, analyzing, and displaying an image of the magnetic field distribution at the surface of a material, specifically the magnetic field of a periodic recording on magnetic tape. Data are presented in various forms, limitations in the collection of

* Deceased.

data are discussed, and possible methods of improving the power of this technique are suggested.

II. Basic Description of the Method

While the emission of X rays, backscatter, and secondary electrons will all be influenced by the chemistry and structure of the material being bombarded by the electron beam, only the secondary electrons will be appreciably affected by electric or magnetic fields present at the surface of the material. A variation in electrical potential of the surface will vary both the energy a secondary has when it arrives at an electron collector surface and the trajectory by which it reaches the surface. A magnetic field will only influence the trajectory or the effective angle of emission. It is the measurement of this change in the angle of emission that forms the basis of this technique for measuring magnetic fields in the scanning electron probe.

An electron traveling in a magnetic field experiences a force perpendicular to the magnetic field and its direction of motion. The acceleration a_θ of such an electron with charge to mass ratio e/m with radial velocity v_r in the yz plane, in a magnetic field B, parallel to the x axis as illustrated in Fig. 1 is given by

$$a_\theta = (e/m)v_r B. \tag{1}$$

If y_0 is the point of impact of a beam of high energy electrons incident normal

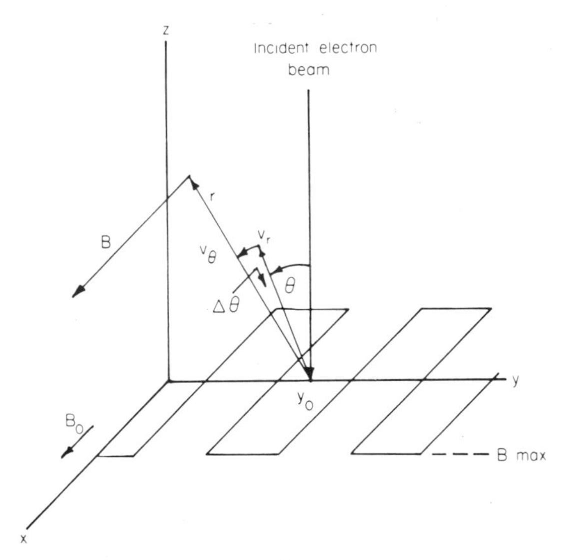

Fig. 1. Basic geometry.

to the specimen surface in the xy plane, then low energy or secondary electrons are emitted in all directions from that point with a distribution of radial velocities. A secondary electron emitted with radial velocity v_r at an angle θ with respect to the normal to the specimen surface will obtain a component of velocity v_θ, perpendicular to the original direction of travel such that for small changes in direction $\Delta\theta$,

$$v_\theta = \int_0^R a_\theta / v_r \, dr \tag{2}$$

where R is the distance from y_0 to the collector surface. When R is large compared to the wavelength of the periodic magnetic field, the field is only strong enough to cause an appreciable a_θ while the electron is very near the surface of the specimen. Therefore, $\Delta\theta$ may be approximated simply as

$$\Delta\theta \approx v_\theta / v_r . \tag{3}$$

The backscatter electrons, on the other hand, will have such high energy that they will not be deflected an amount great enough to be of any significance.

III. Magnetic Field Calculation

If the horizontal component on the magnetic field at the surface of the specimen, B_0, is expressed as a Fourier series

$$B_0 = \sum_{n=0}^{\infty} B_{0n} \tag{4}$$

where each term in the series is given by

$$B_{0n} = a_n \cos[2\pi y/(\lambda/n)] + b_n \sin[2\pi y/(\lambda/n)], \tag{5}$$

then the magnetic field above the surface may be expressed as the series

$$B = \sum_{n=0}^{\infty} B_n \tag{6}$$

where each term in the series is given by

$$B_n = B_{0n} \exp\left(\frac{-2\pi r \cos\theta}{\lambda/n}\right). \tag{7}$$

In Eqs. (5) and (7), λ is the fundamental wavelength of the recording.

$$a_0 = \lambda^{-1} \int_0^\lambda B_0 \, dy \qquad \text{and} \qquad b_0 = 0 \qquad \text{for} \quad n = 0;$$

$$a_n = 2/\lambda \int_0^\lambda B_0 \cos[2\pi y/(\lambda/n)]\, dy \quad \text{and} \quad b_n = 2/\lambda \int_0^\lambda B_0 \sin[2\pi y/(\lambda/n)]\, dy$$

$$\text{for } n \geq 1.$$

The attenuation factor $\exp[(-2\pi r \cos\theta)/(\lambda/n)]$ is based on the attenuation with distance of the maximum in the horizontal component of magnetic field given by Mee [1] for a sinusoidal recording. Equation (6) is merely the superposition of sinusoidal components given in Eq. (5), each with its own respective wavelength and attenuation factor.

IV. Angular Deflection of Secondaries

Using the relationships

$$y = y_0 - r \sin\theta$$

where y_0 is the distance from the origin of the system and the point of impact of the incident beam and

$$v_r = (2eV/m)^{1/2}$$

where V is the energy in electron volts of the secondary electron, where the height of the collector surface $R\cos\theta$ is much greater than $\lambda/2\pi$ and y_0, and by combining Eqs. (5)–(7) and (1)–(3), the angle of deflection is found to be

$$\Delta\theta = \frac{\lambda}{4\pi}\left(\frac{2e}{mV}\right)^{1/2} \sum_{n=0}^{\infty} \frac{a_n}{n}\cos\left(\frac{2\pi y_0}{\lambda/n} - \theta\right) + \frac{b_n}{n}\sin\left(\frac{2\pi y_0}{\lambda/n} - \theta\right). \tag{8}$$

Notice that this expression is a function of the fundamental wavelength of the recording, the position from which the secondary electron originates, its energy and angle of emission. An approximate expression for the maximum angle of deflection is

$$\Delta\theta_{\max} \approx (\lambda/4\pi)(2e/mV)^{1/2} B_{\max} = 4.73 \times 10^{-2} \lambda B_{\max}/V^{1/2} \tag{9}$$

in practical units.* This is the $\Delta\theta_{\max}$ at an emission angle of 0° for a pure sine wave recording. It is now possible to calculate, for example, the maximum change in direction that a secondary electron will experience for a recording wavelength of 10 mils, a secondary electron energy of 4 eV, a maximum magnetic field at the specimen surface of 100 G, and an emission angle from the normal to the surface of 0°. This angle is 0.059 rad, or 3.4°. Under these conditions, the small $\Delta\theta$ approximation used in Eqs. (2) and (3) is easily justified.

* Radians, centimeters, gauss, volts, or electron volts.

V. Detected Signal Due to Magnetic Field

For an angular distribution of secondary electrons as given by Jonker [2],

$$dI/d\theta \sim \cos\theta, \tag{10}$$

and two collector surfaces subtending arcs in the yz plane as shown in Fig. 2, the current intercepted by either collector surface in the absence of a magnetic field is

$$I(\theta_1, \theta_2) \sim \sin\theta_2 - \sin\theta_1. \tag{11}$$

Of course, an actual collector surface has finite width in the x direction, but this may be kept small enough to reasonably satisfy this two dimensional treatment. The ratio of the change in current intercepted by one of the collectors due to the deflection of secondaries emitted at θ_1 and θ_2 to the secondary current intercepted in the absence of deflection is given by

$$\frac{\Delta I(\theta_1, \theta_2)}{I(\theta_1, \theta_2)} = I^1\left[\frac{dI}{d\theta}(\theta_1)\,\Delta\theta_1 - \frac{dI}{d\theta}(\theta_2)\,\Delta\theta_2\right] \tag{12}$$

$$= \frac{(\cos\theta_1)\,\Delta\theta_1 - (\cos\theta_2)\,\Delta\theta_2}{\sin\theta_2 - \sin\theta_1}. \tag{13}$$

Because of the symmetry of the system, the current gained by one collector should approximately equal the current lost by the other. Therefore, if the signal from one collector is subtracted from the other, the backscatter and most of the gross secondary signal should be eliminated, leaving twice the change in signal from either collector.

A quantity S, independent of the beam current, may be defined as the ratio of this difference signal to the signal from either detector in the absence of deflection as given by

$$S = \Delta I(\theta_1, \theta_2) - \Delta I(-\theta_2, -\theta_1)/I(\theta_1, \theta_2). \tag{14}$$

Using Eq. (8) in order to evaluate $\Delta\theta$ at θ_1 and θ_2 and using Eqs. (13) and (14), S may be evaluated as

$$S = \frac{\lambda}{2\pi}\left(\frac{2e}{mV}\right)^{1/2}(\sin\theta_2 + \sin\theta_1)\sum_{n=0}^{\infty}\frac{B_{0n}(y_0)}{n} \tag{15}$$

where

$$B_{0n}(y_0) = a_n \cos[2\pi y_0/(\lambda/n)] + b_n \sin[2\pi y_0/(\lambda/n)]. \tag{16}$$

$B_{0n}(y_0)$ is simply the nth Fourier term of the magnetic field at the surface of the specimen evaluated at the point of impact of the incident electron beam. We thus have a means of measuring that surface field with resolution limited only by the effective beam size and the sensitivity of the detection system. Here, S_{max} may be evaluated in terms of $\Delta\theta_{max}$ of Eq. (9) and is given by

$$S_{max} \approx (\lambda/2\pi)(2e/mV)^{1/2}(\sin\theta_2 + \sin\theta_1)B_{max}$$
$$\approx 9.46 \times 10^{-2}\lambda B_{max}(\sin\theta_2 + \sin\theta_1)/V^{1/2}. \tag{17}$$

For the same λ, B_{max}, and V used previously to calculate the $\Delta\theta_{max}$ of 3.4°, and a collector θ_1 and θ_2 of 10 and 60°, S_{max} is approximately 12%.

VI. Detected Signal Due to Surface Relief

Up to now, we have considered that the specimen was a smooth surface normal to the incident beam. As illustrated in Fig. 2, if this is not the case, then the distribution of secondary electrons will shift with respect to the

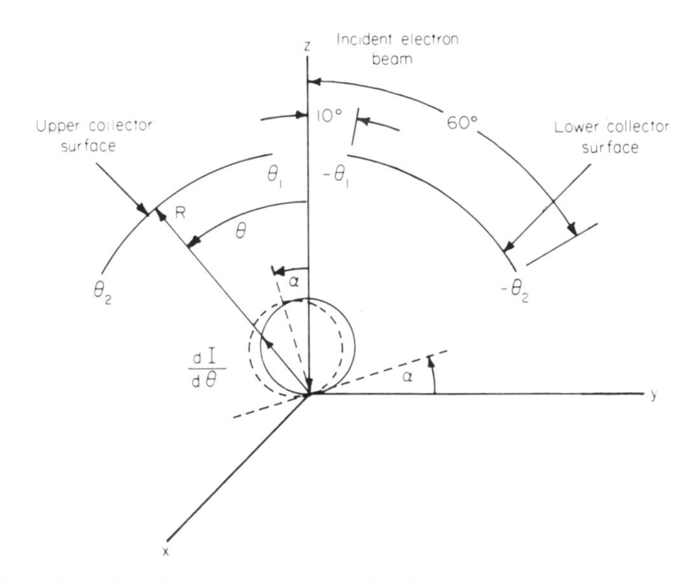

FIG. 2. Basic collector geometry and distribution of secondary electrons.

beam and the collectors by an angle α equal to the shift in the normal to the surface [2]. A calculation of S due to such a change in the specimen surface may be made using Eqs. (13) and (14). Where $\Delta\theta_1 = \Delta\theta_2 = \alpha$,

$$S = 2(\cos\theta_1 - \cos\theta_2)\alpha/(\sin\theta_2 - \sin\theta_1). \tag{18}$$

For θ_1 and θ_2 equal to 10 and 60°, $S_{max} = 1.4\alpha$. This is 1.4% for every 10^{-2} rad or 0.57° shift in the normal of the specimen surface. Contrast due to such surface relief is illustrated in Fig. 3 in the form of a scanning electron probe image of a 200 mesh grid. Although this contrast will certainly interfere with the measurement of magnetic fields, there was no serious interference at the 10 μ resolution level on the magnetic tape used for this work.

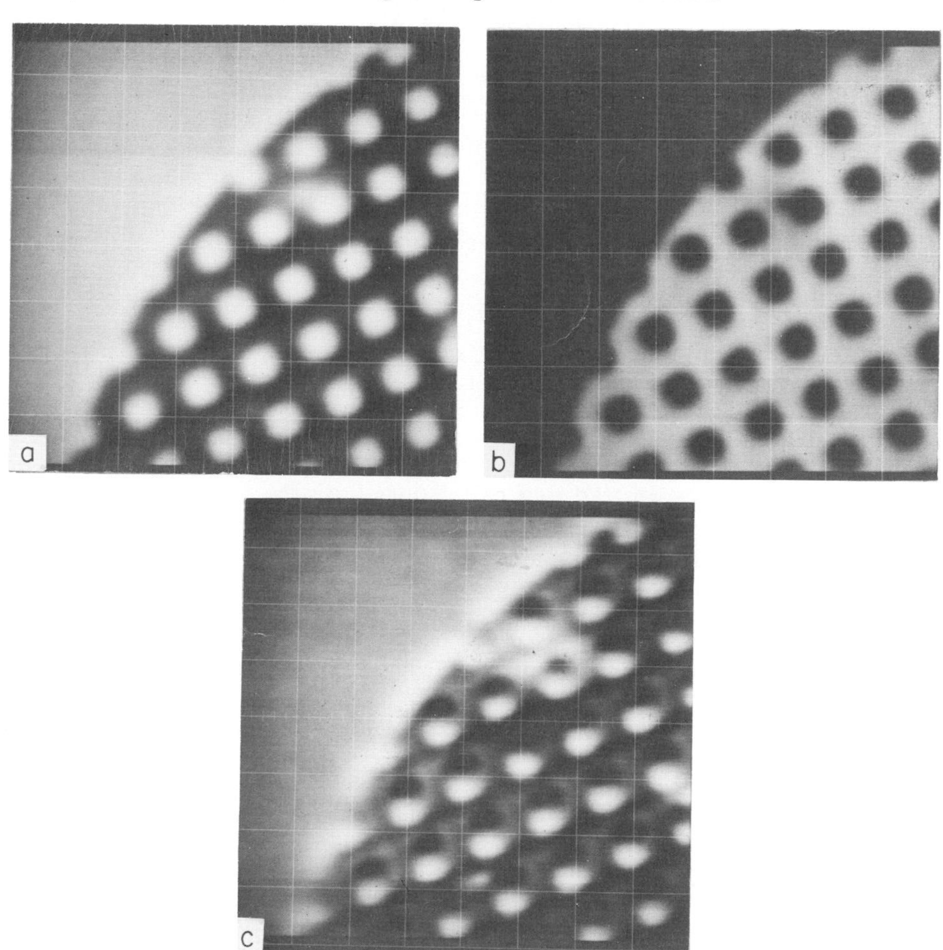

FIG. 3. Contrast due to surface relief on a 200 mesh grid. (a) upper collector, (b) lower collector, (c) upper–lower collector.

VII. Other Limitations

We have also ignored in our calculations the fact that the secondary electrons are emitted with a distribution of energies. A detector was constructed to measure this distribution, as illustrated in Fig. 4. However, no

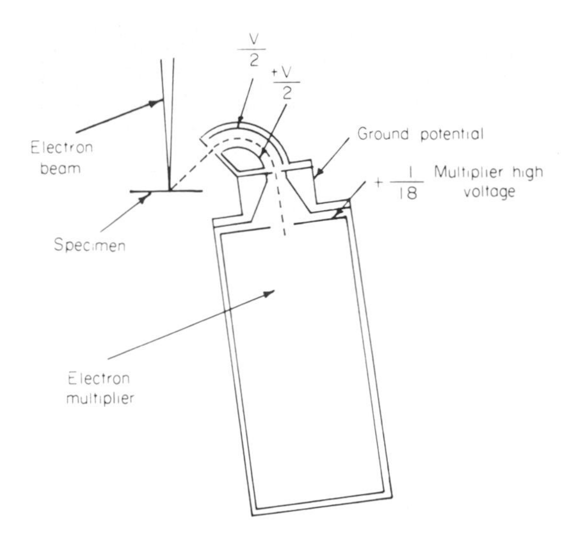

FIG. 4. Secondary electron detector with energy selection.

attempt has been made as yet to integrate the combined signal due to all secondary energies. Instead, our calculations have the energy of 4 eV corresponding to the peak in the signal displayed in Fig. 5. The peak for most metals is generally between 2 and 5 eV [3]. The display in Fig. 5 is a plot of current x energy because the width of the energy window in this type of detector is proportional to the energy being selected.

Figure 6 shows a scanning electron probe or microscope image of a magnetic recording using this technique. The signal recorded on the tape was a 100 cycle/in. square wave. The distortion due to the attenuation of the high frequency components by this method of detection is evident in the quantitative signal superimposed on the image. The measured S of 2.1% indicates a maximum horizontal component of surface field of 14.8 G.

VIII. DESCRIPTION OF DETECTOR

The actual collector geometry used to generate these data is shown in Fig. 7. Although a narrow detector following the arc shown in Fig. 2 was the first design used, it was more desirable for economy of space to use a flat surface. This surface is designed such that the curved edges are at θ_1 and θ_2, respectively, with subtend arcs such that $\omega \sin \theta$ is a constant. The straight edges approximate the same conditions, and also approximate hyperbolas formed by cones at angles of $\pm 10°$ out of the yz plane having axes coincident with the x axis of the system. The reason for wanting to hold

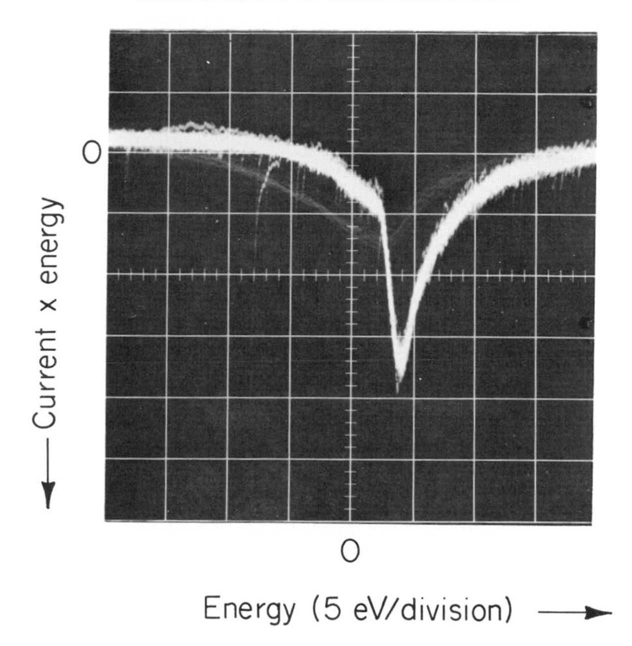

FIG. 5. Measured secondary electron energy distribution.

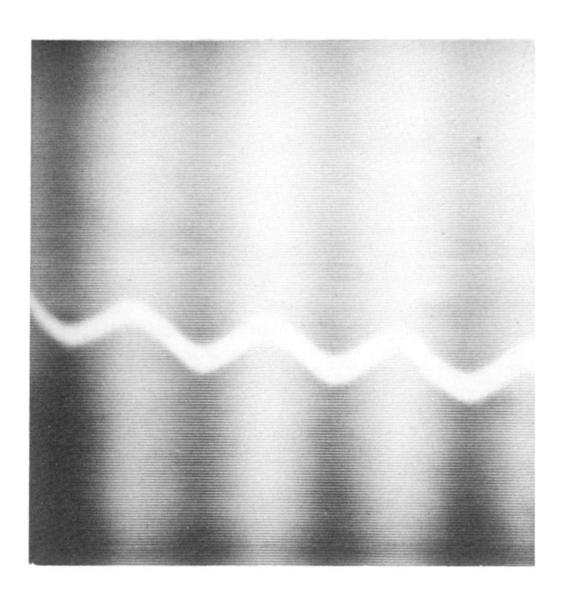

FIG. 6. Scanning electron probe image of a 100 cycle/in. square wave magnetic recording.

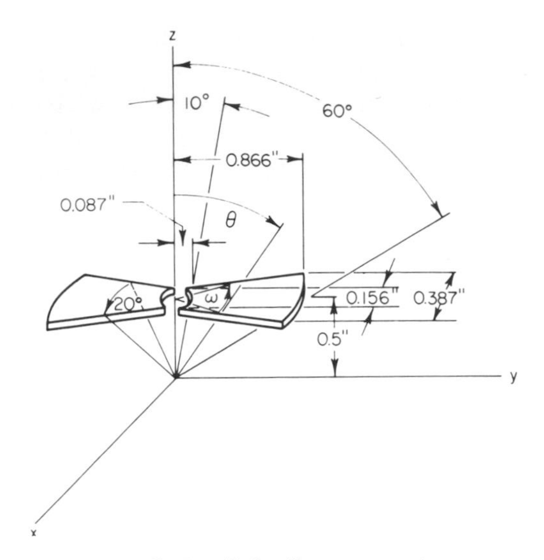

FIG. 7. Detailed collector geometry.

$\omega \sin \theta$ constant is that the current intercepted by the collector per unit increment of θ be proportional to $\cos \theta$, as in Eq. (10). That is, since the current per unit solid angle for a secondary distribution is given by

$$dI/d\Omega = I_t/\pi \cos \theta, \tag{19}$$

then since

$$dI/d\theta = \omega \sin \theta \, dI/d\Omega, \tag{20}$$

$$dI/d\theta = CI_t/\pi \cos \theta \tag{21}$$

where

$$\omega \sin \theta = C. \tag{22}$$

Since the angle subtended by the collector out of the yz plane is roughly constant, Eq. (13) will be valid in that the change in collected current will only occur at θ_1 and θ_2 rather than along the straight edges of the collector as well. Aside from the considerations involved in satisfying the two-dimensional nature of the theoretical treatment used to develop S in Eq. (15), θ_1 and θ_2 should be chosen to give the largest possible signal or $SI(\theta_1, \theta_2)$. Combining Eqs. (15) and (20),

$$SI(\theta_1, \theta_2) = (CI_t \lambda/2\pi^2)(2e/mV)^{1/2}(\sin^2 \theta_2 - \sin^2 \theta_1) \sum_{n=0}^{\infty} \frac{B_{0n}(y_0)}{n}. \tag{23}$$

This indicates that the maximum signal occurs when θ_1 and θ_2 are separated by 90°. Our choice for θ_1 and θ_2 of 10 and 60° provides a signal which is 70% of this maximum value and an S which is 3% higher. The collectors themselves are made of aluminum, attached from above to an insulator with no direct line of sight to the specimen and are connected via short leads to the amplifier shown in Fig. 8. Aluminum was chosen as a general purpose collector material

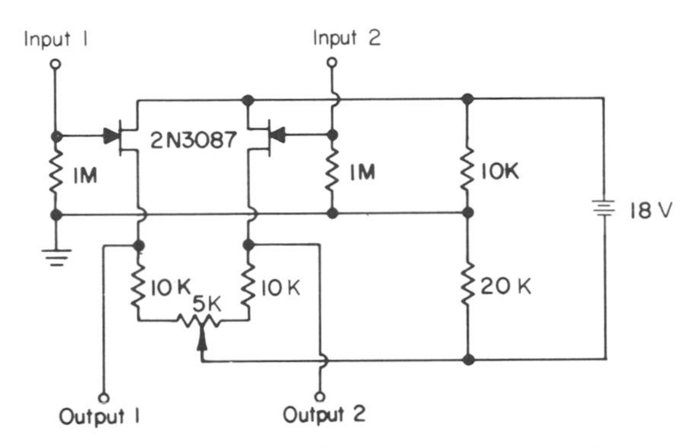

FIG. 8. Differential source follower input impedance amplifier. Input impedance 1 MΩ; output impedance 10 kΩ; bandwidth, dc to 50 kHz; effective rms noise current at input $= 10^{-10}$ A $\times$ 3 $\times$ 10^{-11} A at 5 kHz.

because of its low backscatter yield although a high backscatter yield material might be better for this application in order to reduce unwanted signals due to collected backscatter electrons from the specimen. The amplifier is mounted inside the vacuum chamber along with the collector plates since the bandwidth is limited primarily by the input capacity and resistance of the amplifier. The amplifier outputs are connected to a commercial differential amplifier with a voltage gain of up to 10^4. Other applications of this detector are discussed in Appendix I.

IX. Specimen Preparation

The magnetic recording tape was prepared for examination by first evaporating several hundred angstroms of gold on the surface. No attempt to smooth the surface was made although this has been found by others to be necessary at higher resolutions in the electron mirror microscope. The tape was then mounted on a glass slide and covered by a mask of tin foil in which a rectangular window had been cut. A conducting paint was then applied at the edge of the window in order to make good contact to the gold surface. The foil was then connected to ground or to a bias voltage supply. Although this procedure gave the best results, occasionally the signal from such a sample was considerably smaller than that usually obtained. It is not clear what may be causing this unless discontinuities in the gold film or the formation of a dielectric contaminant allow regions of the surface to charge and either reduce the effective secondary yield or deflect the distribution away from the collector plates. When the specimen was grounded, there should not have been any appreciable electric field between it and the collectors.

X. Measurement Techniques, Results and Discussion

Typical results of measurements made of the saturated 100 cycle/in. recording on 3M951 magnetic recording tape shown in Fig. 6 are outlined in Table I. With a 20 kV 1.35×10^{-7} A electron beam, the secondary current

TABLE I

A Typical Set of Measurements on the 100 cycle/in. Square Wave Recording Shown in Fig. 6

Measured quantities			
Specimen bias (V):	+12	0	−12
Specimen current (A):	1.35×10^{-7}	1.1×10^{-7}	1.0×10^{-7}
Collector current (A):	5×10^{-10}	4×10^{-9}	4×10^{-9}
Peak-to-Peak Signal (A):	0	1.5×10^{-10}	4×10^{-10}
Calculated quantities			
Secondary current (A):	$1.35 \times 10^{-7} - 1.0 \times 10^{-7} = 3.5 \times 10^{-8}$		
Collected secondary current $I(10°, 60°)$ (A):	$4 \times 10^{-9} - 5 \times 10^{-10} = 3.5 \times 10^{-9}$		

collected on each detector plate, $I(10°, 60°)$, was within 10% of 3.5×10^{-9} A. The maximum difference signal $S_{max}I(10°, 60°)$ was 7.5×10^{-11} A from a peak to peak signal of 1.5×10^{-10} A. This corresponds to an S_{max} of 2.1% and a maximum horizontal component of magnetic field of 14.8 G.

The beam current was taken to be the absorbed or specimen current with a positive 12 V bias on the specimen (see Appendix II). An approximate value for the secondary current was obtained by measuring the difference between the absorbed current with a bias of positive 12 V and negative 12 V. This difference was 3.5×10^{-8} corresponding to a secondary yield of 26%. The secondary current collected on each collector plate was measured by taking the difference between that collected with no specimen bias and that collected with a positive 12 V bias on the specimen. The 3.5×10^{-9} A, thus determined, should not contain the component of collected current due to backscatter electrons from the specimen. This collected secondary current corresponds to 10% of the total secondary yield. The collector plate subtends a solid angle of 4.4% of the 2π sr above the surface of the specimen, and for the assumed cosine distribution of secondaries, the collector plate should intercept 8.4% of the total secondary yield.

A very useful maximum gain in signal was obtained by putting a negative 12 V bias on the specimen. The peak to peak signal increased by a factor of 2.7, from 1.5×10^{-10} to 4×10^{-10} while the current intercepted by each collector increased by less than 10%, and the current absorbed by the specimen was reduced by 10%. The calculation in Appendix III indicates that this increase in peak to peak signal should be by a factor of 2. In some "poor" samples where the zero bias signal was unusually low, the increase with positive bias was as much as an order of magnitude, accompanied by a factor of 2 increase in collected secondary current and a 30% decrease in absorbed current. Better agreement for magnetic field calculations in such samples was obtained by using the collected secondary current, calculated by taking the difference between the collected current with negative bias on the specimen and the collected current with positive bias on the specimen instead of the usual no bias to positive bias difference. The reason for the decrease in absorbed current at the specimen is unclear, but it may be due to the elimination of secondaries created by backscattered electrons which have struck the collectors or the repulsion of secondaries that are otherwise attracted back to the specimen under zero bias condition by a positive surface charge. Because of the long working distance of about 5 in. from the magnetic lens in the system used for these measurements, the magnetic field due to the lens is less than 0.1 G. Since the radius of curvature of a 4 V electron traveling perpendicular to earths field of 0.5 G is about 13 cm, it is not possible that any of these electrons could be returned to the specimen. Electrons with lower energy than 4 eV, however, may be returning by this means to the specimen under the zero bias condition as the radius of curvature is proportional to the square root of the energy. This disproportionate increase in S must be due to a condensation of the secondary distribution toward the normal to the specimen surface. The theoretical calculations for this negative bias situation show that the 4 eV electrons emitted parallel to the specimen surface are deflected by the electric field well onto the collector plates so current is no longer lost at the high angle edge due to magnetic deflection. Therefore, magnetic contrast is solely due to the deflection and secondary current density at the low angle edge of the detector.

Since, in general, it will be desirable to use a standard specimen in order to make other than relative field measurements and calibration at several wavelengths will have to be done anyway, there is considerable advantage in calibrating for the negative bias case. Such a calibration was made using sine wave recordings of known vertical field strength as measured by a recording head. Using a sensitivity coefficient k, defined by the equation

$$S = k\lambda B, \tag{24}$$

the k was determined for a 5, 10, and 20 mil wavelength with a 10 keV incident electron beam to be 10.5×10^{-2}, 8.12×10^{-2}, and 6.94×10^{-2} $(\text{cm G})^{-1}$. 9.5×10^{-2} $(\text{cm G})^{-1}$ is the sensitivity calculated from Eq. (17) using $V=$

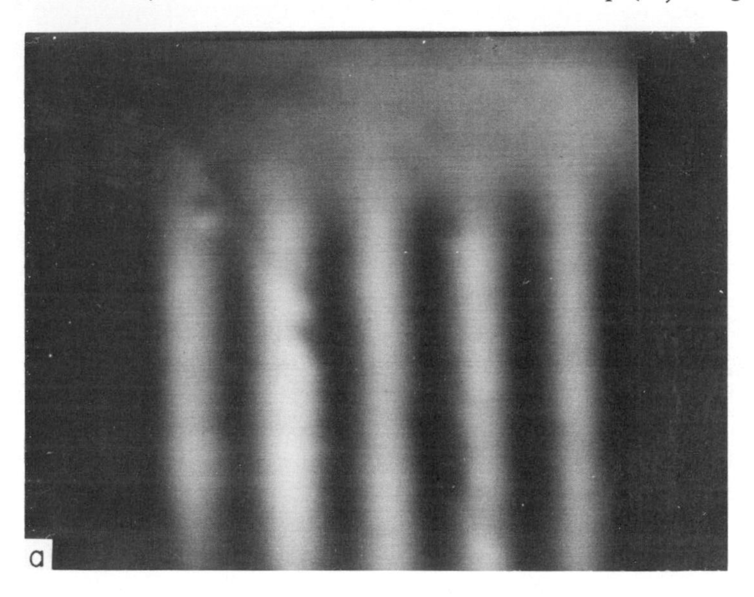

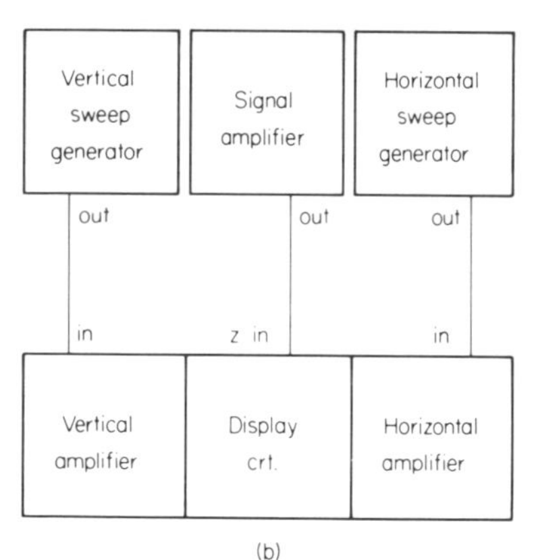

FIG. 9a, b.

FIG. 9. Images of a 10 mil sine wave magnetic recording illustrating two methods of display.

4 eV, θ_1 and θ_2 equal to 10 and 60°, respectively, and multiplying by a factor of 2 due to the negative specimen bias.

XI. Image Displays

The display of such magnetic information as the scanning electron probe provides may be accomplished in a number of ways. Figure 9 contains illustrations of two such methods. At the left-hand side of the figure, (a) and (b) illustrate the manner in which a conventional scanning microscope

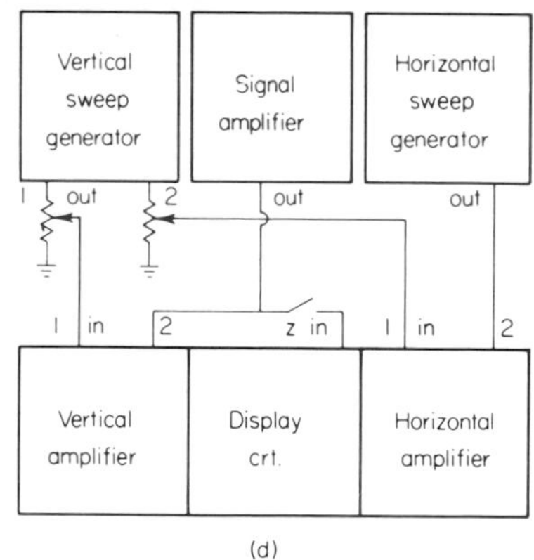

Fig. 9c, d.

image is obtained. The electron beams in the probe and the display cathode ray tube (crt) are made to sweep synchronously while the signal being generated from each point on the specimen is being used to modulate the intensity of the beam in the display crt via its z axis input. The same 10 mil sine wave recording that is displayed in (a) is also displayed in (c) but in a slightly different fashion. While there is still a one-to-one correspondence between a point on the specimen and a point on the display tube, the position of the

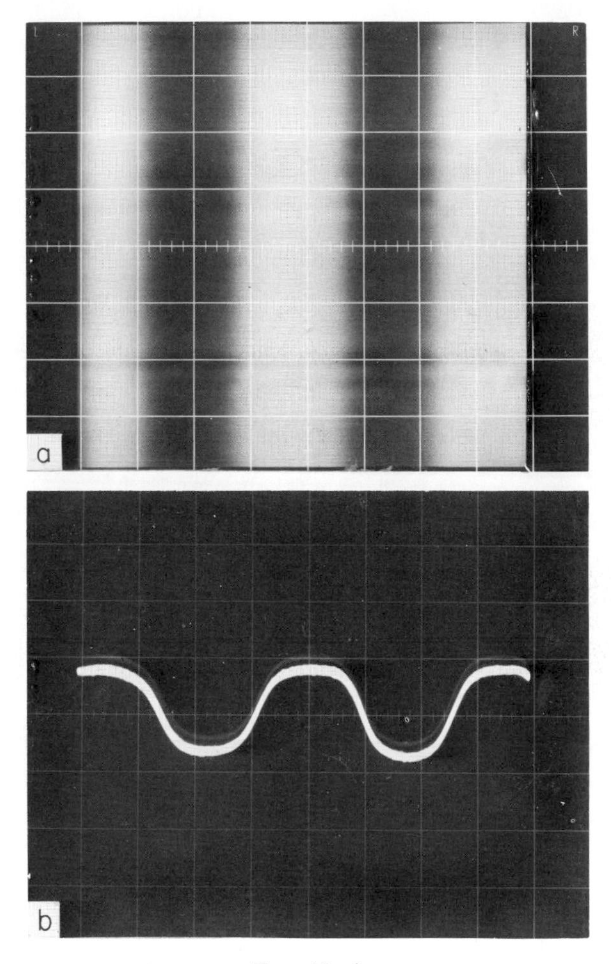

FIG. 10a, b.

FIG. 10. Various types of displays of a 20 mil square wave magnetic recording. (a) conventional scanning microscope image, (b) quantitative signal generated by a single line scan, (c) tilted view of scanning microscope image with resolved scan lines, (d) signal displayed as a function of position with intensity modulation.

beam in the display tube is elevated above or below that point by an amount proportional to the magnitude of the signal. To accomplish this, a fraction of the vertical sweep voltage is added to the signal in the vertical amplifier of the display crt while another fraction of the vertical sweep voltage is added to the horizontal sweep signal in the horizontal amplifier of the display crt. By changing the fractions of the vertical sweep voltage used, it is possible to change the angle of view of the family of curves representing the magnitude of the signal as a function of position on the specimen. It is also possible to amplitude modulate this display, as is done in (c) of Fig. 9. The image in this figure is a region of a recording near the edge of the recording track. In Fig. 10 are various combinations of the two basic types of displays illustrated

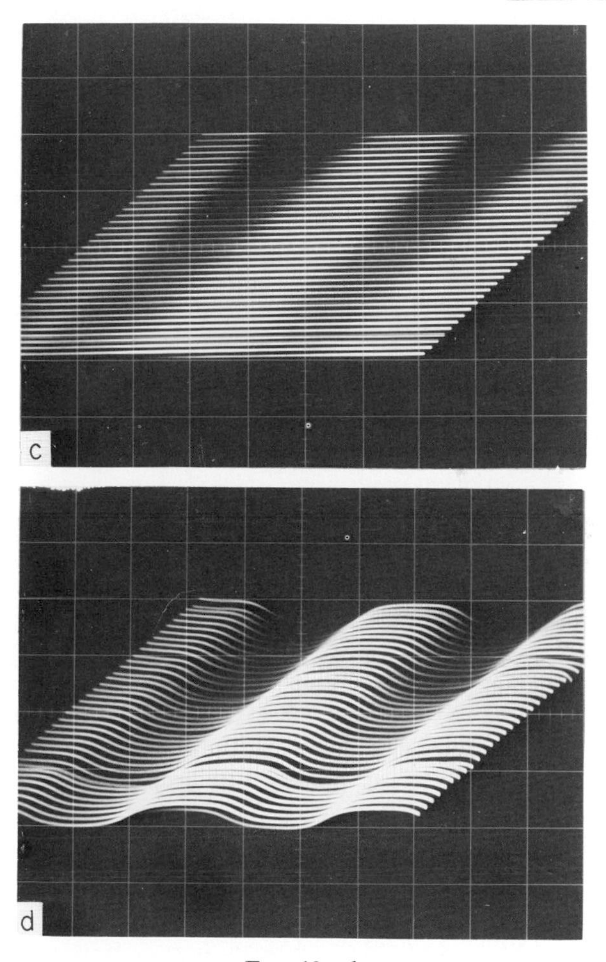

FIG. 10c, d.

in Fig. 9. The specimen in Fig. 10 is a 20 mil square wave recording. It is possible in the conventional scanning microscope image (a) to see the variation in magnetic field across the recording track due to the 3 mil laminations in the recording head.

XII. Conclusions

While the technique for making magnetic field measurement described in this paper is not expected to be a method of reading out magnetic recordings anytime in the near future, it may be developed into a useful technique for studying the quality of such recordings and the characteristics of the recording materials. For that matter, it may be used to detect, if not measure, magnetic field configurations of any type which have a transverse or horizontal component. In order to improve the value of the technique, one will certainly want to use a smaller beam size than 10 μ. This, however, will require smoother specimen surfaces to prevent interfering contrast due to surface relief. The surface must be electrically conducting to avoid local charge effects; and if a film is deposited in order to satisfy this requirement, it must be nonmagnetic and thin compared to the wavelength of the high frequency components expected to be resolved. The surface should also be one with a high secondary yield, but one with low backscatter yield, if possible. The most significant improvement in this technique will probably be in the use of some kind of large collection angle detector with current multiplication, perhaps a semiconductor detector at a high positive potential behind a grid at a potential of about 12 V. However, at the moment, electron energies of over 10 keV at the detector would be necessary to achieve significant gain in presently available semiconductor detectors.

Appendix I. Other Detector Applications

The detector described in this paper to measure magnetic fields, as has already been stated, is sensitive to surface relief. Since the signal from one detector surface is subtracted from the other, the differential signal is relatively independent of atomic number or density variations which would modulate the backscatter [4] or secondary [5] yields. These variations could be detected using the same collector geometry, however, simply by adding instead of subtracting the two signals. This would attenuate the variations in collected current due to surface relief and yield a signal relatively proportional to the backscatter and secondary yields.

It is possible to separate the signals due to backscatter electrons from those due to secondary electrons by applying a bias voltage between the specimen surface and the detector. By making the specimen positive with respect to the detector it is possible to eliminate the signal due to secondaries with energies up to that corresponding to the potential difference applied. There is no way with this detector geometry to eliminate the signal due to backscatter electrons while retaining that due to secondaries. The signal due to secondaries, however, may be enhanced, as we have shown, by making the specimen negative with respect to the detector.

A possible application that immediately comes to mind for this type of detector would be the measurement of current flow in microelectronic circuits. The small fields available over short distances above such circuits probably preclude such an application. As a means of estimating the sensitivity for such an application, one may use the field above a circular wire as given by

$$B = 0.2I/R$$

in practical units.*

Appendix II. Specimen and Collector Currents under Bias Conditions

If I_p is the primary current incident on the specimen, I_b is the current leaving the specimen due to backscattered primaries, I_s is the current leaving the specimen in the form of secondary electrons, and fI_{bs} is the secondary current incident on the specimen which is generated when the backscattered primaries strike the collector surfaces, then the current absorbed by the specimen is approximately

$$I_a = I_p - I_b - I_s + fI_{bs},$$

and the current absorbed by the collector is given by

$$I_c = f_b I_b + f_s I_s - I_{bs}$$

where f_b is the fraction of the backscattered electrons leaving the specimen that are absorbed by the collector, f_s is the fraction of the secondary electrons leaving the specimen that are absorbed by the collector, and I_{bs} is the secondary current leaving the collector which is generated by the incident backscatter electrons. Therefore, I_{bs} is proportional to $f_b I_b$. The backscattered electrons that are backscattered from the collector are ignored since they will be unaffected by the bias. The secondary electrons from the collector that are

* Gauss, amperes, centimeters.

generated by secondaries from the specimen are omitted because they are so relatively few in number.

If the specimen is made positive with respect to the collector, the current absorbed by the specimen is

$$I_{a(+)} = I_p - I_b + fI_{(+)}I_{bs}$$

where $f_{(+)}$ may be larger than f because of increased collection efficiency on the part of the specimen depending on its shape and size, etc. Under these same conditions, the collector current is

$$I_{c(+)} = f_b I_b - I_{bs}.$$

If the specimen is made negative with respect to the collector, the current absorbed by the specimen is

$$I_{a(-)} = I_p - I_b - I_s$$

and the collector current is

$$I_{c(-)} = f_b I_b + f_{s(-)} I_s$$

where $f_{s(-)}$ may be larger or smaller than f_s depending on collector geometry, etc.

Therefore

$$I_{a(+)} - I_a = I_s + (f_{(+)} - f)I_{bs},$$

$$I_a - I_{a(-)} = fI_{bs},$$

$$I_{a(+)} - I_{a(-)} = I_s + f_{(+)}I_{bs},$$

and

$$I_{c(+)} - I_c = f_s I_s,$$

$$I_{c(-)} - I_c = (f_{s(-)} - f_s)I_s + I_{bs},$$

$$I_{c(-)} - I_{c(+)} = f_{s(-)}I_s + I_{bs}.$$

The best measurement of the effective incident beam current is obtained by measuring $I_{a(+)}$. Although I_b may be appreciable, it does not produce secondaries and hopefully $f_{(+)} I_{bs}$ is small. The best value for the secondary current is obtained by measuring $I_{a(+)} - I_a$ unless one has reason to believe that part of I_s may be returning to the specimen under zero bias conditions. In this case $I_{a(+)} - I_{a(-)}$ is probably better. A rough estimate of the ratio of backscatter to secondary yield may be obtained by the ratio of $I_{c(+)}$ to $I_{c(+)} - I_c$, $I_{c(+)} - I_c$ being the collected secondary current under zero bias conditions. Again, if there is reason to believe that part of I_s is returning to the specimen, collected secondary current might better be determined by $I_{c(-)} - I_{c(+)}$.

Appendix III. Magnetic Signal under Bias Conditions

The application of a negative bias to the specimen will compress the distribution of secondary emission in the direction of the normal to the specimen in such a fashion that

$$\frac{dI}{d\theta'} = \frac{\sin\theta}{\sin\theta'}\frac{dI}{d\theta}.$$

Here, θ' is the angle from the normal to the specimen to the line joining the origin to the point of incidence of the electron at the collector. The relationship between θ and θ' is given by

$$1 = \frac{\tan\theta'}{\tan\theta} + \frac{V_b}{4V}\frac{\tan^2\theta'}{\sin^2\theta}$$

where V_b is the amount of negative bias. At small angles

$$\theta' = \frac{2V}{V_b}\left[\left(1 + \frac{V_b}{V}\right)^{1/2} - 1\right]\theta$$

or

$$= \tfrac{2}{3}\theta$$

for $V = 4$ eV and $V_b = 12$ V. A change in θ' with respect to a change in θ is likewise reduced at small angles. The θ' corresponding to a θ of 90° is given by

$$\theta' = \tan^{-1} 2[V/V_b]^{1/2}$$

or

$$= 49.1°$$

for the above values of V and V_b. As this is well onto the collector plates, there is no longer any loss of signal due to the deflection of the secondary distribution off the high angle edge of the collectors. Following the procedure used in the derivation of Eq. (23), the signal due to a magnetic field with negative specimen bias is given by

$$S_{(-)} = \frac{\cos^2\theta_1}{\sin^2\theta_2' - \sin^2\theta_1'}\left(\frac{\sin\theta_1}{\sin\theta_1'}\right)S$$

or

$$= 2.0S$$

for $V = 4$ eV, $V_b = 12$ V, $\theta_1' = 10°$, and $\theta_2' = 60°$.

REFERENCES

1. C. D. Mee, "The Physics of Magnetic Recording," p. 136. North-Holland Publ., Amsterdam, 1964.
2. J. L. H. Jonker, *Philips Res. Repts.* **6**, 372 (1951).
3. H. Bruining, "Physics and Applications of Secondary Electron Emission," p. 105. McGraw-Hill, New York, 1954.
4. G. D. Archard, *J. Appl. Phys.* **32**, 1505 (1961).
5. H. Bruining, "Physics and Applications of Secondary Electron Emission," p. 84. McGraw-Hill, New York, 1954.

Nondispersive X-Ray Emission Analysis for Lunar Surface Geochemical Exploration

J. I. TROMBKA and I. ADLER

Goddard Space Flight Center
National Aeronautics and Space Administration
Greenbelt, Maryland

I. Introduction

Our laboratory is presently investigating possible instrumentation for geochemical exploration of lunar or planetary surfaces for both manned and unmanned missions. The studies presently going forward will hopefully be used to define a total system having as its objectives: (1) obtaining operational numbers to help an astronaut in sample selection, and (2) obtaining geochemical information for mapping and preliminary surface analysis for manned and unmanned missions.

During the NASA 1965 Summer Conference on Lunar Exploration and Science, a considerable amount of attention was given by the Geochemistry Working Group to problems of surface exploration. While it was the consensus that the lunar environment was not particularly conducive to performing detailed geochemical analysis, there was, however, a great need for diagnostic tools to help the astronaut in selecting samples to be returned to earthside laboratories for comprehensive studies.

Consideration of diagnostic tools in relationship to the mission's objectives leads to a number of requirements that must be satisfied. These requirements range from geochemical parameters to be tested, optimum instruments to be employed, and the vital problem of data acquisition and processing in very real time. Additionally, although the proposed instrumentation would primarily enable the astronaut to make decisions about sample selection, the design is sufficiently flexible to be useful for either manned or unmanned missions.

There are certain obvious characteristics that such instrumentation must meet such as small size and weight, reliability, stability, and specificity. Because some of these requirements are, at least in part, conflicting, one must accept some tradeoffs. In this paper, one such instrument will be described as well as a computer program for data reduction.

A. Instrumentation

Obviously, instrumental and system design is strictly determined by the mission profile. If, for example, one established a criterion that the probe is needed to help an astronaut in sample selection, by permitting him to perform relatively simple *in situ* analysis, then "nondispersive" X-ray emission* analysis becomes attractive. Such a device lends itself to portability as well as simple and rugged construction having minimal space and power requirements. In fact, such instrumentation has been proposed by a number of investigators [1–4]. One of the most desirable features of such instrumentation is that X-ray spectra can be excited effectively by means of radioactive isotopic sources. The use of this type excitation has been discussed in a number of publications [5–10]. Most applications have involved the use of gamma, beta, or bremsstrahlung sources, but only recently has a systematic study of alpha excitation been undertaken [2, 7, 8, 10]. The advantages of alpha particle excitation will be discussed in additional detail in Section I,B.

In the program to be described, attention has been directed to the common rock forming elements such as Mg, Al, Si, K, Ca, and Fe. The characteristic K X rays emitted by the majority of these elements may be classed as soft X rays and, with the exception of Fe, range from 3 to approximately 12 Å. In this range one encounters problems of excessive absorption in the detector windows as well as low fluorescence yields. The instrument therefore requires an efficient design with emphasis on optimum excitation and geometry. It is also essential to use the thinnest detector windows compatible with mechanical

* The term nondispersive is actually a misnomer. This expression which is in common usage means simply pulse height analysis as opposed to Bragg-crystal dispersion.

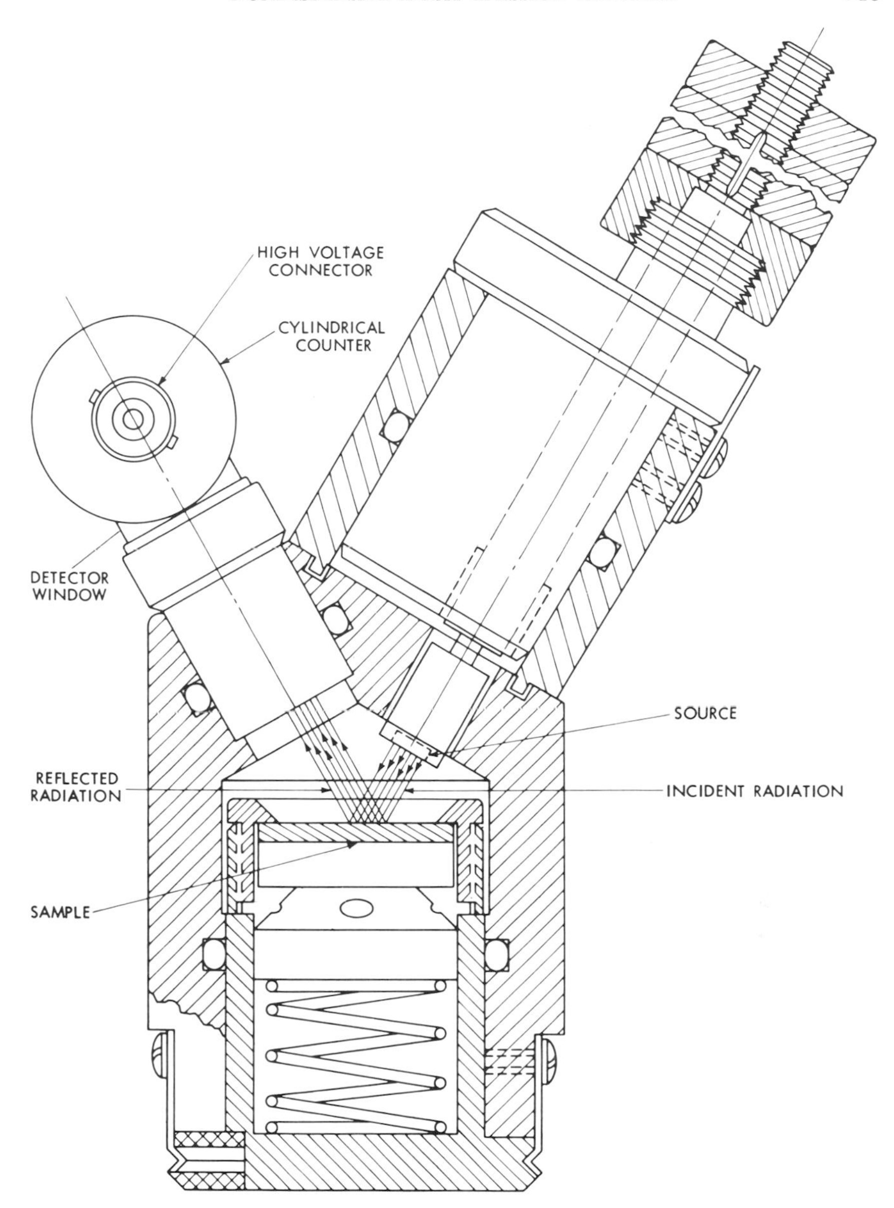

FIG. 1. An α source X-ray analyzer.

strength and low counter gas leakage. An additional, but very important, requirement is to obtain the best possible detector resolution.

Figure 1 shows the rather simple instrumentation used in the laboratory*

* This is not proposed as a flight instrument but rather is a breadboard for laboratory evaluation of the technique.

for analysis. The apparatus consists of a radioactive source, a means for holding the sample, a proportional detector, and an inlet for introducing helium into the chamber. Although a modest vacuum is desirable, helium is used at present in order to preserve the integrity of the alpha source window.

The samples to be analyzed may be introduced as powders, briquets, or small flat sections of rock. The various components of the instrumentation will now be examined more closely.

B. Excitation Sources

The obvious advantage in using radioisotopes for exciting X rays is that the source is small, requires no power, and is the ultimate in reliability. It has already been amply demonstrated that radioactive sources of relatively low output will generate very adequate X-ray spectra in time periods as short as minutes.

Various types of isotopes have been found useful for X-ray emission analyses. These are beta sources, bremsstrahlung sources, K capture X-ray sources, and alpha emitters. It has been shown by Sellers and Ziegler [2] that alpha sources are particularly useful for light element analysis. These advantages can be tabulated as follows.

(1) The cross section for ionization of the K X rays increases approximately as Z^{-12}. Thus, the rapidly increasing ionization cross section with decreasing atomic number more than compensates for the decreasing fluorescence yield.

(2) In comparison to electron excitation the alpha bremsstrahlung produced continuum is reduced by a factor of $(m/M)^2$ where m and M are, respectively, the masses of the electron and alpha particles. As a consequence the continuum is reduced by a factor of about 10^6 to 10^7.

(3) Although alpha emission is usually associated with gamma ray emission, it is possible to select nuclides from which the gamma emission is both a small fraction of the alpha emission and the gamma rays are sufficiently removed from the excited X rays to be easily discriminated against.

(4) It is possible to obtain sources of high specific activity emitting very energetic alpha particles having energies of 4 to 6 MeV such as ^{210}Po and ^{242}Cm.

While the advantages listed above make alpha emitters attractive sources for light element analysis, there are a number of problems encountered in practice. Alpha sources are a serious potential health hazard and must be handled with extreme care. Because of the phenomenon of aggregate recoil, one should work either with sealed sources or in adequately ventilated hoods or glove boxes. As yet, it is difficult to obtain sealed sources that can be operated in vacuum without using windows of such thickness as to seriously degrade the energy of the emitted alpha particles. In the work to be described here, measurements were made in helium at atmospheric pressure. The source holder is shown in Fig. 2. The ^{242}Cm source was approximately 15 mCi

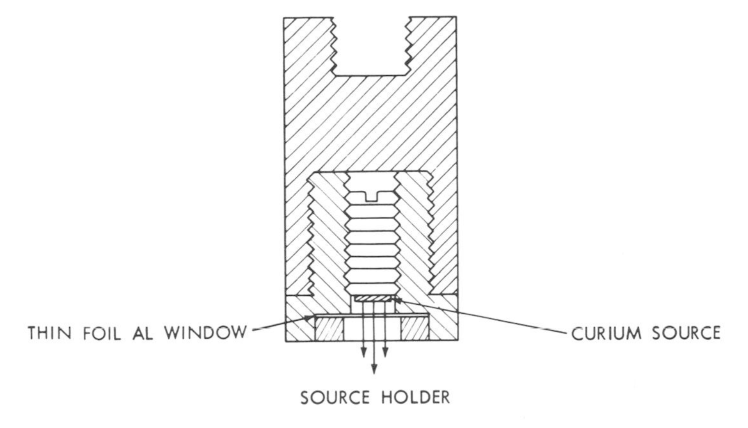

FIG. 2. An α emitter source holder.

deposited by coating on a stainless steel foil. The window in this instance was 0.0005 in. aluminum. Although no direct measurement of the energy distribution of the alpha particles was made, it was estimated that the effect of the window and helium atmosphere reduced the 6 MeV alphas about 3 MeV with an associated broadening of the energy peak.

Although it is true that alpha bombardment excites virtually no bremsstrahlung, one must expect some background. This background may come from the source container as well as from X rays excited by processes such as internally converted gamma rays and due to contaminating radioactive species. Figure 3 shows the background radiation from the ^{242}Cm source scattered from a boric acid sample. One can identify at the very least an Al K line from the source window, an Fe K line from the substrate on which the ^{242}Cm is deposited, a W line from the hevi-met (sintered W) container, as well as lines in the 13–18 keV region which may be attributed to the daughter product Pu as well as other contaminants.

C. Detector

A major problem in light element analysis is associated with detection. The ideal would be a windowless detector with high efficiency and resolution. This is unfortunately beyond the state of the art at this time, particularly in

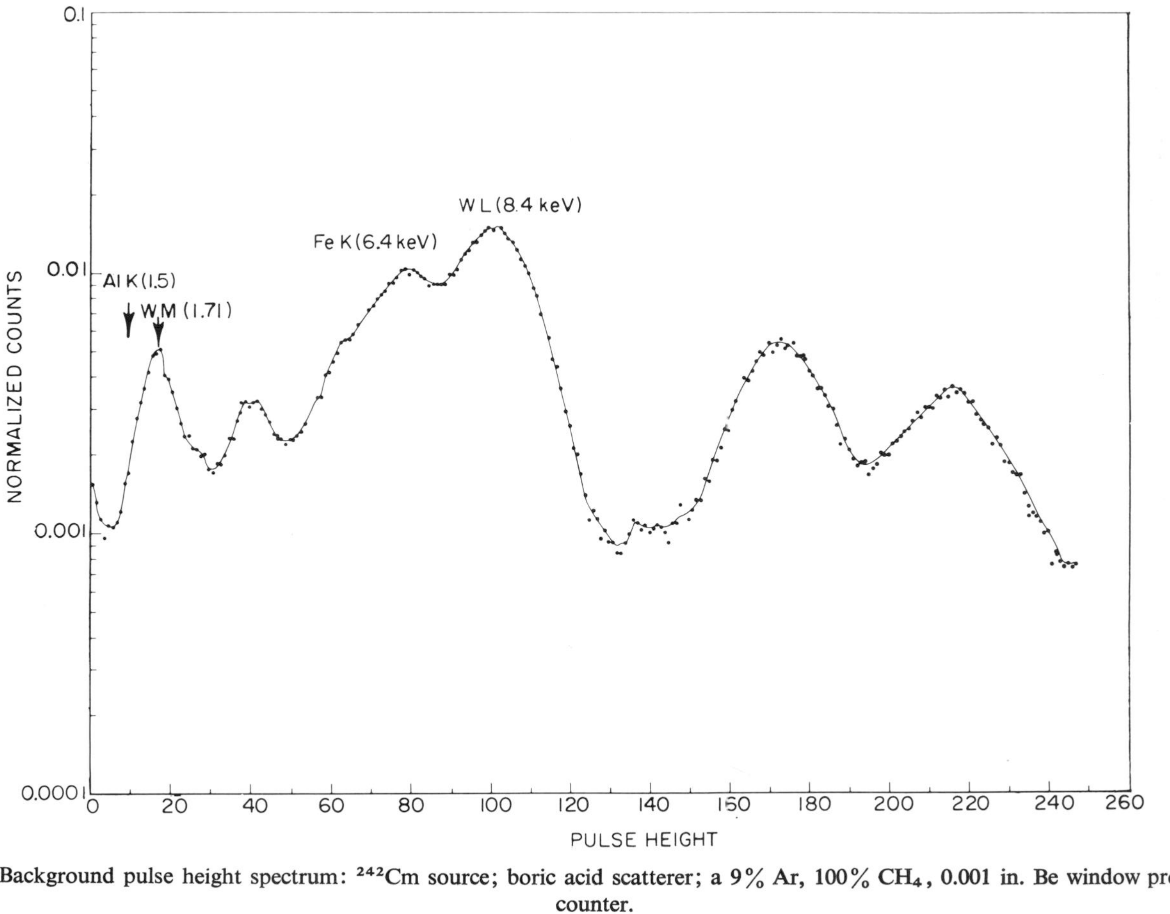

FIG. 3. Background pulse height spectrum: ^{242}Cm source; boric acid scatterer; a 9% Ar, 100% CH_4, 0.001 in. Be window proportional counter.

the region of from 7 to 1 keV. The best detector available at this time is the thin window proportional counter, either flow or sealed. With flow counters, one can work easily with 0.001 in. Be or 0.00025 in. Mylar. Figure 4 shows a typical efficiency curve for a 2 cm detector using an argon–methane mixture.

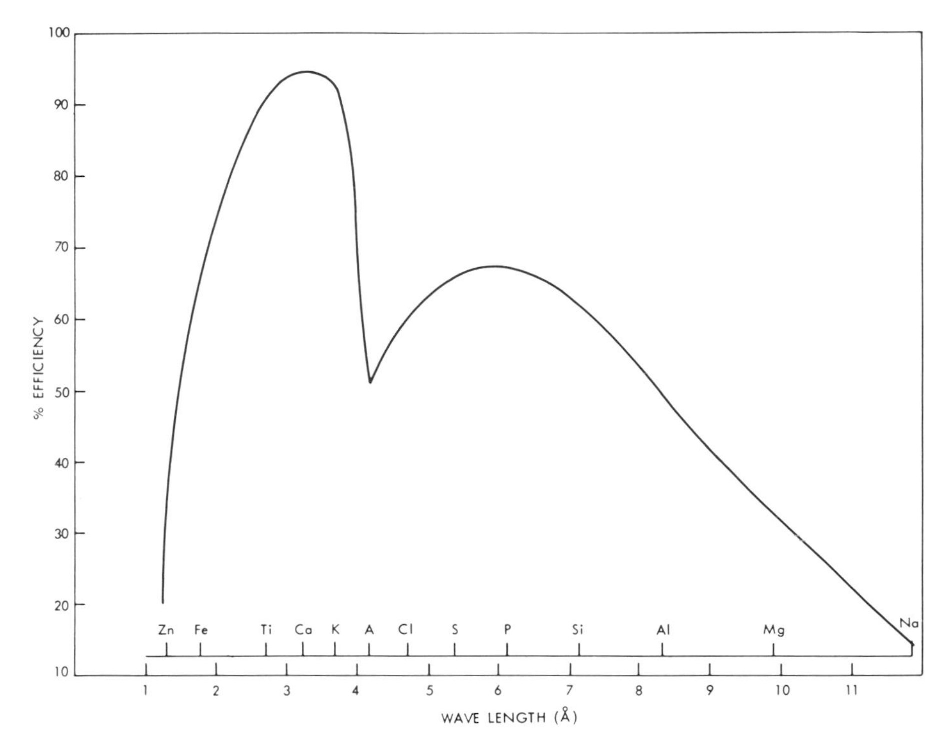

FIG. 4. Counter tube efficiency: 90% Ar, 10% CH_4 filling; 0.001 in Be window; 2 cm absorption path.

This curve represents the product of window transmission and gas absorption. It can be seen that counter efficiency drops to about 10% at 11 Å. When one combines this with phenomenon with the low fluorescence yield for the low *Z* elements, then the high ionization cross section of the alpha particles becomes particularly significant. The detector resolution used in this study was of the order of 17% for ^{55}Fe.

D. Electronics

This electronic arrangement was conventional consisting of a charge sensitive preamplifier feeding a high quality multichannel analyzer. Although greater capabilities were available, 256 channels of the analyzer were found to be sufficient. Data readout was on perforated tapes which were then converted to punch cards for computer processing.

II. Data Analysis

A. Analytical Approach

In the previous section, the design of the X-ray emission system was shown to be greatly influenced by the mission objectives. Similarly, the analytical method is also closely related to these objectives. Methods can be developed for analysis of the measured spectra which at one extreme can compare spectral patterns, and at the other extreme decompose the observed spectrum into monoelemental components for detailed analysis. The simpler pattern recognition techniques are now being investigated for their possible application to problems involving sample selection, but since in almost all cases the more detailed analysis will be required, we will address ourselves to this problem.

The problem of determining qualitative and quantitative information from a nondispersive measurement of the X-ray emission can be divided into two parts. First, the differential X-ray energy spectrum (true energy spectrum incident upon the proportional counter) must be determined from the measured pulse height spectrum. The second step is then to deduce the chemical composition of the irradiated sample from the differential energy X-ray spectrum. The first step will be described in detail in this paper, and possible methods for performing the second step will be discussed by example at the end of this section.

B. Nature of the Pulse Height Spectrum

We shall begin by defining what is meant by a pulse height spectrum. Consider the case of an ideal spectrometer while looking at the Fe K spectrum that consists of a number of monochromatic energies. We would observe a spectral distribution consisting of a number of discrete lines as shown in Fig. 5. If, in practice, the X-ray energies are examined by means of a proportional counter detector and a pulse height analyzer, the spectrum becomes a continuous distribution as shown in Fig. 5. Here we see the effect of instrumental smearing due to such factors as resolution, a statistical fluctuation in the detection process, detector geometry, electronic noise, etc. In many cases the spectrum is further complicated by an escape peak phenomenon due to

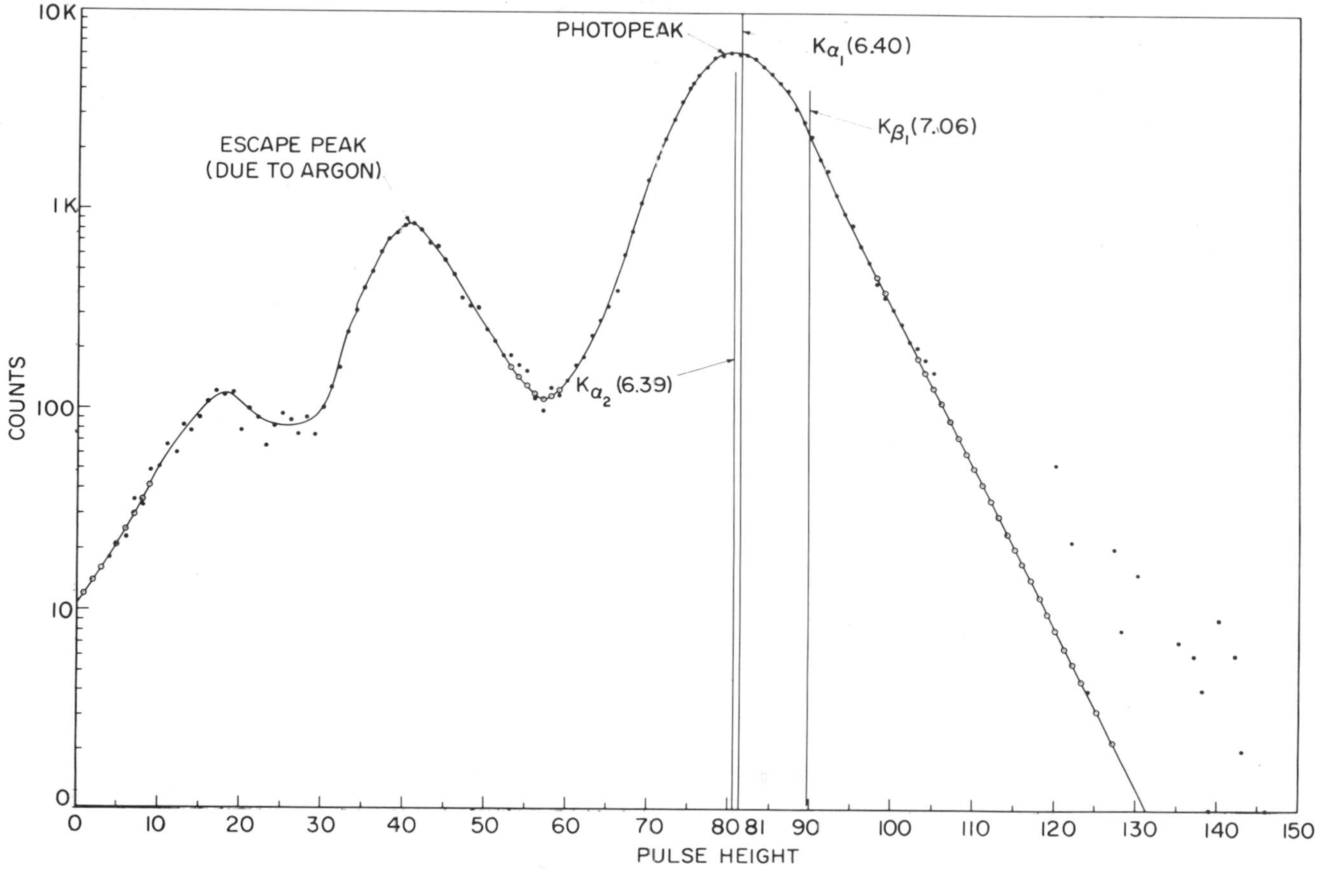

FIG. 5. Discrete energy spectrum and corresponding pulse spectrum of the characteristic Fe X-ray lines. Pulse height spectrum measured with a sealed 90% Ar, 10% CH_4 proportional counter.

partial absorption of the incident energies. It is obvious that any program of reducing a pulse height distribution to pulse height energies must concern itself with and remove these instrumental factors. This necessary approach will now be discussed in some detail.

It has been observed that the X-ray photopeaks show a nearly Gaussian distribution, and that the resolution of the peaks is a function of energy, the resolution R being defined as

$$R = W_{1/2}/P_m, \tag{1}$$

where $W_{1/2}$ is the width of the photopeak at half the maximum amplitude, and P_m is the position of the pulse height maximum, where the width and peak are expressed in the same units (energy or channel window). The resolution is inversely proportional to some power of the pulse height or energy

$$R \sim E^{-n}. \tag{2}$$

Theoretically [11] $n = 0.5$ although experimentally n is usually greater or equal to 0.5. In the cases we have examined, n has varied from 0.5 to 0.67 and appears to be affected by the resolution of a given detector.

If one irradiates a mixture of elements with alpha particles and records the pulse height spectrum, the total observed polyelemental spectrum will be made up of a summation of the monoelemental components as shown in Fig. 6.

If ρ_i is taken as the measured count in channel i, then

$$\rho_i = \sum_{\lambda} C_{i_\lambda}, \tag{3}$$

where C_{i_λ} is the total number of counts in channel i due to the λth element. The pulse height spectra of the monoelemental components can be normalized as follows:

$$C_{i_\lambda} = \beta_\lambda A_{i_\lambda}, \tag{4}$$

where A_{i_λ} is the normalized number of counts in channel i due to the element λ, and β_λ is the relative intensity of the λ component, then

$$\rho_i = \sum_{\lambda} \beta_\lambda A_{i_\lambda}. \tag{3a}$$

Another component of the measured pulse height spectra which must be considered is the background. In the case of alpha induced fluorescence, there are two separate sources of radiation contributing to the background spectrum (see Figs. 3 and 6). As indicated above, one component can be attributed to the natural background and to gamma rays from the source striking the detector. The second component can be attributed to the X rays

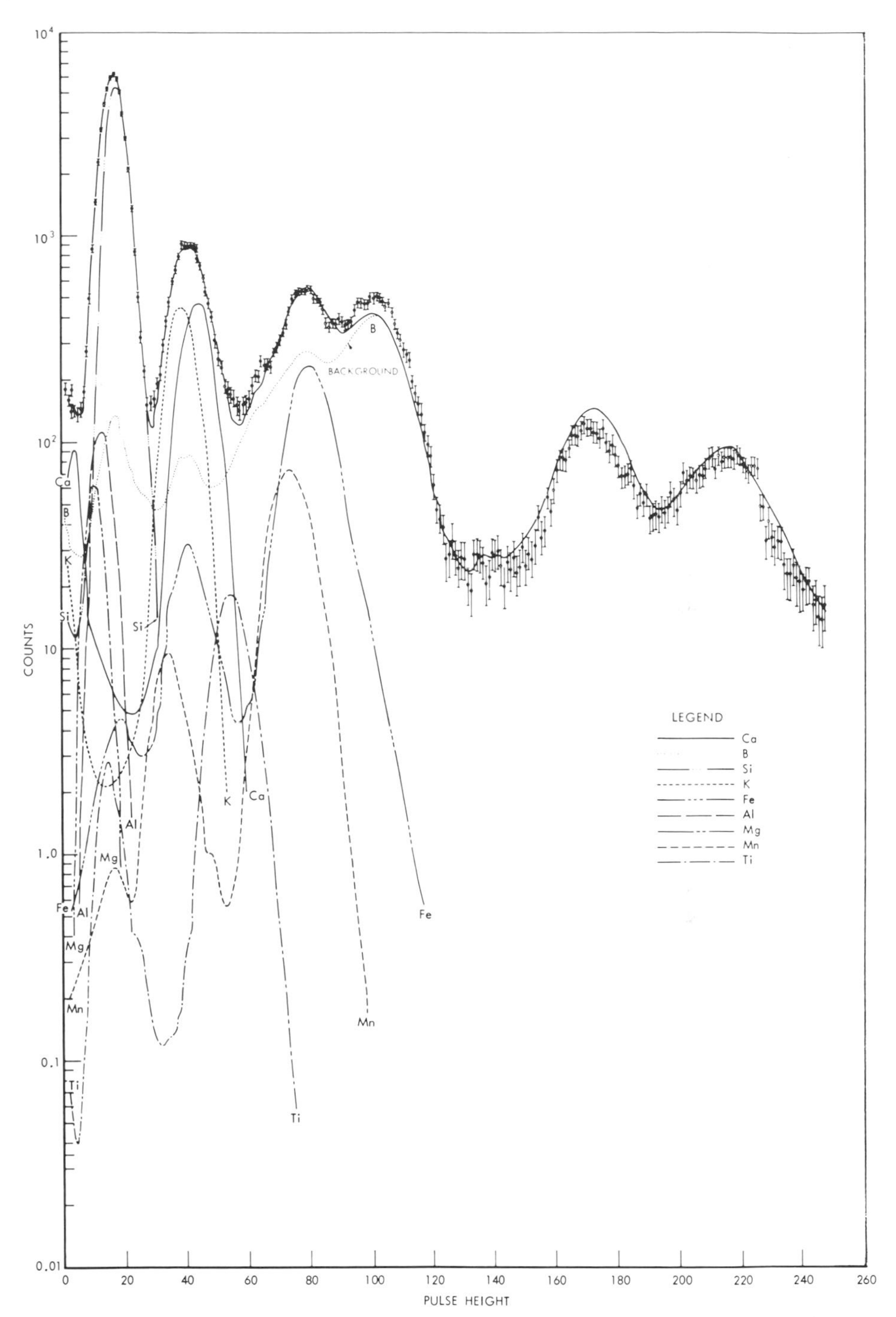

FIG. 6. Pulse height spectrum of basalt measured with a proportional counter filled with 90% Ar and 10% CH_4, showing the monoelemental components; the envelope due to the sum of the monoelemental components; and the measured pulse height spectrum with statistical variations included.

emanating from the source being scattered by the sample. We have found that in this instance the major contribution to the background seems to be due to this second scattered component. The amount of scatter will depend on the average atomic number of the sample and thus changes from sample to sample. Furthermore, we have found from our measurements that while the shape of the background pulse height spectrum does not change significantly from sample to sample, the intensity does. Further study of the shape of the background as a function of composition is underway. The problem of subtracting background is not a simple one. Our approach to this problem will be described in detail, in the final section of this paper in which actual spectral analyses will be presented. Figure 6 shows the measured pulse height spectrum of a basalt sample with the background included. Equation (3a) should include the background component B_i, where B_i (see Fig. 3) is the intensity of the background in channel i

$$\rho_i' \approx \sum_{\lambda} \beta_\lambda A_{i_\lambda} + B_i. \tag{3b}$$

C. *Formulation of the Linear Least-Square Analysis*

1. *Matrix Equations*

If Eq. (3b) were truly an equality the relative intensities, β_λ could easily be determined by a simple matrix inversion of (3b), that is,

$$\boldsymbol{\beta} = (A^{-1})\boldsymbol{\rho}, \tag{4a}$$

where $\boldsymbol{\beta}$ is a vector of the relative intensities, A is an $m + n$ matrix of the monoelemental components, and $\boldsymbol{\rho}$ is a vector of the measured polyelemental spectrum minus background $(\rho_i' - B_i)$.

The problem in performing this inversion arises when one considers the statistical variance in the measurement A_{i_λ} and ρ_i. Because of this, the equality (3b) is not true, and a unique solution to (4a) does not exist. Equation (4a) is overdetermined, therefore an infinite set of equally good solutions is possible, and there are no criteria for selecting the best one. As a consequence of this and because of the statistical variance involved, only the most probable values of β_λ can be determined by using the following criterion:

$$M = \sum_i \omega_i \left((\rho_i' - \sum_\lambda \beta_\lambda A_{i_\lambda} \right)^2 \Rightarrow \text{minimum}, \tag{5}$$

where ω_i is a statistical weighting function and is proportional to σ_i^{-2}, and σ_i^2 is the statistical variance in the measurement in channel i.

In formulating the linear least-square analysis method, we begin by assuming that the pulse height scale is invariant (i.e., there is no gain shift or zero drift) and, therefore, the minimum can be found by taking a partial derivative with respect to some relative intensity β_γ. The derivative is then set equal to zero

$$\partial M/\partial\beta_\gamma = \sum_i \omega_i\left(\rho_i - \sum_\lambda \beta_\lambda A_{i_\lambda}\right)A_{i_\gamma} = 0. \tag{6}$$

This can be written in matrix form as a no longer overdetermined set of equations

$$\tilde{A}\omega\rho - (\tilde{A}\omega A)\beta = 0, \tag{6a}$$

where the definition given in (4) still holds, and ω is a diagonal matrix of the weighting function ω_i, and $\tilde{A}$ is the transpose of the A matrix. Solving for the relative intensity vector $\boldsymbol{\beta}$, we find that

$$\boldsymbol{\beta} = (\tilde{A}\omega A)^{-1}\tilde{A}\omega\rho \tag{6b}$$

where $(\tilde{A}\omega A)^{-1}$ is the inverse matrix of $(\tilde{A}\omega A)$. Essentially, it is the method of attaining the solution of Eq. (6b) which will concern us in the rest of the discussion. This equation has been applied to the solution of a number of spectroscopic problems involving pulse height analyses [12–18]. There are a number of questions which must be answered in determining the algorithm (i.e., the method of solution) to be used for the given problem:

(1) Is it possible to invert the matrix $(\tilde{A}\omega A)$?
(2) Are the components of the A matrix linearly independent or do they correlate?
(3) How does one calculate this effect?
(4) How does resolution effect the correlation?
(5) What is the nature of the ω matrix?
(6) What is the effect of background subtraction?
(7) Is the system linear?
(8) If there are nonlinearities in the system, how does one compensate for nonlinearities in the application of this method?

2. *Library Functions*

Let us consider the X-ray emission method, and determine the algorithm for obtaining β from Eq. (6b). We start by considering the nature of the A matrix. The rows of the matrix can consist of either the monoelemental pulse height spectra (i.e., spectra characteristic of all the possible elements found in the irradiated sample) or the monoenergetic pulse height spectra (i.e., the pulse height spectra characteristic of all the possible monoenergetic X rays striking the detector). Since we are interested in qualitative and quantitative elemental composition, we will consider the problem using the monoelemental pulse height spectra.

The monoelemental pulse height spectra are determined by irradiating pure samples of the elements in the same geometric configuration as that used for the unknowns. Because these measurements have an inherent statistical error and because of the dynamics of the measurement system (possible gain drifts and zero drifts), these monoelemental pulse height spectra can be determined to only two or perhaps three significant figures with any great certainty. In the solution proposed in this paper, the statistical errors in the measurement of the monoelemental pulse height spectra are kept much smaller than the error in the measurement of the unknown pulse height spectra by extended and repeated measurements and therefore can be ignored. The difficulty comes then when an attempt is made to invert $(\tilde{A}\omega A)$. This inversion requires obtaining differences involving the third, fourth, and even fifth significant figure. As indicated above, the measurements of the monoelemental spectra are only good to two or at the most, three significant figures. Thus, unless some additional constraints are introduced, serious oscillations in the solution will result, and in fact one cannot successfully invert the matrix. Two necessary constraints that can be imposed are symmetry conditions of the $(\tilde{A}\omega A)$ matrix and also a "nonnegativity option." The symmetry constraint follows from matrix algebra and is accomplished in the program which examines the symmetric off diagonal elements, i.e., the A_{ij} and A_{ji}, where $i \neq j$ and sets them equal. The use of the nonnegativity option which proves very useful to the X-ray case will be further amplified.

3. *Nonnegativity Constraint*

It has been shown above that an exact solution cannot be obtained because of the statistical variation in the measurement. If such exact solutions were possible, the relative intensity of the various components would either be positive or zero. Now we are looking for the most probable solution, and in this case negative values can appear. When such negative values are observed, there must been associated error that is as great or greater than the absolute magnitude of the value obtained. Negative values with errors significantly smaller than the absolute value do appear, however, in the application of least-square analysis to the reduction of pulse height spectra. It is this effect which produces oscillations in the solution described above.

The oscillation in the solution can be described in the following way. A given relative intensity can appear as a negative value, and possibly more negative than it should be because of the errors in the inverse transformation just discussed. The overestimation of a negative component by the analytical method will tend to make some other component more positive to compensate. This in turn will tend to make a third component smaller than it should be in order to compensate, and thus an oscillation in the solutions will be produced. By imposing a nonnegativity constraint, one essentially damps out these oscillations in the solution due to such factors as errors in the determination of the A matrix, and errors in improperly subtracting the background. What is meant by the nonnegativity constraint then, is that when determining the relative intensities of the $\boldsymbol{\beta}$ vector (6b), only positive values of monoelemental components are allowed, and negative values are set equal to zero.

The use of physical constraints in the solution of linear least-square analyses was first discussed by Beale [19], and suggested for use in the analysis of pulse height spectra by Burrus [13]. A method using the nonnegativity constraint for the analysis of gamma ray pulse height spectra [12, 13] is now being applied to the problem of X-ray nondispersive analyses and has been discussed briefly in other papers [1, 4].

Because of the necessarily empirical nature of the method of solution, the absolute minimum obtained from Eq. (6b) is not necessarily the best solution. One must find a minimum in the domain described by other physical constraints. In addition to nonnegativity, one should consider the following constraints, for example:

(1) If the relative intensities of two species are known, then the ratio of intensities is invariant in the least-square analysis.

(2) The sum of the relative intensities once determined for various elements always equals some constant.

Later in the discussion concerning limits of resolution and correlation, the latter condition will be invoked to help in the solution. Continuing with the discussion of a nonnegativity constraint, we shall give only a brief and nonrigorous physical argument to show how one obtains a solution (since a detailed description of this solution is given elsewhere [12, 13, 19]). We first assume that we have a linearly independent set of monoelemental functions (function library A_{ij}), and we are required to find the relative intensities (β_λ) of these library functions which combined will yield the best fit to the experimentally determined polyelemental pulse height spectrum (raw data spectrum ρ). The library must include all possible components. We now assume that all but two components from the library are zero. A least-square fit (6b) to the total spectrum ρ is used to determine the relative intensity of these two components. Since the number of components in the measured spectrum ρ is generally equal, or larger than, two, the estimates of the relative intensities of these two components will be greater than they should be. As components from the library are added, the relative intensities of the individual components will grow smaller. Thus if the least-square fit gives a negative relative intensity for either of these two components, that component is eliminated from the library of elements and not used in the analyses (i.e., its relative intensity is set equal to zero). If both relative intensities are positive, both components are kept. Additional library components are added, one at a time, a least square fit is made, and all negative components are again set equal to zero. The process is continued until all components have been tested. This then will yield the desired solution which is the least-square solution for relative intensities in the positive domain. If the library elements are linearly independent, then the solution obtained will be completely independent of the order in which the monoelemental components are added.

The question may be asked concerning the necessity to perform the itera-tive process described above in order to find the solution to the minimum in the positive domain. Why not, for example, find the absolute minimum with all the components at once, and then adjust the negative values to zero? The answer is that because of the possibility of oscillation in the solution, it is difficult to tell whether the negative values result from the oscillations in the solution or because of true statistical variation. The nonnegativity, as we have indicated, tends to damp this oscillation out, and the algorithm described above finds the desired solution.

One must be rather careful in using this nonnegativity principle for, strictly speaking, negative solutions should be allowed in least-square analysis. What the nonnegativity constraint does is to increase the mathematical error because one does not find the absolute minimum as required by the least-square method. Furthermore, by rejecting certain components, estimates of possible minimum detectable limits for these components are not determined. In order to obtain the minimum detectable limits in the program prepared for this analysis, both the absolute minimum and the minimum in the positive domain can be found and compared.

4. *Error Calculation*

We now consider the problems of calculating the error in the relative intensity β_λ due to the statistical variation in ρ_i, the counts in channel i, contributed by the components of the polyelemental pulse height spectrum. Let us consider Eq. (6b), which shows that β_λ can be written as a linear combination of the ρ_i's

$$\beta_\lambda = \sum_i \sum_\nu C_{i\lambda}^{-1} A_{i_\nu} \omega_i \rho_i . \tag{6c}$$

We define $C = (\tilde{A}\omega A)$, where $(\tilde{A}\omega A)$ is a symmetric matrix whose elements $C_{\nu\gamma}$ of C are given by

$$C_{\nu\gamma} = \sum_i \omega_i A_{i_\nu} A_{i_\gamma} . \tag{7}$$

Remembering that ω_i is the statistical weight, and that

$$\omega_i = \sigma_i^{-2} \tag{8}$$

where σ_i^2 is the variance of the measurement of ρ_i in channel i, and C^{-1} is the inverse of matrix C, e.g.,

$$CC^{-1} = I, \tag{9}$$

and where I is the identity matrix (i.e., all off diagonal elements are zero and the diagonal elements are one), or

$$I_{\nu\lambda} = 1 \quad \text{if} \quad \nu = \lambda, \qquad I_{\nu\lambda} = 0 \quad \text{if} \quad \nu \neq \lambda. \tag{10}$$

The elements of the $I_{\nu\lambda}$ can be given

$$I_{\nu\lambda} = \sum_{\gamma} C_{\nu\gamma} C_{\gamma\lambda}^{-1}. \tag{11}$$

From Eq. (6c), it is seen that β_λ is a linear homogeneous function of the counts ρ_i under the assumption that there is no significant error in the A_{ij} compared to the measurement of ρ_i. Thus, the mean square deviation $\sigma^2(\beta_\lambda)$ corresponding to the variation in ρ_i can be written as a linear sum of the variances on the ρ_i's. Thus from Eq. (6c), we get the variance for β to be

$$\sigma^2(\beta_\lambda) = \sum_i \sum_\nu \sum_\gamma C_{\nu\lambda}^{-1} C_{\gamma\lambda}^{-1} A_{i_\nu} A_{i_\gamma} \omega_i^2 \sigma^2(\rho_i), \tag{12}$$

and from (8),

$$\sigma^2(\beta_\lambda) = \sum_i \sum_\nu \sum_\lambda C_{\nu\lambda}^{-1} C_{\gamma\lambda}^{-1} A_{i_\nu} A_{i_\gamma} \omega_i$$

or

$$\sigma^2(\beta_\lambda) = \sum_\nu \sum_\lambda C_{\nu\lambda}^{-1} C_{\gamma\lambda}^{-1} \sum_i \omega_i A_{i_\nu} A_{i_\gamma}.$$

Then from (7),

$$\sigma^2(\beta_\lambda) = \sum_\nu \sum_\gamma C_{\nu\lambda}^{-1} C_{\gamma\lambda}^{-1} C_{\nu\gamma} \quad \text{or} \quad \sigma^2(\beta_\lambda) = \sum_\nu C_{\nu\lambda}^{-1} \sum_\gamma C_{\nu\gamma} C_{\gamma\lambda}^{-1}$$

and from Eq. (11)

$$\sigma^2(\beta_\lambda) = \sum_\nu C_{\nu\lambda}^{-1} I_{\nu\lambda}.$$

Finally, from (10)

$$\sigma^2(\beta_\lambda) = C_{\lambda\lambda}^{-1}, \tag{12a}$$

that is, $\sigma^2(\beta_\lambda)$ can be found from the diagonal elements of the C^{-1} matrix. Equation (12a) is true if χ_i^2 is equal to one, and χ_i^2 is defined as follows:

$$\chi_i^2 = \frac{\sum_i \omega_i(\rho_i - \sum_\lambda \beta_\lambda A_{i_\lambda})^2}{n - m} \tag{13}$$

where $n - m$ are the number of degrees of freedom, n is the number of channels, and m is the number of library components. If $\chi_i^2 \neq 1$, then

$$\sigma^2(\beta_\lambda) = \chi_i^2 C_{\lambda\lambda}^{-1}. \tag{12b}$$

Later χ_i^2 and its utilization will be discussed in slightly more detail.

5. *Correlation*

Let us now consider what we mean by linearly independent library functions. It was shown that $C_{\lambda\lambda}^{-1}$ is the variance on β_λ. It can be further shown that $C_{\lambda\gamma}^{-1}$ is the covariance of λ'th and γ'th component [20, 21]. A measure of interferences $F_{\lambda\gamma}$ between the λ'th and γ'th component is given by

$$F_{\gamma\lambda} = \frac{(C_{\gamma\lambda}^{-1})^2}{(C_{\lambda\lambda}^{-1})(C_{\gamma\gamma}^{-1})} \times 100\%. \tag{14}$$

Equation (14) is a measure of how different, or how well one component or library spectrum can be resolved from another. It is also a measure of whether or not the components in the library of spectra can be considered to be linearly independent.

As an illustration of how Eq. (14) is used, let us assume that the library function or monoelemental functions can be described by Gaussians

$$A_{i_\lambda} = \exp\left[-(i - P_\lambda)^2/2\alpha_\lambda^2\right], \tag{15}$$

where A_{i_λ} is the normalized counts in channel i due to the λ'th component, α_λ^2 is a constant, and is a measure of the width of the Gaussian for energy λ, and P_λ was defined in Eq. (1).

Now let us calculate the percent inference between two monoenergetic pulse height spectra with Gaussian form using Eq. (14) and a shape given by (15). If we assume no statistical error, Eq. (14) can be written

$$F_{\gamma\lambda} = \frac{\sum_i A_{i_\gamma} A_{i_\lambda}}{(\sum_i A_{i_\gamma})^2(\sum_i A_{i_\lambda})^2}. \tag{14a}$$

Using equations of the form given in (15) for A_{i_γ} and A_{i_λ}, and replacing the summation by an integral from −infinity to +infinity, (14a) becomes

$$F_{\gamma\lambda} = \frac{\int_{-\infty}^{+\infty}\left[\exp - \frac{(i-P_\gamma)^2}{2\alpha_\gamma^{\,2}} \exp - \frac{(i-P_\lambda)^2}{2\alpha_\lambda^{\,2}}\right] di}{\left[\int_{-\infty}^{+\infty}(di)\exp - \frac{(i-P_\gamma)^2}{2\alpha_\gamma^{\,2}}\right]^2\left[\int_{-\infty}^{+\infty}\exp - \frac{(i-P_\lambda)^2}{2\alpha_\lambda^{\,2}}(di)\right]^2}$$

$$= \frac{2\alpha_\lambda\alpha_\gamma}{\alpha_\lambda^{\,2}+\alpha_\gamma^{\,2}}\exp - \frac{P_\lambda - P_\gamma}{-\alpha_\lambda^{\,2}+\alpha_\gamma^{\,2}}. \tag{14b}$$

The constant α_γ can be written in terms of the resolution R_γ as

$$\alpha_\gamma = R_\gamma P_\lambda/2(2\ln 2)^{1/2}. \tag{16}$$

Remember that P_γ stands for the pulse height portion of some energy E_γ and is proportional to E_γ. If we now assume that Eq. (2) is applicable, and let $n = 0.5$, we get

$$R_\gamma/R_\lambda = (P_\lambda/P_\gamma)^{1/2}. \tag{17}$$

For this special case, we can substitute Eq. (17) into Eq. (14b)

$$F_{\gamma\lambda} = \frac{2(P_\lambda P_\gamma)^{1/2}}{P_\lambda + P_\gamma}\exp - \frac{P_\gamma(P_\lambda - P_\gamma)^2}{(P_\lambda - P_\gamma)\alpha_\lambda^{\,2}}. \tag{14c}$$

Figure 7 is a plot of Eq. (14c) for various detector resolutions as a function of the percent separation $(P_\gamma - P_\lambda)/P_\lambda$ for decreasing P_γ, where P_γ and P_λ are the axis for two adjacent energies. If $P_\gamma = P_\lambda$, the Gaussians are identical and the percent interference is 100%. As $(P_\gamma - P_\lambda)$ increases the percent interference decreases, and this is then a measure of how well the two Gaussians can be resolved. For example, consider the following case. Given:

(1) Two X-ray energies separated by 10%.
(2) The detector resolution for the higher energy is 20%;
(3) The relationship given in Eq. (17) holds; and
(4) The photopeak is described by a Gaussian which fully described the pulse height spectrum of the monoelemental component (e.g., no escape peak or continuum), there will be a 50% interference between these energies.

The sum $\beta_\gamma + \beta_\lambda$ can be determined and will be constant, but both β_γ and β_λ can be varied by 50% keeping their sum constant, and the fit based on the least-square criteria will be just as good. In general, given a library of functions A, the percent interference between various components can be determined by forming $(\tilde{A}A)^{-1}$, and calculating $F_{\gamma\lambda}$ from (14). This point will be described further in Section III, in which a solution of an actual experimental problem is presented.

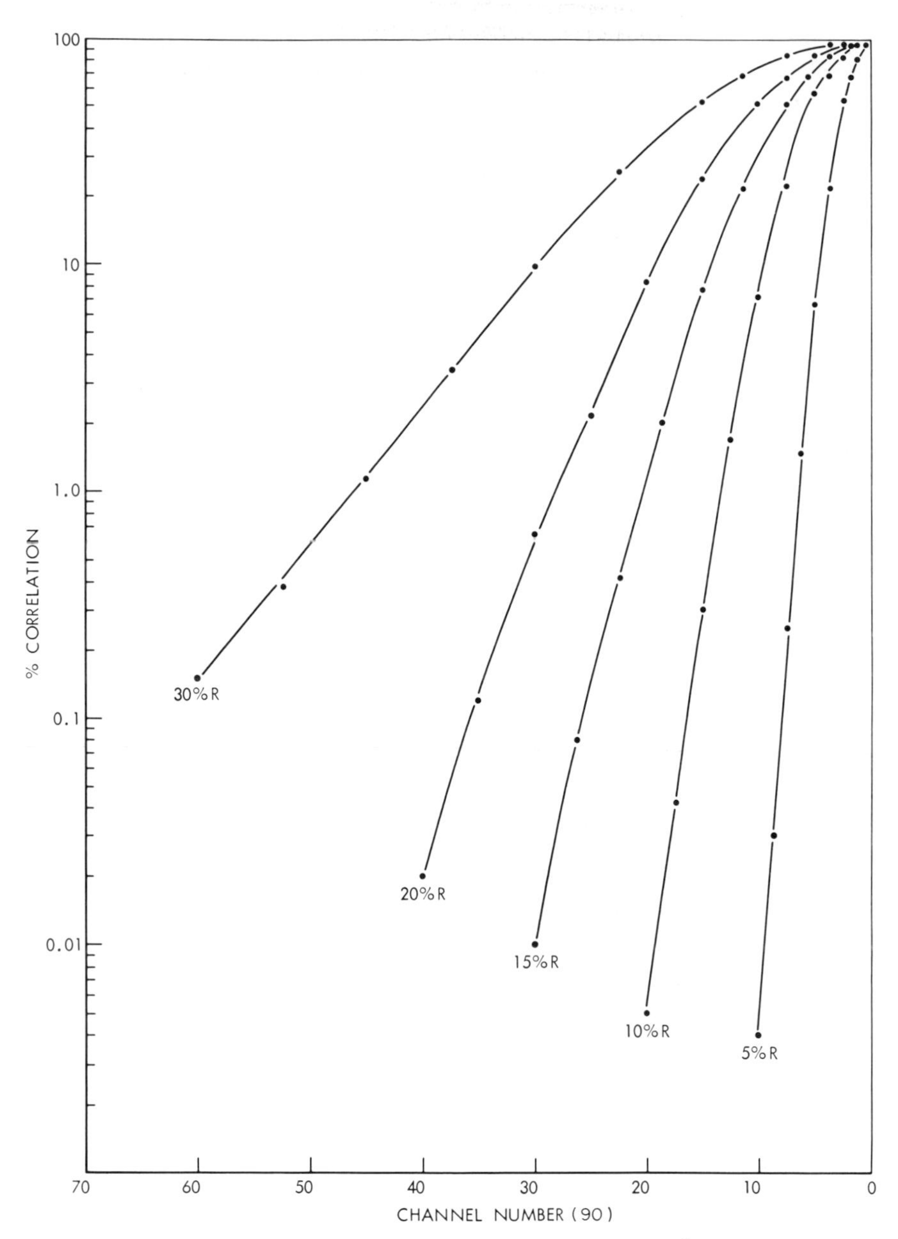

FIG. 7. Percentage of interference for various resolutions, $(P\delta - Px)/Px$.

If two functions do interfere strongly, it is sometimes possible to impose a physical constraint that can eliminate the interference. For instance, in the problem of gamma ray spectroscopy with a mixture of isotopes with varying half-lives, the pulse height spectra can be followed until one of the species has decayed out. The second specie is then followed and the intensity determined. Knowing the half-life of the second species, the intensity of the second component can be determined and extrapolated back to the earlier measurements when the correlation was large. The second component can then be stripped out and the analysis repeated for the first element with the correlation due to the second element eliminated. In X-ray fluorescence, on the other hand, it is possible to use filters, and determine either the intensity of one energy or determine the ratio of one energy with respect to another. Using the intensity of one component or the ratio of one component to another as a physical constraint, the interference can be eliminated.

The percent interference described above has only considered the problem of resolution, but there are two other effects which will produce strong interference. In one case the statistical error due to counting and background subtraction may be so high that the percent interference will be large. This will be related to the statistical weight ω in the $(A\omega A)^{-1}$ equation. Second, interference can occur between two elements if the library is incomplete and an element is missing. Then the two components on either side of the missing element will be under or overestimated in order to make up for the missing element, and the two adjacent elements will correlate. When one calculates the difference between the measured spectrum and the calculated spectrum using least squares, a negative or positive peak will appear at the position of the missing element. By adding in the missing component, the interference and χ_i^2 can be greatly reduced.

6. *Chi Square*

In the method developed for this problem a χ_i^2 test is performed (see Eq. (13)) for goodness of fit. The closer χ_i^2 is to unity the better the fit.

7. *Background Correction*

At first thought, this problem should be rather simple to handle. For example, a background is measured for either the same time or a time longer than the measurement of the unknown system, and then the background is subtracted with adjustment being made for the counting times. In X-ray fluorescence a major part of the background is due to coherent scattering from the sample of the incident radiation. Thus, the fraction of background to be subtracted will be affected by the nature of the sample (average atomic number, density, etc.). However, as an alternative approach the following method was used. Measurements of the raw data spectrum ρ are made to extend to energies (or pulse height channels) higher than the highest energy expected in the pulse height spectrum due to the sample. A background spectrum using a strong scatterer such as boric acid or plastic in the sample position is taken. Characteristic X rays from these substances are so soft as to remain undetected by our present instrumentation. As was pointed out in the section on the shape of the pulse height spectra, the shape of the background spectrum does not seem to be strongly affected by the scatterer although the intensity is greatly affected. The background is then included as a library component and subjected to the least-square treatment, already described, which yields a measure of the background intensity included in the total spectrum. In practice, this latter approach has proved to be feasible because in the measurement of the raw data spectrum ρ, there is a higher energy portion of the background spectrum which is free of lines generated in the specimen. If this were not the case, the background spectrum would strongly correlate with each of the monoelemental components, and a unique solution would not be obtained. In this event one goes back to the subtraction technique. This second approach has also been used, but since an iterative process is necessary, it is more time consuming. This follows because the fraction of background to be subtracted is not known and can only be arrived at by trial and error, e.g., various fractions are subtracted, subjected to least-square analysis and a minimum χ_i^2 sought. In this study, the results presented in the paper used the method of including the background as a library component although the computer program developed has the subtract option included. Included as part of this second option is an error calculation due to this subtract mode and a proper evaluation of the weighting function ω used in the least-square analysis.

8. *Gain Shift*

Up to this point it has been assumed that the pulse height scale for the library spectra is the same as that for the measured raw data spectrum ρ. Because of the possibility of gain shifts and zero drift, the pulse height scale can be compressed or expanded and linearly shifted. Two approaches can be used to compensate for this effect. Nonlinear least-square analysis methods may be used. That is, in Eq. (5), the derivative with respect to β_γ and A_{i_γ} can be obtained. This technique is rather complex, but a number of methods have been developed [22, 23]. Because of these complexities, we use the linear method and make corrections to the raw data ρ_i before performing the analysis. Other similar techniques have been developed [13, 15, 18, 24].

Let us first consider the problem of gain shift. It must be remembered that the measured pulse height spectrum is a histogram, that is, it is a measure of all pulses in some increment Δi about i as a function of i. The total sum $\sum_i \rho_i$ is equal to the total number of interactions between the incident X-ray flux and detector. This number must remain a constant for a given measurement and is independent of a compression, extension, or linear translation along the pulse height axis. If we assume that there is a gain shift G on the pulse height axis between library spectra A_{ij} and the raw data spectrum ρ, the pulse scale i will have to be multiplied by G, and the count rate scale ρ_i divided by G to compensate for the shift. This keeps the area a constant.

It is important to point out that the library spectra histograms are only included for integer values of pulse height which are separated by $\Delta i = 1$. The gain is shifted by some value G which will produce values at fractional pulse heights and cause Δi to become either greater or less than unity. For example, consider the case for $G = 0.98$ or $G = 2.0$. Channels 10 and 11 will become channels 9.8 and 10.78 for $G = 0.98$, and channels 20 and 22 for $G = 2.0$. For the first case $G = 0.98$, we must find the value at channel 10, 11, 12, etc., and for the second $G = 2.0$, we would have to find the value at channel 20, 21, 22, etc. The value at channel 10 ($G = 0.98$) or at channel 21 ($G = 2.0$) is found in our method by linear extrapolation which is an integral part of our program. Only two points are used at a time in this extrapolation. Higher order methods can also be used, but have not been found necessary in this application.

Two choices are possible: shifting the library spectra to correspond to the raw spectrum or vice versa. The second alternative has been found more economical in terms of time. The results obtained using either alternative are the same.

In practice, either a given gain shift or a range in gain shift (G_{min} and G_{max}) is assumed. A series of least-square fits between the assumed G_{min} and G_{max} are made, χ_i^2 calculated and the minimum χ_i^2 found for this range. Figure 8 shows a measured spectrum and the gain shifted spectrum. The spectrum has been shifted 30%.

A similar method compensating for zero drift is now being developed for inclusion in the program. Here a minimum ε_{min} and maximum ε_{max} is introduced. The pulse height scale is shifted by a value ε, the intercepts at integer pulse height values are then determined, and again least-square fits are made until a minimum χ_i^2 is determined in the range ε_{min} to ε_{max}.

9. *Computer Program*

The computer program flow diagram which has been developed to perform the above described calculations is shown in Fig. 9. The zero shift part of the program is not included in the flow diagram. There are four flows

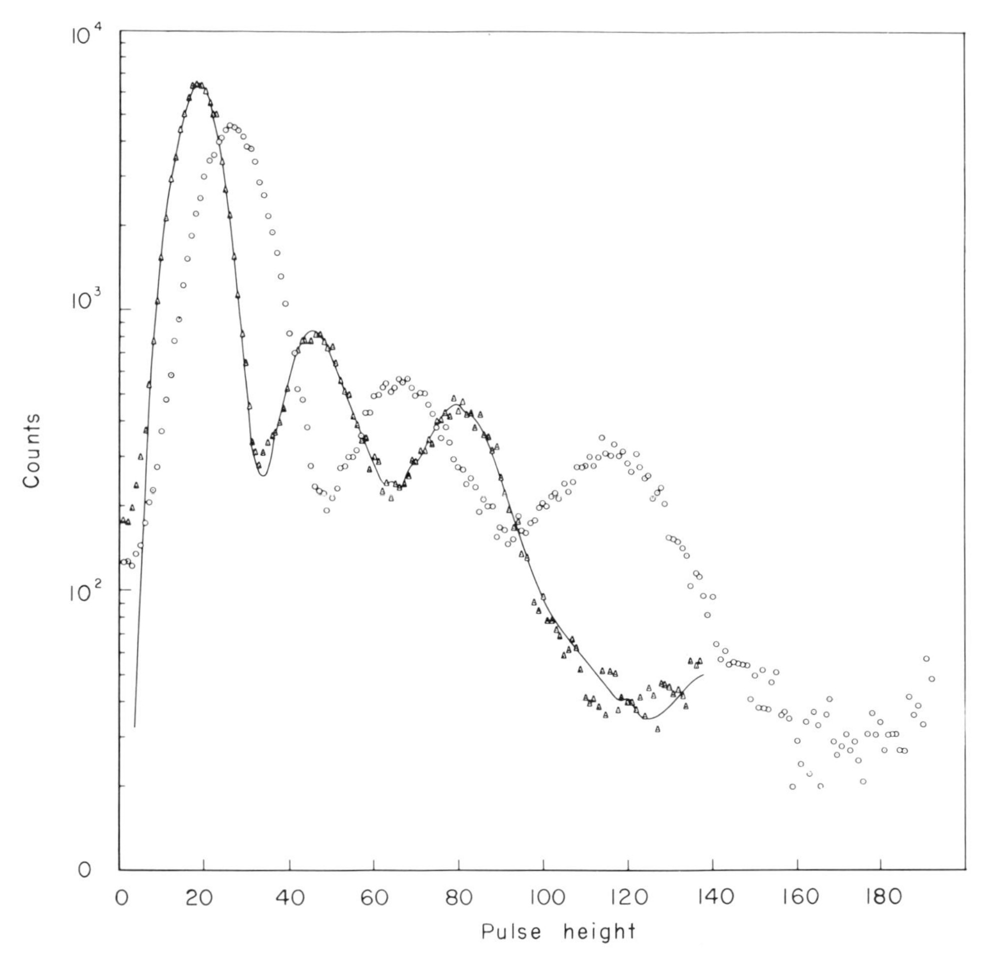

FIG. 8. The effect of gain shift on silty sand from Hoppe Butte, Arizona.

available in this program. Flow 1 allows the stripping of one or more components from the spectrum before analysis. Flow 2 allows background subtraction, gain shifting of the spectrum, and contains the nonnegative constraint as an option. Flow 3 is just a flow for solving Eq. (5) without gain shift and without subtraction of background, and without the nonnegativity constraint. Flow 4 is similar to flow 2, but does not allow for background subtraction. Calculation for errors introduced due to stripping of components and background subtraction are included in the appropriate flow. A number

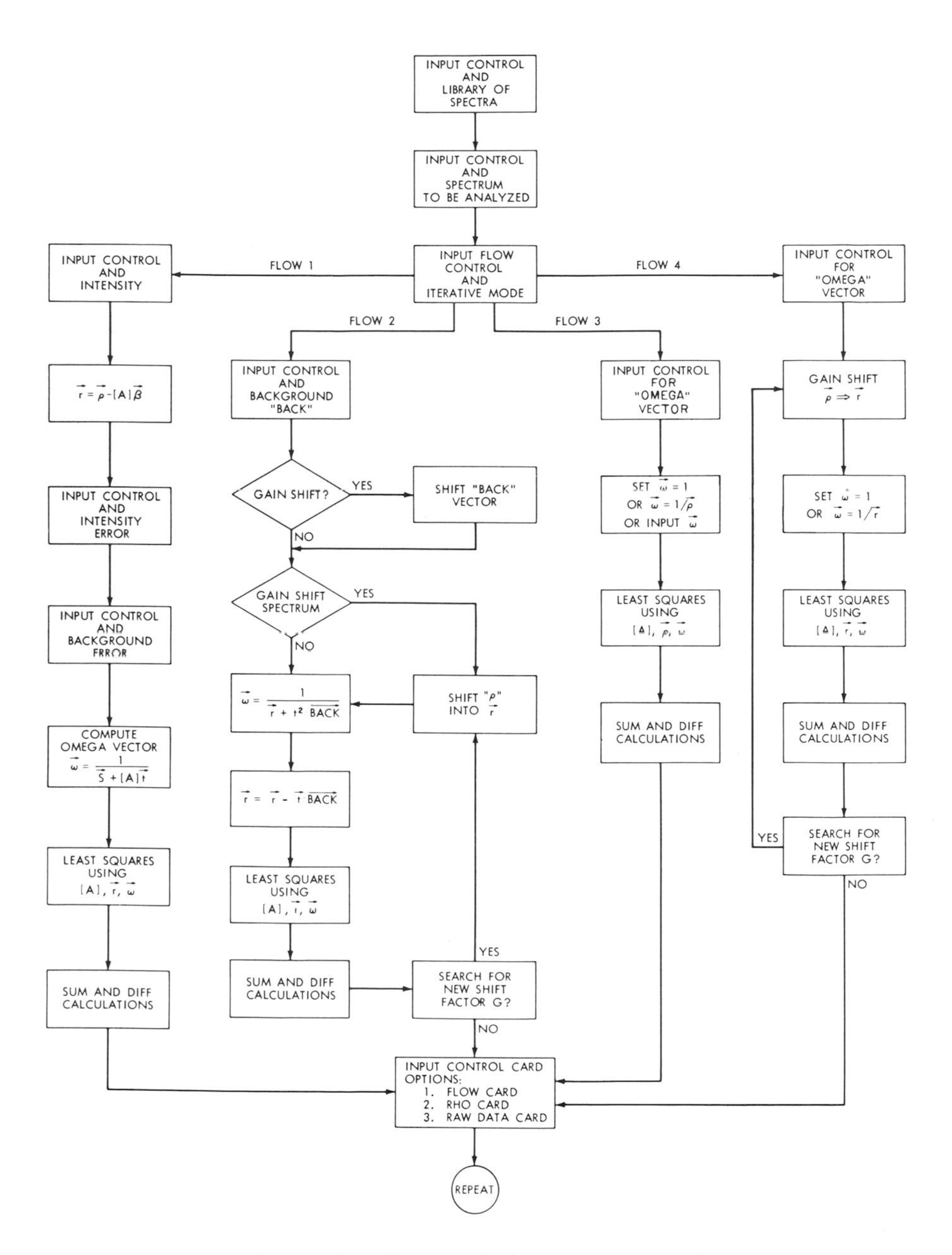

FIG. 9. Flow diagrams for least-squares analysis.

of options are allowed for calculation of ω. The statistical weight matrix, can be either (1) set equal to the unity matrix, (2) set equal to ρ^{-1}, (3) calculated internally in Flows 1, 2, and 4, or (4) read in separately according to the analyst's desire. The output from the computer program yields the following information: (1) the flow used, (2) the gain for minimum χ_i^2 if the option is chosen, (3) whether or not the nonnegativity constraint was used, (4) the relative intensities for β, (5) the monoelemental library functions used, (6) the elements rejected if nonnegativity is used, (7) the standard deviation corresponding to each of the relative intensities calculated, (8) the percent interference between library elements, (9) the actual spectrum analyzed, (10), the synthesized spectrum obtained using the relative intensities determined by least square, (11) the difference between the actual and synthesized spectrum, (12) and the χ_i^2 value. A more detailed description of the program can be found in the work of Imamura *et al.* [10], Curran [11], and Poulson *et al.* [22].

III. Application of Least-Square Technique to an Experimental Problem

A. *General Experimental Results*

The procedures which have been described in some detail above have been applied to the problem of the semiquantitative determination of the composition of a variety of rock types ranging from ultrabasic rocks such as dunite and peridotite through acidic rocks such as granite. For this purpose a suite of six rocks, carefully prepared by the US Geological Survey and presently being circulated for comparative analysis were used. All rock samples were finely divided and homogenized.

Library spectra were obtained using oxides and carbonates of the various elements to be determined such as Mg, Al, Si, K, Ca, Ti, Mn, and Fe.

For the measurements reported here, a sealed Ar–CH_4 proportional counter was used which gave a resolution of 17.5% of the Mn K_α line from ^{55}Fe.

Background was determined using a boric acid briquet as a scatterer. Figures 10–15 show the observed spectra for dunite, peridotite, basalt, andesite, granodiorite, and granite. These experimentally measured data are shown as points superimposed on the solid line curves, synthesized by the computer using the least-square technique proposed and the library elements.

In every instance the spectra have been gain shifted so that they are on a common pulse height energy scale. On this basis, one can state that up to approximately channel 90, covering the elemental range from Mg to Fe, the

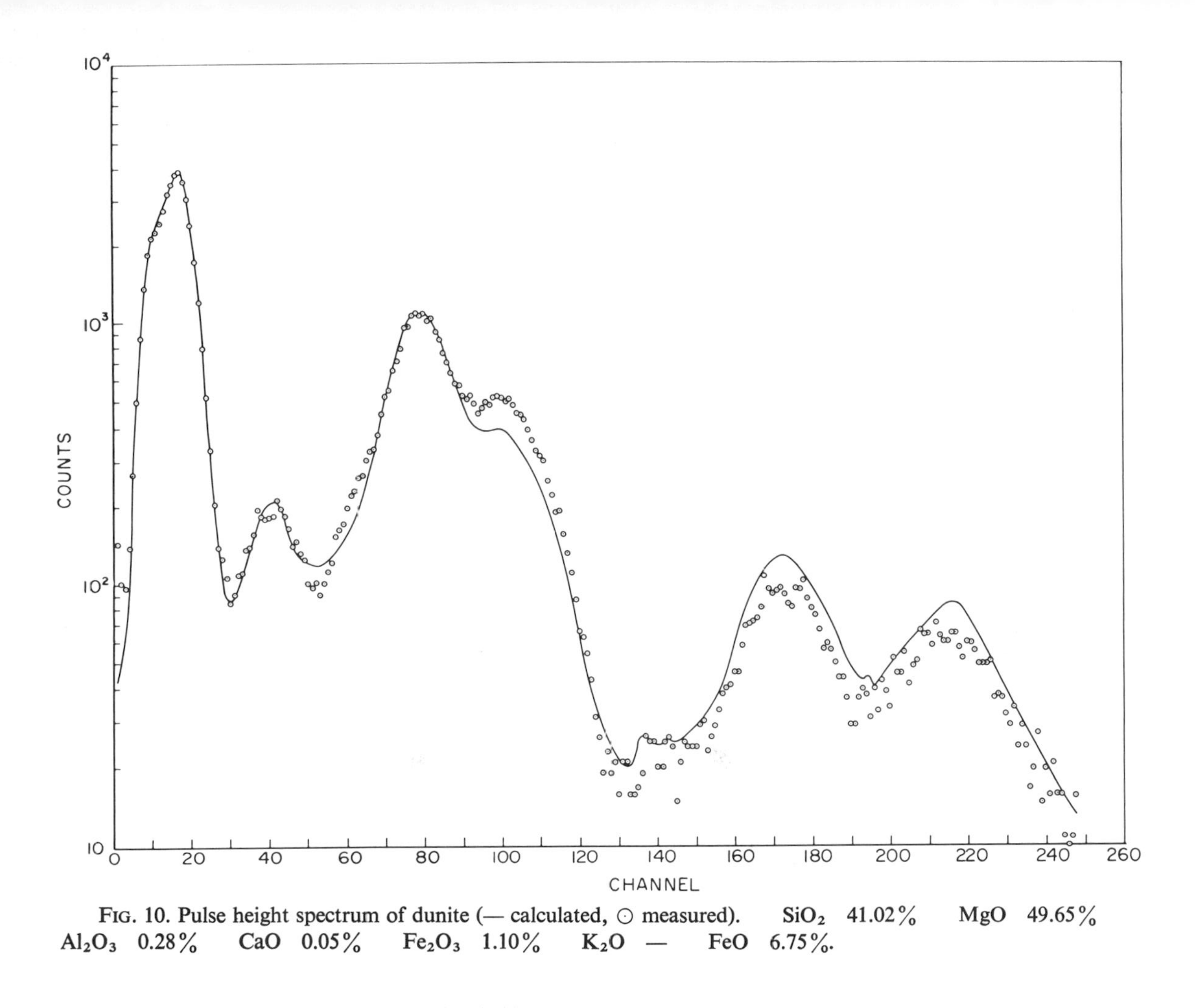

FIG. 10. Pulse height spectrum of dunite (— calculated, ⊙ measured). SiO_2 41.02% MgO 49.65% Al_2O_3 0.28% CaO 0.05% Fe_2O_3 1.10% K_2O — FeO 6.75%.

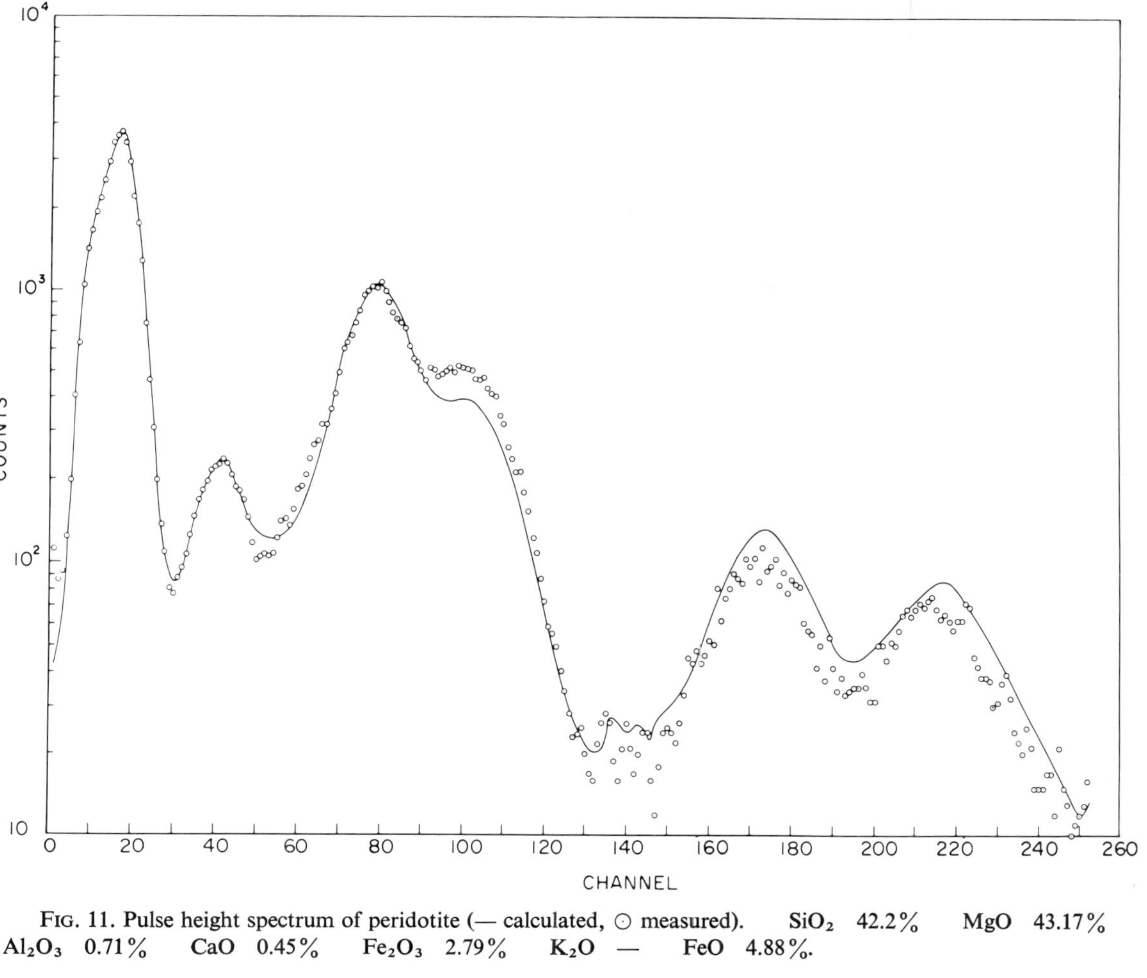

FIG. 11. Pulse height spectrum of peridotite (— calculated, ⊙ measured). SiO_2 42.2% MgO 43.17% Al_2O_3 0.71% CaO 0.45% Fe_2O_3 2.79% K_2O — FeO 4.88%.

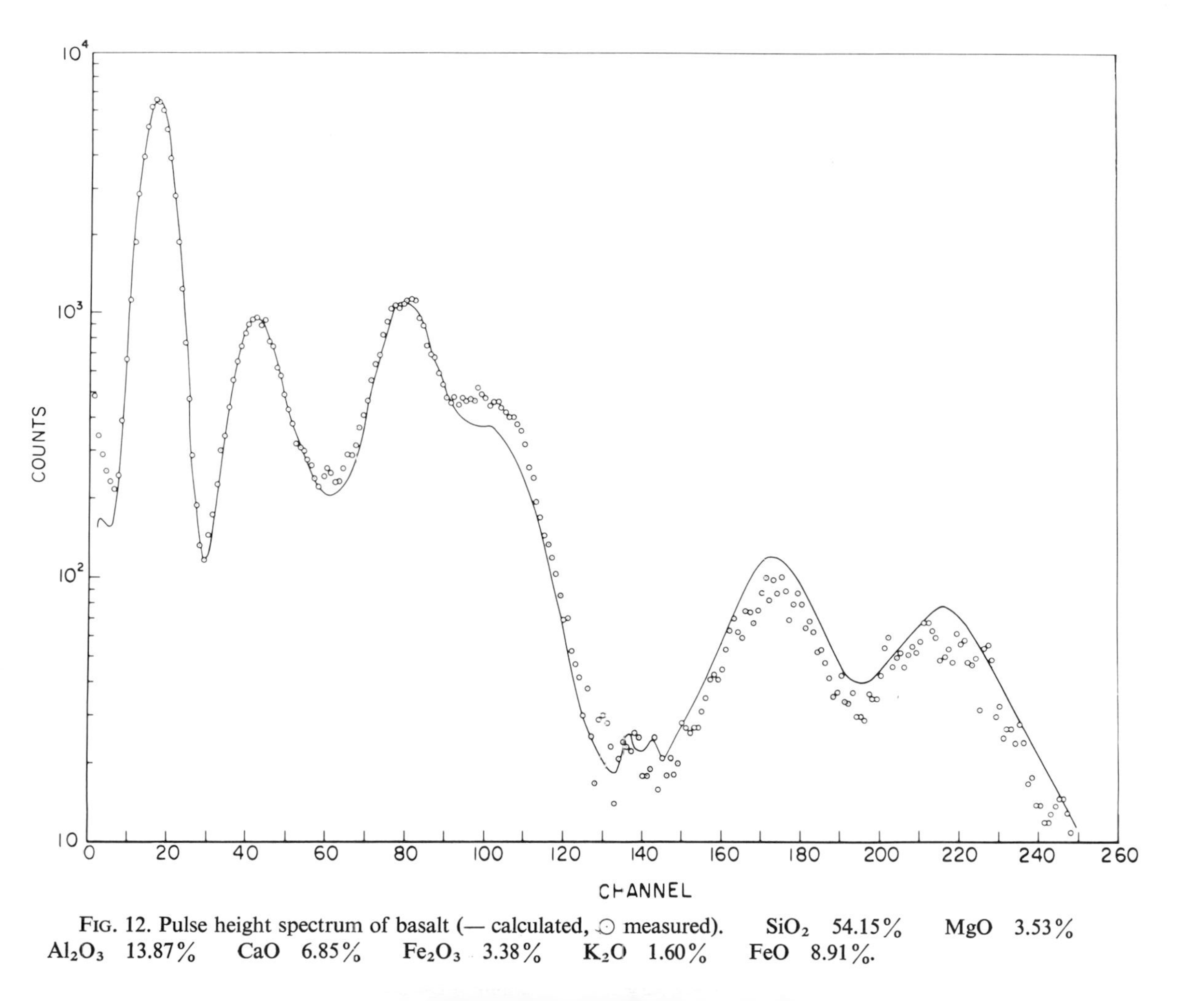

FIG. 12. Pulse height spectrum of basalt (— calculated, ○ measured). SiO_2 54.15% MgO 3.53% Al_2O_3 13.87% CaO 6.85% Fe_2O_3 3.38% K_2O 1.60% FeO 8.91%.

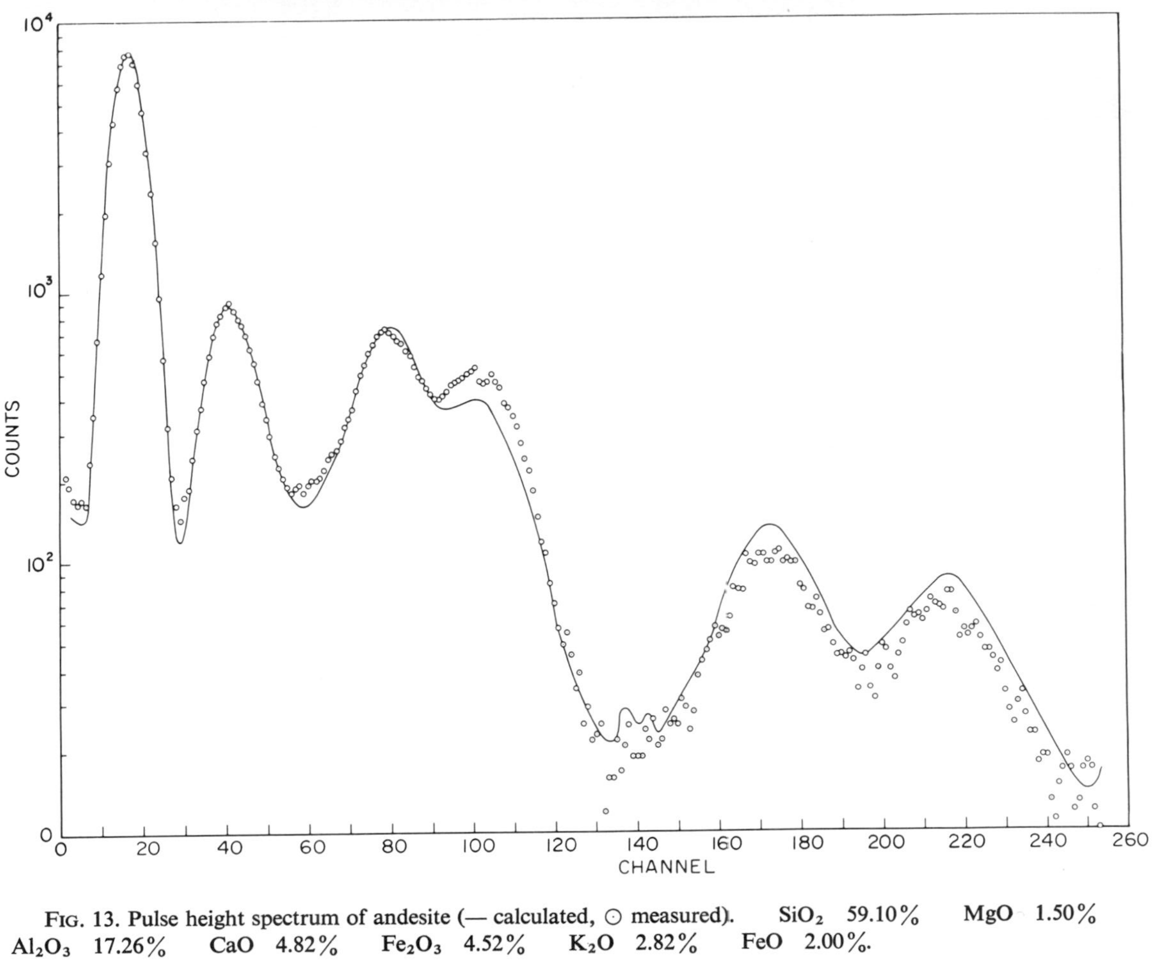

FIG. 13. Pulse height spectrum of andesite (— calculated, ⊙ measured). SiO_2 59.10% MgO 1.50% Al_2O_3 17.26% CaO 4.82% Fe_2O_3 4.52% K_2O 2.82% FeO 2.00%.

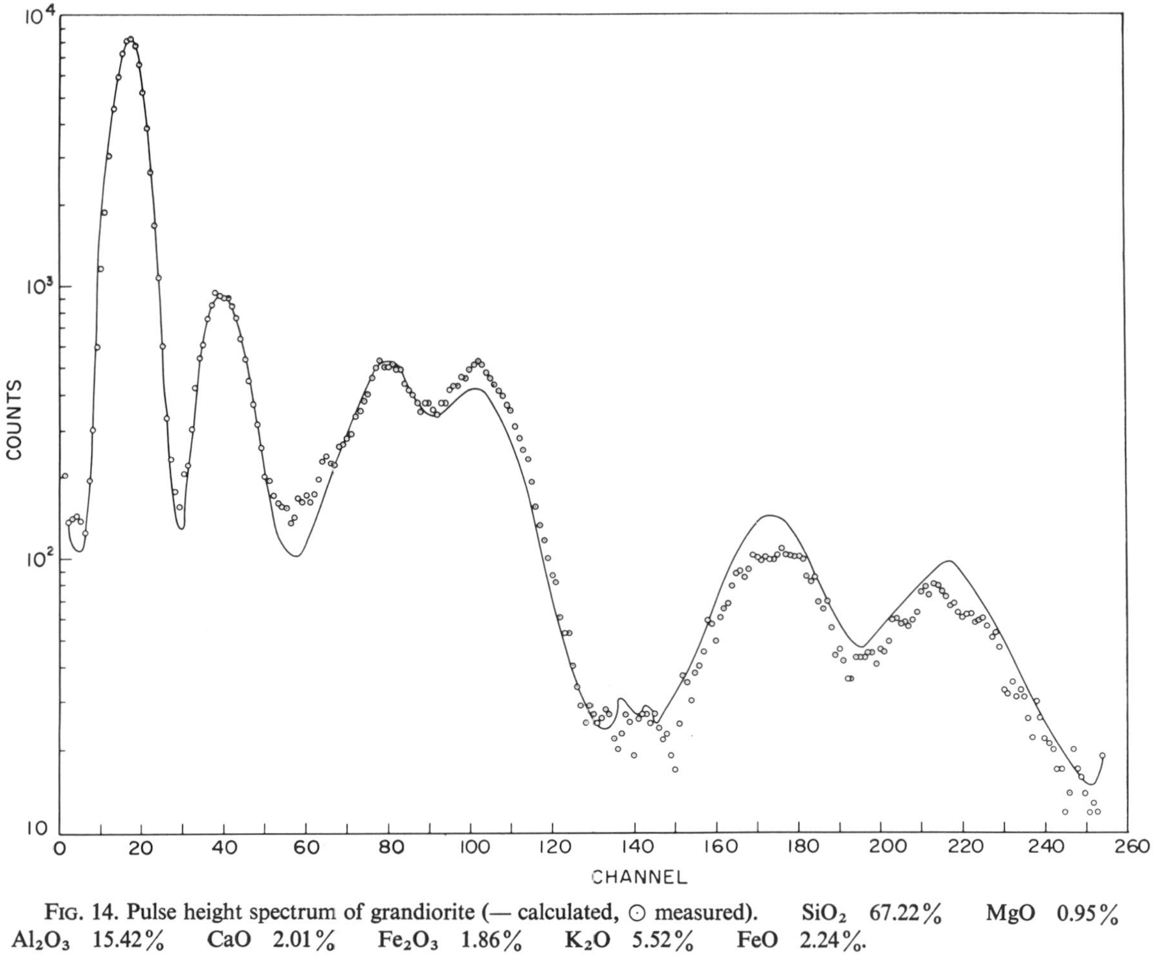

FIG. 14. Pulse height spectrum of grandiorite (— calculated, ⊙ measured). SiO_2 67.22% MgO 0.95% Al_2O_3 15.42% CaO 2.01% Fe_2O_3 1.86% K_2O 5.52% FeO 2.24%.

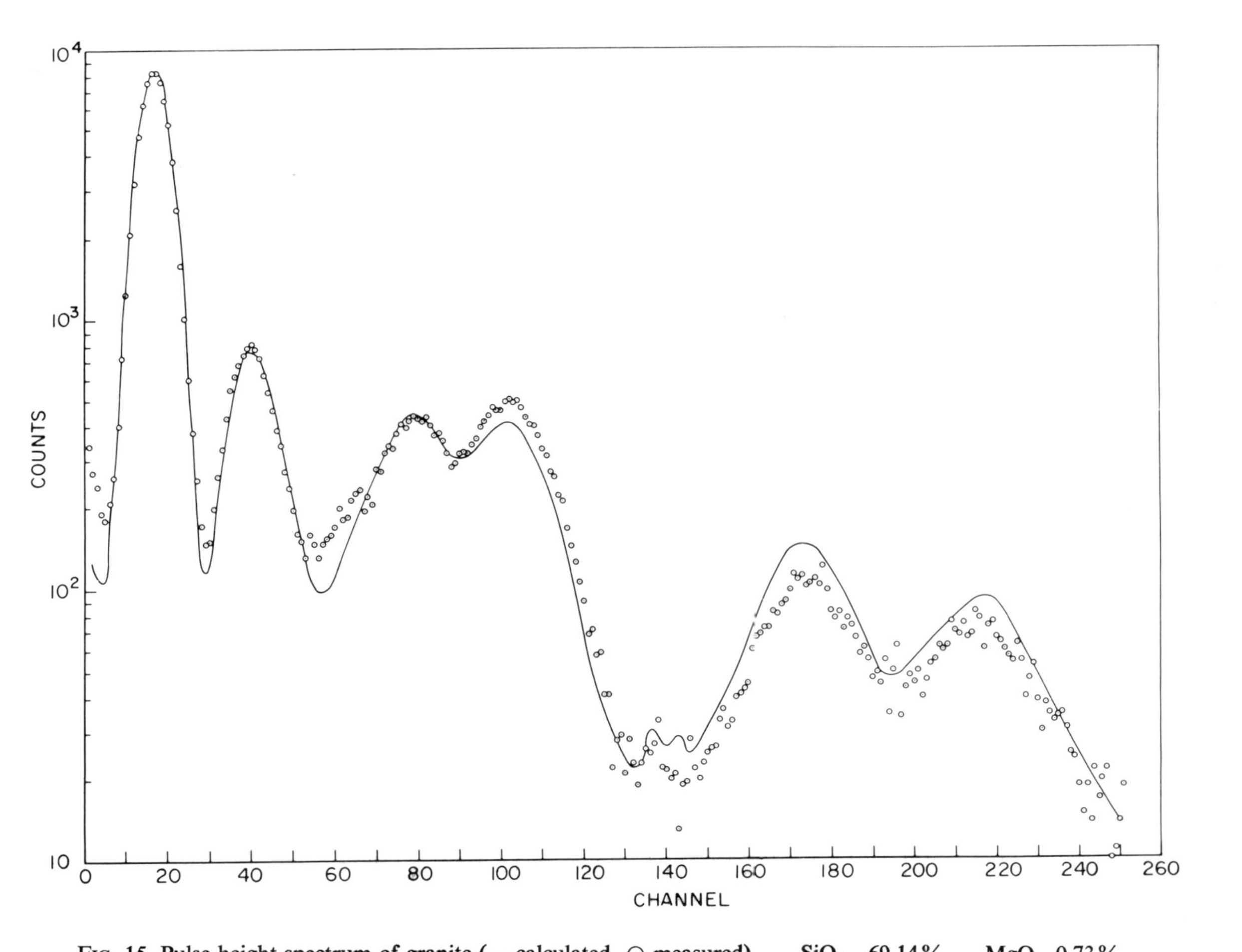

FIG. 15. Pulse height spectrum of granite (— calculated, ⊙ measured). SiO_2 69.14% MgO 0.73% Al_2O_3 15.49% CaO 1.93% Fe_2O_3 1.13% K_2O 4.50% FeO 1.43%.

agreements are excellent and, in fact, within the statistical error. Beyond channel 90 where one observes a number of discrete lines in the background spectrum, there is a slight divergence, very likely due to the difference in scattering between the samples measured and the boric acid pellet. These divergences shown in Figs. 10–15 would actually appear smaller if the statistical spread for the data points were plotted (see Fig. 6). Nevertheless these differences are real and do increase the statistical error in the determination of the relative intensities. These increased errors are reflected in an increase of χ_i^2. The increase on the variance $\sigma^2(\beta_\lambda)$ due to increase in χ_i^2 is then given by Eq. (12b) above.

In order to produce the synthesized curves shown in Figs. 10–15, the relative intensities of each of the library components has been determined by using the least-square criteria and the nonnegativity constraint. These relative intensities are shown in Table I together with the statistical error. Blank portions in the table indicate components showing negative or zero intensity values. These are automatically rejected in the analysis and are not included as part of the library. This is an indication that a particular component is either absent or that its intensity contribution is indistinguishable statistically from the noise. The actual composition for each sample is shown in the corresponding figure (Figs. 10–15). The percentage interference between elements (i.e., those greater than 10%) are given in Table II. It can be seen that the elements fall into three groups when interference is strong. These are: (1) Si, Mg, Al, (2) Ca, K, and (3) Fe, Mn. In these groups, unless an additional constraint is used, such as that described below, it is very difficult to determine the concentrations of the individual components even though the sums of the relative intensities are correct.

B. Resolving Correlations

We now shall consider the case of the Si–Al–Mg correlation in detail. As indicated above, another physical constraint is required in order to resolve the interferences. Since aluminum has a strong absorption edge lying between the silicon and the aluminum K lines; a 0.5 mil aluminum foil was used in front of the detector window to preferentially absorb the silicon K radiation. Figures 16 and 17 show, for example, the pulse height spectra of dunite and granite taken under these conditions. In comparison to the spectra for these rocks previously shown in Figs. 10 and 15, the dunite now shows a distinct Mg peak, and the Al line is predominant in the granite.

The least-square analysis technique using a nonnegativity constraint and an aluminum foil filter was applied to the same suite of six rocks shown in

TABLE I

RELATIVE INTENSITIES OBTAINED FROM THE LEAST-SQUARES ANALYSIS

Element	Rock type					
	Dunite	Peridotite	Basalt	Andesite	Granodiorite	Granite
Si	0.26×10^5 $\pm 0.006 \times 10^5$	0.25×10^5 $\pm 0.006 \times 10^5$	0.41×10^5 $\pm 0.009 \times 10^5$	0.50×10^5 $\pm 0.008 \times 10^5$	0.56×10^5 $\pm 0.008 \times 10^5$	0.56×10^5 $\pm 0.01 \times 10^5$
Corrected value[a]	0.301×10^5 $\pm 0.018 \times 10^5$	0.30×10^5 $\pm 0.017 \times 10^5$	0.47×10^5 $\pm 0.011 \times 10^5$	0.52×10^5 $\pm 0.012 \times 10^5$		
Al	0.18×10^4 $\pm 0.08 \times 10^4$	0.30×10^4 $\pm 0.07 \times 10^4$	0.13×10^5 $\pm 0.009 \times 10^5$	0.13×10^5 $\pm 0.009 \times 10^5$	0.12×10^5 $\pm 0.006 \times 10^5$	0.13×10^5 $\pm 0.01 \times 10^5$
Corrected value[a]	0	0	0.721×10^4 $\pm 0.085 \times 10^4$	0.111×10^5 $\pm 0.014 \times 10^5$	0.105×10^5 $\pm 0.013 \times 10^5$	0.780×10^4 $\pm 0.093 \times 10^4$
Fe	0.13×10^5 $\pm 0.005 \times 10^5$	0.12×10^5 $\pm 0.005 \times 10^5$	0.14×10^5 $\pm 0.006 \times 10^5$	0.79×10^4 $\pm 0.05 \times 10^4$	0.40×10^4 $\pm 0.04 \times 10^4$	0.24×10^4 $\pm 0.05 \times 10^4$
Mg	0.12×10^5 $\pm 0.005 \times 10^5$	0.89×10^4 $\pm 0.04 \times 10^4$	0.47×10^3 $\pm 0.39 \times 10^3$	0.28×10^3 $\pm 0.33 \times 10^3$		0.93×10^3 $\pm 0.42 \times 10^3$
Corrected value[a]	0.972×10^4 $\pm 0.12 \times 10^4$	0.74×10^4 $\pm 0.08 \times 10^4$	0.252×10^4 $\pm 0.151 \times 10^4$	0.94×10^3 $\pm 0.99 \times 10^3$	0.194×10^4 $\pm 0.149 \times 10^4$	0.364×10^4 $\pm 0.280 \times 10^4$
Ca	0.33×10^3 $\pm 0.20 \times 10^3$	0.67×10^3 $\pm 0.19 \times 10^3$	0.75×10^4 $\pm 0.03 \times 10^4$	0.58×10^4 $\pm 0.04 \times 10^4$	0.31×10^4 $\pm 0.03 \times 10^4$	0.30×10^4 $\pm 0.03 \times 10^4$
K	0.49×10^2 $\pm 1.8 \times 10^2$	0.18×10^3 $\pm 0.17 \times 10^3$	0.38×10^4 $\pm 0.04 \times 10^4$	0.50×10^4 $\pm 0.03 \times 10^4$	0.76×10^4 $\pm 0.04 \times 10^4$	0.61×10^4 $\pm 0.04 \times 10^4$
Ti	0.55×10^3 $\pm 0.14 \times 10^3$	0.50×10^3 $\pm 0.13 \times 10^3$	0.18×10^4 $\pm 0.02 \times 10^4$	0.81×10^3 $\pm 0.18 \times 10^3$		
Mn	0.39×10^4 $\pm 0.03 \times 10^4$	0.37×10^4 $\pm 0.04 \times 10^4$	0.33×10^4 $\pm 0.05 \times 10^4$	0.16×10^4 $\pm 0.04 \times 10^4$	0.10×10^4 $\pm 0.04 \times 10^4$	0.10×10^4 $\pm 0.04 \times 10^4$

[a] Corrected values are those with interference removed.

TABLE II

CORRELATION GREATER THAN 10% BETWEEN LIBRARY COMPONENTS

	Si	Al	Fe	Mg	Ca	K	Ti	Mn	Back
Si		45%		16%					
Al				55%					
Fe								42%	
Mg									
Ca						40%			
K									
Ti									
Mn									
Back									

Figs. 10–15. Because of the strong silicon absorption, the silicon component was eliminated from the analysis. The relative intensities of the Al and Mg obtained were corrected for absorption due to the Al filter. While a strong correlation is possible between these elements, this was not a significant factor in this instance because where the Mg concentration was high, the Al concentration was very low and vice versa.

If Al and Mg were both present in large amounts or if it is required that the Al and Mg be known with great accuracy, then a magnesium filter could be used to selectively absorb the Al radiation, and resolve this correlation.

Since the Al and Mg relative intensities had been determined, these components were then stripped out of the original spectrum, and the Si relative intensity determined using flow 1 of the computer program. These results are listed in Table I as corrected values for Si, Al, and Mg and can be compared to the uncorrected values within the same column.

Figure 18 is a plot on a log–log scale of the relative intensities versus chemical composition for Si, Al, Mg, K, Ca, and Fe. The solid lines are least-square fits to the data points obtained from the computer analysis. There is a good approximation to a linear relationship between relative intensity and percentage composition, consistent with the small concentration range. No line was drawn for the Mg because the lower Mg concentrations are associated with high Al concentrations in these samples, and the strong correlation between these two elements makes it difficult to determine the Mg component without very large errors. It is significant that one can easily distinguish between a 10% relative concentrational variation in the 30% range. The error bars in these plots are strictly fluctuations due to counting statistics and background correction. These can be partially reduced by increased counting times and background reduction.

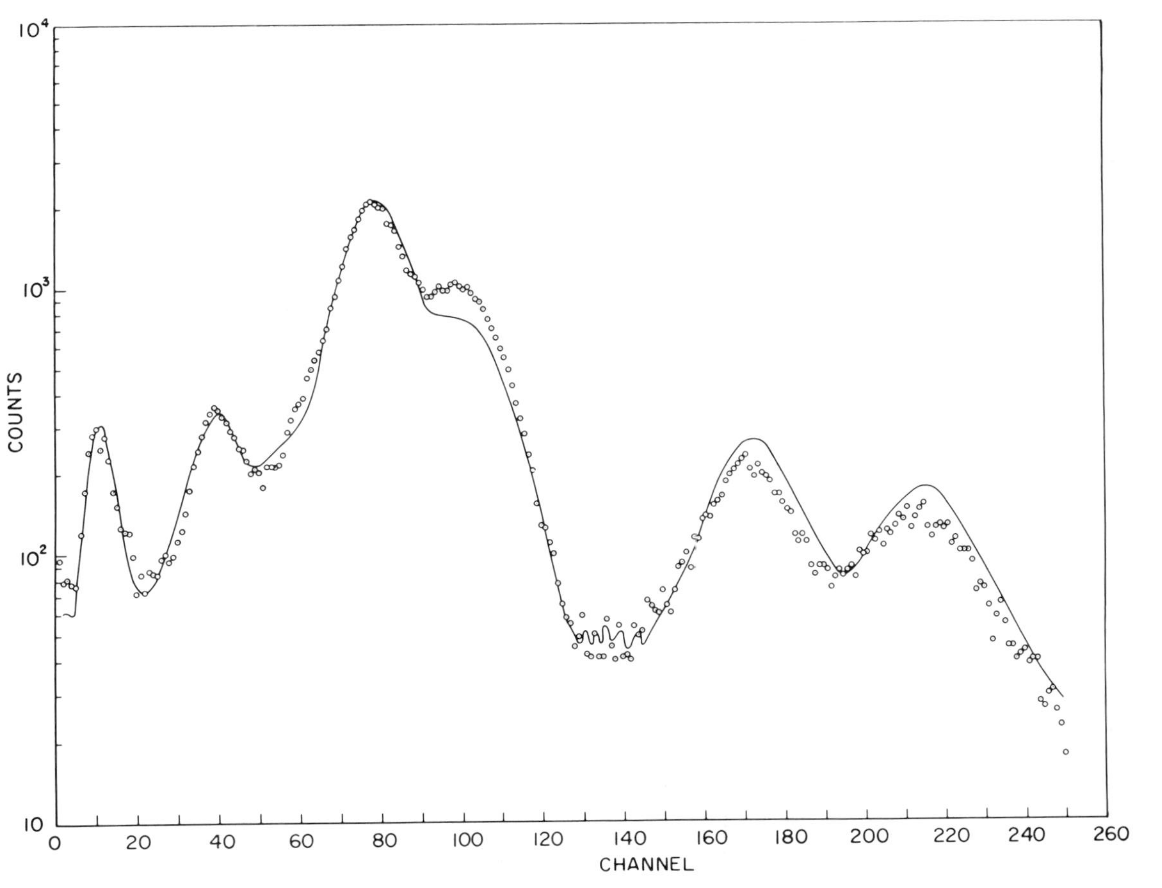

FIG. 16. Dunite spectrum with $\frac{1}{4}$ mil aluminum absorber (— calculated, ⊙ measured).

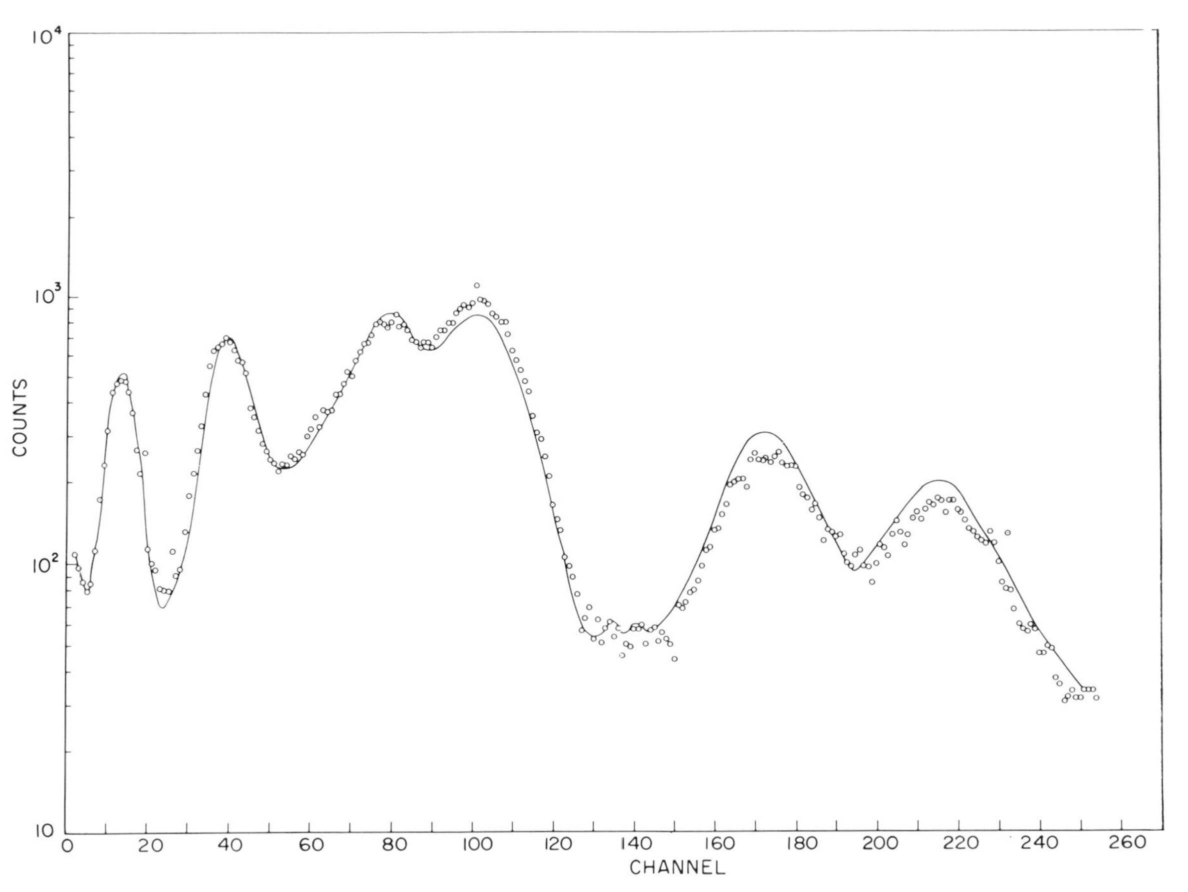

FIG. 17. Granite spectrum with ½ mil aluminum absorber (— calculated, ⊙ measured).

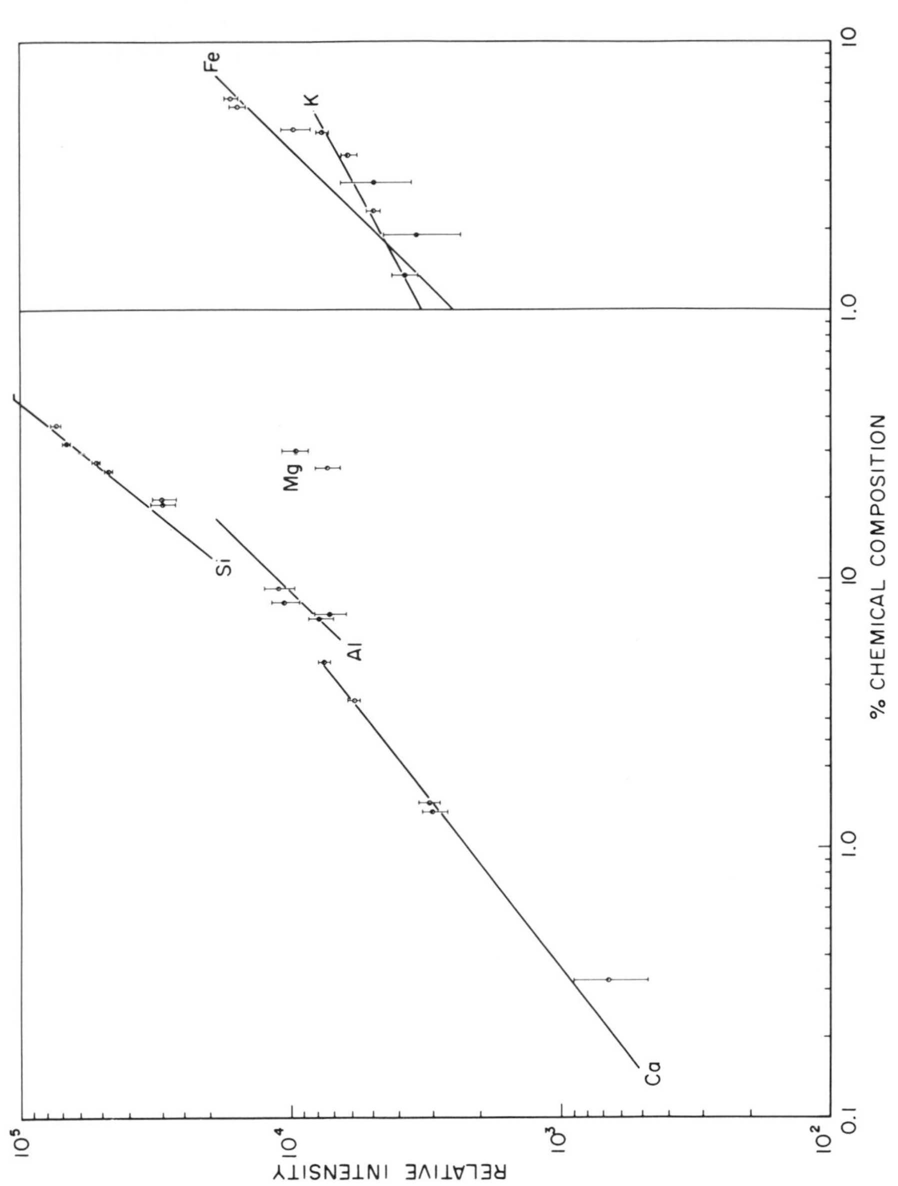

FIG. 18. Measured relative intensity as a function of chemical composition.

C. Classification of Rock Types

Although the curves in Fig. 18 show a reasonable linear approximation between relative intensity and concentration, one does not intend to infer that this method can yield precise chemical analysis. No attempt has been made at careful sample preparation or to correct for matrix effects such as absorption or enhancement. In fact, in terms of a lunar surface mission, a serious attempt has been made to determine how much chemical information can be obtained without careful control of the above factor. Using these ground rules, the data will be used in an attempt at broad classification of rock types. The approach is similar to those in the studies performed using neutron methods [25–29].

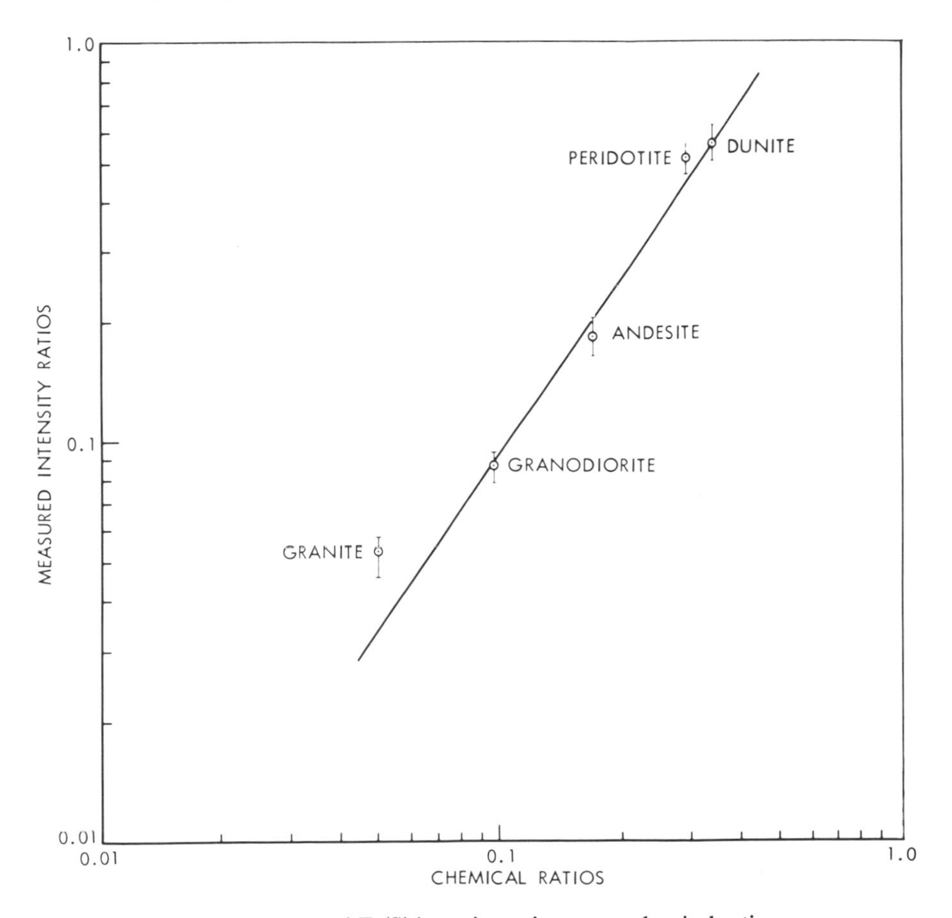

FIG. 19. Measured Fe/Si intensity ratios versus chemical ratios.

A preliminary classification of the rocks according to type is made on the basis of the presence or absence of certain chemical elements simply determined from the computer output, for example the presence or absence of large amounts of Mg, Al, Ca, and K. For example the ultrabasic rocks are rich in Mg with negligible amounts of Al, Ca, and K (see Table I and Figs. 10–15).

Following this, an approach similar to that described for the neutron-gamma techniques [25–29] is used. Rather than attempting to carefully control the sample geometry, ratios of various elements have been calculated as shown in Figs. 19–25. One observes again a good approximation to a

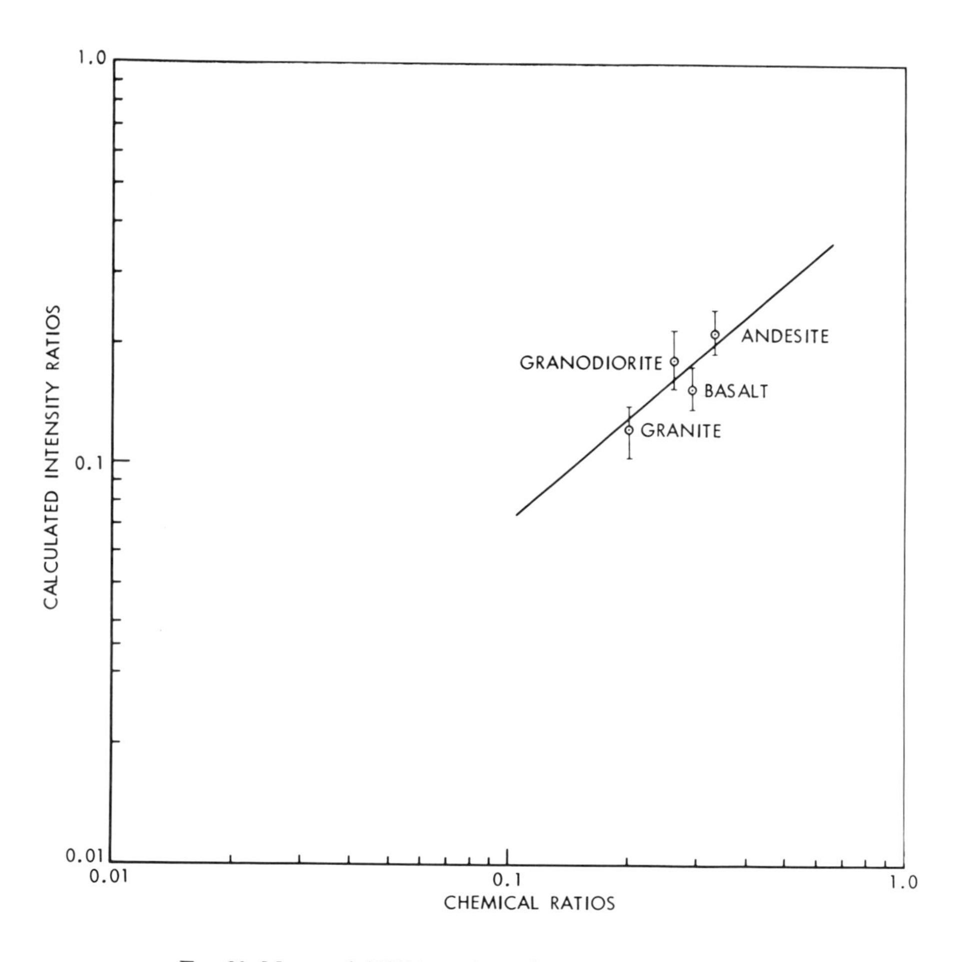

FIG. 20. Measured Al/Si intensity ratios versus chemical ratios.

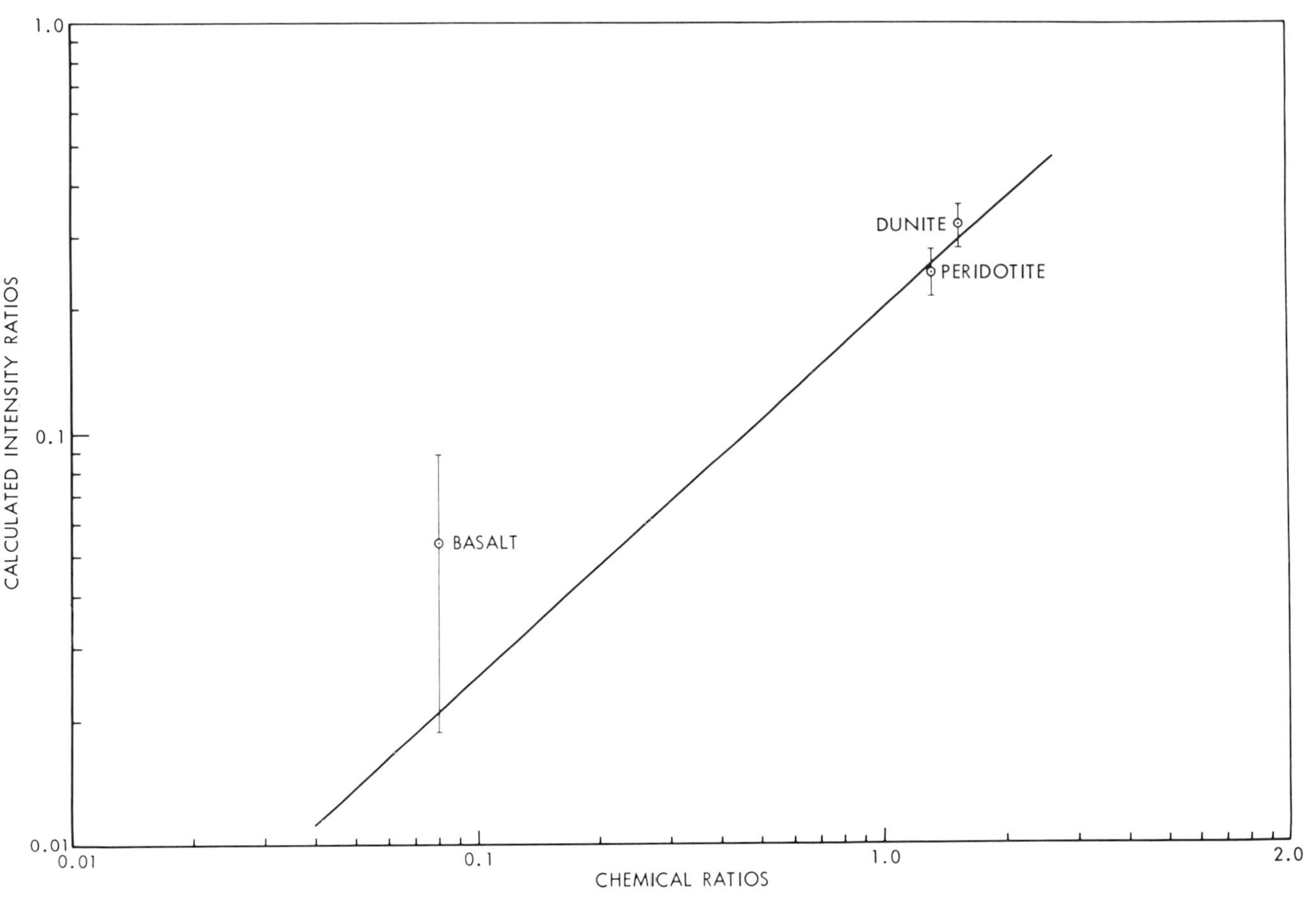

FIG. 21. Measured Mg/Si intensity ratios versus chemical ratios.

linear relationship between chemically determined and calculated ratios. As pointed out by Waggoner and Knox [29], it is obvious that no one element or ratio of elements can be used to classify a rock type. However, if one combines a number of these factors, then it is possible to categorize a rock as being somewhere on the scale between the ultrabasic and acidic rocks. This can, of course, only be done to the extent that a chemical analysis is meaningful in this context.

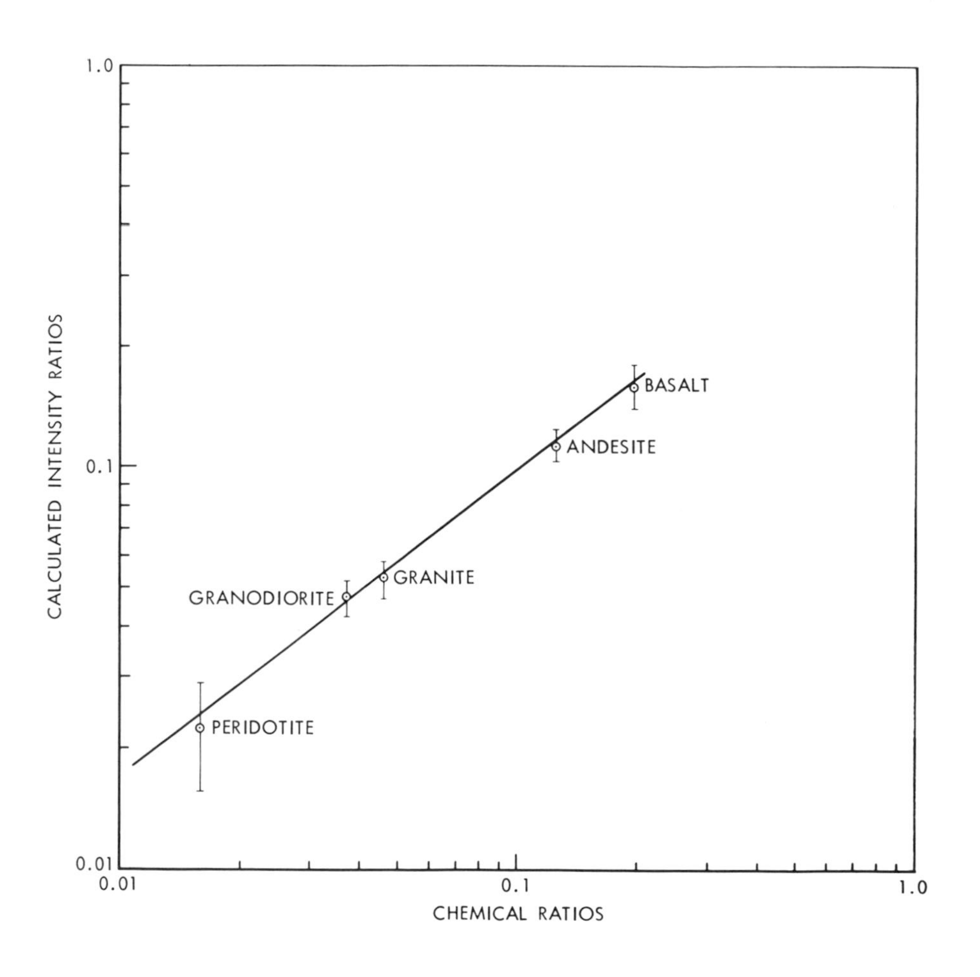

FIG. 22. Measured Ca/Si intensity ratios versus chemical ratios.

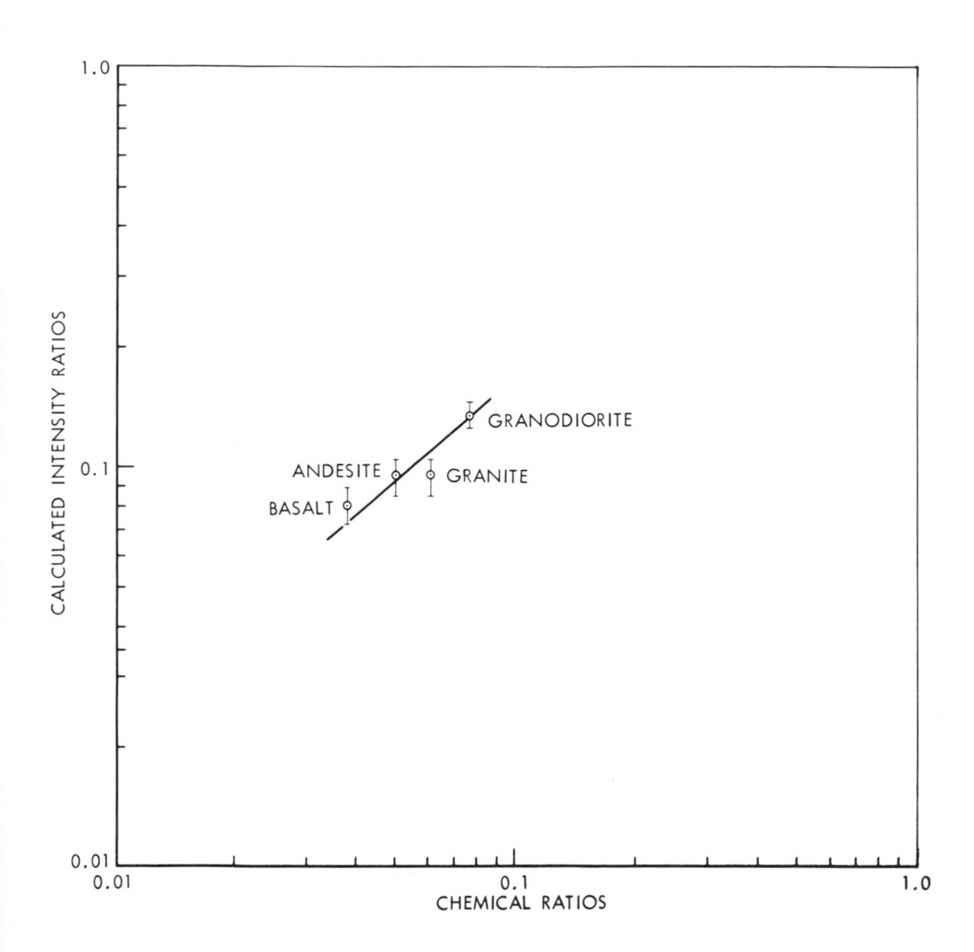

FIG. 23. Measured K/Si intensity ratios versus chemical ratios.

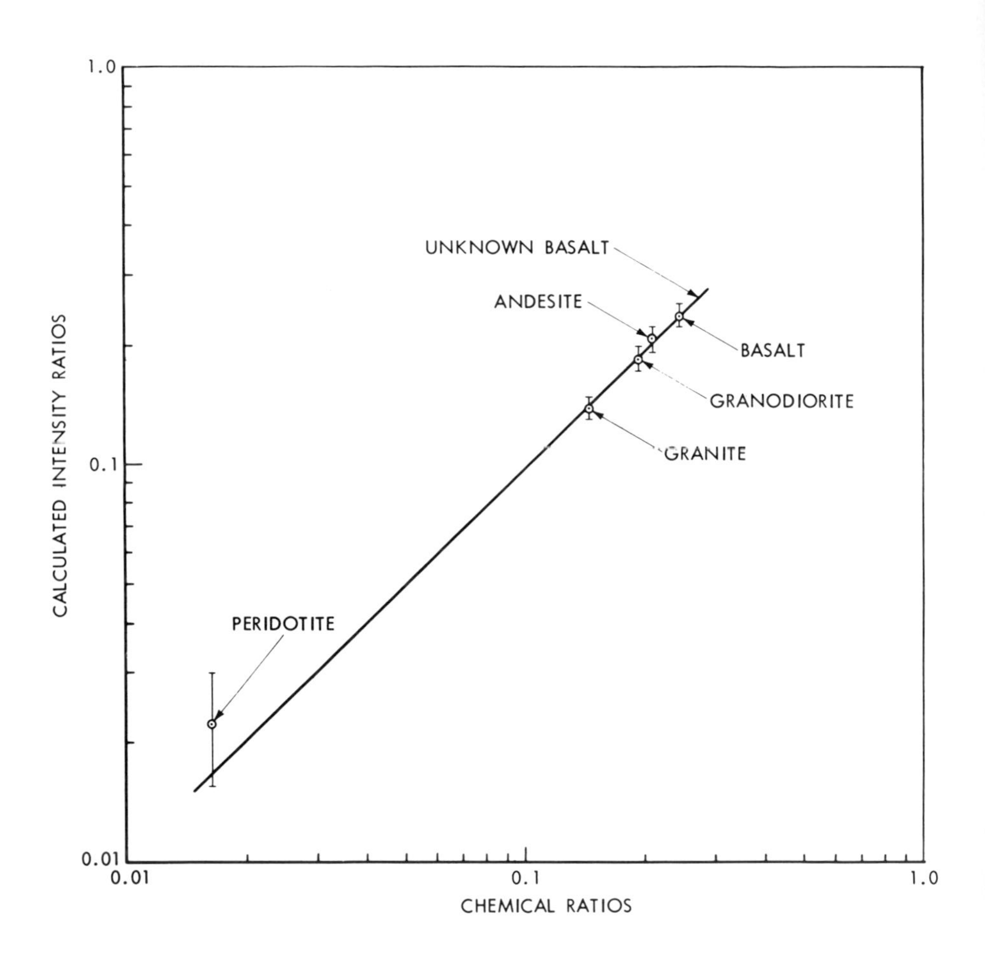

FIG. 24. Measured (Ca + K)/Si intensity ratios versus chemical ratios.

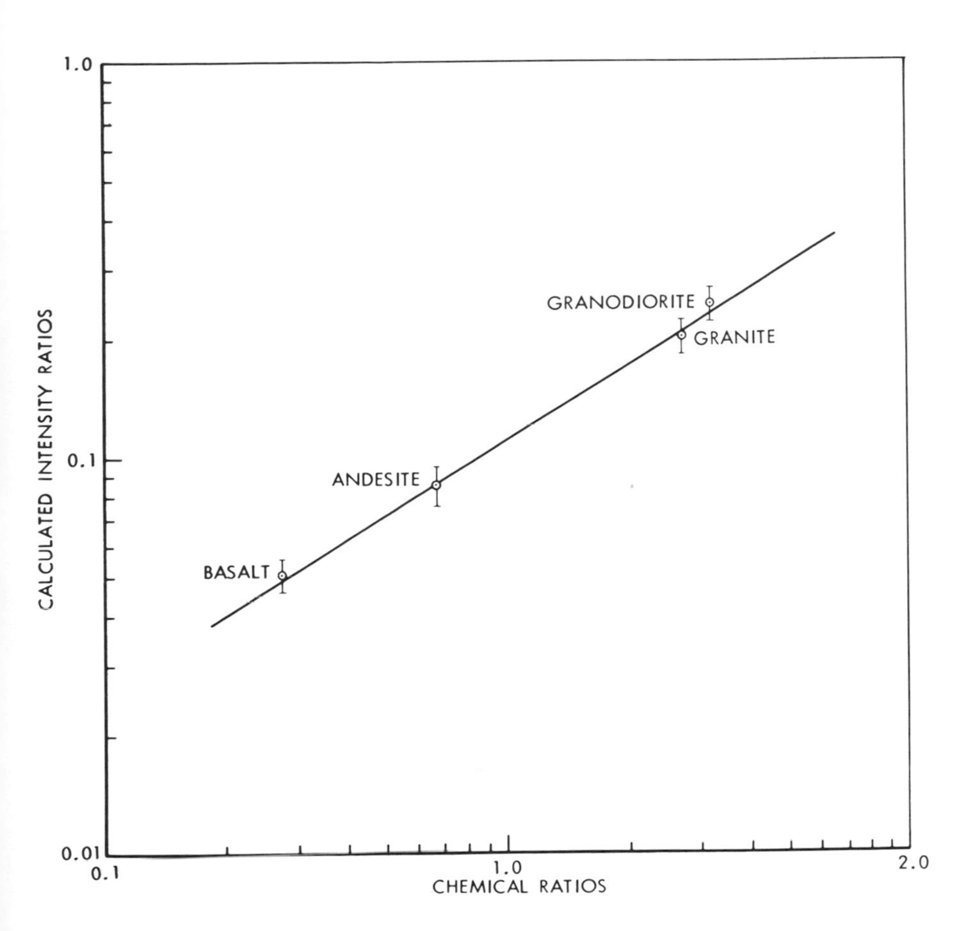

FIG. 25. Measured K/Ca intensity ratios versus chemical ratios.

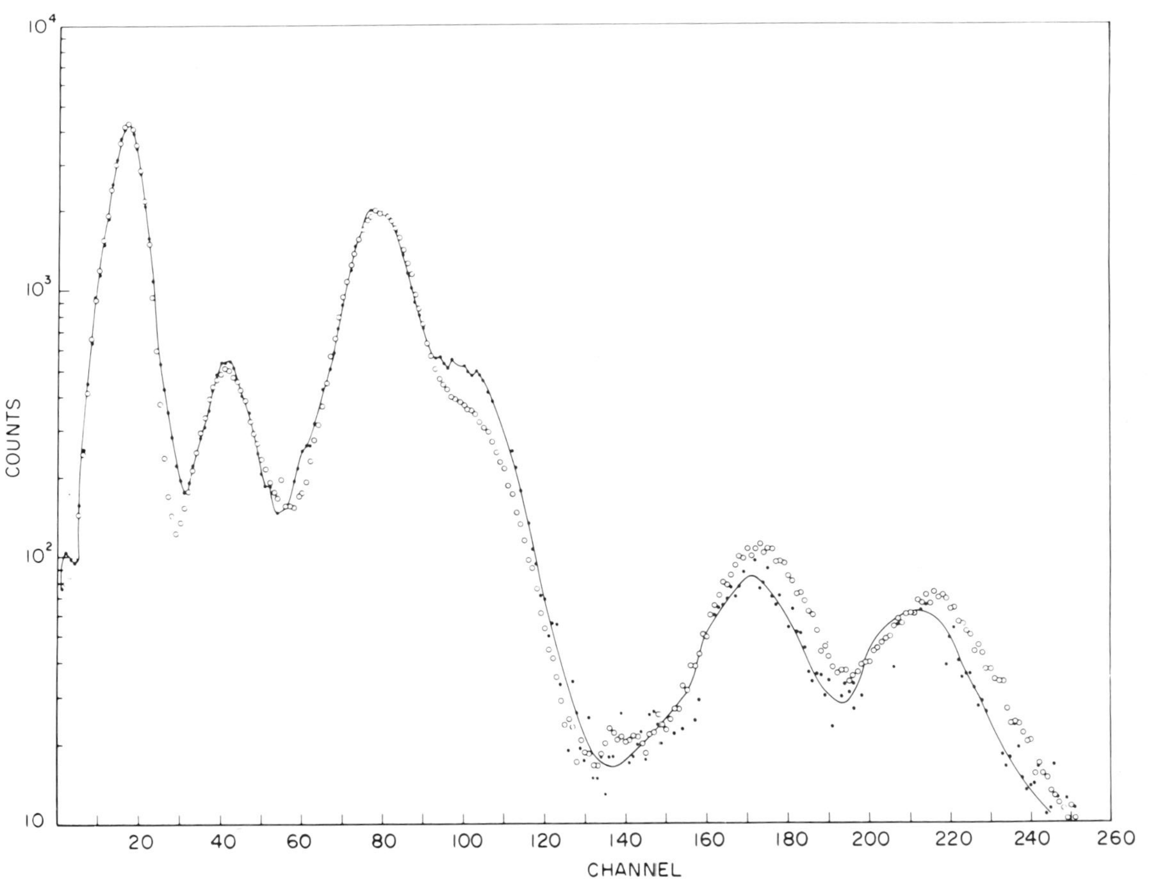

FIG. 26. Spectrum of Plainview (bronze) meteorite (— calculated, ⊙ measured).

D. Quantitative Analysis

The performance of quantitative analysis normally requires extreme care in sample preparation, the use of appropriate standards and adjustment for matrix effects. Such requirements are sometimes difficult to fulfill even in a laboratory environment. One cannot even hope to control these parameters in a remote exploration program. Keeping these constraints in mind, an attempt was made, using these rough procedures to perform an analysis on a meteorite to obtain some idea of the possibility for performing a quantitative analysis. Figure 26 shows the measured pulse height spectrum obtained for the Plainview meteorite, a bronzite-chondrite. Table III summarizes the results obtained using the curves of Fig. 17 as calibration curves. The agreements can be considered as quite satisfactory for a first cut chemical analysis. These results combined with the data used to prepare the calibration curves show that one can expect to obtain useful chemical values, satisfying the goals for a lunar geochemical exploration device.

TABLE III
RESULTS OBTAINED FOR PLAINVIEW METEORITE USING THE LEAST-SQUARES ANALYSIS

Element	Relative intensities from least-squares analysis	% Composition from Fig. 17	Reported chemical analysis[a]
Fe	$0.36 \times 10^5 \pm 0.02 \times 10^5$	14.1 ± 2.8	17.24
Si	$0.31 \times 10^5 \pm 0.01 \times 10^5$	17.5 ± 0.7	17.25
Al	$0.25 \times 10^4 \pm 0.09 \times 10^4$	3.2 ± 1.5	1.08
Mg	$0.54 \times 10^4 \pm 0.05 \times 10^4$	18.0 ± 2.5	13.71
Ca	$0.24 \times 10^4 \pm 0.03 \times 10^4$	1.08 ± 0.21	1.19
K	$0.95 \times 10^3 \pm 0.30 \times 10^3$	0.1 ± 0.03	0.066

[a] Quart. Rept., US Geological Survey, April–June 1965.

REFERENCES

1. A. E. Metzger, R. E. Parker, and J. I. Trombka, *IEEE* (*Inst. Elect. Electron. Eng.*) *Trans. Nucl. Sci.* **13**, (1) (1966).
2. B. Sellers and C. A. Ziegler, *Symp. Low Energy X and Gamma Sources and Applications, Chicago, Illinois, 1964*. U.S. At. Energy Comm., Washington, D.C.
3. J. O. Karttunen *et al.*, *Anal. Chem.* **36**, 1277 (1964).
4. J. I. Trombka and I. Adler, *Natl. Conf. Electron Probe Microanalysis, 1st, Univ. of Maryland, College Park, Md., 1966.*
5. J. F. Cameron and J. R. Rhodes, *Intern. J. of Appl. Radiation Isotopes* **7**, 244–250 (1960).
6. J. F. Cameron and J. R. Rhodes, *Nucleonics* **19**, (6), 53–57 (1961).

7. A. Robert, Contributions to the analysis of light elements using X-fluorescence excited by radioelements. Comm. Energie Atomique, Rappt. CEA-R2539 (1964).
8. H. Friedman, *Advan. Spectros.* (1964).
9. A. Robert and P. Martinelli, Paper No. SM 55/76, *IAEA Meeting, Salzburg, Austria, 1964.*
10. H. Imamura, K. Vehida, and H. Tominaya, *Radioisotopes* **11**, (4) (1965).
11. S. C. Curran, *in* "Handbuch der Physik" (S. Flügge, ed.), Vol. XLV. Springer, Berlin, 1958.
12. J. I. Trombka, Least-squares analysis of gamma-ray pulse height spectra. NAS-NS-3107, pp. 183–201 (1963).
13. W. R. Burrus, *IRE* (*Inst. Radio Engrs.*) *Trans. Nucl. Sci.* **7**, (23) (1960).
14. M. E. Rose, *Phys. Rev.* **91**, 610 (1953).
15. W. A. Hestin, R. L. Heath, and R. D. Helmer, Quantitative analysis of gamma ray spectras by the method of least squares. IDO-16781 (1962).
16. R. L. Heath, Data analysis techniques for scintillation spectrometry. IDO-16784 (1962).
17. E. Schoenfeld, private communication.
18. A. Turkivitch and E. Franzyrote, private communication.
19. E. Beale, *Naval Res. Logistics Quart.* **6** (September 1959).
20. H. Scheffe, "The Analyses of Variance." Wiley, New York, 1959.
21. C. A. Bennett and H. L. Franklein, "Statistical Analysis in Chemistry and the Chemical Industry." Wiley, New York, 1954.
22. P. Poulson, R. E. Parker, and J. I. Trombka, Computer Program Rept. Linear least square analysis of radiation spectra. Interoffice Memo, Jet Propulsion Lab. (1965).
23. J. I. Trombka, Dissertation. Univ. of Michigan, 1962.
24. J. I. Trombka and A. Metzger, *in* "Analysis Instrumentation 1963" (L. Fouler, R. D. Eanes, and T. J. Kehoe, eds.), pp. 237–256. Plenum Press, New York, 1963.
25. C. D. Schroder, J. A. Waggoner, J. A. Benger, E. F. Martina, and R. J. Stinner, *ARS* (*Am. Rocket Soc.*) *J.* **32**, 631 (1962).
26. R. C. Greenwood and J. H. Reed, *Proc. Intern. Conf. Mod. Trends Activation Anal., A and M College of Texas, College Station, Texas, 1961.*
27. L. E. Fite, E. L. Steele, and R. E. Wainerdi, An investigation of computer coupled automatic activation analysis and remote lunar analyses. Quart. Rept. TID-18257 (1963).
28. A. E. Metzger, Some calculations bearing on the use of neutron activation for remote compositional analysis. Jet Propulsion Lab. Tech. Rept. No. 32-386, Jet Propulsion Lab., Pasadena, California (1962).
29. J. A. Waggoner and R. J. Knox, Elemental analysis using neutron inelastic scatter. UCRL-14654-T.

The Divergent Beam X-Ray Technique

HARVEY YAKOWITZ

Metallurgy Division, Institute for Materials Research
National Bureau of Standards, Washington, D.C.

List of Symbols

- ϕ angle between diffracting plane and the crystal surface nearest the X-Ray source
- θ Bragg angle
- ζ Mosaic spread
- d Planar spacing
- (hkl) Miller indices of a plane

D	Perpendicular distance from X-Ray source through the photographic film or plate
e	Electron charge
m	Electron mass
c	Velocity of light
λ	Wavelength in vaccuum
λ_0'	Wavelength in crystal
F	Structure factor for plane of interest
N'	Number of unit cells per unit volume
τ	Primary extinction attenuation coefficient
μ	Ordinary linear X-ray attentuation coefficient
I_r	Reflected intensity
I_0	Incident intensity of primary beam
u	A small angular quantity
μ'	Attentuation coefficient in the presence of secondary extinction
Q	Volume reflecting power of mosaic crystal
Q'	Volume reflecting power of mosaic crystal reduced by the presence of primary extinction, i.e., the actual reflecting power
g	Secondary extinction coefficient
$G'(u)$	$=gQ'$
x	Specimen thickness
W	Deficiency line breadth
σ	Photographic contrast of unresolved deficiency conics
K	Film response factor for entire incident X-ray spectrum (takes into account the ratio of energies of characteristic to continuum radiation)
$\exp-(\bar{\mu}x)$	Fictitious average transmission for the entire radiation spectrum passing through the sample
s	X-ray source size
P	Integrated intensity of a diffraction conic observed on a transmission pattern
P_B	Integrated intensity of a diffraction conic observed on a back-reflection pattern
S_T	Total spread of a diffraction conic observed on a transmission pattern
(ξ,y)	Fractional coordinate of the exit range of multiple reflections
y	$=(1-\xi)$
$T(\xi)$	Intensity of transmitted beam issuing from perfect nonabsorbing crystal at point ξ
$R(\xi)$	Intensity of reflected beam issuing from perfect nonabsorbing crystal at point ξ
q	Amplitude of beam reflected by one plane for an incident beam of unit amplitude
A'	q multiplied by the number of reflecting planes intersected by a line parallel to the reflected beam
A''	q multiplied by the number of planes intersected by the incident ray
J_0	Zero order Bessel function
J_1	First order Bessel function
a	Lattice parameter
h_0	$=(h^2+k^2+l^2)^{1/2}$
H_0	$=(H^2+K^2+L^2)^{1/2}$
ω	Angle between two crystallographic poles
Y'	Perpendicular distance measured between conics on the film in the case of near approach

X' Perpendicular distance measured between conics on the film in the case of overlapping conics
t' Point at which true tangency of two conics would occur on a film
R_K $=X'/Y'$
γ Half-angle of a lens-shaped intersection of two conics on the film
ϕ' Angle from pattern center to a given pole of interest
Δp Angular separation between α_1 and β
(B, F, M) Coordinate position for the exact intersection of three or more conics (origin at pattern center)
(X, Y) Coordinate position of a point on any conic with respect to rectangular axes having an origin at the pattern center
Z is equivalent to D in Morris' method obtaining lattice spacing values
ρ Density
Z_A Atomic number
ε Strain
T_s Symmetric strain tensor
S_I Principal strain
l_{ij} Direction cosine for principal strain axis
ζ' $=(\varepsilon_{11}, \varepsilon_{22}, \varepsilon_{33}, \varepsilon_{23}, \varepsilon_{31}, \varepsilon_{12})$
tE Exposure time in minutes for a Kossel pattern
i_s Electron current rejected to ground by the specimen
η Half-angle of Kossel cone intercepted by the film
0 (subscript) Number of photons produced per incident electron per unit of solid angle
r Fraction of incident electrons which are backscattered from a target.

I. Introduction

Propagation of a diverging X-ray beam through a single crystal produces diffraction cones occurring in directions governed by both the lattice symmetry and the crystal orientation. If these cones are intercepted by a photographic film, conic sections corresponding to each monochromatic component in the primary beam will be recorded, i.e., cones for say K_{α_1}, K_{α_2}, and K_β will appear together with a general background blackening.

Information which can be obtained from the photograph includes lattice spacings, crystal orientation, and data concerning the degree of crystal perfection. In particular, lattice spacing data precise to two or three parts per million are possible to obtain. Orientation may be carried out to accuracies approaching 0.1° of arc, and information relating to internal strains may be obtained from line profiles.

Despite the fact that it was known in 1938 that such information was obtainable, the divergent beam technique did not achieve wide usage until 1962. Among the reasons for this lack of popularity were the relatively large size of sources of X-radiation available; the effect of the large source size was to broaden the resultant X-ray lines and hence impair measurement precision. Furthermore, line contrast was reduced and exposure times lengthened by source spread. The advent of the electron probe microanalyzer provided the investigator with an almost ideal point source of X-rays as well as a means for viewing the actual spot irradiated. The so-called capillary X-ray tube which may also be used for preparing divergent beam patterns provides a slightly larger source, requires a larger single crystal, and has no means for viewing the irradiated area. Nevertheless, it has been used with great success in a number of cases.

Another difficulty was that choosing a radiation source capable of guaranteeing the investigator the possibility of either a highly precise lattice spacing measurement and/or a convenient means for orientation was a matter of chance. This difficulty was not overcome until 1965.

Yet another problem was the means to obtain lattice spacing data for both cubic and noncubic materials from the film. New criteria for these procedures have been developed since 1962. The use of digital computer techniques has been of great value in this area.

Still other problems concerned specimen preparation, photographic technique and instrumentation.

Thus the development of the technique itself has been of prime importance in the past few years although applications have by no means been neglected. In effect, the widespread work on the technique largely completes the systematic studies begun by Kossel and his co-workers in 1934. At this point in time, the technique for obtaining and interpreting divergent beam X-ray photographs is nearly complete. Refinements and extensions will certainly be made, but it appears that the "technique phase" is essentially over and that the "applications phase" is about to begin.

Accordingly, this survey will deal largely with the technique of preparing and interpreting divergent beam X-ray patterns. Early work will for the most part be reviewed only briefly. Finally, some of the applications to date will be mentioned.

II. Early Work

The first successful experimental demonstration of a divergent beam diffraction technique was made by Rutherford and Andrade in 1914. These workers used an external radium source of γ rays and a cleaved rocksalt crystal as the target. From the photographically recorded diffraction pattern, they were able to determine the monochromatic radiation components of the source [1].

In the period 1914 to 1934, little was done owing to the necessity of constructing special wide angle X-ray tubes [2, 3]. While attempts to use conventional equipment to produce divergent beam patterns were made, the experimental arrangement was very complex [4]. Thus, twenty years after its demonstration, the divergent beam diffraction technique was still largely a laboratory curiosity.

It was at this point that Kossel and his co-workers constructed the first truly practical divergent X-ray beam generator. With the aid of this generator, a systematic experimental study of the divergent beam technique was carried out [5–10]. The results inspired von Laue to study the theory of divergent beam diffraction especially from the point of view of the interaction of the beam with the crystal planes of the sample and resultant intensity of the diffracted waves [11].

During the course of these investigations, Kossel demonstrated that the divergent beam technique could be utilized to make extremely precise lattice spacing determinations in cubic crystals [8]. The work of van Bergen, using the method outlined by Kossel, was the first experimental demonstration of the potential practical power of the divergent beam technique [12, 13].

It was in this period (1937–1941) that Fujiwara and his co-workers [14], in Japan, were developing the prototype of the capillary X-ray tube. While World War II completely ended divergent beam X-ray studies in Germany, studies in Japan were only temporarily suspended. Thus, the capillary X-ray tube was improved [15, 16], and Imura [17], became the first to study the effects of lattice imperfections on the divergent beam conics.

In England, Lonsdale studied the effect of secondary extinction on the patterns as well as the geometry of the cones and specimen preparation. Lonsdale was also the first to use the method to precisely measure a wavelength value [18–20]. Peace and Pringle [21] extended Lonsdale's work on extinction effects and used the divergent beam method to measure mosaic spread in crystals.

The first work in the United States appears to be that of Geisler, Hill, and Newkirk who were able to use standard X-ray equipment to obtain patterns. They placed a source foil ahead of the crystal and used fluorescent radiation diverging from it to prepare their patterns. Laue spots were also superimposed [22, 23]. A similar method was also developed by Borrmann [24].

In 1959, Schwarzenberger [25] used divergent beam techniques to determine the lattice spacings of beryllium. Her work appears to have been the first determination of noncubic lattice spacings by this means.

The value of the electron probe microanalyzer as a source for Kossel patterns was recognized by Castaing [26] as soon as his first instrument was completed. He presented patterns for Al and Al–Cu alloy in his Thesis in order to demonstrate the capability of the microprobe in this area.

This brief summary brings us to 1962. Some of this early work will be referred to again in presenting and extending methods for obtaining data from divergent beam patterns. Much of the work prior to 1950 was reviewed by

James [27]. In fact, it was he who referred to the divergent beam conic array as Kossel lines in honor of the pioneering aspect of Kossel's work. It is clear that James intended Kossel lines to mean those conics produced when the X-ray source was within the crystal itself. When the source is external to the crystal, Imura *et al.* [28] have referred to the conics as pseudo-Kossel lines. Usage has virtually made the names "divergent beam X-ray diffraction" and "Kossel technique" synonomous, and hereinafter they will be used interchangeably.

III. Formation of the Observed X-Ray Conic Sections

The Kossel and pseudo-Kossel methods are employed for generating divergent beam X-ray patterns. In the Kossel method, the X-ray source is excited within the specimen itself by a focused electron probe or a collimated X-ray beam of suitable energy or radioactivity. In this case, the atoms of the crystal act as independent sources of monochromatic radiation, and the waves diverging from these excited atoms will be diffracted by atom planes of the lattice. The effect of such diffraction is to give rise to a wavefield which exits from the crystal and in which the radial directions lying at the Bragg angle, θ, with respect to the diffracting planes form right circular cones. Thus, the external wavefield may be considered to lie on the loci of a set of cones rigidly fixed to the source of X-ray excitation. A flat film intercepting this system of waves shows a set of conic sections superimposed upon a general background blackening. The axis of any specific cone is the normal to its plane of diffraction; the semiapex angle of such a cone is $(90° - \theta)$, where θ is the Bragg angle.

The attenuation of the diverging primary beam is often greatly increased along the set of cone elements by the X-ray extinction effects produced within the crystal. Whether the appropriate extinction effect is purely due to primary or secondary extinction or, as is most likely, a combination of both, depends upon the state of perfection of the specimen crystal. This point will be examined in detail later. For the present discussion, it is sufficient to state that the primary beam radiation transmitted by the crystal may be deficient in intensity along the cone elements. Therefore, a film placed to intercept the transmitted radiation will often show conic sections lighter than the general background; which are called deficiency conics.

Furthermore, diffracting planes will lie at some angle, Φ, with respect to the crystal surface nearest the source (the entrance surface). If Φ is greater than the Bragg angle θ, then a diffraction conic containing an excess of energy with respect to the background may be observed in the transmission pattern. Hence, both dark and light conics may be expected on a transmission pattern.

In back-reflection, the film is placed on the same side of the crystal as the source. Diffraction conics darker than background will be observed for planes whose Bragg angle θ, is greater than the angle Φ in the case of crystals having an appreciable mosaic spread, ζ. For a perfect crystal in which ζ is negligible, conics or portions of conics may appear lighter than background as predicted by dynamical X-ray theory.

Figure 1, taken from Sharpe [29], shows the geometry for the formation of both diffraction and deficiency conics for the case of a mosaic crystal. Note

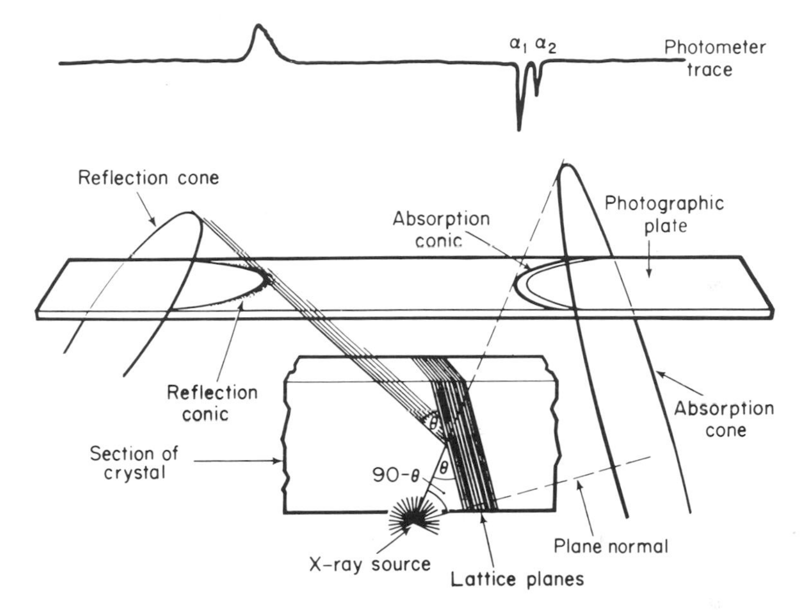

FIG. 1. Formation of Kossel conics (after Sharpe [29]).

that the diffraction conic is represented as being somewhat broadened with respect to the deficiency conic. Broadening occurs since reflection occurs from a large number of parallel planes. For each plane, there is a cone of semiapex angle $(90 - \theta)$ but all cones have vertices displaced by one interplanar distance, d. Therefore, reflection conics are generally not as sharp as deficiency conics, all of which have the X-ray source as a common vertex [27]. However, Lonsdale [19] has pointed out that reflection conics recorded on transmission patterns are nearly as sharp as deficiency conics. Reflection cones from planes near the entrance surface are lost by emergence from the surface or are lost by multiple reflection and absorption. Therefore, only cones reflected from planes near the exit surface are recorded by the film, thereby reducing the vertex displacement problem by decreasing the number of vertices.

Another extremely important feature is that each separate plane of a family is represented by a separate conic, i.e., there is no averaging effect as in a diffractometer. Thus, even in a cubic material, if (*hkl*) is seen, then ($\bar{h}\bar{k}\bar{l}$) also appears as a separate entity.

The pseudo-Kossel case is entirely analogous to the foregoing except that the source is external to the sample, i.e., the apices of the cones are external to the parent crystal. In the pseudo-Kossel method, the source is usually a thin foil of the material whose radiation is desired. This source foil is either placed directly upon or slightly above the crystal of interest; the 180° divergence of X-rays in the sample direction gives rise to the same behavior of the diffracted energy as in the Kossel case.

To continue this qualitative discussion of the divergent beam pattern, consider now the geometry of the conics observed on a film. Maier [30] gave the conic shape as a function of the Bragg angle and the angle Φ. These results suggest the type of pattern expected in that θ, Φ, and the observed conics are mutually dependent.

In the case where Φ is greater than θ, a hyperbola is formed. Diffracted radiation from planes lying such that Φ is greater than θ will not exit the crystal on the same side as the source. Hence, these conics will never be observed on a back-reflection pattern, but rather only on a transmission photograph. Figure 1 is seen to conform to this case. ($\Phi > \theta$) A deficiency conic from planes (*hkl*) is also formed and appears as a mirror image of the diffraction conic on the film. (In practice, a true mirror image may not be observed due to projection distortion effects on a flat film.) Lonsdale has pointed out that if both light and dark conics from planes (*hkl*) are recorded on the film, then complementary light and dark conics from planes ($\bar{h}\bar{k}\bar{l}$) will also be recorded. The dark line from (*hkl*) is parallel to the light line from ($\bar{h}\bar{k}\bar{l}$) and vice versa. The physical separation between the parallel complementary light and dark conics depends upon the distance from the X-ray source to the opposite side of the crystal taken in a direction perpendicular to the film [19].

For Φ equal to θ, a parabola results. In this case, both the back-reflection and transmission patterns will show diffraction conics and in the transmission pattern a complementary deficiency conic will also be present. The probability for Φ equal to θ is extremely small for any given case.

When Φ is less than θ, ellipses occur. Any diffraction conics will exit only on the same side of the crystal as the source. Thus, in back-reflection, a series of ellipse-like figures will be observed on the pattern. In transmission, the corresponding deficiency ellipses will result; they will be unaccompanied by complementary diffraction conics. As Ellis and Weissmann [31] point out, the curves are of higher order than ellipses, but the deviation is negligible.

For Φ equal to zero, a circle is obtained. This is a special case of Φ less than θ and the comments for that situation apply. It should be noted that a circular trace indicates immediately that some plane is parallel to the crystal surface which is usually placed parallel to the film.

Summarizing, a back-reflection pattern may be expected to contain largely elliptical conics with perhaps some circular traces all of which are darker than background. A transmission pattern may contain complementary light and dark hyperbolic traces, light elliptical traces, and perhaps some light circular traces. The chances of finding a parabolic trace in either type of pattern are very slight. Figure 2 shows a positive print of a transmission Kossel pattern of iron–3 wt % Si.

To this qualitative view of the divergent beam photograph must be added considerations concerning the intensity and contrast of the conics on the film with respect to background. Contrast is spoken of rather than intensity for deficiency conics since these conics as such do not represent energy deposited per unit area. Contrast is taken to mean that portion of the peak appearing above or below the general background on the film. When discussing the diffraction cones, intensity will be used as the descriptive term.

Two classes of crystals are to be considered: those having a finite mosaic spread, ζ, and those for which ζ is zero. The former will cause secondary extinction of the incident divergent X-ray beam; primary extinction may be present to some degree. The latter will cause only primary extinction to be operable and may be called "perfect" crystals for purposes of this discussion. Extinction exists when the measured integrated intensity is less than that predicted by the mosaic formula [32].

The following brief discussion of extinction is greatly simplified. The concept of primary and secondary extinction is itself rather idealized [32]. An extensive discussion and development of the topic will be found in James [27].

Primary extinction is usually said to operate when diffraction occurs from large "perfect" regions of the crystal. These regions may be called coherent domains or large mosaic blocks. In this case, the integrated diffracted intensity is not proportional to the volume of the coherent domain. The attenuation of the primary beam is large in such a block due to multiple reflections of the beam by the perfect lattice. Instead of the usual X-ray absorption coefficient, the crystal is said to have an extinction coefficient, τ, such that $I_r/I_0 = \exp -(\tau x)$. The value of τ may found from:

$$\tau = \tfrac{1}{2}\pi(l^2/mc^2)FN'\lambda. \tag{1}$$

Calculation will show that τ is usually two to five orders of magnitude greater than the ordinary linear X-ray attenuation coefficient μ.

Secondary extinction is said to be operable when the coherent domains are small enough so that in a single block, the effect of τ is negligible. However, the crystal is considered to be composed of a number of these smaller

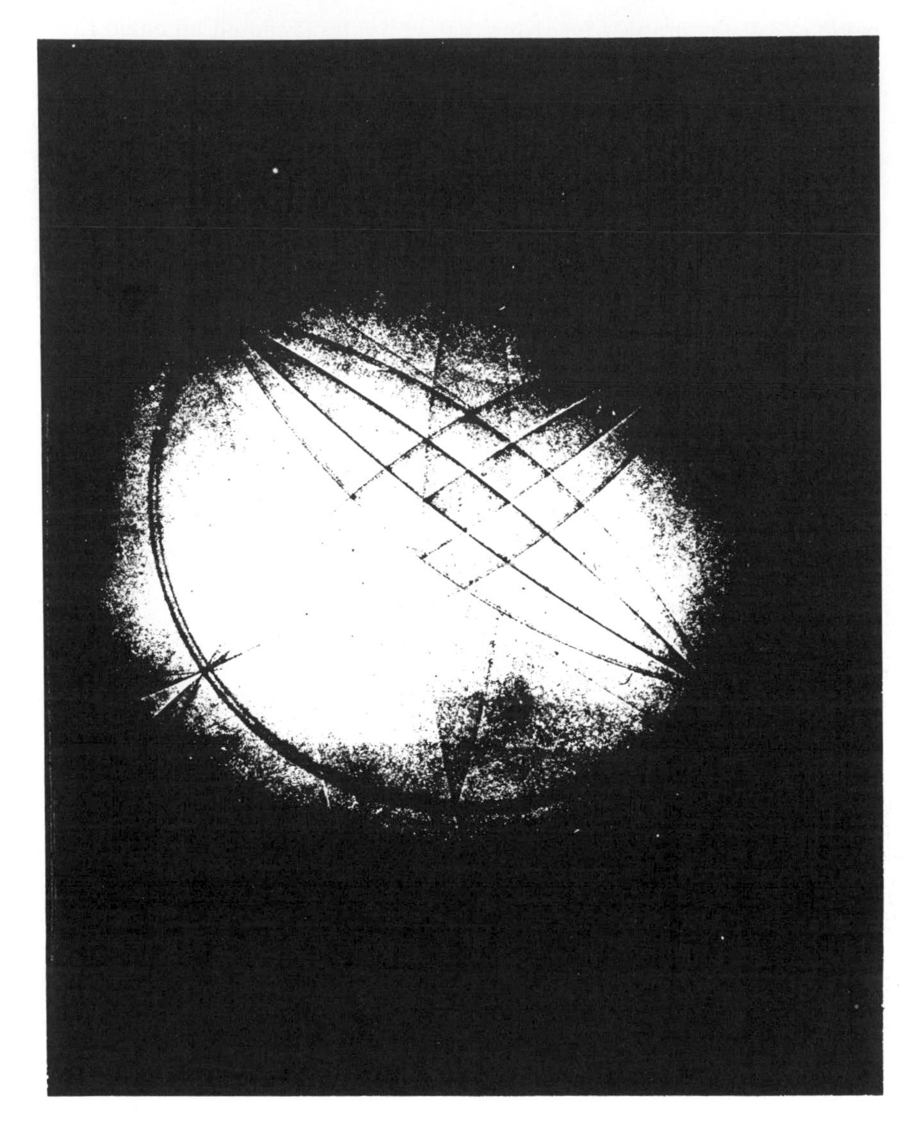

FIG. 2. Kossel transmission pattern of Fe–3 wt % Si using Fe radiation. The 110 plane is parallel to the film.

blocks most of which are closely parallel to one another. Hence, a loss of intensity in the primary beam is experienced by a robbing due to diffraction by properly oriented domains above the domain of interest [32].

When secondary extinction is dominant, the attenuation of a monochromatic ray of the primary beam incident upon the crystal at an angle $(\theta \pm u)$ where u is a small angular quantity is apparently enhanced. Peace and Pringle [21] have expressed this attenuation as

$$\exp -(\mu' x) = \exp -\{[\mu + G'(u)]x\}. \tag{2}$$

Equation (2) is in effect the probability of survival of a packet of energy [21]. It is possible to define a secondary extinction coefficient g such that

$$g \equiv [1/2(2\pi)^{1/2}] \exp -(u^2/2\zeta^2)\,. \tag{3}$$

The total amount of secondary extinction present is then given by

$$G'(u) = gQ'. \tag{4}$$

It should be noted that g is peculiar to the sample at hand and is not a general physical property. Furthermore, the mosaic spread, ζ, which is the root-mean-square of random rotations of small magnitude of the coherent domains about a common axis (assuming a normal distribution) will usually vary for different planes of the same crystal. Thus g may be expected to vary slightly from plane to plane within the crystal. The value of Q' decreases as planar index increases. Therefore, $G'(u)$ decreases as planar index increases.

Peace and Pringle deduced that the deficiency line breadth W depends upon the amount of secondary extinction present rather than being proportional to the mosaic spread. Their expression for the line breadth is

$$W = \int_{-\infty}^{+\infty} (1 - \exp[-2G'(u)x \sec\theta]\, du. \tag{5}$$

The value of W/ζ can then be plotted against $Q'x \sec\theta/\zeta$ as shown in Fig. 3. Among other things, this curve shows that the contrast of unresolved lines or the breadth of resolved lines increases very slowly with Q' where a large amount of secondary extinction is present. By resolved and unresolved is meant the angular structure of the deficiency conic being observed or not observed [21].

The observation of the angular structure depends on ζ, the crystal to film distance, and the source size. Under the usual experimental conditions for transmission Kossel photography, the crystal to film distance is about 10 cm, and the X-ray source size is about a 3 to 30 μ diameter hemisphere. For a value of ζ of 1 min of arc, the angular structure width is 30 μ at a 10 cm distance. However, Peace and Pringle [21] state that blurring due to multiple reflections also occurs causing a resolution decrease. Furthermore, the film grain size may cause blurring of the angular structure as well. Therefore, the great majority of deficiency conics observed have their angular structure unresolved. For the case of unresolved deficiency conics the contrast σ in the

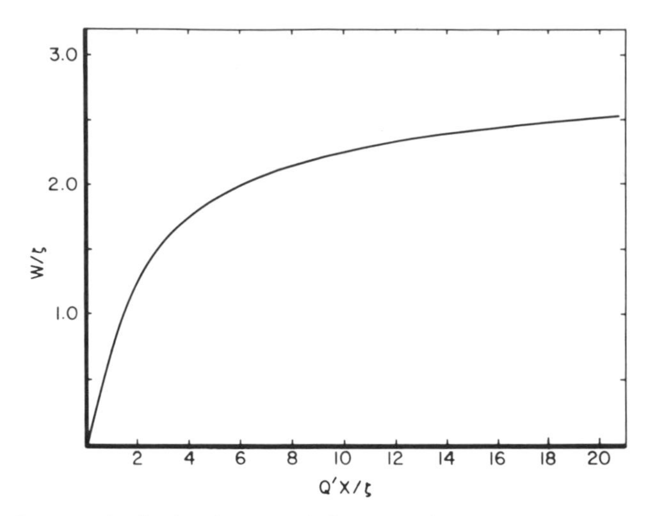

FIG. 3. Integrated reflection in transmission: Darwin solution for a Gaussian mosaic spread ζ in the symmetric case (after Peace and Pringle [21]).

presence of a background due to continuum radiation has been given by Peace and Pringle [21] as

$$\sigma = W/s[1 + (K)\exp(\mu x - \bar{\mu}x)]. \tag{6}$$

The contrast is thus a sensitive function of the crystal thickness x, the operating voltage used to produce the X rays (through $\bar{\mu}$), the X-ray source size s, the mosaic spread, the amount and distribution of both primary and secondary extinction, and, finally of the film response and exposure used. While some conclusions on proper values of x and operating voltage in order to maximize σ can be drawn (see Section VII,B), values of g are not usually known for the crystal of interest. Therefore, measurements of the contrast of deficiency conics may not give unambiguous information concerning the crystal. This point is discussed in detail at the end of Section VI.

It is now necessary to consider the intensity to be expected for diffraction conics found on a transmission Kossel pattern, i.e., from planes where $\Phi > \theta$. Again the crystal is assumed to have finite mosaic spread. Recalling when secondary extinction is dominant that $\mu' = \mu + gQ'$, the integrated intensity P for the transmitted diffraction conics is given by James [27] for the case of $\Phi = 90°$ as

$$P = (Q'x \sec\theta)(\exp -[(\mu + gQ')x \sec\theta]). \tag{7}$$

For cases where Φ is not 90°, the relation for P is of the form of Eq. (7) except that extra geometrical factors must be included [33]. In any case, for a given Q' and θ, departure of Φ from 90° lowers the value of P. Equation (7) is valid for a single reflection; however, diffraction conics are usually composed of multiple reflections overlapping one another due to the previously mentioned cone vertex displacement. Peace and Pringle were able to deduce, by a statistical argument, the intensity issuing from a given point in the range of multiple reflections. The expression depends upon the same variables as Eq. (7) [21]. Furthermore, they were able to show that the total spread S_T of the emergent beam including the direct ray is not less than

$$S_T = (x \sin \theta)/\sin \Phi \sin(\Phi - \theta) \qquad (\Phi > \theta). \tag{8}$$

If this result is coupled with Lonsdale's relation for the separation of black and white conics in transmission, the condition for black and white overlap can be found [19]. Overlap will cause intensity and contrast cancellation which in turn will make meaningful measurements involving the lines difficult if not impossible to obtain. Overlap occurs if the inequality of Eq. (9) is fulfilled.

$$(\sin \theta)/2 \sin \Phi \sin(\Phi - \theta) > (\sin 2\theta)/(\sin^2 \Phi - \sin^2 \theta) \qquad (\Phi > \theta). \tag{9}$$

It is probably most prudent to make only measurements involving deficiency conics for which $\Phi < \theta$ so that possible black line interference is eliminated.

Diffraction conics observed on a back-reflection pattern from a mosaic crystal are entirely analogous to those produced in transmission. The value of P_B is different, however [27]:

$$P_B = Q'/2(\mu + gQ'). \tag{10}$$

Equation (8) is also still applicable so that the lateral line spread can be calculated.

Consider now the mosaic crystal in which the domains are of random orientation. Lonsdale [19] states that all secondary extinction effects will be smeared out in such a case. This is tantamount to setting g equal to zero in the foregoing discussion; Q' may also increase to Q. This will lead to no deficiency conic contrast and an increased background absorption in transmission. Diffraction conics may appear in transmission and in back reflection since their integrated intensity is increased by setting g equal to zero. It might be parenthetically stated that this condition is apparently difficult to obtain since an iron foil broken in torsion gave deficiency conic contrast from a point about 10 μ from the fracture surface.

Having surveyed the intensity relations for mosaic crystals, the case of perfect crystals must be considered. As was previously mentioned, primary extinction is operable but secondary extinction is negligible. Hence, the attenuation coefficient τ is far more important than μ. Typically, the primary beam intensity loss is greater than 90% after only 2 or 3 μ depth within the crystal. Thus, deficiency conic contrast exists for thicknesses greater than this. The continuous background still appears as the primary beam intensity attenuation is exp $\bar{\mu}x$ away from Bragg angles. However, the line is extremely sharp, ranging from far less than 1 sec to perhaps 1 min of arc for a perfect crystal; the higher the planar index, the less the line breadth [27]. In this angular range, essentially total reflection of the primary beam occurs. Hence, one would expect very high deficiency conic contrast. Unfortunately, such contrast is usually not obtained.

The reasons for this lack of contrast may be severalfold. First, the source size s must be as small as practicable in order to prevent blurring of the deficiency conics and direct contrast loss [19, 21]. Second, the smearing function of the film may obliterate the line; or may cause the line to appear broadened by truncating the peak [34]. Therefore, the finest grained film possible should be employed when perfect crystals are to be investigated. Finally, if the crystal is too thick, $\exp(\bar{\mu}x)$ lightens the background and causes contrast loss.

There is the additional complication that mosaic crystals having a very low value of ζ may give essentially the same deficiency conic behavior. Hence, it may become difficult to distinguish between the perfect crystal and an imperfect one [21].

Transmitted diffraction conics will follow the dynamical theory which says that the deficiency conic and its associated diffraction conic have precisely complementary intensities. However, for the case where Φ is not 90° but $\Phi > \theta$, composite cones will result. In this case, a dark line will change to a light line because of the angular range of the emergent radiation. This effect can be quantitatively evaluated using von Laue's relations for the total excess or defect of intensity of a Kossel conic from a perfect crystal [27]. For $90° = \Phi > \theta$, the usual complementary pairs will occur.

Peace and Pringle were able to deduce the defect of transmitted intensity and the reflected intensity by assuming the perfect crystal to be nonabsorbing and that incident intensity multiplied by $(\text{breadth})^2$ could be equated to unity. The relations are for the intensity at the fractional coordinate ξ, of the exit range, in Fig. 4. Using the geometry of Fig. 4 and $y = (1 - \xi)$, they obtained

$$T(\xi) = \left\{\frac{q \sec\theta}{2d}\left(\frac{yx \csc(\Phi - \theta)}{\xi \times \csc(\Phi + \theta)}\right)^{1/2} J_1[2(\xi y A' A'')^{1/2}]\right\}^2 \tag{11a}$$

$$R(\xi) = \left\{\frac{q \sec\theta}{2d} J_0[2(\xi y A' A'')^{1/2}]\right\}^2 \tag{11b}$$

Integration of Eq. (11b) over the range of ξ yields Waller's relation for the power of the reflected beam.

Diffraction conics in back-reflection show a light border on the convex side of the dark line. This border corresponds to the lower limit of the range

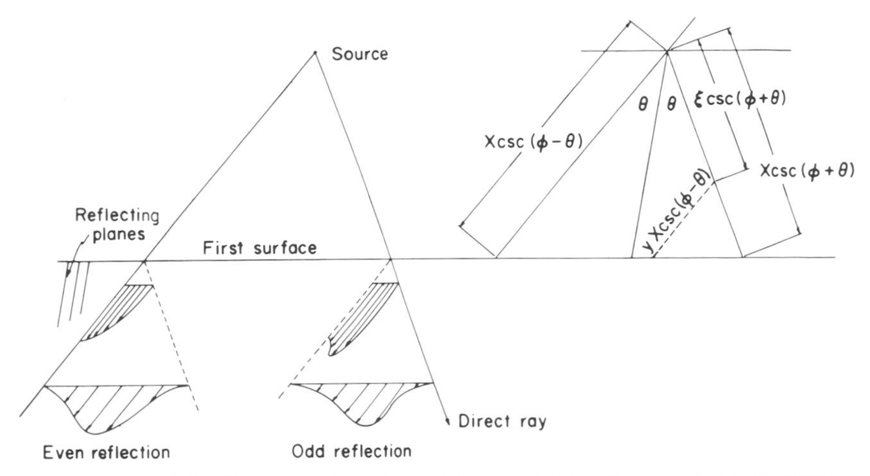

FIG. 4. Multiple X-ray reflections in a thick crystal (after Peace and Pringle [21]).

of total primary beam reflection by the perfect crystal. Photometer traces of such diffraction conics compare very well with the theoretical curves of the dynamical theory showing the variation of intensity with the incidence angle of the radiation [10, 27].

One other comment on contrast is germane: James indicates that the intensity drop-off within the cones should be proportional to D^{-2}. Thus, for equal exposures, the greater D is, the better the contrast for diffraction conics [27]. This is not true for deficiency conics in general; their contrast is relatively independent of D and may, in fact, become worse as D increases [29]. The reason is that the intensity defect of the conic is fixed by the crystal perfection. The lightened background resulting from increasing D may cause less white line contrast.

IV. PATTERN INTERPRETATION

Having considered the formation and contrast of the lines on the pattern, it is necessary to explore the means for indexing patterns, determining lattice spacings and orientations, and to consider the effect of lattice imperfections.

A. Pattern Projection

The pattern which would be observed on a spherical film surrounding the sample consists of circles, the size and position of each of which depends upon Φ, θ, and the angle between a reference diameter on the sphere and the cone axis [19]. However, it is not feasible to use spherical film, and it is almost universal practice to use a flat film placed parallel to the plane containing the X-ray source. Using the perpendicular source to film distance D as a radius, a sphere can be constructed. The film is tangent to this sphere at a point along the normal to the plane of the source and the film plane. The line joining this point and the source of X-rays will be called the projection axis; the point of intersection with the film is, in fact, the geometric pattern center. Although it is also referred to as the "film center," the point does not lie at the geometric center of the sheet of film, in general.

The pattern observed on such a film is a gnomonic projection. The projection of any circle from the sphere onto the film plane is a conic section. Figure 5 suggests the process involved. The gnomonic projection means that

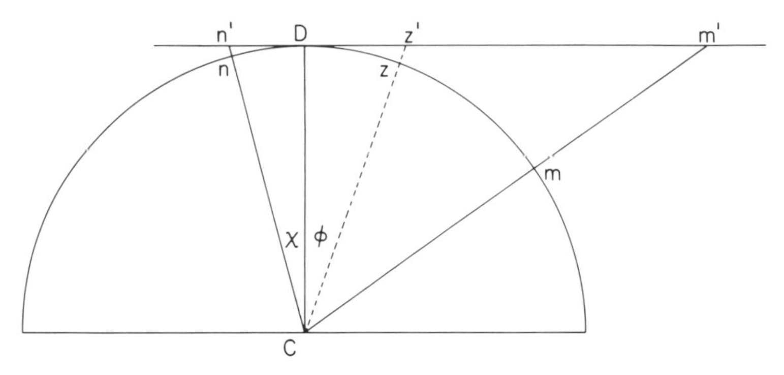

FIG. 5. Gnomonic projection of points m, z, and n into m', z', and n', respectively (after Lonsdale [19]).

the angle χ which is equivalent to arc nD on the sphere is given by $\chi = \arctan(n'D/DC)$ on the flat film.

Lonsdale [19] suggested the use of the stereographic projection to represent the expected pattern. In this case, circles on the sphere project as circles on the projection having angular radii $(90° - \theta)$. However, the projection of the circle from the sphere is not the center of the projected circle; thus, lines which are parallel on the sphere are not parallel in the stereographic projection.

To plot the projection requires that one know λ and the lattice parameters. Then $(90 - \theta)$ is obtained as $\cos^{-1}(\lambda/2d)$. It should be mentioned that no plane for which $d \leq 0.5\lambda$ will give rise to an observable conic. Furthermore, systematic conic absences must be taken into account. The next step is to plot the poles of expected conics with respect to the arbitrarily chosen pole. In the cubic system, these roles are always in the same place; for systems of lower symmetry, polar positions must be calculated for each case. Three $(90 - \theta)$

positions are plotted using the pole as center; then a circle is constructed through these points. This circle is the stereographic representation of a single plane for a single wavelength. An example is shown in Fig. 6 for a hexagonal lattice.

The preparation of such projections by hand is tedious, time consuming, and subject to large errors—especially when the projection contains many conics. This difficulty has been overcome by Frazer and Arrhenius [35] who have programmed a digital computer equipped with an $X-Y$ plotter to prepare stereographic projections for any set of lattice spacings and wavelengths.

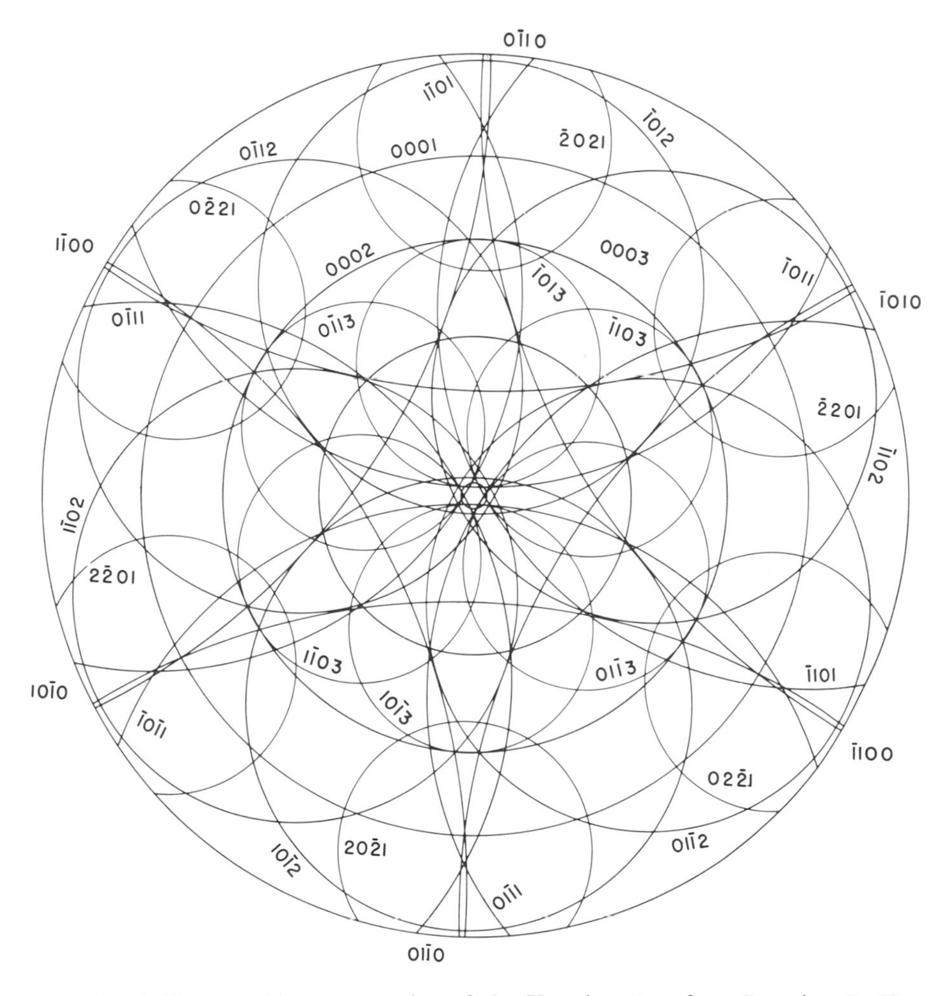

FIG. 6. Stereographic representation of the Kossel pattern from Be using Co–K_{α} radiation, ⟨00.1⟩ pole at center.

B. Indexing the Pattern

If a stereographic projection is available, indexing can be carried out directly by comparison with the film; the gnomonic and stereographic projections are quite similar over the range of angles normally recorded on the film.

In some instances, plotting of the projection can be avoided by tabulating $(90 - \theta)$ values for each conic and then comparing the radii of curvature for the conics actually recorded on the film. The index of a plane is inversely proportional to its radius of curvature. Such comparisons can be reinforced by also noting the α_1–α_2 doublet separation and/or the α–β separation. The planar index is also inversely proportional to these separations.

Yet another method consists of preparing a pseudo-Greninger chart for the source to film distance used [36]. This chart may be overlaid on the pattern and approximate $(90 - \theta)$ values for complete conics obtained directly. Furthermore, incomplete conics can be evaluated by Peters' [36] technique. In this method, coordinates of three or more points on the conics are obtained by using a polar gnomonic net; these points are then transferred to a stereographic projection and a circle is constructed through them. Hence, the pole of the conic and $(90 - \theta)$ can be found. By use of the Bragg law, the d values for all observed conics can be found; indexing is then carried out by the usual procedures. This technique is most useful for completely unknown systems.

C. Determination of Lattice Spacings

It was indicated earlier that extremely precise lattice spacing data can be obtained by the Kossel method. Achievement of the maximum possible precision, e.g., 3–20 ppm, often requires attainment of somewhat special conic intersections on the film. Thus, the geometry of the pattern can play a critical role while the intensity of conics is minor except that maximum contrast is sought.

Measurement of the lattice spacing is done by manipulating Bragg's law; several variations and their precision will be discussed.

Historically, the Kossel technique has been almost exclusively employed to measure cubic lattice spacings. Only recently have noncubic materials been reported [25, 37–39]. However, the advent of the digital computer now makes noncubic spacing determinations more feasible.

The accuracy of the determination has as its largest source of error uncertainties in the wavelength scale values used. Therefore, the quantity λ/d or for a cubic material λ/a may be measured as accurately as the Kossel technique will allow. The question of wavelength errors on the accuracy of the lattice spacing determination will be discussed separately.

1. *The Method of Kossel*

Consider any two conics from any Bravais lattice having indices (*hkl*) and (*HKL*), respectively. The interpolar angle ω between the two is a function of the lattice parameters. The angle ω is also the angle between the cone axes. The conics will intersect in a lens shaped figure on the film, be tangent, or not touch one another as the sum of $(90 - \theta)_{HKL}$ and $(90 - \theta)_{khl}$ is greater, equal to, or less than the angle ω.

For the cubic case in which ω is independent of everything except the planar indices, Kossel pointed out that for tangency to occur, the relation given as Eq. (12) must apply [8].

$$\cos^{-1}(\lambda H_0/2a) + \cos^{-1}(\lambda h_0/2a) = \cos^{-1}\{hH + kK + lL/h_0 H_0\}. \quad (12)$$

Thus, for the case of tangency, Eq. (12) could be solved for λ/a. Unfortunately, the probability for tangency is virtually zero. However, cases for which near tangency occurs between conic pairs from the cubic lattice are not uncommon [40]. The situation may be represented by Fig. 7 in which X' and Y' are the interconic distances for the cases of overlap and near-approach, respectively. The point t' is the point where tangency would occur given the proper geometric conditions. Then depending upon which case is applicable, the

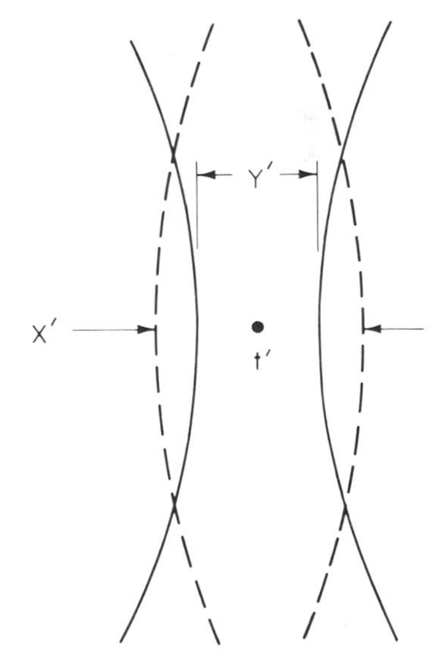

FIG. 7. Construction showing typical conic positions for Kossel's method of lattice spacing determination (after Kossel [8]).

relations of Eqs. (13) may be applied:

$$(90 - \theta) = \omega + \tan^{-1}(X'/2D) = \cos^{-1}(\lambda h_0/2a) \tag{13a}$$

$$(90 - \theta) = \omega - \tan^{-1}(Y'/2D) = \cos^{-1}(\lambda H_0/2a). \tag{13b}$$

Equations (13) are strictly true only when X' and Y' are measured along a line passing through intersection of the projection axis with the film and for conics of the same planar form. If X' or Y' does not lie along a line through the projection axis, then gnomonic projection distortion must be accounted for. If the conics are not of the same planar form, interpolation of the angle from the point of tangency to each conic is required.

The value of D is virtually never known accurately unless very elaborate measures are taken [41]. For this reason, D is usually eliminated in all Kossel lattice spacing determinations by taking the ratio of two variables each containing D as a factor. For example, by solving Eqs. (13) for X' and Y', and forming the ratio, one obtains

$$R_K = \frac{X'}{Y'} = \frac{\tan[\cos^{-1}(\lambda h_0/2a) - \omega]}{\tan[\cos^{-1}(\lambda H_0/2a) + \omega]}. \tag{14}$$

Equation (14) is the typical form in which this method is used.

There is no simple analytical treatment to solve for a nor for $\Delta a/a$ from Eq. (14). In this case, it is necessary to assume a values and compute R_K values. Then R_K versus a can be plotted. Over a reasonable range of R_K and a values such a plot is a straight line.

In order to deduce the values of $\Delta a/a$, one can define a parameter Ψ_R such that

$$\Psi_R = (a/R_K)(\Delta R_K/\Delta a). \tag{15}$$

The slope of the R versus a plot yields $\Delta R/\Delta a$ while a and R are the measured values. Hence

$$\Delta a/a = (\Delta R_K/R_K)\Psi_R^{-1}. \tag{16}$$

The value of $\Delta R/R$ can be obtained [41]. Thus, a good approximation to $\Delta a/a$ can be obtained. It is worthwhile to note that this method for $\Delta a/a$ will work for any ratio method. The parameter Ψ_R is thus, in effect, a sensitivity ratio and should be maximized.

This method was the first used to obtain lattice spacings by Kossel patterns, and it was exploited by Kossel and more extensively by van Bergen [9, 12, 13]. The most recent usage of this technique was made by Potts and coworkers [42, 43] in the study of Ge and GaAs; specific examples may be found in their work. For Ge, a Ψ_R value of 2.6 is obtained from Pott's data [40]. Since $\Delta a/a$ is claimed to be 1.4×10^{-5}, it would have been necessary to measure the ratio of lengths used to calculate a to one part in 27,000, i.e., $\Delta R/R$ of 3.6×10^{-5} for a single measurement. However, about 625 individual measurements were made and a histogram plotted of frequencies R versus a; the indicated ratio error is 1 part in 3400 for a single measurement. Thus, with the low Ψ_R value, probably little was gained from the use of Kossel patterns to obtain a [40].

In general, the precision of any method is governed by the relation

$$\Delta a/a = \cot\theta\,\Delta\theta. \tag{17}$$

which follows from the Bragg law. Therefore, in order to obtain high Ψ_R values, high index planes with large θ values are a necessity. It has been stated that, for high precision lattice spacing determinations, Ψ_R should be at least 100 [40]. This will obviate the need for many measurements in order to obtain high precision. The most desirable experimental conditions would seem to be a combination of a high and a low index conic for use in obtaining the X' and Y' lengths, respectively. Reference to Eq. (17) shows that Ψ_R will be large in such a case.

In a more recent paper, Pietrokowsky [44] has attempted to revive Kossel's method as an approach to lattice parameter determinations in iron. While the equations he used do account for projection distortion and ratios of lengths are used to eliminate D, the problem of extremely cumbersome transcendental relations for the lattice parameter a still remained. The notation in the paper is also extremely complex, thus making it difficult to obtain these relations in terms of a. In any case, the relation R_1 as used by Pietrokowsky yields

$$R_1 = \frac{\sqrt{2}\lambda_1\lambda_2 \tan\{\arccos(\sqrt{3}/3) - \arccos(\sqrt{6}_{\lambda_1}/2a)\}}{\lambda_2(a^2 - 2\lambda_1{}^2)^{1/2} - \lambda_1(a^2 - 2\lambda_2{}^2)^{1/2}}$$

in which λ_1 and λ_2 are the wavelengths of Fe–Kα_1 and Fe–Kα_2 respectively, and R_1 is the ratio of the distances of closest approach of the {112} Fe–Kα_1 to {220} Fe–Kα_1 conics and {220} Fe–Kα_1 to {220} Fe–Kα_2 conics respectively.

Although it is not stated explicitly, the ψ_R method described in detail by Gielen *et al.* [40] was used by Pietrokowsky to evaluate the sensitivity. For Pietrokowsky's R_1 ratio, the value of ψ_R is 72 which is considerably below the requirements for a highly precise lattice spacing determination. It is worth mentioning that Gielen *et al.* [40] investigated the use of Kossel's method for iron–3 wt% Si but abandoned that approach after ascertaining the cumbersome nature of the expressions for a and the low ψ_R values to be expected.

The method of Kossel is not recommended for noncubic cases. The noncubic analog of Eqs. (12)–(14) can be written, but values of the lattice parameters required a priori in order to calculate the ω value. Hence, all reasoning becomes circular.

2. *Lens Methods*

Lens methods make use of the overlapping of conics to form lens-shaped figures on the film. Hanneman [45, 46] was the first to recognize the possibilities of the lens as an aid to determine lattice spacings. Heise [47] later refined Hanneman's method to the use of two lens length measurements in order to eliminate *D*. However, the complete evaluation of the lens in a general form was only later performed by Gielen *et al.* [40].

The general aspect of the lens-shaped intersection is shown in Fig. 8.

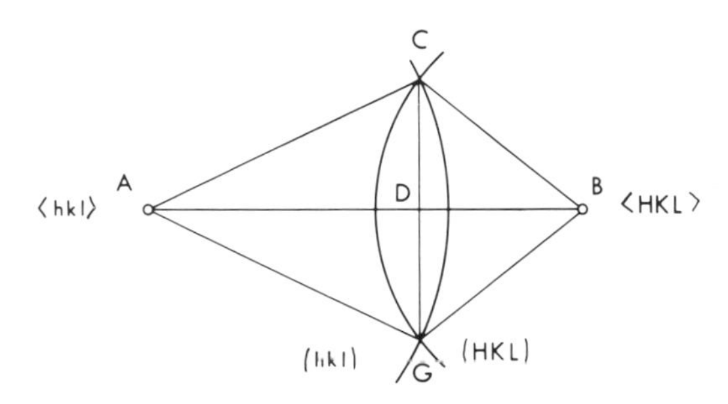

FIG. 8. Typical lens-shaped intersection of conics.

$$\cos^2 CD = \cos^2 \gamma = \cos^2 AC + \left[\frac{\cos BC - (\cos AB)(\cos AC)}{\sin AB}\right]^2$$

The relationship between the angular lens length γ and the component spherical triangles has been obtained [40]

$$\cos^2 \gamma = \cos^2 CD = \cos^2 AC + \left\{\frac{\cos BC - (\cos AB)(\cos AC)}{\sin AB}\right\}^2. \tag{18}$$

In Eq. (18), $\cos AC = \cos(90 - \theta)_{khl}$, $\cos BC = \cos(90 - \theta)_{HKL}$ and AB is the interpolar angle ω. Hence, the half-angular lens length γ can be related to the lattice spacing d for any Bravais lattice. For the general case, let λ be the wavelength associated with (HKL) of spacing d, and λ_1 that associated with (hkl) of spacing d_1. Furthermore, assign $h_0 \geq H_0$; this convention will be retained throughout. By substitution in Bragg's law and then into Eq. (18), the general lens relation can be obtained:

$$\cos^2 \gamma = \left[\frac{d^2\lambda_1{}^2 - 2dd_1\lambda\lambda_1 \cos \omega + d_1{}^2\lambda^2}{4d^2d_1{}^2(1 - \cos^2 \omega)}\right]. \tag{19}$$

Equation (19) contains one unknown only in the cubic case. Relations for noncubic Bravais lattices can be written using Eq. (19). However, unless D is accurately known, one is usually required to have two independent equations for each unknown d; this is also true in the cubic case.

The relations for noncubic systems have been obtained, but they are extremely cumbersome. Furthermore, obtaining the required number of suitable intersections on the same film in a fashion such that gnomonic projection distortion effects can be taken into account may prove difficult. For these reasons, lens methods cannot always be recommended as the best general means for obtaining noncubic lattice spacings by the Kossel technique although at least some noncubic cases have been solved in this fashion [37, 39].

The lens method is, however, particularly suited to the determination of cubic lattice spacings. Several materials such as Ni, Fe–3 wt % Si, LiF and Al have been studied with success [40, 41, 46, 48].

For the cubic case, Eq. (19) reduces to

$$\cos \gamma = [(\csc \omega)/2a](\lambda_1{}^2 h_0{}^2 + \lambda^2 H_0{}^2 - 2\lambda_1 \lambda H_0 \cos \omega)^{1/2}. \tag{20}$$

Hence, the lattice parameter a is expressed in terms of the angle "γ" which is unknown and the wavelengths and indices which are presumed known. Thus,

$$a = f \sec \gamma \tag{21}$$

where f is obtained from Eq. (20) and γ is the half-angle of the lens-shaped intersection on the film [41].

By differentiation of Eq. (21), the quantity $\Delta a/a$ for a given lens is

$$\Delta a/a = \tan \gamma \, \Delta\gamma. \tag{22}$$

In general, Eq. (22) indicates that lens-shaped intersections yielding small γ values should be sought for measuring highly precise cubic lattice parameters.

Equation (21) indicates that all that needs to be done is to obtain "γ" with high precision in order to express the lattice parameter with a high degree of precision. However, obtaining "γ" precisely may present somewhat of a problem.

Any Kossel pattern obtained on a flat film is a gnomonic projection of a reference sphere; such a reference sphere has the source of divergent radiation at its center and has as a radius the source-to-film distance. The film is arranged parallel to a set of crystallographic planes. A measured lens length on the flat film must be related to the true angular lens length on the sphere.

The pertinent goemetrical relations for a lens lying in an arbitrary position with respect to the film center are shown in Fig. 9. The film center is the intersection of a normal to the film from the source point. Therefore, the pole of a crystallographic plane exactly parallel to the film will be at the film center. The conic corresponding to such a plane will be a circle on the film.

The linear distance L in Fig. 9 has been related to the source-to-film

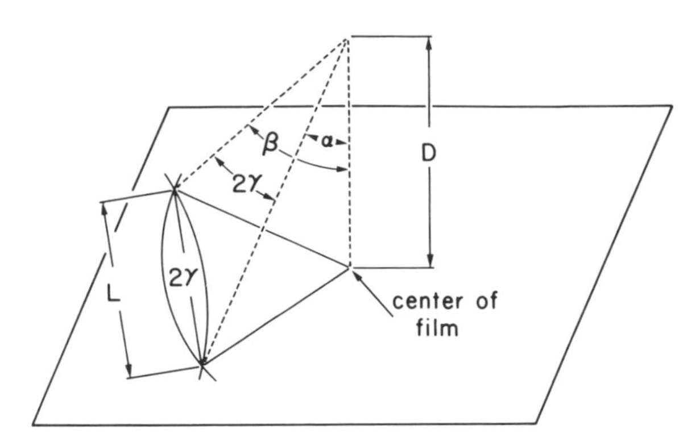

FIG. 9. Gnomonic projection effects: general case ($\alpha \neq \beta \neq \gamma$).

distance D and the angles, α, β, and the angle of interest γ by Gielen *et al.* [40] as:

$$L = D[(\sec \alpha + \sec \beta)^2 - 4 \sec \alpha \sec \beta \cos^2 \gamma]^{1/2}. \tag{23}$$

Unfortunately, in attempting to express the right side of Eq. (23) in terms of the lattice parameter, wavelengths and indices, sets of spherical triangles are obtained which cannot be solved uniquely for γ. Hence, a single lens lying in the general position with respect to the film center is an analytically insoluble case.

For this reason, it is necessary to seek special orientations of lenses with respect to the film center. There are three useful orientations in which the lens length L can be expressed rigorously in terms of the lattice parameter, wavelengths, and indices. Furthermore, in anticipation of these expressions, it may be stated now that the expressions depend on having some pole coincident with the center. Therefore, it is an experimental necessity to provide some means to alter the orientation of the specimen and to accurately mark the film center [41]. These two requirements will be discussed later.

The three soluble cases have been named the tangential case (Fig. 10), the isosceles case (Fig. 11), and the radial case (Fig. 12) [40]. Considering first the tangential case, the film center lies at the midpoint of the line connecting the lens intersections. Under these conditions $\alpha = \beta = \gamma$ in Fig. 9 and L may be expressed trigonometrically as

$$L = 2D \tan \gamma. \tag{24}$$

Using Eq. (21), one obtains

$$L = 2D[f_T^{-1}(a^2 - f_T^2)]^{1/2} \tag{25}$$

where f_T is the f value from Eq. (20) for a tangential lens.

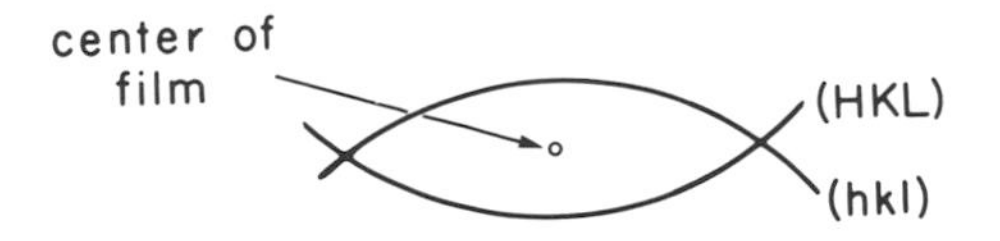

FIG. 10. Gnomonic projection effects: tangential case ($\alpha = \beta = \gamma$).

$$L_{2\gamma} = 2D \tan \alpha, \qquad L_{2\gamma} = 2D \frac{(a^2 - f_T{}^2)^{1/2}}{f_T}$$

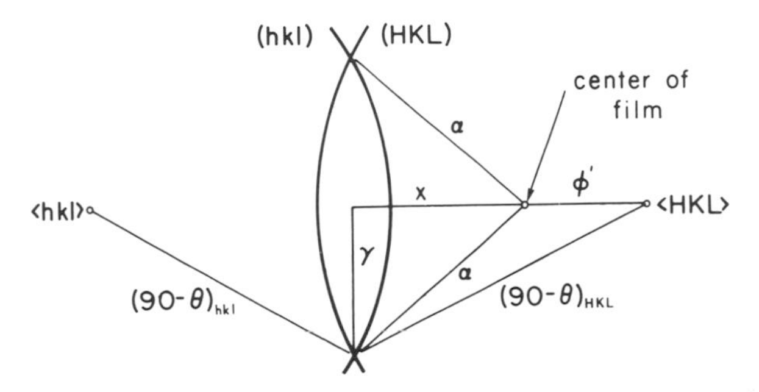

FIG. 11. Gnomonic projection effects: isosceles case ($\alpha = \beta = \gamma$).

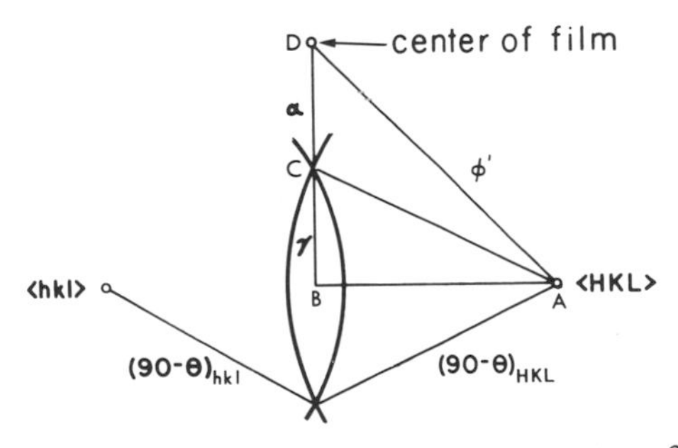

FIG. 12. Gnomonic projection effects: radial case ($\alpha = \beta \neq \gamma$).

Proceeding to the isosceles case (Fig. 11), it is seen that $\alpha \neq \beta \neq \gamma$ in Fig. 9. The trigonometric expression for L in Gielen *et al.* [40] is

$$L = 2D \sec \alpha \sin \gamma. \tag{26}$$

This may be transformed trigonometrically to

$$L = 2D \sec X \tan \gamma. \tag{27}$$

Performing the required algebra and trigonometry leads to the analytical expression of Eq. (27) as

$$L = 2D\left[\frac{2(a^2 - f_{\mathrm{I}}{}^2)^{1/2}}{\lambda H_0 \cos \phi' + \sin \Phi'(4f_1{}^2 - \lambda^2 H_0{}^2)^{1/2}}\right] \tag{28}$$

where f_I is the isosceles lens f. If the requirement that some pole is at the position of the film center, has been satisfied, Φ' is known unambiguously.

The final case of interest is the radial case (Fig. 12). This corresponds to $\beta = (\alpha + 2\gamma)$ in Fig. 9. The trigonometric equation for $L_{2\gamma}$ in Gielen *et al.* [40] is

$$L_{2\gamma} = D\{\tan(\alpha + 2\gamma) - \tan\alpha\}. \tag{29}$$

While it is possible to express $L_{2\gamma}$ analytically, the result is so cumbersome that it is better to express the length associated with the angle α, i.e., the distance between the film center and the lens extremity nearest the film center, analytically. The appropriate trigonometric relation is

$$L_{2\gamma} = D\tan\alpha. \tag{30}$$

Reduction of Eq. (30) to analytical form results in

$$L = D\left[\frac{a^2}{(f_R{}^2 - a^2)(K^2 - 1) + f_R K\{f_R K + 2[(f_R{}^2 - a^2)(K^2 - 1)]^{1/2}\}} - 1\right] \tag{31}$$

where f_R is f for the radial lens, and

$$K = (2f_R \cos\Phi')/\lambda H_0.$$

Again note that there must be some pole at the film center in order that Φ' can be found unambiguously.

Thus, lenses conforming to the tangential, isosceles, and radial cases can be rigorously solved if one can properly orient the sample and accurately mark the film center. The orientation requirement can be met fairly well by means of an ordinary Laue procedure; however, it is convenient and far more accurate if there is a goniometric device mounted directly on the sample holder or stage. The film center can be marked in a simple fashion devised by Ogilvie [40]. In this method a small hole is drilled in the sample mount. With the gun voltage off and the filament current on in the electron beam column, the center of the hole is then brought to the coordinate position of the beam. The light from the filament is allowed to strike the film for a few seconds; this impresses a spot on the film center whose diameter is on the order of a few minutes of arc. It is necessary that the film be perpendicular to the beam path. If a circle is present on the pattern, the spot must lie at its center. A good test is to measure several diameters; if a true circle exists, its center is the pattern center.

The source-to-film distance D appears as a factor in each of the equations relating a measured length L to the lattice parameter. In general, it is not possible to reproduce values of D, in any given experimental arrangement, with the accuracy required although Morris has claimed sufficient accuracy through the use of a depth micrometer [41, 48]. It has also been suggested that a material whose lattice spacing is known be used as standard for D [28].

In order to eliminate the D parameter, one takes the ratio of two lens lengths. As long as both lenses conform to soluble cases, such a procedure is completely rigorous. It might appear that obtaining two lenses conforming to the orientation requirements on the same pattern is unlikely. However, at least several materials yield patterns for which a number of lenses satisfy the requirements. A few examples are Fe under Fe–K radiation, Fe under Mn–K radiation, LiF under Cu–K radiation, Al under Ta–L radiation, and Ni under Ni–K radiation. There are many others.

Taking the ratio of any two of the equations (25), (28), or (31) leads to a relation for the lattice parameter which, when both legs of each of the lenses are formed by wavelengths λ and λ_1, respectively, takes the form given in Eq. (32). Even if four wavelengths are involved in forming the two lenses, a somewhat analogous relation develops which can be handled straightforwardly; the mathematics are, however, somewhat more cumbersome since u', v, and w, in Eq. (32) are functions of wavelength values. This is also true in the radial case.

$$a = (\lambda\lambda_1/u')[(vR^2 - w)/(\lambda^2 R^2 - \lambda_1{}^2)]^{1/2}. \tag{32}$$

Here, R is the ratio of two lens lengths L_1 and L_2, respectively, chosen such that $L_1 > L_2$, i.e., $R > 1$ and u', v, and w are numerical constants dependent upon the planar indices and which come directly from the equations relating the measured length and the lattice parameter. Hence, u', v, and w are calculable. The nature of Eq. (32) is such that as R increases a decreases.

The expected precision may be obtained by differentiating Eq. (32) with respect to R

$$\left|\frac{\Delta a}{a}\right| = \left|(R\,\Delta R)\left[\frac{v}{(vR^2 - w)} - \frac{\lambda^2}{(\lambda^2 R^2 - \lambda_1{}^2)}\right]\right|. \tag{33}$$

In the case in which $\lambda = \lambda_1$ this reduces to

$$\left|\frac{\Delta a}{a}\right| = \left|\frac{R\,\Delta R\,(w - v)}{(R^2 - 1)(vR^2 - w)}\right|. \tag{34}$$

Since $R = L_1/L_2$,

$$\Delta R = \left[\frac{(\Delta L_1\, L_2)^2 + (\Delta L_2\, L_1)^2}{(L_2{}^2)^2}\right]^{1/2}. \tag{35a}$$

For a good comparator or other measuring device $\Delta L_1 \approx \Delta L_2$. Using this approximation

$$\Delta R = (\Delta L/L_2)(1 + R^2)^{1/2}. \tag{35b}$$

In practice, one measures an RMS value of ΔL and combines Eqs. (33) and (35b) to obtain a measure of the sensitivity.

Some general statements about the precision may now be made. A large value of R is desired since Eq. (33) shows that $\Delta a/a$ is approximately inversely proportional to the first power of R. Thus, consistent with Eq. (22), a small value of γ_2 leading to a small L_2 is required.

Equation (33) represents the geometric accuracy of a determination of λ/a, or the precision is determining the lattice parameter a. This assumes that the error due to a slight displacement of a pole from the film center is negligible compared to that derived from Eq. (33). In order to test this assumption, the effect of lateral displacements of the film center can be approximated. The results indicate that the increase in $\Delta a/a$ predicted by Eq. (33) is less than 4% if orientation can be carried out to within $\frac{1}{2}°$, a feasible requirement [41].

A possible increase in the error can be obtained through the actual length measurement of the lens extremities on the film. Such lens intersections are generally not points. In fact, the uncertainty in determining the exact intersection position increases as the contributing conics become more oblique. While Eq. (22) indicates that the Bragg angles of the conics forming the lens do not play a role in the value of $\Delta a/a$, high index, i.e., high θ, conics, in fact, give better results. The high θ conics give "thick" lenses whose extremities are well defined, thus reducing the chances of measurement errors due to an uncertainty in the exact position of intersection.

Hanneman [45, 46] devised a method for the use of a single lens coupled with a measurement standard on the film. Thus, the two lens requirement is obviated. However, the measuring standard must lie very nearly athwart the lens; if not, the use of the standard to remove the projection distortion fails [40].

The usual standard is the angular separation, Δp, between α, and β conics (Fig. 13). The relation for γ is

$$\gamma = (L_{2\gamma}/2L_p)\,\Delta p. \tag{36}$$

The lattice parameter is found from Eq. (20). Then Δp is recalculated using this a value; reiteration proceeds until a converges.

A basic tenet of this method is that Δp must change much more slowly than does γ for a given change in a. The exact relationship can be obtained by differentiation of (Δp) with respect to a combined with Eq. (22).

$$\left|\frac{d\gamma}{d(\Delta P)}\right| = \left|\frac{\cot\gamma}{\cot(90-\theta)_{\alpha_1} - \cot(90-\theta)_{\beta}}\right|. \tag{37}$$

Equation (37) shows that $d(\Delta P)$ is proportional to θ; hence, the lowest possible index conics should be chosen for the measurement standard. Equation (37) reconfirms that a low γ value is also required. For the case of

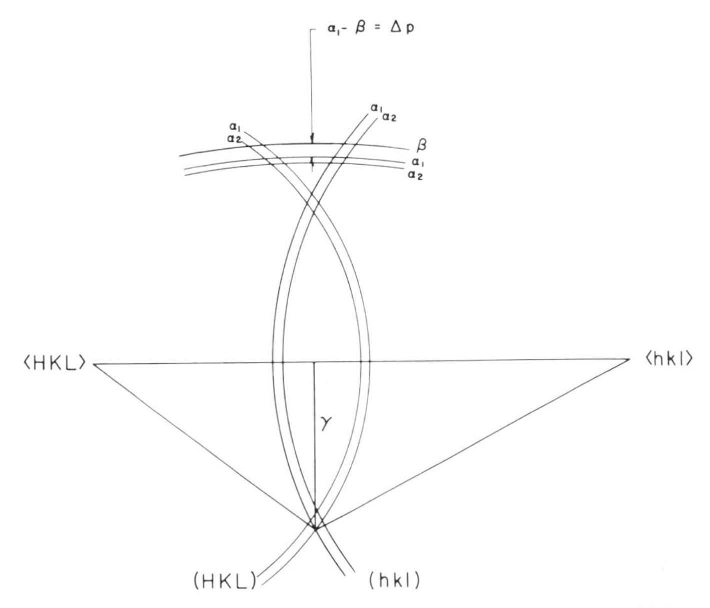

FIG. 13. Lens with measuring standard ΔP (after Hanneman [45]).

Ni that Hanneman used, $d\gamma/d(\Delta p)$ is about 10^3 which justifies the assumption that $d(\Delta P)$ is virtually negligible.

The relation for $\Delta a/a$ is cumbersome since $d(\Delta P)$ appears. Therefore, a plot of $L_{2\gamma}/L_p$ versus a is suggested. From this, Ψ_R can be obtained and $\Delta a/a$ computed from Eq. (16).

3. *Use of Lens Methods for Cubic and Noncubic Lattices*

In order to simplify the lens relations and to make optimum use of the lens method the following technique called the two-circle method has been adopted. Whenever possible, matters are arranged so that a small circle of λ equal to K_{α_1} is placed exactly at the film center. If possible $(90 - \theta)$ should be $5°$ or less. Under these conditions, two tangential relations are usually possible—the diameter of the K_β circle as L_1 and the diameter of the K_{α_1} circle as L_2 or alternately, the diameter of the K_{α_2} circle as L_2. Furthermore, isosceles lenses with the K_{α_1} circle are often formed by other conics, thus permitting several independent determinations of the lattice spacing [41]. It is often possible to find a suitable high θ conic. It is necessary to be able to orient the crystal so that this conic lies at the pattern center. If possible, one arranges matters so that a conic depending only on one of the parameters is the small circle whose center coincides with the pattern center. Then that parameter can be determined as if the material were cubic. Having obtained one parameter, other lenses can usually be used on the same film to obtain the other parameters. Alternatively, the methods of Morris and/or Mackay (pp. 391–394) can be used; there procedures are greatly simplified if one parameter is already available. The application of this method to hexagonal systems using beryllium as an example has been described in detail [39].

Recently, Morris proposed a modification of the lens method suitable for any Bravais lattice. Approximate lattice parameters are required beforehand as well as the exact location of the film center and an accurate value for D. A computer program yields the proper wavelength choice for obtaining a lens having a small γ value. The value of $\sec \gamma$ in terms of the lattice parameters is related to $\sec \gamma_0$, the value obtained from the approximate parameters, by means of a Taylor series expansion. Furthermore, $\sec \gamma$ is determined experimentally from the relation

$$\sec^2 \gamma = \frac{4ab'}{(a + b')^2 - L^2}$$

in which L is the lens length; P_1 and P_2 are the distances from the film center to the two points of the lens extremities, respectively; $a = (P_1{}^2 + D^2)^{1/2}$; and $b' = (P_2{}^2 + D^2)^{1/2}$. Then, the experimental value of $\sec \gamma$ is combined with the Taylor series approximation to yield the difference between the assumed and the measured parameters. Reiteration yields more accurate differences; the measured lattice parameter is then the sum of $a_0 + \Delta a$, $b_0 + \Delta b$, etc. The actual computations are performed by computer [49].

There is no simple method for evaluating the precision. Morris suggests that after determining the lattice parameters, the calculations be repeated using a second set of data representing typical deviations from the original set. This procedure yields a good estimate of the errors.

4. *Intersection Methods*

There are often a number of places on a Kossel pattern at which three or more conics intersect or very nearly intersect in a point (Fig. 6). Such intersections are of two types—invariant and accidental. Invariant intersections depend only upon the crystal geometry. The higher the crystal symmetry the greater the number of possible invariant intersections.

Frazer and Arrhenius [35] have demonstrated that for an invariant intersection to occur a definite proportionality must exist between the Miller indices of the conics participating and their respective interplanar spacings.

$$\frac{h_1 d_1{}^2 - h_2 d_2{}^2}{h_2 d_2{}^2 - h_3 d_3{}^2} = \frac{k_1 d_1{}^2 - k_2 d_2{}^2}{k_2 d_2{}^2 - k_3 d_3{}^2} = \frac{l_1 d_1{}^2 - l_2 d_2{}^2}{l_2 d_2{}^2 - l_3 d_3{}^2}. \tag{38}$$

Equation (38) has two other requirements: namely that the conics intersect in the first place and that the participants come from planes belonging to the same crystallographic zone.

Physically, an invariant intersection occurs when the reciprocal lattice points corresponding to the participating conics and the reciprocal lattice point corresponding to the origin of the Ewald sphere all lie on the circumference of the same circle [50]. Invariant intersections may be of use in providing information about the symmetry or the Bravais lattice of a specimen.

Accidental intersections or near intersections depend strongly upon the exciting wavelength and the interplanar spacings of the participants. By measuring the deviation from exact intersection, the lattice spacings may be determined. The higher the indices of the participants, the greater the motion of the intersection as a function of lattice spacing variation (Eq. (17)). The first to employ this technique was Lonsdale [19] who measured the lattice parameter of diamond with a precision of about 15 ppm. For cubic materials, Geisler *et al.* [23] have tabulated the coordinates of many invariant intersections with respect to the $\langle 001 \rangle$ pole.

Mackay [38, 51] was able to obtain equations relating the indices, lattice spacings, and the coordinate position of accidental as well as invariant intersections. The approach was to obtain the equation of the intersection of the sphere of radius D with that of the plane containing the Kossel circle on the sphere—the Kossel plane which is parallel to the reflecting plane of index (hkl). The form of the equation for orthogonal lattices is

$$\frac{hB}{a} + \frac{kF}{b} + \frac{lM}{c} = \frac{\lambda}{2d^2} \tag{39}$$

in which (BFM) is the coordinate position for exact intersection of three or more conics. Thus, for orthogonal axes

$$B^2 + F^2 + M^2 = 1. \tag{40a}$$

For hexagonal axes

$$B^2 - BF + F^2 + M^2 = 1. \tag{40b}$$

In practice, it is convenient to express Eq. (39) as a determinant whose denominator, δ, is for any Bravais lattice

$$\delta = \begin{vmatrix} h_1 & k_1 & l_1 \\ h_2 & k_2 & l_2 \\ h_3 & k_3 & l_3 \end{vmatrix}. \tag{41}$$

For example, the hexagonal and tetragonal systems yield

$$B = (2ac^{-2})(\delta^{-1}) \begin{vmatrix} k_1 & l_1 & \varepsilon_1 \\ k_2 & l_2 & \varepsilon_2 \\ k_3 & l_3 & \varepsilon_3 \end{vmatrix}, \tag{42a}$$

$$F = (2ac^{-2})(\delta^{-1}) \begin{vmatrix} h_1 & l_1 & \varepsilon_1 \\ h_2 & l_2 & \varepsilon_2 \\ h_3 & l_3 & \varepsilon_3 \end{vmatrix}, \tag{42b}$$

$$M = (2a^{-2}c)(\delta^{-1}) \begin{vmatrix} h_1 & k_1 & \varepsilon_1 \\ h_2 & k_2 & \varepsilon_2 \\ h_3 & k_3 & \varepsilon_3 \end{vmatrix}, \tag{42c}$$

$$\text{hexagonal:} \quad \varepsilon = \lambda[\tfrac{4}{3}(h^2 + hk + k^2)c^2 + a^2l^2], \tag{42d}$$

$$\text{tetragonal:} \quad \varepsilon = \lambda[(h^2 + k^2)c^2 + a^2l^2]. \tag{42e}$$

The extension of Eqs. (42) to the orthorhombic case or their reduction to the cubic case is straightforward.

If there is no intersection of the three conics, B, F, and M are infinite. Furthermore, if the intersection is invariant, B, F, and M are indeterminate, i.e., $B = F = M = 0/0$. Only for an accidental intersection does one obtain real and positive results [51].

Combination of the relations for B, F, and M with Eq. (40) yields an equation involving the lattice parameters and the wavelength for which *exact* intersection *would* occur. As many of these equations as there are lattice parameters are required, i.e., for a tetragonal lattice, at least *two* independent accidental intersections must be analyzed.

The method consists of determining the point where exact intersection would occur. Then using the α_1–α_2 doublet separation and/or α–β separations as measuring standards, in a manner analogous to that used by Hanneman, the wavelength for which exact intersection would have occurred is determined by interpolation. According to Mackay [51], extrapolation leads to unacceptable errors. As many such wavelengths as lattice parameters are required. Then the resulting simultaneous relations are solved for the lattice parameters [25, 38].

In the cubic case, the lattice spacing can be obtained directly by interpolating the exact coincidence point position between the chosen measuring standards. Sharpe gives several examples for the case of MgO under Cu–K radiation [29].

This method has the advantage that projection distortion effects can essentially be neglected since measurement does indeed take place athwart the measuring standard. However, a disadvantage is that there is no simple way to evaluate the general precision to be expected. Qualitatively, it would appear that the participants should contain at least one high index conic in order that the motion of the intersection be large for small changes in lattice spacing.

For this reason, lens methods are probably better suited for the cubic case. However, the intersection method holds promise for the noncubic cases—especially when combined with the Frazer–Arrhenius program for stereographic projection plotting.

A means to obtain noncubic lattice spacings can be listed:

(1) Plot the projection for a given set of radiations by computer—approximate lattice parameters are needed from an outside source.

(2) Locate good accidental intersection points on the plot; these can be tested by Eq. (38) or the matrices.

(3) Enlarge the regions of interest by computer replotting.

(4) Perform the experiment and obtain the λ values for exact intersection.

(5) Using Eq. (42) or others as appropriate, calculate the lattice parameters.

(6) Replot using the parameters from (5) and the wavelengths from (4)—check for exactness of intersection.

(7) Replot using the parameters from (5) with the actual λ values used. Photographic pattern should match results in all symmetry regions.

The great power of the computer plotting program is seen in that it allows a search for radiations yielding the desired intersections and the proper orientation required to place the intersections on the film automatically.

Mackay's relation (Eq. (39)) has been generalized by Morris [49]. The equation of any conic on a flat film obeys the relation

$$l_1 X + l_2 Y + l_3 Z = \sin\theta (X^2 + Y^2 + Z^2)^{1/2}. \tag{43}$$

The pattern center is taken to be the origin, i.e., $X = Y = 0$, while Z is equivalent to the source-to-film distance D. The direction cosines of the pole of the diffracting plane with (X, Y, Z) axes are (l_1, l_2, l_3), respectively. A traveling microscope is used to measure the (X, Y) coordinates of three or more points lying on a conic; then l_1, l_2, l_3 and $\sin\theta$ are obtainable as is d if λ is known. If more than three (X, Y) values are available, a least squares fit gives the l, θ, and d values. This regression method is programmed for computer solution.

It is convenient to express Eq. (43) as a determinant with denominator δ'

$$\delta' = \begin{vmatrix} p_1' & q_1 & r_1 \\ p_2' & q_2 & r_2 \\ p_3' & q_3 & r_3 \end{vmatrix} \tag{44}$$

in which $p_1' = X_1/(X_1^2 + Y_1^2 + Z_1^2)^{1/2}$, $q_1 = Y_1/(X_1^2 + Y_1^2 + Z_1^2)^{1/2}$, and so on to $r_3 = Z_3/(X_3^2 + Y_3^2 + Z_3^2)^{1/2} = D/(X_3^2 + Y_3^2 + D^2)^{1/2}$.

For any conic of interest

$$l_1 \csc \theta = U\delta' = \begin{vmatrix} q_1 & r_1 & 1 \\ q_2 & r_2 & 1 \\ q_3 & r_3 & 1 \end{vmatrix}, \tag{45a}$$

$$l_2 \csc \theta = V\delta' = \begin{vmatrix} p_1' & r_1 & 1 \\ p_2' & r_2 & 1 \\ p_3' & r_3 & 1 \end{vmatrix}, \tag{45b}$$

$$l_3 \csc \theta = W\delta' = \begin{vmatrix} p_1' & q_1 & 1 \\ p_2' & q_2 & 1 \\ p_3' & q_3 & 1 \end{vmatrix}. \tag{45c}$$

Since θ is identical for the same conic, Eqs. (45) can be solved for d using the Bragg law and the fact that for orthogonal axes

$$l_1^2 + l_2^2 + l_3^2 = 1. \tag{46}$$

The value of d is

$$d = (\lambda/2)(U^2 + V^2 + W^2)^{1/2}. \tag{47}$$

Morris' method requires the exact position of the pattern center to be known since all (X, Y) coordinate values depend upon that point. Furthermore, the value of D is required as accurately as is obtainable. The only other source of uncertainty is the length measurements for X and Y. Note that potential gnomonic projection distortion problems are eliminated by choosing the pattern center as the origin.

This method offers another means for obtaining noncubic lattice spacings by means of Kossel patterns. A great potential advantage is that d values can be obtained for an unknown Bravais lattice without prior knowledge of some crystallographic data such as the indices of the conics. The value of this capability in the study of small phases *in situ* is immediately apparent. Morris' method is the analytical analogue of Peters' graphical method for obtaining d values from completely unknown systems and the end results of Maier's [30] original attempts to write general equations for conics on a flat film.

Isherwood and Wallace [52] have described a means for determining precise lattice spacings by means of the divergent beam method which appears to show its greatest promise for the study of perfect cubic crystals. The method makes use of the fact that if a primary beam is reflected from a set of planes $h_1k_1l_1$, it is sometimes possible for the reflected beam to have the exact direction to be reflected from another set of planes, $h_2 k_2 l_2$. Furthermore, the direction of this doubly reflected beam is the same as that of a beam reflected directly by the planes $h_3 k_3 l_3$ where $h_3 = h_1 + h_2$, $k_3 = k_1 + k_2$, and $l_3 = l_1 + l_2$. In addition, in the divergent beam method, the primary beam may also reflect directly from $h_2 k_2 l_2$.

If $h_3 k_3 l_3$ has a very small or zero structure factor, then the double diffraction scheme produces an apparent reflection from $h_3 k_3 l_3$. According to Isherwood and Wallace [52], these reflections give high contrast and are readily obtained from perfect crystals. In addition, the direction of the $h_3 k_3 l_3$ doubly produced reflection is exactly defined; if it were not, the reflection would not appear since the structure factor of $h_3 k_3 l_3$ was constrained to be zero or nearly so. Isherwood and Wallace [52] have called such double diffraction effects UMWEGANREGUNG (literally: indirectly produced) reflections after Renninger's notation.

Use is made of the separation of planes of a family by the divergent beam technique to obtain an angular separation of the indirectly produced individual reflections. The angular separation is calculated with the boundary conditions: (1) That the wave vectors $\vec{k}$ must terminate on the Brillouin zone boundaries corresponding to the reciprocal lattice vectors $(h_2 k_2 l_2)$, $(k_2 h_2 l_2)$, and $(h_3 k_3 l_3)$. (Here $k_2 h_2 l_2$ is simply another plane in the general $h_2 k_2 l_2$ family.) (2) That the wave vectors also terminate on the sphere $|\vec{k}| = 1/\lambda_0'$ [52].

Measurement conditions are chosen so that the angular separation between the reflections is very small (about 10 minutes of arc). This small value results from (a/λ_0') being nearly equivalent to a triple diffraction situation, i.e., the exact intersection of $(h_2 k_2 l_2)$, $(k_2 h_2 l_2)$, and $(h_3 k_3 l_3)$. The angular separation is a measure of the difference in (a/λ_0') from the triple diffraction point [52].

Isherwood and Wallace [52] have developed the foregoing explicitly for the case of a highly perfect silicon single crystal. In this case, $h_2 k_2 l_2$ is 313 while $k_2 h_2 l_2$ is 133 and $h_3 k_3 l_3$ is 222. An equation for a/λ_0' in terms of the angular separation of the doubly diffracted 222 reflections due to 313 and 133 respectively was derived in explicit form. In the apparatus, the source to film distance D was 116 cm while the distance between reflections L_R was small. Hence, the required angular separation was (L_R/D); a densitometer-comparator was used to measure L_R. From this, using Cu–Kα_1 radiation, the value of a at 25° was determined as $5.43060 \pm \sigma = 0.000033$ Å.

5. *Multiple Exposure Method*

This method was essentially developed as a device to eliminate the parameter D from consideration [31, 53]. Instrumentation of the type described by

Fujiwara and co-workers [14], as improved by Imura [17], is usually used. In this instrumentation, the film may be placed at any distance from the source virtually at will.

Figure 14 shows two views of a nearly elliptical conic taken at two values of D in the back reflection mode. It will be recalled that only ellipselike

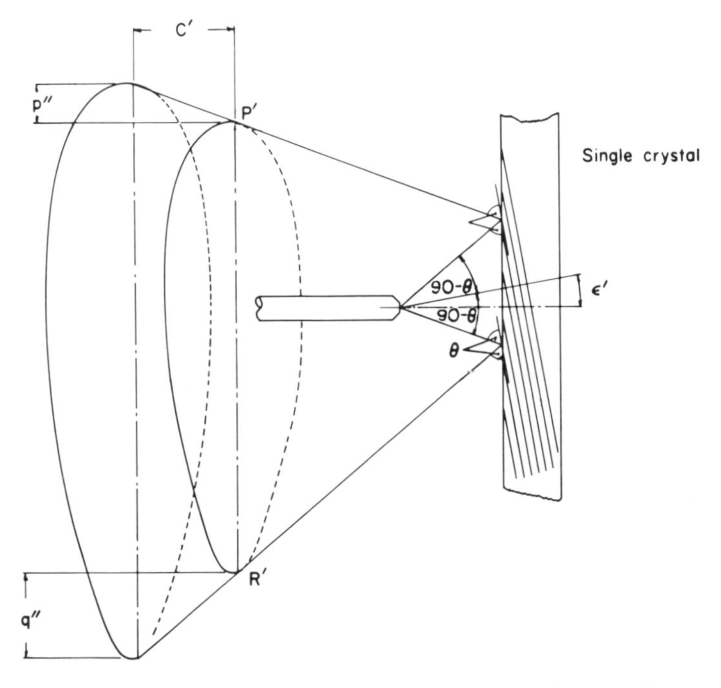

FIG. 14. Formation of conics for multiple exposure method (after Ellis and Weissman [31]).

figures appear in this mode. The slopes m_1 and m_2 of the diffracted rays are given by

$$m_1 = P''/c' = \tan(90 - \theta + \varepsilon'), \tag{48a}$$

$$m_2 = q''/c' = \tan(90 - \theta - \varepsilon'). \tag{48b}$$

The value of c is $(j_1 - j_2)$, the distance between two consecutive film positions.

The method consists of a determination of the slopes by exposing seven or eight times on the same film at different D values. Precision spacers are used so that each c' value is known very well. Thus a family of ellipses corresponding to one (hkl) reflection is obtained. It should be mentioned that the film is kept in the same position by evacuating the back of the cassette [31, 53].

Figure 15 shows a "side view" of this method. In it, t_7 and t_8 are the

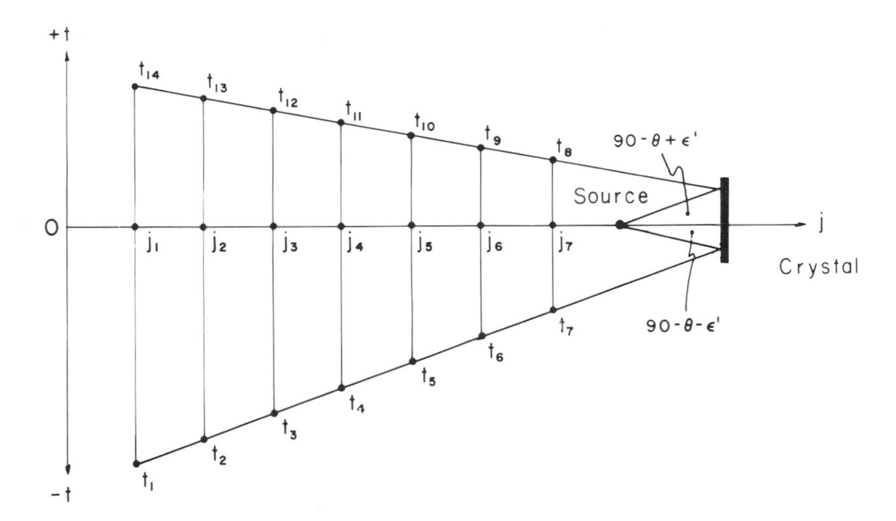

FIG. 15. Schematic of multiple exposure method (after Ellis and Weissman [31]).

major axis of the ellipse produced at the lowest D value (j_7). The film is then moved to j_6, and so on.

The equation of line t_1t_7 is

$$t_1t_7 = m_1 j + B_1'. \tag{49}$$

By a least-squares fit

$$m_1 = \frac{\sum t_i(j_i - \bar{j})}{\sum (j_i - \bar{j})^2}.$$

However, the following is also true

$$K'm_1 = \tan(90 - \theta - \varepsilon') \tag{50}$$

where K' is the film shrinkage factor.

In a like manner for $t_{14}t_8$

$$K'm_2 = \tan(90 - \theta + \varepsilon'). \tag{51}$$

Combining Eqs. (45) and (46), one obtains

$$\theta = 90 - \tfrac{1}{2}[\tan^{-1}(K'm_1) + \tan^{-1}(K'm_2)]. \tag{52}$$

The value of d may then be found from Bragg's law.

This method has the advantages that any Bravais lattice can be studied and that no search for suitable wavelengths is necessary. It was observed that a high θ conic gave higher precision than did low θ conics. Consequently, it was stated that it is highly desirable to record a great many complete ellipses in the vicinity of the film center [31, 53].

The center of the film must be accurately known in order to determine the t values properly. This is done by connecting the points of intersection of two families of ellipses and extrapolating. If several sets of such ellipses are used, the intersection of the extrapolated lines is claimed to be the film center [31, 53]. In general, these are curved lines and so some subjective judgment is called for in the determination of the film center.

Another basic tenet is that the film shrinkage be uniform. After study, Ellis and Weissmann [31] state that a special Cronar base single emulsion graphic arts film was found to satisfy the requirements for isotropic film shrinkage. In a transmission pattern this is probably true although other films seem to meet the requirements as well [40, 46]. Witt [54] found the resolution of Cronar base film to be inferior to X-ray emulsions.

In the back-reflection mode, a hole is normally punched in the film near the center to allow the electron beam to pass. Jellinek [55] has reported experimental results which show that a large shrinkage occurs where a hole is punched and that on either side of the hole, there is a region where the film stretches slightly followed in turn by a region of shrinkage. Therefore, the punching of a hole would seem to automatically preclude the possibility of isotropic film shrinkage. The effect is apparently much worse near the film center than elsewhere; hence, measurements in this region would be most in doubt. For such reasons as avoiding nonuniform film shrinkage, some investigators have advocated the use of glass plates for taking Kossel patterns even in back-reflection. The difficulty of routinely obtaining glass plates of the proper size with a hole drilled to match the apparatus in the center is apparent. On the other hand, glass plates can routinely be used in the transmission mode.

Shrier *et al.* [56] have described a simplification of the multiple exposure method. In this case only two exposures are required; the source to film difference between the film positions for the exposures Δj is measured as accurately as possible. The coordinates of at least three pairs of corresponding points (those from the same reflected ray) are measured on the film.

The position of correspondence is established by placing a square grid of 1 mil diameter tungsten wire with grid spacing of 3/4 inch between the film and the sample. The grid size is that of the film and it is fixed as far from the specimen as possible. The double exposure shows thin breaks in the conic traces due to the wire. Correspondence can be found by matching the breaks in the conics [56].

The method is the multiple exposure analogue to Morris [49] general technique. Hence, the relation for d is analogous to that of Eq. (47). In this case [56]:

$$d = 0.5\lambda \begin{vmatrix} \Delta X_1 & \Delta Y_1 1 \\ \Delta X_2 & \Delta Y_2 1 \\ \Delta X_3 & \Delta Y_3 1 \end{vmatrix}^{-1} \left\{ \begin{vmatrix} t_1 & \Delta Y_1 1 \\ t_2 & \Delta Y_2 1 \\ t_3 & \Delta Y_3 1 \end{vmatrix}^2 + \begin{vmatrix} \Delta X_1 & t_1 1 \\ \Delta X_2 & t_2 1 \\ \Delta X_3 & t_3 1 \end{vmatrix}^2 + \frac{1}{(\Delta j)^2} \begin{vmatrix} \Delta X_1 & \Delta Y_1 & t_1 \\ \Delta X_2 & \Delta Y_2 & t_2 \\ \Delta X_3 & \Delta Y_3 & t_3 \end{vmatrix}^2 \right\}^{1/2} \tag{53}$$

where $t_n = [(\Delta X_n)^2 + (\Delta Y_n)^2 + (\Delta j)^2]^{1/2}$ and ΔX_n, ΔY_n are the differences between the X and Y coordinates, respectively, of the corresponding points on the double exposure.

All of the general comments applied to Morris' method and to the back reflection method apply to the technique of Shrier *et al.* However, this technique appears to be the simplest and most straightforward for the measurement of lattice spacings by means of back reflection.

D. Choice of Radiation

Originally, the radiation chosen was usually a matter of chance or convenience. In the true Kossel method one is limited to the radiation of the sample. However, the work of Hanneman *et al.* [46], Heise [47], and Gielen *et al.* [40], showed that the choice of radiation was often critical for obtaining lenses or intersections capable of yielding highly precise lattice spacing data.

The ability to vary the radiation is an extremely powerful device since if n is the number of wavelengths available, the probability for finding a suitable intersection using the pseudo-Kossel technique is $n^2/2$ larger than in the true Kossel case [40]. Furthermore, it will be shown that it is most desirable to have the X-ray source away from contact with the specimen.

The usual method of testing for a suitable wavelength has been to plot stereographic projections. The hand plotting of the projection is tedious and time consuming; errors are also common [40]. The machine method of Frazer and Arrhenius removes this difficulty and is recommended if a computer is readily available. It has the potential disadvantage that a large number of wavelengths cannot be tested simultaneously.

For the cubic system, Gielen *et al.* [40] have reported a computer program for the lens method. This program allows any number of wavelengths to be investigated simultaneously. Morris has extended the Gielen program to provide low γ intersection data for noncubic systems. The Frazer–Arrhenius stereographic plotting program has also been reprogrammed by Morris [49].

If a computer is not available, one can solve Eq. (20) directly in the search for low γ lenses. A graphical method for the cubic case in which λ versus $\cos(90 - \theta)$ is plotted for a given a value has also been described [57]. This method gives conic overlap values and may be used for either the lens case or Kossel's method.

COMPARISON OF LATTICE SPACING METHODS

Method	Comment
Tangency	(a) cubic only, (b) sensitivity poor, (c) cumbersome to use
Lenses	(a) special orientations required to eliminate projection distortion effects, (b) lattice parameters only, (c) primarily for cubic materials
Intersections	(a) cannot predict existence of suitable intersection, (b) intersection motion unpredictable, (c) sensitivity unpredictable
Regressive Analysis	(a) Gives d_{hkl} spacings, (b) any crystal, (c) gives orientation information

The regressive analysis method can be summarized as follows: An orthogonal X-Y-Z coordinate system is set up where Z is the X-ray source to film distance and the center of the pattern is the X-Y plane origin $X_0 Y_0$. The equation of any conic on the film is

$$UX + VY + WZ = (X^2 + Y^2 + Z^2)^{1/2} = M$$

with $d = 0.5\lambda(U^2 + V^2 + W^2)^{1/2}$.

Next, we call the direction cosines of the conic having Miller indices (hkl) with respect to the X-Y-Z coordinate system, l_1, l_2, l_3, respectively. Then

$$l_i = \frac{U, V, W}{(U^2 + V^2 + W^2)^{1/2}},$$

$$\cos(90° - \theta) = \frac{l^1}{U} = \frac{l_2}{V} = \frac{l_3}{W}$$

and

$$l_1^2 + l_2^2 + l_3^2 = 1.$$

In principle, we could obtain the d values for all conics on the film by measuring the X–Y coordinates of only three points on each conic. In practice, we measure 10 to 15 points on each conic and use regressive analysis. In this case, the normal equations for U, V, W are:

$$\begin{bmatrix} [X^2] & [XY] & Z[X] \\ [XY] & [Y^2] & Z[Y] \\ Z[X] & Z[Y] & nZ^2 \end{bmatrix} \begin{bmatrix} U \\ V \\ W \end{bmatrix} = \begin{bmatrix} [Xn] \\ [Yn] \\ Zn \end{bmatrix}$$

where $[X]$ means $\sum_i^n X_i$, etc. n is the number of coordinate measurements for the conic.

These equations are conveniently solved by a digital computer. Thus, the d values for conics on the film can be obtained. In fact, we get all planes of a multiplicity set separated and so get d values for each such separate plane. In other words, if (hkl) appears, $(\bar{h}\bar{k}\bar{l})$, etc. will appear as well. This separation is of key importance when the Kossel method is used for strain analysis.

As a by-product of the regressive analysis method, we get orientation data. It has already been shown that the direction cosines of the conic with respect to the coordinate system are obtained. The angle α between any two planes (called 1 and 2 respectively) can be gotten by:

$$\cos\alpha = l_{11}l_{12} + l_{21}l_{22} + l_{31}l_{32}.$$

E. Temperature Effects Due to Electron Beam Heating

This topic has been covered in detail by Morris [58] for several specimen geometries. For true Kossel patterns, it has been stated that one should use as large a spot as possible consistent with the maximum resolution required. One should also use a moderate specimen current and accept a longer exposure time [40, 41].

For lattice spacing measurements, pseudo-Kossel patterns can be used with a foil source placed above the sample and not in thermal contact with the sample. Direct vapor deposition should be used only as a last resort. A thermocouple can be attached to the sample and a bank of thermocouples can record the ambient temperature.

The value of $(\Delta a/a)_T = \alpha\,\Delta T$ in a cubic sample, where α is here the linear coefficient of thermal expansion. Using Morris' relations, one finds that for a true Kossel pattern using a bulk sample that

$$(\Delta a)_T = 0.286 S\alpha/d_B \tag{54}$$

where

$S = (\text{keV})(\mu A)/\text{thermal conductivity}$ (cgs units), $d_B = \text{beam diameter } (\mu)$.

Using Eq. (33), one sets d_B such that $(\Delta a/a)_T$ is less than $(\Delta a/a)$ from all other sources. Since, for metals $\alpha \approx 10^{-5}/°\text{C}$, ΔT should be $\leq 0.1°\text{C}$.

For orthogonal noncubic materials, the volume coefficient of expansion, α_v, can be obtained. Then, expansion of a given axis can be determined from the relation for the cell volume in terms of the cell axes if the thermal expansion is isotropic, i.e., if $\Delta a/\Delta T = \Delta b/\Delta T = \Delta c/\Delta T$. In such a case for a tetragonal system

$$\Delta a/\Delta T = \Delta c/\Delta T = \alpha_v/a(a + 2c). \tag{55}$$

If the thermal expansion is anisotropic, one is left with an indeterminate case. For the tetragonal system in this situation

$$2(\Delta a/a) + (\Delta c/c) = (\alpha_v/a^2 c)\, \Delta T. \tag{56}$$

Thus, if anisotropic thermal expansion is suspected, it is absolutely imperative to keep temperature fluctuations absent.

It is noteworthy that the emitting volume in a true back-reflection Kossel exposure is almost certainly at a higher temperature than ambient. The effect is much less important in the transmission method where the emitting volume is often 100 times as large as in the corresponding back-reflection case.

In any case, when a true Kossel pattern must be prepared, the probe diameter should be made as large as is prudent within the requirements of the experiment. Furthermore, the specimen current should be maintained at a moderate value and the longer resultant exposure time accepted.

F. Effect of Wavelength Uncertainty on the Accuracy of Lattice Spacing Measurements by the Kossel Technique

It has been shown that the useful range of precision using the Kossel technique to obtain lattice spacing values is 3–15 ppm. The question may now be raised as to what accuracy one may expect. Unfortunately, the accuracy is somewhat poorer than this. The primary reason for this is wavelength uncertainty; the effect of such a $\Delta\lambda$ term can be seen by obtaining Eq. (33) with wavelength considered as being variable.

A correction which is normally made in transmission Kossel work is the refractive index correction [19]. Usually, the classical equation is used [27]:

$$\lambda_0/ = \lambda(1 - \delta'') = 1 - 2.72 \times 10^{-6}(\lambda_0{}^2 \rho Z_A/A). \tag{57}$$

This equation is not particularly accurate. Often measured values of δ'' disagree with the classical expression. Recent work has also not offered completely satisfactory results [34]. Therefore, this effect may lead to errors in wavelength of about 1 to 10 ppm. It is worthwhile mentioning that surface refraction effects can be shown to be negligible [40, 50].

The precision of the wavelength measurements themselves enters here.

Bearden [59] shows this to be 1–20 ppm depending on the λ value. Furthermore, he gives the conversion error for kX to Å units as being in doubt by 5 ppm. However, the kX unit is being superseded by the use of Å* units and the conversion Å*/Å is more relevant at present; it is also in doubt by 5 ppm.

Finally, the fact that even the refractive index corrected wavelength may not lie at the centroid of the emission peak coupled with line symmetry contributes an error. Superimposed upon this is a line broadening due to the small but finite X-ray source size. The error associated with these effects may be estimated at 10 to 20 ppm.

The temperature can probably be controlled in the specimen chamber to the point where temperature errors are 0–1 ppm.

Assuming a Cauchy distribution of errors, we get an accuracy of 17 to 71 ppm. A reasonable "average" accuracy is about 40 ppm or 1 part in 25,000. It is probably unsafe to claim higher accuracy without carefully documented proof.

V. Orientation of Crystals by Means of the Kossel Technique

Orientation procedures utilizing Kossel patterns are entirely analogous to the well-documented Laue procedures [60]. The major advantage of the Kossel technique is that a microsource of X-rays is employed, thus enabling twin grains, matrix–precipitate relations, grain misorientation, and polycrystalline materials to be examined directly. Either the transmission or back-reflection mode is useful; the choice is largely a matter of convenience.

The orientation of a crystal is adequately described by the stereographic projection of the poles of the diffracting planes. Thus, if a stereographic projection for the material of interest is at hand, the Kossel pattern of the material can be directly compared to the projection in order to determine the relative orientation of the crystal. Comparison is possible since when a Kossel pattern is prepared on a film parallel to some crystal face (as is the usual case), the face normal will intersect the film center. A reference direction is established on the film by fiducial marks. Thus, the indices of the face and a particular crystallographic direction within it can be found [36, 61].

More accurate than comparison is, of course, direct plotting of the poles. The Kossel analog of a Greninger chart used for Laue work can be prepared. Morris has prepared such a chart and has described in detail and with solved examples its use in solving orientation problems in iron metal whiskers [58, 62]. Such a chart is prepared using exactly the relations Greninger [63] employed. The chart is also used in the same fashion as for Laue patterns which requires the investigator to know the lattice spacings, the Bravais lattice, the value of D, and the wavelength responsible for the conics.

Ogilvie and Bomback [64] have described a simplified method for stereographically plotting poles if all of the foregoing crystallographic data are known. The steps in plotting the pole of one plane are as follows:

(1) Place a vertical fiduciary mark on the film during exposure.
(2) Mark the film center during exposure.
(3) Choose a conic as close as possible to the film center.
(4) Erect a perpendicular from the film center to the conic.
(5) Measure the angle between the fiduciary mark and the line in (4).
(6) Lay off the angle in (5) with respect to the vertical direction on the stereographic projection along a line through the center of the projection.
(7) Calculate $(90 - \theta)$ for the conic chosen.
(8) Measure the length of the line in (4) as carefully as possible. Call it L.
(9) Compute the angle between the cone axis and the film center as: $G = (90 - \theta) \pm \arctan(L/D)$. The plus sign is taken when the conic is convex to the film center; the minus sign is used when the conic is concave to the film center.
(10) Lay off G along the line in (6) using a Wulff net. This intersection is the required stereographic projection of the conic pole.

We may investigate the expected orientation accuracy by determining the error in G. It is assumed that errors in $(90 - \theta)$ and in plotting will be smaller than the error in G. We obtain

$$\Delta G = (D\,\Delta L + L\,\Delta D)/(D^2 + L^2). \tag{58}$$

Assuming $D = 10$ cm, $\Delta D = 1\%$ of 10 cm, $L \approx 1$ cm, $\Delta L \approx 10^{-2}$ cm, we get ΔG about 7 min of arc. All other errors probably approach this value. It is probably safe to say that orientation can be carried out to $\pm\frac{1}{4}^\circ$ if the crystal exhibits sharp conics. When heavily strained crystals, such as Kamacite plates in meteorites, are studied, the orientation accuracy may be worse due to line broadening or splitting [65]. In addition, orientation accuracy is promoted if measurements can be made with respect to conics having large Bragg angles.

While two such pole-orientation-plots completely determine the orientation, it is probably best to plot three as an internal check. After this, if the Bravais lattice and the lattice spacings are known, the positions of all other poles can be calculated with standard interplanar angle relations [60].

A more sophisticated procedure developed by Peters can be used when there is insufficient information to use the previous method. In this procedure, it is not even necessary to know the Bravais lattice. A polar gnomonic net is used to measure the angular coordinates of three points on the conics. These points are then plotted on the stereographic projection and the circle which they define drawn. With a Wulff net, the true center of this circle is found. This point is the stereographic projection of the conic pole [36, 61].

As indicated (p. 394) Morris [49] has written the analytical analogue to the Peters procedure. Morris' method gives the direction cosines l_1, l_2, l_3 which the Kossel cone axis makes with the (X, Y, Z) axes, respectively. Here the coordinates of the pattern center are taken to be $X = Y = 0, Z = D$. Hence, the orientation of any plane is determined with respect to the pattern center.

Since the cone axis is the pole of the plane, the orientation information can be plotted stereographically to give a permanent record. Plotting, especially for non-cubic lattices, is simplified since the angle ω between two planes, p and q, can be found from:

$$\cos \omega_{pq} = l_{1p} l_{1q} + l_{2p} l_{2q} + l_{3p} l_{3q}. \tag{59}$$

Morris states that orientation can be carried out to within 0.1° if the values of X_n, Y_n and D can be determined to 0.5%. Orientation accuracy is relatively insensitive to D but is especially sensitive to errors to determining the pattern center [49].

Heise [66] has offered an orientation procedure using Kossel lines when the film is placed cylindrically around the specimen. It is complex and the orientation accuracy is only $\pm 1.6°$.

Peters and Ogilvie [61] have compared the Kossel and Laue methods for orientation. They point out that unless several Laue spots representing major poles are present on the Laue film, analysis is difficult. Although the usual Kossel film intercepts a solid angle of about $\pi/3$, which is comparable to the standard back-reflection Laue film, the poles are randomly distributed over the entire film thus providing a larger percentage of the important low index poles. The striking example that a Kossel pattern of copper contains 41% of the permissible cubic poles while the corresponding back-reflection Laue photograph contains about 15% of the major poles and zones is presented. It is concluded that this less limited distribution of poles provides a statistically more accurate orientation capability.

Bevis and Swindells [67] have developed a method for determining the orientation which eliminates stereographic plotting. To do this, a digital computer is used to prepare tables giving the indices of the directions represented by the intersection of all possible Kossel cones with any other given Kossel cone. The appropriate Bravais lattice and the corresponding lattice parameters are required to prepare the table [67].

The method involves measuring the lens angle between two intersections. Then, if the position of center of the film, and the value of D are known accurately, the tables described above can be used to determine the Miller indices of the plane of the crystal surface. The usual fiducial marks both on the film and the crystal are required [67]. The precision of the method as shown in Bevis and Swindells' example is about $\pm 1°$.

This technique appears especially valuable if a great number of orientation determinations are to be made on specimens of the same Bravais lattice and which have the same lattice parameters. However, if such a series of orientations is not to be determined, it seems that using the computer time to prepare a stereographic projection, according to available methods [35, 49], is indicated. Then, as before, a crude orientation can be carried out by inspection; or alternatively the method of Ogilvie and Bomback [64] can be applied if better accuracy is desired.

Another area in which the Kossel method proved its superiority was in the determination of orientation and common plane in platelets of two different phases having different Bravais lattices, each of which was approximately 50–100 μ in breadth [36]. The standard procedures for common plane determination [60] could be carried out directly since the X-ray source was placed directly in each phase. Much complicated and tedious labor was thereby saved since some variant of the rotating-crystal method would have been otherwise required to make the common plane determination [60].

Bevis and Swindells [67] have developed a clever method of determining the orientation of a planar interface without using two surfaces. The method involves polishing the surface and then, assuming the removal to be uniform, determining the amount removed. The angle of the normal to the interface plane can then be calculated by trigonometry from measurements of the interface motion between successive positions of the surface [64].

Microhardness marks are used to relocate the area of interest after polishing; the depth of the removed surface layer is found from the change in dimensions of the microhardness marks. As a test for uniformity of removal, it is required that the sides of the microhardness indentation remain plane [67].

By combining, orientation information obtained by the divergent beam method on the original and new surface, the pole of the common plane can be found. This method should be especially valuable in the study of small included phases.

VI. Use of the Kossel Technique to Study Mechanical and Thermodynamic Characteristics of Crystals

The internal crystalline perfection may affect the appearance of a Kossel pattern in several ways. Internal strains may cause line broadening if the strain is distributed nonuniformly according to the relation given by

Imura [17]

$$\Delta W = \Delta \varepsilon \tan \theta \tag{60}$$

Furthermore, as ε varies, the value of ΔW will vary, and hence a single conic may show some regions much broader than others. A crystal yielding such a pattern is anisotropically strained.

Figure 3 shows the expected line width as a function of ζ, for a transmission pattern. The value of W passes through a maximum as ζ is varied [21]. This maximum may be fairly flat if the crystal lies on the plateau region of Fig. 3. However, the maximum may be sharp if the crystal is in the steeply rising region of the curve. This in turn affects the contrast t, σ, of a transmission pattern. Hence, if external stress is applied, the contrast of a transmission Kossel photograph may pass through a definite maximum [68]. Imura's experimental results show this contrast change dramatically for the $(1\bar{1}1)$ plane of an aluminum crystal which was strained in tension. Series of photometer traces clearly show both the broadening as well as the contrast changes as a function of elongation. Similar effects were observed in back-reflection patterns but to a lesser extent [17].

Another effect which may be observed is line broadening, kinking or actual line splitting. These effects are caused by lattice bending or local lattice rotation and by fragmentation into small crystallites mutually misoriented with respect to one another. Imura observed these effects in tensilely deformed Al [12]. Le Hazif *et al.* found that the fragmentation effect led only to line broadening in beryllium compressed parallel to the *c* axis. This effect was attributed to a fine substructure. In the same material, large subgrains having a misorientation of several degrees were also found; these gave an actual line splitting [69]. The diffracting volume apparently contained two "grains" from which two patterns were recorded displaced by the misorientation angle. In such a case, the sample is in effect a bicrystal; the lattice misorientation can be estimated by determining the angular displacement between the two patterns.

When the strain is so great that fragmented crystallites become misoriented by many minutes of arc, line kinking takes place as opposed to line broadening. Imura found that kinking occurred with block misorientations of greater than 15 min of arc in strained Al. He observed that the conics became very broad and diffuse at the kinks but were comparatively sharp elsewhere [17]. This observation is to be expected since the block itself will diffract in the usual fashion and if the atoms of the block are arranged in a near perfect manner, then a fairly sharp conic will result.

The condition of extremely minute fragments mutually misoriented with respect to one another leads to nearly 100% of kinks. This is the explanation for Lonsdale's case of the "perfectly imperfect" crystal from which no observable pattern is obtained [19]. Achieving this condition seems to depend entirely on the crystal. Fisher and Harris [70] have observed patterns from commercial 18–Cr–8Ni stainless steel after permanent elongation of 1.5%. Patterns have been obtained near fracture surfaces as well. But well annealed indium gave no pattern; dipping in liquid nitrogen did not help the situation.

Only Weissmann and his co-workers have, to date, attempted to use Kossel techniques to obtain more than semiquantitative data concerning a strain distribution. These workers have studied ordering strains in Cu–Au alloys and the effect of compression on W and Ta crystals as well as strains in Borrmann crystals of Ge and W [28, 31, 53, 71–76]. The back-reflection method has been employed exclusively; lattice spacing values have, in most cases, been determined by means of the multiple exposure method described previously.

A basic premise of the strain analysis is that the Kossel technique separates all planes of a family yielding a separate conic for each. Thus, d values can be obtained for each such plane separately. The strain is taken to be $\Delta d/d_0$ where d_0 is the lattice spacing for the unstrained crystal. Separate strain values for each reflection enable the investigator to avoid the complex line averaging required in other X-ray strain-analyses such as the Warren–Averbach method [77].

The divergent X-ray beam pattern supplies information about the strain state of the crystal in a finite but small neighborhood of a point. Hence, it has been found necessary to use a statistical algorithm to infer an average state of strain from the measured $(\Delta d/d_0)$ values [28]. This algorithm is used in order to obtain the symmetric strain tensor T_s which completely describes the state of the solid [78, 79].

In the cubic system, a point within the irradiated volume of the crystal is taken as reference; three mutually perpendicular axes corresponding to [100], [010], and [001] pass through this point. Then the components of the vector $\mathbf{H} = (hkl)$ are the direction numbers for the plane form H. It can be shown that for a general plane form H_r

$$|H_r|^2(\Delta d/d_0)_r = \alpha_r'\zeta' \tag{61}$$

in which α_r' and ζ' are 6-vectors

$$\alpha_r' = (h_r^2, k_r^2, l_r^2, k_r l_r, l_r h_r, h_r k_r), \qquad \zeta' = (\varepsilon_{11}, \varepsilon_{22}, \varepsilon_{33}, \varepsilon_{23}, \varepsilon_{31}, \varepsilon_{12}). \tag{62}$$

If the six components of ζ' can be determined, T_s can be found [28]. Therefore, it is necessary to measure lattice spacing values in at least six crystallographic directions.

From these measurements one obtains a 6 × 6 matrix A with components α_r' as the elements of its rth row, which when multiplied by ζ' equals a 6-vector whose components are $|H_r|^2\ (\Delta d/d_0)_r$. The rank of A must be 6, i.e.,

Det $A \neq 0$ in order to compute a unique set of strains. There is no simple way to decide whether a particular set is suitable without constructing A and testing it [28].

Even if A is of rank 6, the computed strain system would probably be in error. To minimize this error, more than six strains are recorded.The least squares method is then used to determine an average value of ζ'. This yields values for the average strains, $\langle\varepsilon_{11}\rangle$, $\langle\varepsilon_{22}\rangle$, $\langle\varepsilon_{33}\rangle$, $\langle\varepsilon_{23}\rangle$, $\langle\varepsilon_{31}\rangle$, $\langle\varepsilon_{12}\rangle$. From elastic theory [79]

$$\mathrm{Det}(T_s - SI) = 0 \tag{63}$$

where S corresponds to the principal strains with $S_1 \geq S_2 \geq S_3$ and I is the invariant of Eq. (63) and which has components:

$$I_1 = S_1 + S_2 + S_3$$
$$I_2 = S_2 S_3 + S_1 S_3 + S_1 S_2$$
$$I_3 = S_1 S_2 S_3 .$$

The principal strain axes are obtained from

$$(T_s - S_r I)l = 0; \qquad r = 1, 2, 3. \tag{64}$$

In explicit form, the determinant of Eq. (63) is written from its characteristic matrix as:

$$\begin{vmatrix} \langle\varepsilon_{11}\rangle - S & 0.5\langle\varepsilon_{12}\rangle & 0.5\langle\varepsilon_{31}\rangle \\ 0.5\langle\varepsilon_{12}\rangle & \langle\varepsilon_{22}\rangle - S & 0.5\langle\varepsilon_{23}\rangle \\ 0.5\langle\varepsilon_{31}\rangle & 0.5\langle\varepsilon_{23}\rangle & \langle\varepsilon_{33}\rangle - S \end{vmatrix} = 0. \tag{65}$$

This determinant reduces to the following cubic equation in S, the roots of which are the latent roots of Eq. (65):

$$\begin{aligned} &S^3 - S^2[\langle\varepsilon_{11}\rangle + \langle\varepsilon_{22}\rangle + \langle\varepsilon_{33}\rangle] + S[\langle\varepsilon_{22}\rangle\langle\varepsilon_{33}\rangle + \langle\varepsilon_{11}\rangle\langle\varepsilon_{33}\rangle + \langle\varepsilon_{11}\rangle\langle\varepsilon_{22}\rangle \\ &\quad - 0.25(\langle\varepsilon_{12}\rangle^2 + \langle\varepsilon_{23}\rangle^2 + \langle\varepsilon_{32}\rangle^2)] \\ &\quad - \{\langle\varepsilon_{11}\rangle\langle\varepsilon_{22}\rangle\langle\varepsilon_{33}\rangle - 0.25[\langle\varepsilon_{11}\rangle\langle\varepsilon_{23}\rangle^2 + \langle\varepsilon_{22}\rangle\langle\varepsilon_{31}\rangle^2 + \langle\varepsilon_{33}\rangle\langle\varepsilon_{12}\rangle^2] \\ &\quad + 0.25[\langle\varepsilon_{12}\rangle\langle\varepsilon_{23}\rangle\langle\varepsilon_{31}\rangle]\} = 0. \end{aligned} \tag{66}$$

Equation (66) can be solved by standard cubic equation formulae [79]. Computer programs have been written which reduce lambda matrices of the type just discussed to their latent roots directly [28]. All of the latent roots of Eq. (65) are real [79].

In practice, Eq. (66) is solved for its roots by means of a computer program [28]. The results give $S_1 \geq S_2 \geq S_3$, the principal strains. With these available,

Eq. (64) can be written in explicit form in order to obtain the principal strain axes.

$$\begin{pmatrix} \langle\varepsilon_{11}\rangle - S_1 & 0.5\langle\varepsilon_{12}\rangle & 0.5\langle\varepsilon_{31}\rangle \\ 0.5\langle\varepsilon_{12}\rangle & \langle\varepsilon_{22}\rangle - S_1 & 0.5\langle\varepsilon_{23}\rangle \\ 0.5\langle\varepsilon_{31}\rangle & 0.5\langle\varepsilon_{23}\rangle & \langle\varepsilon_{33}\rangle - S_1 \end{pmatrix} \begin{pmatrix} l_{11} \\ l_{21} \\ l_{31} \end{pmatrix} = 0 \tag{67a}$$

$$\begin{pmatrix} \langle\varepsilon_{11}\rangle - S_2 & 0.5\langle\varepsilon_{12}\rangle & 0.5\langle\varepsilon_{31}\rangle \\ 0.5\langle\varepsilon_{12}\rangle & \langle\varepsilon_{22}\rangle - S_2 & 0.5\langle\varepsilon_{23}\rangle \\ 0.5\langle\varepsilon_{31}\rangle & 0.5\langle\varepsilon_{23}\rangle & \langle\varepsilon_{33}\rangle - S_2 \end{pmatrix} \begin{pmatrix} l_{12} \\ l_{22} \\ l_{32} \end{pmatrix} = 0 \tag{67b}$$

$$\begin{pmatrix} \langle\varepsilon_{11}\rangle - S_3 & 0.5\langle\varepsilon_{12}\rangle & 0.5\langle\varepsilon_{31}\rangle \\ 0.5\langle\varepsilon_{12}\rangle & \langle\varepsilon_{22}\rangle - S_3 & 0.5\langle\varepsilon_{23}\rangle \\ 0.5\langle\varepsilon_{31}\rangle & 0.5\langle\varepsilon_{23}\rangle & \langle\varepsilon_{33}\rangle - S_3 \end{pmatrix} \begin{pmatrix} l_{13} \\ l_{23} \\ l_{33} \end{pmatrix} = 0. \tag{67c}$$

A computer routine [28] is used to solve Eq. (67).

The eigenvalues of the l eigenvector determined from Eq. (67) may be written as the orthogonal matrix [28]:

$$\begin{pmatrix} l_{11} & l_{21} & l_{31} \\ l_{12} & l_{22} & l_{32} \\ l_{13} & l_{23} & l_{33} \end{pmatrix} - l.$$

At this point, it is worthwhile to recall that for any orthogonal matrix such as l, the transpose l^T is equivalent to the reciprocal l^{-1}.

$$l^T = l^{-1} = \begin{pmatrix} l_{11} & l_{12} & l_{13} \\ l_{21} & l_{22} & l_{23} \\ l_{31} & l_{23} & l_{33} \end{pmatrix}.$$

Furthermore, the sum of the squares of any row or column is unity and $l_{11}l_{12} + l_{21}l_{22} + l_{31}l_{23} = 0$, etc.

The row vectors in the l matrix determine the principal axes for the strains. Explicitly l_{11}, l_{21}, l_{31} are the direction cosines of the prinipal axis associated with principal strain S_1 while l_{12}, l_{22}, l_{32} and l_{13}, l_{23}, l_{33} are the direction cosines of the principal axes associated with principal strains S_2 and S_3 respectively. By determining the direction numbers A, B, C of the l matrix row vectors, a nearly orthogonal set of crystallographic directions having small Miller indices is found which closely represents the principal axes [72]. Explicitly:

$$\left.\begin{array}{l} \dfrac{A_1}{l_{11}} = \dfrac{B_1}{l_{21}} = \dfrac{C_1}{l_{31}} \\[2ex] \dfrac{A_2}{l_{12}} = \dfrac{B_2}{l_{22}} = \dfrac{C_2}{l_{32}} \\[2ex] \dfrac{A_3}{l_{13}} = \dfrac{B_3}{l_{23}} = \dfrac{C_3}{l_{33}} \end{array}\right\} \rightarrow \begin{bmatrix} A_1 & B_1 & C_1 \\ A_2 & B_2 & C_2 \\ A_3 & B_3 & C_3 \end{bmatrix}.$$

Thus, the principal strains and their axes are determined. With these available, the extremely important question of the maximum shear strain in any (hkl) plane can be investigated [72]. In principle, this means taking a plane slice through the strain ellipsoid and determining the shear strain distribution in the slice. This may be done most simply by the method of Lagrangian multipliers [79]. The result is that the shear strain will be zero normal to some point in the plane and that the direction of maximum shear strain in the plane will lie at 90° to this position. The magnitude of the maxi-maximum shear strain $|\varepsilon_{lm}|_{max}$ in any (hkl) plane is given by [72]:

$$|\varepsilon_{lm}|_{max} = 2[(S_2 - S_3)^2 m_2^2 m_3^2 + (S_3 - S_1)^2 m_1^2 m_3^2 + (S_1 - S_2)^2 m_1^2 m_2^2]^{1/2} \tag{68}$$

where

$$m_1^2 = 1/ho^2[hl_{11} + kl_{21} + ll_{31}]^2,$$

$$m_2^2 = 1/ho^2[hl_{12} + kl_{22} + ll_{32}]^2,$$

and

$$m_3^2 = 1/ho^2[hl_{13} + kl_{23} + ll_{33}]^2.$$

Further analysis, in which the maximum of all $|\varepsilon_{lm}|_{max}$ values was found, led to the conclusion that, in the cubic system, the maximum shearing strains occur on planes of type (101) referred to the principal strain axes, independent of strain configuration [72].

Finally, the stored elastic energy W_E, in a cubic crystal was deduced; the result is based on the previously determined $\langle\varepsilon_{ij}\rangle$, and l values [72]:

$$W_E = \tfrac{1}{2}C_{11}(\varepsilon_{11}^2 + \varepsilon_{22}^2 + \varepsilon_{33}^2) + C_{12}(\varepsilon_{22}\varepsilon_{33} + \varepsilon_{33}\varepsilon_{11} + \varepsilon_{11}\varepsilon_{22}) + \tfrac{1}{2}C_{44}(\varepsilon_{23}^2 + \varepsilon_{31}^2 + \varepsilon_{12}^2) \tag{69}$$

where C_{11}, C_{12}, and C_{44} are the elastic constants. In Eq. (69), W_E is referred to cubic axes. Transformation from the principal axis frame to the cubic frame is carried out by the matrix operation:

$$\begin{pmatrix} l_{11} & l_{12} & l_{13} \\ l_{21} & l_{22} & l_{23} \\ l_{31} & l_{23} & l_{33} \end{pmatrix} \begin{pmatrix} S_1 & 0 & 0 \\ 0 & S_2 & 0 \\ 0 & 0 & S_3 \end{pmatrix} \begin{pmatrix} l_{11} & l_{21} & l_{31} \\ l_{12} & l_{22} & l_{32} \\ l_{13} & l_{23} & l_{33} \end{pmatrix} = \begin{pmatrix} \varepsilon_{11} & 0.5\varepsilon_{12} & 0.5\varepsilon_{13} \\ 0.5\varepsilon_{21} & \varepsilon_{22} & 0.5\varepsilon_{23} \\ 0.5\varepsilon_{31} & 0.5\varepsilon_{32} & \varepsilon_{33} \end{pmatrix} \tag{70}$$

that is

$$l^T T_s l = \varepsilon' = (\varepsilon_{11}, \varepsilon_{22}, \varepsilon_{33}, \tfrac{1}{2}\varepsilon_{23}, \tfrac{1}{2}\varepsilon_{31}, \tfrac{1}{2}\varepsilon_{12})$$

since

$$\varepsilon_{31} = \varepsilon_{13}; \qquad \varepsilon_{21} = \varepsilon_{12}; \qquad \text{and} \quad \varepsilon_{23} = \varepsilon_{32}.$$

It is possible to partition W_E into a combination of elastic energy due to shear W_S elastic energy due to a mixture of normal and shear strains W_{NS} and elastic energy due to normal strains W_N. The value of W_S can be found explicitly [72]:

$$W_S = \tfrac{1}{2}\{C_{44} - \tfrac{1}{4}\nu[(l_{11}{}^2 - l_{13}{}^2)^2 + (l_{21} - l_{23})^2 + (l_{31}{}^2 - l_{33}{}^2)^2]\}(S_1 - S_3)^2$$

where (71)

$$\nu = C_{12} + 2C_{44} - C_{11}.$$

The energy partition is only valid when $W_S/W_E = 0$ in the spherically symmetrical strain configuration case and $W_S/S_E = 1.0$ in the case of simple shear. W_N and W_{NS} cannot be determined explicitly but W_E and W_S can be found and the sum $(W_N + W_{NS})$ can be obtained [72].

In order to express the anisotropy of deformation, a purely geometrical measure of distortion has been presented as

$$D_G = \frac{2}{11}\left(\frac{4I_1{}^2 - 11I_2 - 3I^{2/3}_3}{I_1{}^2 - 2I_2}\right) \tag{72}$$

where I_1, I_2 and I_3 are expressed in terms of S_1, S_2, S_3 as before. For the spherically symmetric case $D_G = 0$ while for the simple shear configuration, $D_G = 1$. However, D_G and W_S/W_E do not coincide except at the extremes since D_G is independent of the strain frame [72].

In addition to the strain configuration, the stress has been included by involving the general Hooke's law matrix of elastic constants [72]. The principal stresses and their associated axes can then be found by solving the stress analogues to Eqs. (63) and (64) [79].

The methods just described enable the investigator to obtain both the stress and strain tensors in the elastic region. Hence, the entire stress-strain configuration can be found in quantitative terms.

As Weissmann and his co-workers have stated, only the beginnings of the straining process can be observed by the means described. When the strain becomes large, line broadening, splitting and kinking may occur, thus making highly precise $(\Delta d/d_0)$ values impossible to obtain. In this case, Fourier methods can be applied to the conic profiles in order to augment the strain analysis [72, 80]. The procedure is somewhat analogous to that for elastic strains in that six separate directions must be sampled.

The analysis can be made only on the elastic portion of the strain configuration; no information is obtained about plastic strain components. Thus, if the crystal exhibits any recovery characteristics near room temperature, the results of the strain analysis may change with time. The sampling procedure adopted, i.e., irradiation of a number of plane faces of the specimen, has occasioned the comment that homogeneous strain is, in effect, assumed present in the crystal since an average strain for the entire crystal is computed. Unless external constraints are maintained, a body cannot be homogeneously strained. The effect of removing the constraints leaves residual macrostresses which may be different on one crystal face with respect to another. Thus, a residual strain gradient is expected to be present throughout the crystal [81]. The question seems to revolve around whether or not it is possible to obtain mechanically isotropic single crystals. It seems clear that it will not be possible to use Wiessmann's experimental methods when studying single grains in polycrystalline material due to inhomogeneity of the strain distribution.

In order to study polycrystalline samples, Morris' method (p. 394) can be used to obtain reliable measurements of the several d values. A loading device which fits into the Kossel camera [92] will ensure that the constraints are in place at the time of measurement. As in any measurement of d values, the most sensitive determinations arise from conics with high θ values. Therefore, an effort should be made to obtain as many of these sensitive conics on the film as possible.

The question of deficiency line contrast in transmission Kossel patterns as related to crystal perfection has arisen in several cases. Morris concluded that the Borrmann effect was present in iron whiskers with an orientation [111], (110), and which had a thickness such that $\mu x > 2$, because deficiency conic contrast dropped while transmitted diffraction conic intensity increased. The effect increased with increasing whisker thickness. It was also concluded that a 100-μ-thick whisker exhibiting the Borrmann effect could be expected to be dislocation free [58, 62]. However, another possible explanation is that Eq. (6) can be used to show that increasing the sample thickness at a given operating voltage will always decrease deficiency line contrast owing to the nature of W and b. At some point, deficiency conic contrast may become zero even if the film is exposed properly due to the exponential increase of the denominator of Eq. (6) [68]. Furthermore, all planes in an iron Kossel pattern prepared using Fe K radiation will have $\Phi > \theta$, except for $\{220\}$, if the crystal has a $\langle 110 \rangle$ pole perpendicular to the film. Thus, all planes whose conics are seen except $\{220\}$ will exhibit normal diffraction lines in the forward (transmission) direction.

Potts found that deficiency conics from GaAs quenched from 1100°C after heating in an excess of As vapor were much sharper and more faint than those from similar samples heated without excess As. It was concluded that the crystals heated in excess As vapor were more perfect [43].

Frazer and Arrhenius [35] postulated that changes in atomic position associated with changes in the physical state of a crystal could be traced by consecutive measurements of intensity changes in a critical set of deficiency conics [35]. However, the contrast of a transmission Kossel pattern is a critical function of the X-ray source size, crystal thickness, and mosaic spread, extinction characteristics of the crystal, operating voltage, exposure time, development technique, and the exciting wavelength spectrum. The overall deficiency of intensity is strongly dependent on the crystal perfection which is apt to vary with temperature and with any small change in strain state;

less strong is the dependence on atom position changes. In fact, according to Fig. 3 and Eq. (6), in some cases, the apparent contrast for a given plane form can vary by a factor of 2 for a relatively small change in crystal perfection. However, a similar change in a thicker sample of the same crystal may lead to a contrast difference so small as to be nearly undetectable (plateau region of Fig. 3). For these reasons, measurements of the contrast of deficiency conics are probably not useful as *quantitative* indicators of the physical state of the specimen crystal.

VII. Experimental Methods in Divergent Beam Diffraction

Three major factors affect the photographic quality and overall usefulness of a Kossel pattern. These factors are the specimen preparation, photographic exposure time, and the type of instrumentation used to obtain the pattern.

A. Specimen Preparation for the Back-Reflection Mode

The pattern obtained in back-reflection usually originates only a few microns below the surface and represents only a few hundred cubic microns of material. It is requisite that the surface be truly representative of the bulk material in such a case.

In order to check on the surface, a Laue photograph may be made before any other specimen preparation is undertaken. After preparation, a second Laue pattern can be prepared. There should be no change in the spot shapes or sharpness if the specimen preparation has been properly carried out.

Ordinary metallographic or petrographic polishing may be sufficient; a worked surface layer may be removed by etching. Alternatively, electropolishing can be used to avoid surface strains. In addition, spark erosion techniques have been used [74].

For small inclusions or precipitates, electropolishing may be a very useful technique. In such a procedure, the anodic (less noble) phase will be preferentially atttacked; hence, the matrix could be attacked at the expense of the inclusion thereby giving a better opportunity to obtain a good pattern. Extraction replica techniques may also be used to isolate inclusions.

For some materials microtomy may lead to satisfactory surfaces. Brittle materials are probably more amenable to this technique than are ductile substances. Ductile substances may smear when subjected to microtomy [82, 83].

B. Specimen Preparation for the Transmission Mode

Deficiency conics are most prevalent in a transmission pattern. However, experimentally it has been observed on Kossel photographs that deficiency conics are often of weak contrast and occasionally even absent [19, 43, 58, 62]. This problem has led several workers to consider the question of contrast optimization particularly with respect to the sample thickness.

Lonsdale [19] was the first to study this problem and presented data for "optimum thickness" values in NaCl and diamond. Later in a somewhat more general treatment, Peace and Pringle [21] were led to an optimum sample attenuation corresponding to $\mu x = 1$.

Subsequently, Hanneman *et al.* [46] proposed on an empirical basis that μx of 5 was an optimum value. More recently, Potts [42] stated that the contrast was relatively independent of the sample thickness. All of these workers appear to have adopted the same definition of contrast. However, none of these workers investigated the effect of the operating voltage on the contrast. Equation (6) has been used to estimate the thickness-operating voltage relationship [68].

In Kossel transmission work, the radiation used is generally chosen on the basis of crystallographic considerations. For example, a particular wavelength which yields sensitive intersections for precise lattice spacing measurements will often be chosen. This fixes the value of the linear absorption coefficient. Furthermore, the values of Q' and ζ are also predetermined. The value of s is fixed by the capabilities of the electron probe forming column. Thus, the only variables left with respect to which contrast may be adjusted are the sample thickness and the operating voltage. The operating voltage enters since "$\bar{\mu}$" is a sensitive function of the continum distribution which is manifestly a function of operating voltage. These considerations suggest that there may be a thickness at a given voltage which will yield a maximum contrast or a voltage at a given thickness which will do likewise.

Consider the denominator of Eq. (6) which needs to be minimized in order to increase contrast. If $\bar{\mu}x > \mu x$, then σ can be raised by merely raising the thickness or $\bar{\mu}$. However, the total intensity I reaching the film is given by $I = I_0 e^{-\bar{\mu}x}$. Therefore, if $\bar{\mu}x$ is increased much beyond 3, there will be little blackening of the film with time, i.e., exposure times become prohibitively long. In fact, one would tend to prefer $\bar{\mu}x \approx 1$ in order to maximize any diffraction conics intercepted and to reduce the exposure time. It would seem, therefore, that values of $\bar{\mu}x = 1 > \mu x$ would almost always yield high contrast patterns. Unfortunately, the nature of the numerator of Eq. (6) precludes this simple solution.

Peace and Pringle have plotted W/ζ versus $Q'x/\zeta$; this plot is reproduced as Fig. 3. At values of $Q'x/\zeta$ greater than 6, increasing the thickness has little effect on the values of W, and hence upon the contrast. However, if x is too low, the value of $Q'x/\zeta$ may drop below 6, especially for high index planes for which Q' may have a small value [27] which will make the contrast a critical function of x and may well cause the loss of some high index lines. Therefore, since $Q'x/\zeta$ is generally unknown, it is necessary to increase x to insure that the region in which $6 > Q'x/\zeta$ is avoided for all lines to be recorded on the pattern. If this is done a pattern of nearly uniform contrast can be obtained owing to the nature of the function in Fig. 3.

The next step in the evaluation of the situation is to estimate values of "$\bar{\mu}$." Let $\bar{\mu}$ be defined as $b\mu$ where b is a factor to be determined. The factor b depends upon the short wavelength limit of the continuum, λ_c, the emitted intensity as a function of wavelength $I_0(\lambda)$, the thickness x, the long wavelength limit λ_f of the spectrum, and the transmission of the sample as a function of wavelength, $\exp -[\tilde{\mu}(\lambda)x]$.

An approximate relation giving the value of b can be obtained for incident electrons of given energy by using Kramers' [84] approximation to the shape of the continuum band from a thick target coupled with the function $[\tilde{\mu}(\lambda)]$ which is tabulated [85].

It is found that b increases with decreasing thickness at a given operating voltage and also that b increases with decreasing operating voltage at a given thickness [68]. As an illustration, Fig. 16 shows b values for Cu irradiated with Cu X-rays.

Three considerations are important in choosing a suitable thickness–operating voltage relationship. When the overvoltage ratio (E/E_c), denoted by U, is decreased below 2, characteristic peak to background ratios decrease rapidly as does the absolute spectral intensity. Furthermore, it should be recalled that the thickness chosen should yield a $Q'x/\zeta$ value greater than 6. Opposing this is the fact that a $(\bar{\mu}x)$ value greater than three gives prohibitively long exposure times coupled with an increased K value in Eq. (6). It therefore appears that for most cases, a choice of $1.5 \leq \mu x \leq 2.5$ combined with $2.5 \leq U \leq 3.0$ will yield transmission Kossel or pseudo-Kossel photographs having nearly optimum contrast characteristics. Such a choice reduces the need for extremely thin, hard to prepare samples, yet gives a high enough over-voltage ratio to produce enough X-ray intensity for good photographic characteristics. Furthermore, such thickness values tend to help the value of $(Q'x/\zeta)$ rise above 6 [68].

Thus, the thickness required can be chosen and specimen preparation carried out accordingly. While a transmission pattern is representative of the entire specimen thickness used, it is desirable to carry out specimen preparation so that the inherent crystal perfection is not disturbed. It is possible to do this for many metals. One can often cut a 10–15 mil thick strip from a slab sample, pickle it in an appropriate solution and then electropolish to the final

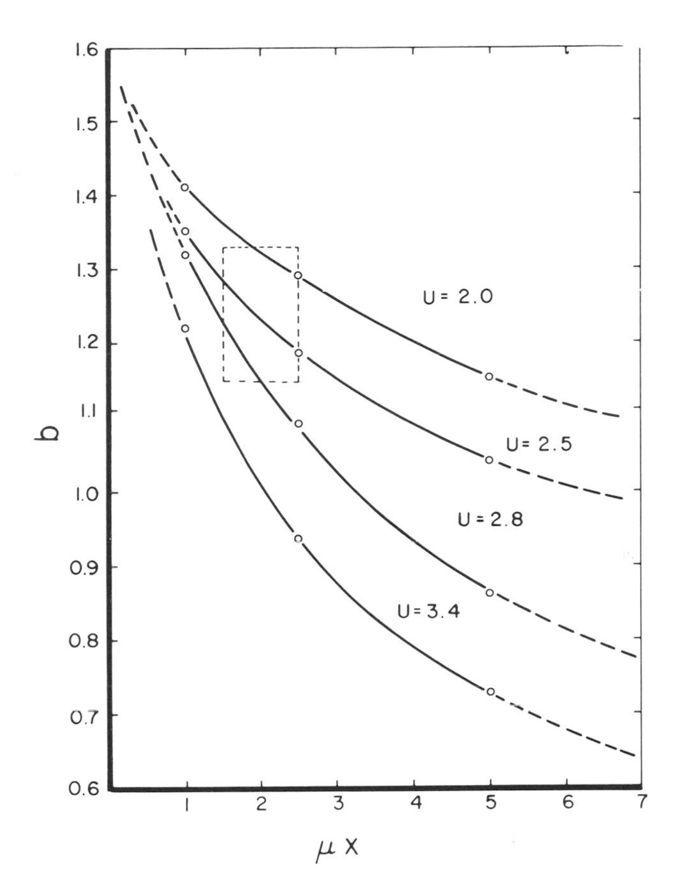

FIG. 16. Value of $b = \bar{\mu}/\mu$ as a function of μx for copper irradiated with Cu–K at various overvoltage ratios $U = E/E_c$. The dotted rectangle indicates the most useful experimental region.

surface finish and thickness desired. Appropriate electropolishing solutions are available for a great many metals. Often, pickling alone is sufficient to obtain the desired result. A perfect metallographic finish is not really required for good transmission Kossel work.

When brittle materials such as semiconductors must be investigated, the crystal may be cut and etched to nearly the right thickness. Mechanical polishing is carried out by cementing the crystal onto an optically flat glass disk; the crystal is centered between four slightly thicker brass disks which are also cemented to the glass. The brass disks are used to make certain that the polishing pressure is uniform and hence that the crystal thickness is uniform. With practice, a specimen such as AlSb can be prepared in a reasonable time [86]. A deep etch may be used as the final step in order to remove disturbed surface material.

If no Kossel transmission pattern is obtained, a Laue photograph should tell by inspection whether or not the crystal was too perfect or too imperfect. If a pattern is desired, appropriate steps can be taken in either case, e.g., carefully annealing the imperfect crystal, bending the perfect crystal slightly, or, if brittle, by irradiating it or abrading it. Dipping in liquid air, He, or nitrogen also often helps [19]. Peace and Pringle [21] have discussed the effect of such dipping in detail.

C. Exposure Time of Kossel Photographs

According to Lonsdale [19], the exposure time of a Kossel photograph has an optimum value. Depending upon the experimental conditions, the exposure time may vary from a few seconds to several hours [87, 62]. Therefore, it is desirable to be able to estimate the exposure time.

Only a few investigators have discussed the exposure time in more than a cursory manner. Morris and Ogilvie [62] plotted the exposure time for Fe whiskers at 30 kV, a 5 cm value of D, and a specimen current of 90 nA for single emulsion Kodak M film. An equation can be fitted to this curve:

$$t_E = 5.78\{\exp[0.14 + 383x]\} \tag{73}$$

Peters and Ogilvie [61] were able to give an approximation to the exposure time for M film for a specific set of conditions. Beyond this, only approximate times for specific photographs are usually given.

Yakowitz and Vieth [88] obtained semiempirical relations for the exposure time which are valid to about ± 10 or 15%. For the case of the film being in vacuum and a pseudo-Kossel experiment, they obtained for a transmission pattern as shown in Fig. 17

$$t_E = \frac{2.67 \times 10^{-15} E_0 D^2 (1 - r)(\sec^2 \eta')}{n_0 i_s} \{\exp -[\bar{\mu}_f x_f + \bar{\mu}_s x_s]\}. \tag{74}$$

Equation (74) takes into account the number of photons produced per incident electron, the operating voltage (through n_0) and the fact that the exposure is greatest at the central portion of the film and least at the edges. The latter is the result of a combination of two effects, the first of which is an increasing absorption path within the sample with increasing emergence angle. The other effect is geometrical in that a unit of solid angle subtends a larger area on the film at the edge than at the center.

The quantity E_0 is that exposure density yielding the maximum contrast

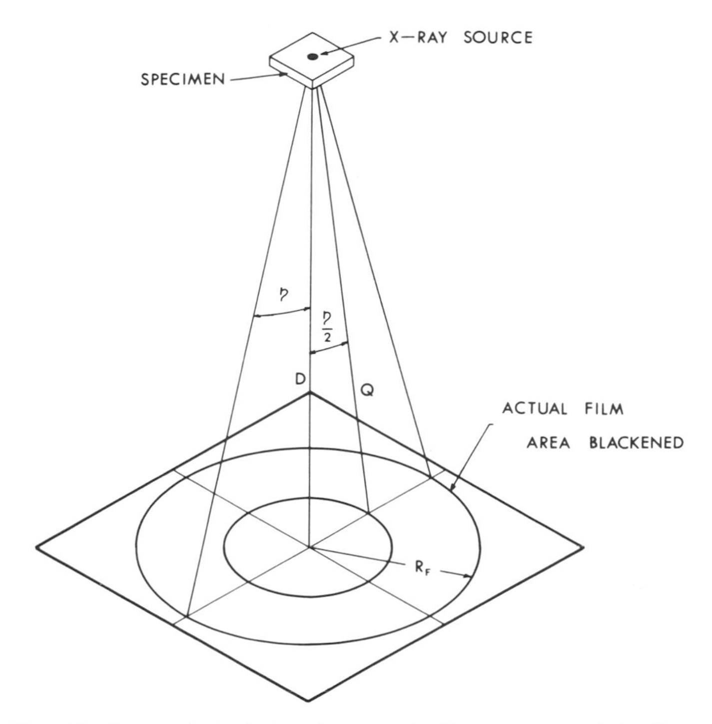

FIG. 17. Geometrical relations between the X-ray source and the film.

between the Kossel conics and the background blackening for a given film at a given distance, D. It must be determined empirically. Once this is done, it is to be expected that E_0 will be essentially constant for a given film type independent of other camera parameters [88]. The E_0 value represents a compromise between the exposure at the edge and central portions of the film.

In the work of Yakowitz and Vieth, it was assumed that b was equal to 2/3 in order to calculate $\bar{\mu}$ values and hence to obtain E_0. At that time, the relations for b as a function of the operating voltage and sample thickness were not available. Now better $\bar{\mu}$ values can be obtained and used [68].

In the case of back-reflection, a relation of the form of Eq. (74) holds except that x_s is undefined. An approximation using Il'in's derivation of an effective thickness, d_f, of the absorbing layer chosen such that exp $-(\mu d_f)$ gives total attenuation of the characteristic line of interest in the working volume of the specimen, i.e., at $\eta = 90°$, is useful [89]. This leads to

$$\rho d_f = H \tag{75}$$

in which H is then a constant for a given value of η'. Figure 18 shows H as a function of η.

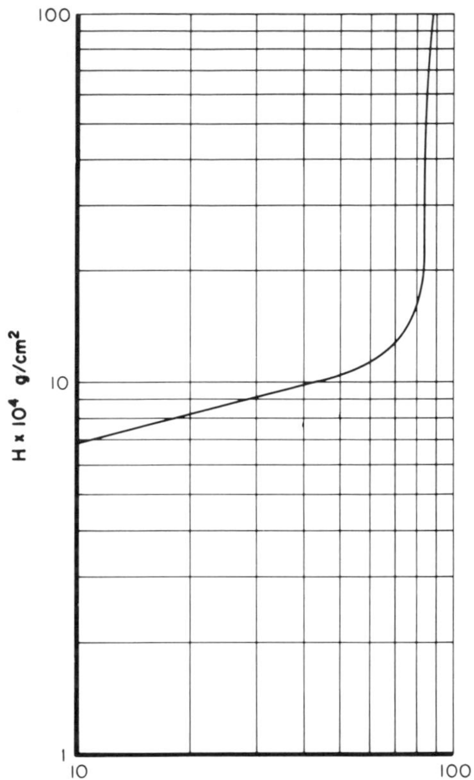

FIG. 18. Parameter H as a function of the half-angle subtended by the film η.

The analog of Eq. (74) for back-reflection is

$$t_E = \frac{2.67 \times 10^{-15} E_0 D^2 (1 - r)(\sec^2 \eta')}{n_0 \, i_s} \left\{ \exp - \left[\bar{\mu}_f x_f + H \left(\frac{\bar{\mu}}{p} \right)_s \right] \right\}. \tag{76}$$

Tests show Eq. (74) and (75) to be valid to within 10 or 15%. These equations are a definite improvement over a pure trial and error method. Furthermore, as more exposures are taken E_0 values can be refined.

D. Instrumentation

All present-day Kossel instrumentation utilizes a focused electron beam to excite the required microsource of X-radiation. While commercial Kossel camera attachments for electron probe microanalyzers represent the most readily available instrumentation, investigators have successfully prepared Kossel cameras for use in point-projection X-ray microscopes and in electron microscopes. This is done because all of the above-named instruments provide convenient electron optics.

However, unless a completely separate Kossel module can be inserted without impairing any other instrument capabilities in a changeover time of about 2 min, the use of such modules is an inefficient means to prepare Kossel patterns. As an example, consider a commercial electron probe microanalyzer which has two full-time professional employees assigned to its operation and maintenance. If such an instrument is amortized in ten years, the total cost per week approaches $1000. Normally, when Kossel photographs are being prepared, equipment such as scanning displays, X-ray spectrometers, and X-ray read-out channels is inoperative. Therefore, only in cases in which it is possible and desirable to simultaneously chemically analyze that portion of the sample for which a Kossel photograph is being prepared is the use of an electron probe microanalyzer fully justified. The only other justification is to have a separate Kossel module used for important incidental tasks such as orientation relations in a microprobe sample or to determine qualitatively if strains are present.

It is suggested that any Kossel camera requiring existing electron column components to be adjusted should not be used. In the long run, it will probably prove more efficient to construct a separate Kossel generator, i.e., electron column including light optical capability and Kossel camera.

This gain in efficiency is possible since the electron optical requirements for virtually all Kossel work are less stringent than those for electron probe microanalysis, and electron or X-ray microcopy. A single lens system with an inexpensive flat grid electron gun is entirely satisfactory [90]. Furthermore, the gun voltage supply stability requirements are much less rigid than those for electron microscopy or microanalysis. Thus, a relativelyinexpensive power supply can be used. Other requirements are a small vacuum system, vacuum gauges, light optics and a current meter whose range is 10–1000 nA [91].

The heart of this system or any modular design for another electron optical column is the Kossel camera itself.

At first, it is necessary to decide whether a transmission or back-reflection camera is to be used. In making this decision, a number of factors merit consideration. First, in order to obtain the maximum precision in a measurement by the Kossel technique, the transmission mode should be adopted as stated previously. The primary reason is that deficiency conics all have precisely the same vertex while diffraction conics do not. Each plane contributing to the diffraction conic corresponds to a separate vertex; hence, there are formed a set of cones all having the same semiapex angle but with vertices each displaced by one interplanar distance. This displacement results in an uncertainty as to the true position of the X-ray source within the sample. In addition, a diffraction conic shows a somewhat greater angular divergence for this reason than does a corresponding deficiency conic. Therefore, measurements involving diffraction conics are to some degree less precise than those involving deficiency conics.

Nevertheless, the back-reflection method is a useful tool that can provide a great deal of information concerning orientation and lattice spacings of small phases in bulk samples which would be difficult to obtain by any other method. Micron sized particles can be studied nondestructively *in situ*. Furthermore, in some cases it is undesirable to prepare a thin specimen suitable for transmission work. Therefore, it is advisable to have both back-reflection as well as transmission capabilities.

It has been asserted that the major features and physical capabilities which are necessary to allow the investigator to obtain the maximum amount of information about the specimen are:

(1) Proper choice of transmission or back-reflection mode for the required study;
(2) Movement of the specimen in orthogonal directions in its own plane;
(3) Capability of observing the specimen;
(4) Capability of orienting the specimen by means of a microgoniometer and facilities for controlling the goniometer from outside the vacuum system;
(5) Capability of precisely determining and marking the position of the center of the film;
(6) A specimen to film distance of 5–11 cm;
(7) An X-ray divergence angle of at least 56°; and
(8) A camera which maintains the specimen in vacuum and the film in air.

Each of these eight requirements is discussed in detail by Vieth and Yakowitz [48]. Their arguments may be summarized as follows:

(1) In order to take full advantage of the technique, both transmission and back-reflection capabilities should be available;
(2) Specimen motion is required to find the region of interest;
(3) Inclusions, different phases, grain boundaries, etc., can only be located with the aid of an optical microscope or alternatively, a target current or backscattered electron beam scanning image;
(4) Orientation capability is required for cubic lattice spacing determination, orientation studies, and for noncubic lattice spacing studies in order to eliminate projection distortion effects, and to obtain all of the desired information on one film;
(5) The film center position is required for almost all lattice spacing studies and as an origin for orientation studies;
(6) Very small values of D may cause contrast loss while very large D values may increase exposure times to prohibitive values;

(7) The angle between the important cubic poles ⟨100⟩ and ⟨111⟩ is just under 55°; and

(8) Operator convenience and control are greatly increased if an air path camera is obtainable.

Another desirable feature would be the means to view the sample and then to insert a pseudo-Kossel source foil. In the transmission mode, one can conceive of a rotating table controlled from outside the Kossel camera in which several suitable foils and an open hole have been provided. This device would be situated just above the specimen. The open position is used to position the specimen with respect to the electron beam axis; then a suitable foil could be placed in position and the exposure made.

In back-reflection, less versatility is possible. A capillary tube or X-ray funnel can be attached to the electron optics with the source foil attached. The electron beam passes through the tube striking the foil and exciting the desired radiation. The diameter of such a tube must be made as small as practicable in order that the tube will not interfere with the X-rays diverging from the sample face. Such a device usually precludes viewing the sample surface except obliquely as the tube is usually fixed.

The capillary tube conception is, in fact, that of Fujiwara and co-workers [14, 15]. Transmission patterns may be obtained by making the foil at the tube end the specimen. A commercial version of this device is available. It utilizes a single lens electron optics column which is focused by obtaining a shadow image of an electron microscope grid placed upon the source foil. When the shadow image is sharpest, the column is focused (as in an X-ray microscope).

The instrument requires a single crystal having a minimum size of about 4 × 4 mm. There is no light optical provision. Usual operation is in the back-reflection mode; the film is, of course, in air. The sample may be mounted on a goniometer and the source to sample and source to film distances may be chosen at will by the operator. The author has found by experience that changing the source foils is difficult and time consuming; a butt solder joint of a 3-mm-diameter foil about 50 μ thick is required. It is not advisable to use source foils or transmission specimens less than about 50 μ thick since the high current electron beam may puncture the foil during operation resulting in a number of undesirable consequences.

Camera modules suitable for adaption to commercial electron probe microanalyzers or to other electron beam columns have been described by various authors [93–97]. Of these, only the camera described by Lewis [93] has goniometric capabilities at the present time. However, Fisher and Harris [70] have modified the Openshaw-Swindells design [94] to include a crude goniometer.

A few current and projected instrumental advances are worth mentioning. Peters [98] was the first to propose the use of Polaroid X-ray films for Kossel patterns. Coupled with an air path camera, such film enables orientation and preliminary study to be carried out much more rapidly than would otherwise be possible.

In the same vein, Vieth [92] has suggested a direct projection of the Kossel pattern onto a fluorescent screen for preliminary studies in much the same manner as suggested by Taylor for Laue photographs.

Finally, Deslattes [34] has suggested a circular electronic counter on a traveling stage as means to obtain direct count rates from planes parallel to the counter face.

VIII. Contribution of Divergent Beam X-Ray Diffraction to Scientific Research

The Kossel technique has been applied to the study of effects causing small lattice spacing changes as a function of sample treatment. It has also been employed in the study of orientation relations in local areas; conventional methods would have been difficult or impossible to apply in such instances. Lattice imperfections in a number of cubic materials have been investigated. Some examples of each type of study will be outlined.

A. Lattice Spacing Changes and Measurements

1. *Lattice Spacings of Cubic Materials*

The lattice parameters of iron, copper, aluminum, rock salt, and calcite were determined by van Bergen [12, 13] using the compensation method described earlier in Section C-1. Lonsdale used intersection methods to obtain the lattice parameter of diamond. Hanneman *et al.* [45] measured a value for Ni using a reiterative process. Lens ratio methods have been used to determine the lattice parameters of Ni, LiF, Al, and Fe–3 wt % Si [40, 41, 47, 48, 99]. The intersection method has been used recently to obtain the lattice spacing of MgO [29], diamond and nickel [51].

Kossel's method was used by Potts [42] to obtain values for Ge and GaAs. All of these determinations except those of van Bergen were performed in the transmission mode. Recently, the back-reflection mode was used to obtain the lattice parameter of tungsten [31, 53]. The unique feature common to the above determinations is that the precision of the measured lattice parameter ranges from 2.5 to 15 ppm.

Ullrich [100, 101], using back reflection methods, has reported lattice parameter values for several materials such as Fe_3Al and $MgCu_2$ among others in recent publications. He also found a lattice parameter varying with position in a grain of polycrystalline $MgCu_2$ [100].

2. *Lattice Spacings of Noncubic Materials*

Only a few materials have been reported upon. Schwarzenberger [25] has determined *c* and *a* values in Be by the intersection method. Subsequently, Mackay [38] using purer crystals and a superior means of analysis redetermined the lattice parameters of Be by means of the intersection method.

The lens method was used to obtain the lattice parameters of hematite [37]. Recently, Be alloys were successfully reinvestigated from the point of view of the lens method [39].

Morris [49] was able to obtain the *c* and *a* values in PbO which is tetragonal. Ullrich [100] reported *c* and *a* values for $MgZn_2$.

3. *Studies Involving Small Lattice Spacing Changes*

Hanneman *et al.* were able to deduce the nature of the radiation damage produced by 3 MeV electrons in Ni by studying the expansion of the Ni lattice as a function of integrated flux (Fig. 19). The conclusions were that irradiation produced Frenkel defects, most of which remained immobile at 30°C. It was also found that complete annealing of irradiation damage in Ni occurs below 400°C [45, 46].

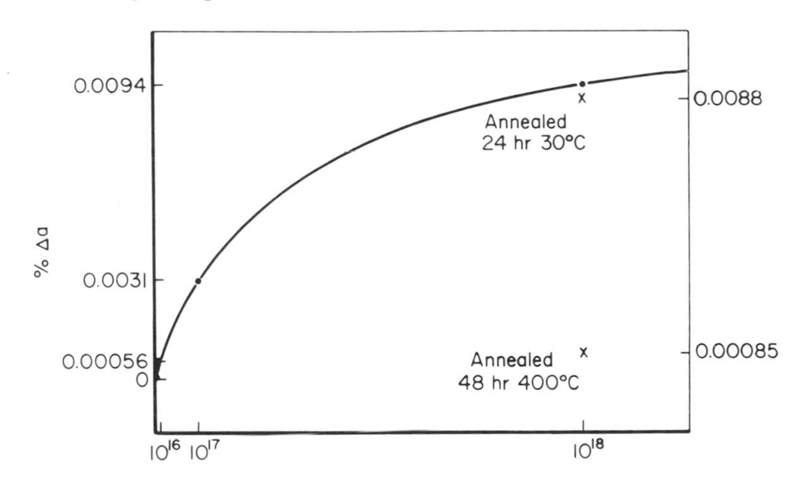

FIG. 19. Change in lattice parameter *a* of Ni as a function of integrated flux of 3 MeV electrons from a van de Graaf generator (after Hanneman [45]).

Sharpe studied effects of irradiation by neutrons on MgO; doses of 3.2×10^{19} nvt and 4.3×10^{20} nvt were used. At the latter value, deficiency conics were so broad as to preclude measurements. The lattice was expanded by nearly 0.1% at the lower neutron dose and the conics were broadened. The damage could be annealed out by holding one hour at 1200°C. Annealing reduction of irradiation damage was observed at 200°C [29].

Potts determined the change in lattice spacing as a function of point defects quenched into GaAs and Ge from high temperatures. No effect was found in Ge but the lattice parameter of GaAs increased after holding for 24 hr at the desired temperature followed by a rapid quench to 0°C. The slope of the $\Delta a/a$ versus T^{-1} curve shown in Fig. 20 gives the enthalpy of the

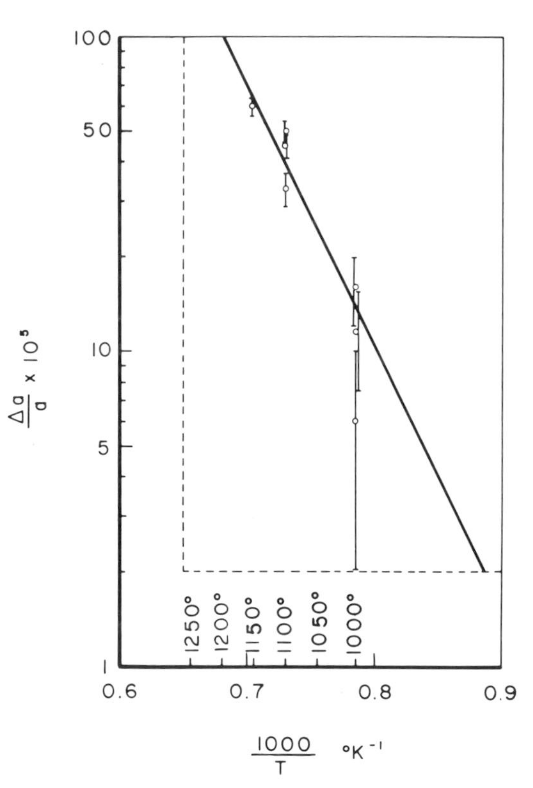

FIG. 20. Value of $\Delta a/a$ in GaAs as a function of reciprocal temperature; Quenched in defects in GaAs (after Potts [42]).

activation of the defect as about 2.0 eV. This effect is attributed to an excess concentration of interstitial atoms leading to Frenkel pairs and Schottky vacancies with the primary defect species being As monovacancies [42, 43].

Weissmann and his co-workers have used the $\Delta d/d_0$ values obtained from pseudo-Kossel patterns as input data for the study of stress-strain configurations in a number of cubic materials such as Cu–Au Cu–Cl and Cu–Ge alloys, as well as W, and Ta. From these data and in conjunction with transmission electron microscopy complete mechanical deformation mechanisms have been postulated for these materials [28, 71–74]. Particular attention has been devoted to the ductile brittle transition in tungsten [31, 76].

B. Orientation Studies

Peters was able to establish that the common plane between the α and κ phases in Cu–Si alloys was the (111)α and (00.1)κ. The orientation relationship was proved to be $(111)_\alpha \parallel (00.1)_\kappa$; $[110]_\alpha \parallel [11.0]_\kappa$ for the alloys as well [36, 61].

Morris determined the orientation of a number of iron whiskers which were too small to be subjected to conventional techniques. He found that the orientation of the α-iron whiskers was predominantly [111](110). For iron oxide whiskers, [110](111) was the orientation [58, 62].

The method of Ogilvie and Bomback ([64], p. 49) has been used to study orientation relations in iron meteorites. In particular, the orientation relationship between kamacite and taenite phases in the Carlton iron has been investigated. The common plane was found to be (111); line broadening precluded a positive determination as to whether the Nishyama or the Kurdjimow–Sachs relations were obeyed.

Bevis and Swindells [67] demonstrated their orientation scheme using a fully austenitic FCC alloy of 20% Ni, 0.6% C, bal Fe. They showed the orientation of several annealing twins and that the habit plane for twinning is 111.

It is interesting to note that by 1967, the Kossel method of orientation was being used as a standard for orientation by selected area electron diffraction [102, 103].

C. Lattice Perfection Studies

Imura [17] showed qualitatively the changes in conics to be expected on straining Cu and Al single crystals in tension. He also showed that an analysis of the line broadening could be used to estimate the density of edge dislocations present as a function of strain. Similar studies by Umeno *et al.* of Al and by Ichinokawa of Fe–3 wt % Si have been performed recently as well [104–107].

Peace and Pringle [21] were able to deduce the mosaic spread in a crystal of pentaerythritol by measuring line breadth and contrast as a function of source to film distance. Solution of Eq. (6) was then made for ζ, assuming Q to be known.

Weissmann and Kalman [75] have recently utilized the divergent beam technique to study lattice defects produced by elastic bending in Ge, Cu, and W crystals. The capillary X-ray device is used and the patterns are prepared in the transmission mode; the specimen and film are rigidly coupled and move synchronously back and forth during the exposure. Anomalous transmission intensity is used to indicate the effect of lattice distortion.

In ductile material such as Cu, slight bending generated dislocations throughout the crystal thereby destroying all anomalous transmission. Tungsten, severely deformed at 77°K, was shown to have the dislocations so produced restricted to a surface layer about five microns in depth. For Ge bent at room temperature, only elastic strains were found until fracture occurred [75, 108].

Acknowledgments

The author wishes to thank Professor R. E. Ogilvie of MIT, in whose laboratory he was a guest worker for one year. Much of the basic material presented in this paper was obtained at that time. Thanks are also given to D. C. Ganow, W. G. Morris, E. T. Peters, P. Gielen, and R. D. Deslattes for many helpful discussions. Special thanks are due to D. L. Vieth for comments and suggestions concerning the manuscript.

Closing Note (July 19, 1967)

At the Second National Conference on Electron Probe Microanalysis (June 14–16, 1967, Boston, Massachusetts), some five papers were given in a session entitled Divergent Beam (Kossel) Diffraction, chaired by Dr. R. E. Hanneman. The titles and authors are as follows:

"An Examination of Factors Limiting Accuracy in Divergent Beam X-ray Diffraction" by K. J. H. MacKay.

"Lattice Parameters of Noncubic Crystals from Kossel Lines" by W. G. Morris.

"Lattice Parameters and Symmetry Information from Kossel Lines: Non-Cubic Crystals" by J. Z. Frazer and A. Reid.

"Grain Boundary Behavior of Macrocrystals in Plastic Deformation" by M. Umeno, N. Gennai, and G. Shinoda.

"A Novel Single Lens Kossel Pattern Generator" by D. L. Vieth, and H. Yakowitz.

In addition, a two day Colloquium on Divergent Beam X-ray Diffraction was held at MIT on June 12 and 13, 1967. This was attended by 22 workers representing the United Kingdom, Japan and the United States, respectively.

References

1. E. Rutherford and E. N. da C. Andrade, *Phil. Mag.* **28**, 263 (1914).
2. H. Seeman, *Ann. Physik* **7**, 633 (1930).
3. W. Hess, *Z. Krist.* **97**, 197 (1937).
4. W. Linnik, *Nature* **124**, 946 (1929).

5. W. Kossel, V. Loeck, and H. Voges, *Z. Physik* **94**, 139 (1935).
6. W. Kossel and H. Voges, *Ann. Physik* **23**, 677 (1935).
7. H. Voges, *Ann Physik* **27**, 694 (1936).
8. W. Kossel, *Ann. Physik* **26**, 533 (1936).
9. W. Kossel, *Ann. Physik* **25**, 512 (1936).
10. W. Kossel, *Ergeb. Exakt. Naturw.* **16**, 295 (1937).
11. M. von Laue, *Ann. Physik* **23**, 705 (1935); **28**, 528 (1937).
12. H. van Bergen, *Ann. Physik* **33**, 737 (1938).
13. H. van Bergen, *Ann. Physik* **39**, 553 (1941).
14. T. Fujiwara, *J. Sci. Hiroshima Univ. Ser. C* **7**, 179 (1947).
15. T. Fujiwara and D. Onoyama, *J. Sci. Horishima Univ. Ser. A* **9**, 115, 125 (1939); **10**, 261 (1940); **11**, 93 (1941); **11**, 325 (1942).
16. T. Imura, *Bull Naniwa Univ. Ser. A* **1**, 52 (1952); **2**, 51 (1954).
17. T. Imura, *Bull. Univ. Osaka Prefect. Ser A* **5**, 99 (1957).
18. K. Lonsdale, *Mineral. Mag.* **28** (196), 14 (1947).
19. K. Lonsdale, *Phil. Trans. Roy. Soc. London, Ser. A* **240**, 219 (1947).
20. K. Lonsdale and J. L. Amoros, *Anales Real. Soc. Espar. Fis. Quim. Ser. A* **47**, 215 (1951).
21. A. G. Peace and G. E. Pringle, *Phil. Mag.* **53**, 1227 (1952).
22. A. H. Geisler, J. K. Hill, and J. B. Newkirk, *Phys. Rev.* **72**, 983 (1947).
23. A. H. Geisler, J. K. Hill, and J. B. Newkirk, *J. Appl. Phys.* **19**, 1041 (1948).
24. G. Borrmann, *Ann. Physik* **27**, 669 (1936).
25. D. R. Schwarzenberger, *Phil Mag.* **4**, 1242 (1959).
26. R. Castaing, Thesis, Univ. of Paris (1951).
27. R. W. James, The optical principles of the diffraction of X-rays, *in* "The Crystalline State" (L. Bragg, ed.), Vol. II. Bell, London, 1950.
28. T. Imura, S. Weissmann, and J. J. Slade, *Acta Cryst.* **8**, 786 (1962).
29. R. S. Sharpe, *Appl. Mater. Res.* **4**, 74 (1965).
30. W. Maier, *Ann. Physik* **40** [5], 85 (1941).
31. T. Ellis and S. Weissmann, A.F. Matls. Lab. Technol. Documentary Rept. ML-TDR-64-220 (1964).
32. B. E. Warren, MIT Physics Dept. course entitled X-Ray Diffraction.
33. R. W. James, G. King, and H. Horrocks, *Proc. Roy. Soc.* (*London*) **A153**, 230 (1935).
34. R. D. Deslattes, private communication (1966).
35. J. Z. Frazer and G. Arrhenius, *in* "Optique des Rayons X et Microanalyse" (R. Castaing, P. Deschamps, and J. Philibert, eds.)., p. 516. Editions Techniq, Paris, 1966 (in press).
36. E. T. Peters, Sc.D. Thesis, MIT (1964).
37. R. E. Ogilvie, ONR Rept. NR031-686 (1962).
38. K. J. H. Mackay, Tube Investments Res. Lab. Technol. Rept. 149 (1962).
39. F. Witt, H. Yakowitz, and D. L. Vieth (to be published).
40. P. Gielen, H. Yakowitz, D. Ganow, and R. E. Ogilvie, *J. Appl. Phys.* **36**, 773 (1965).
41. H. Yakowitz, *in* "The Electron Microprobe" (T. D. McKinley, K. F. J. Heinrich, and D. B. Wittry, eds.), p. 417. Wiley, New York, 1966.
42. H. R. Potts, Stanford Electron. Lab. Rept. SU-SEL-64-075 (1964).
43. H. R. Potts and G. L. Pearson, *J. Appl. Phys.* **37**, 2098 (1966).
44. P. Pietrokowsky, *J. Appl. Phys.* **37**, 4560 (1966).
45. R. E. Hanneman, S. M. Thesis, MIT (1961).
46. R. E. Hanneman, R. E. Ogilvie, and A. Modrzejewski, *J. Appl. Phys.* **33**, 1429 (1932).
47. B. H. Heise, *J. Appl. Phys.* **33**, 938 (1962).
48. D. L. Vieth and H. Yakowitz, *Rev. Sci. Instr.* **37**, 206 (1966).

49. W. G. Morris, Gen. Elect. Res. Develop. Ctr. Rept. 66-C-217, 1966 [*J. Appl. Phys.* **39**, 1813 (1968)].
50. R. deWit, private communication (1965).
51. K. J. H. Mackay, *in* "Optique des Rayons X et Microanalyse" (R. Castaing, P. Deschamps, and J. Philibert, eds.), p. 544. Hermann, Paris, 1966.
52. B. J. Isherwood and C. A. Wallace, *Nature* **212**, (#5058), 173 (1966).
53. T. Ellis, L. F. Nanni, A. Shrier, S. Weissmann, G. E. Padawer, and N. Hosokawa, *J. Appl. Phys.* **35**, 3364 (1965).
54. F. Witt, private communication (1966).
55. M. H. Jellinek, *Rev. Sci. Instr.* **20**, 368 (1949).
56. A. Shrier, Z. H. Kalman, and S. Weissmann, US Gov't. Rept. AD 631 179 (1966).
57. H. Yakowitz *in* "The Electron Microanalyzer and its Applications" (Notes of a special course taught at MIT), p. 101, 1965.
58. W. G. Morris, S. M. Thesis, MIT (1963).
59. J. A. Bearden, A. Henins, J. G. Marzolf, W. C. Sauder, and J. S. Thomsen, *Phys. Rev.* **135**, A899 (1964); [See also J. A. Bearden, U.S. AEC Rept. NYO 10586 (1964)].
60. B. D. Cullity, "Elements of X-ray Diffraction." Addison-Wesley, Reading, Massachusetts, 1956.
61. E. T. Peters and R. E. Ogilvie, *Trans. AIME* **233**, 89 (1965).
62. W. G. Morris and R. E. Ogilvie, A. F. Mat'ls. Lab. Rept. RTD-TDR-63-4198 (1964).
63. M. Greninger, *Trans. AIME* **117**, 61 (1935).
64. R. E. Ogilvie and J. Bomback *in* "The Electron Microanalyzer and its Applications" (Notes of a special course taught at MIT) Lab. Section, p. 10, 1965.
65. J. S. Duerr, S. B. Thesis, MIT (1965).
66. B. H. Heise, *J. Appl. Phys.* **33**, 697 (1962).
67. M. Bevis and N. Swindells, *Phys. Status Solidi* **20**, 197 (1967).
68. H. Yakowitz, *J. Appl. Phys.* **37**, 4455 (1966).
69. R. LeHazif, J. M. Dupouy and Y. Adda, *Conf. Intern. Met. Beryllium, Grenoble*, 1965.
70. D. G. Fisher and N. Harris, Inst. Metals Mtg., Session IV, paper V, London 1967.
71. J. J. Slade, Jr., S. Weissmann, K. Nakajima, and M. Hirabayashi, Dept. Army Rept. DA-ARO(D)-31-124-G300 (1963) (First Rept.).
72. J. J. Slade, Jr., S. Weissmann, K. Nakajima, and M. Hirabayashi, *J. Appl. Phys.* **35**, 3373 (1965).
73. K. Nakajima, J. J. Slade, Jr., and S. Weissmann, *Trans. ASM.* **58**, 14 (1965); See also Dept. Army Rept. DA-ARO(D)-31-124-G300 (1964) (Second Rept.).
74. S. Weissmann and N. Hosokawa, *J. Australian Inst. Metals* **8**, 25 (1963).
75. S. Weissmann, and Z. H. Kalman, *Phil. Mag.* **15** [8], 539 (1967).
76. T. Ellis and S. Weissmann, A.F. Matls. Lab. Doc. Rept. ML-TDR-64-220, Part III (1966).
77. B. E. Warren, *Progr. Metal Phys.* **8**, 147 (1959).
78. R. V. Southwell, "An Introduction to the Theory of Elasticity." Oxford Univ. Press, London and New York, 1949.
79. V. V. Novozhilov, "Theory of Elasticity," Israel Program for Scientific Translations, available from U.S. Dept. Com. as OTS 61-11401.
80. J. J. Slade, Jr. Conf. Intern. *Union Crystallographers, 6th, Rome*, 1963; *Acta Cryst Abs.* 108.
81. R. E. Ogilvie, private communication (1966).
82. V. A. Phillips, ASTM *Spec. Tech. Publ.* 317, p. 34 (1962).
83. V. A. Phillips *in* "Imperfections in Crystals" (H. G. Van Bueren, ed.), 2nd ed. Wiley (Interscience), New York, 1961.

84. H. A. Kramers, *Phil. Mag.* [6] **46**, 836 (1923).
85. H. M. Stainer, U. S. Bur. Mines Inform. Circ. 8166 (1963).
86. P. Gielen, S. M. Thesis, MIT (1964).
87. J. Colby, private communication (1965).
88. H. Yakowtiz and D. L. Vieth, *J. Res. Natl. Bur. Std. C.* **69**, 213 (1965).
89. N. P. Il'in, *Izv. Akad. Nauk* SSSR *Ser. Fiz.* **25** (8), 929 (1961) Transl. *Bull. Acad. Sci.* USSR *Phys. Ser.* **25** (8), 940 (1962).
90. V. E. Cosslett, W. C. Nixon and H. E. Pearson, *in* "X-ray Microscopy and Microradiography" (V. E. Cosslett, A. Engström, and H. H. Pattee, Jr., eds.), p. 96. Academic Press, New York, 1957.
91. D. L. Vieth and H. Yakowitz, *J. Res. Natl. Bur. Std. C.* **71**, 313 (1967).
92. D. L. Vieth and H. Yakowitz, *Rev. Sci. Instr.* **39**, 1929 (1968).
93. R. K. Lewis, Submitted to *Advan. X-Ray Anal.*
94. I. K. Openshaw and N. Swindells *in* "Optique des Rayons X et Microanalyse" (R. Castaing, P. Deschamps, and J. Philibert, eds.), p. 555. Hermann, Paris, 1966.
95. F. Witt, V. V. Damiano, and G. London, *Rev. Sci. Instr.* **38**, 1069 (1967).
96. E. Davidson, W. E. Fowler, H. Neuhaus, and W. G. Shequen, Paper #188, *Pittsburgh Conf. Anal. Chem. Appl. Spectro., 1964*, available from Appl. Res. Labs. Glendale, Calif.
97. Norelco Bull. TC on Kossel Diffraction Camera, available from Philips Elcct. Instr., Mt. Vernon, N.Y.
98. E. T. Peters, private communication (1963).
99. B. H. Heise, *ASTM Spec. Tech. Publ.* **317**, 182 (1962).
100. H. J. Ullrich, *Phys. Status Solidi* **20**, K113 (1967).
101. H. J. Ullrich, *Acta Cryst.* **19**, *Suppl.* A235 (1966).
102. L. Moudy and C. W. Laakso, *Natl. Conf. Electron Probe Microanalysis, 1st, College Park, Md.*, 1966.
103. P. L. Ryder and W. Pitsch, *Phil. Mag.* **15** [8], 437 (1967).
104. M. Umeno, H. Kawabe, and G. Shinoda, *Advan. X-Ray Anal.* **9**, 24 (1966).
105. T. Ichinokawa, *Jap. Appl. Phys. Suppl.* **5**, 36 (1966).
106. M. Umeno, H. Kawabe, and G. Shinoda in "Optique des Rayons X et Microanalyse" (R. Castaing, P. Deschamps, and J. Philibert, eds.), p. 534. Hermann, Paris, 1966.
107. T. Ichinokawa and S. Shirai *in* "Optique des Rayons X et Microanalyse" (R. Castaing, P. Deschamps, and J. Philibert, eds.), p. 561. Hermann, Paris 1966.
108. S. Weissmann, Dept. Army Rept. DA-31-124-ARO(D)-414, 2nd Tech. Rept., (1966).

Author Index

Numbers in parentheses are reference numbers and indicate that an author's work is referred to, although his name is not cited in the text. Numbers in italics show the page on which the complete reference is listed.

Subject Index

F

L

M

N

O

T

U

V

W